EXACT ANALYSIS OF DISCRETE DATA

EXACT ANALYSIS OF DISCRETE DATA

Karim F. Hirji

Boca Raton   London   New York

Published in 2006 by
Chapman & Hall/CRC
Taylor & Francis Group
6000 Broken Sound Parkway NW, Suite 300
Boca Raton, FL 33487-2742

© 2006 by Karim F. Hirji.
Chapman & Hall/CRC is an imprint of Taylor & Francis Group

No claim to original U.S. Government works
Printed in the United States of America on acid-free paper
10 9 8 7 6 5 4 3 2 1

International Standard Book Number-10: 1-58488-070-8 (Hardcover)
International Standard Book Number-13: 978-1-58488-070-7 (Hardcover)
Library of Congress Card Number 2005053883

This book contains information obtained from authentic and highly regarded sources. Reprinted material is quoted with permission, and sources are indicated. A wide variety of references are listed. Reasonable efforts have been made to publish reliable data and information, but the author and the publisher cannot assume responsibility for the validity of all materials or for the consequences of their use.

No part of this book may be reprinted, reproduced, transmitted, or utilized in any form by any electronic, mechanical, or other means, now known or hereafter invented, including photocopying, microfilming, and recording, or in any information storage or retrieval system, without written permission from the publishers.

For permission to photocopy or use material electronically from this work, please access www.copyright.com (http://www.copyright.com/) or contact the Copyright Clearance Center, Inc. (CCC) 222 Rosewood Drive, Danvers, MA 01923, 978-750-8400. CCC is a not-for-profit organization that provides licenses and registration for a variety of users. For organizations that have been granted a photocopy license by the CCC, a separate system of payment has been arranged.

Trademark Notice: Product or corporate names may be trademarks or registered trademarks, and are used only for identification and explanation without intent to infringe.

Library of Congress Cataloging-in-Publication Data

Hirji, Karim F.
   Introduction to the exact analysis of discrete data / Karim F. Hirji.
     p. cm.
   Includes bibliographical references and index.
   ISBN 1-58488-070-8 (alk. paper)
    1. Discrete groups. 2. Computer science--Mathematics. I. Title.

QA178.H57 2005
512'.2--dc22
                                                   2005053883

Taylor & Francis Group
is the Academic Division of Informa plc.

Visit the Taylor & Francis Web site at
http://www.taylorandfrancis.com

and the CRC Press Web site at
http://www.crcpress.com

For

Emma, Rosa, Farida and Sakarma

Contents

List of figures	xiii
List of tables	xv
Abbreviations	xix
Foreword	xxi

1 Discrete Distributions   1
- 1.1 Introduction   1
- 1.2 Discrete Random Variables   1
- 1.3 Probability Distributions   5
- 1.4 Polynomial Based Distributions   6
- 1.5 Binomial Distribution   10
- 1.6 Poisson Distribution   12
- 1.7 Negative Binomial Distribution   14
- 1.8 Hypergeometric Distribution   15
- 1.9 A General Representation   17
- 1.10 The Multinomial Distribution   18
- 1.11 The Negative Trinomial   20
- 1.12 Sufficient Statistics   21
- 1.13 The Polynomial Form   23
- 1.14 Relevant Literature   24
- 1.15 Exercises   25

2 One-Sided Univariate Analysis   29
- 2.1 Introduction   29
- 2.2 One Parameter Inference   29
- 2.3 Tail Probability and Evidence   31
- 2.4 Exact Evidence Function   35
- 2.5 Mid-p Evidence Function   35
- 2.6 Asymptotic Evidence Function   37
- 2.7 Matters of Significance   37
- 2.8 Confidence Intervals   41
- 2.9 Illustrative Examples   44
- 2.10 Design and Analysis   47
- 2.11 Relevant Literature   50
- 2.12 Exercises   52

3 Two-Sided Univariate Analysis — 55
- 3.1 Introduction — 55
- 3.2 Two-Sided Inference — 55
- 3.3 Twice the Smaller Tail Method — 59
- 3.4 Examples — 61
- 3.5 The Likelihood Function — 62
- 3.6 The Score Method — 64
- 3.7 Additional Illustrations — 66
- 3.8 Likelihood Ratio and Wald Methods — 68
- 3.9 Three More Methods — 70
- 3.10 Comparative Computations — 73
- 3.11 The ABC of Reporting — 75
- 3.12 Additional Comments — 78
- 3.13 At the Boundary — 80
- 3.14 Equivalent Statistics — 82
- 3.15 Relevant Literature — 82
- 3.16 Exercises — 83

4 Computing Fundamentals — 87
- 4.1 Introduction — 87
- 4.2 Computing Principles — 87
- 4.3 Combinatorial Coefficients — 88
- 4.4 Polynomial Storage and Evaluation — 93
- 4.5 Computing Distributions — 97
- 4.6 Roots of Equations — 103
- 4.7 Iterative Methods — 106
- 4.8 Relevant Literature — 112
- 4.9 Exercises — 113

5 Elements of Conditional Analysis — 117
- 5.1 Introduction — 117
- 5.2 Design and Analysis — 117
- 5.3 Modes of Inference — 122
- 5.4 The 2×2 Table — 123
- 5.5 The One Margin Fixed Design — 123
- 5.6 The Overall Total Fixed Design — 126
- 5.7 The Nothing Fixed Design — 130
- 5.8 A Retrospective Design — 132
- 5.9 The Inverse Sampling Design — 134
- 5.10 Unconditional Analysis — 135
- 5.11 Conditional Analysis — 139
- 5.12 Comparing Two Rates — 144
- 5.13 Points to Ponder — 147
- 5.14 Derivation of Test Statistics — 149
- 5.15 Relevant Literature — 151
- 5.16 Exercises — 153

6 Two 2 × 2 Tables — 159

- 6.1 Introduction — 159
- 6.2 Sources of Variability — 159
- 6.3 On Stratification — 160
- 6.4 Data Examples — 162
- 6.5 Statistical Models — 165
- 6.6 Conventional Analysis — 170
- 6.7 Conditional Analysis — 173
- 6.8 An Example — 174
- 6.9 A Second Example — 176
- 6.10 On Case-Control Sampling — 177
- 6.11 Anatomy of Interactions — 181
- 6.12 Relevant Literature — 183
- 6.13 Exercises — 183

7 Assessing Inference — 189

- 7.1 Introduction — 189
- 7.2 Exact Unconditional Analysis — 189
- 7.3 Randomized Inference — 194
- 7.4 Exact Power — 196
- 7.5 Exact Coverage — 204
- 7.6 The Fisher and Irwin Tests — 206
- 7.7 Some Features — 210
- 7.8 Desirable Features — 214
- 7.9 On Unconditional Analysis — 217
- 7.10 Why the Mid-p? — 218
- 7.11 Relevant Literature — 219
- 7.12 Exercises — 221

8 Several 2 × 2 Tables: I — 227

- 8.1 Introduction — 227
- 8.2 Three Models — 227
- 8.3 Exact Distributions — 229
- 8.4 The COR Model — 234
- 8.5 Conditional Independence — 237
- 8.6 Trend In Odds Ratios — 241
- 8.7 Recommendations — 246
- 8.8 Relevant Literature — 246
- 8.9 Exercises — 247

9 Several 2 × 2 Tables: II — 253

- 9.1 Introduction — 253
- 9.2 Models for Combining Risk — 253
- 9.3 Testing for Homogeneity — 256
- 9.4 Test Statistics — 258
- 9.5 A Worked Example — 260
- 9.6 Checking the TOR Model — 261
- 9.7 An Incidence Density Study — 263
- 9.8 Other Study Designs — 266
- 9.9 Exact Power — 267
- 9.10 Additional Issues — 269
- 9.11 Derivation — 271
- 9.12 Relevant Literature — 274
- 9.13 Exercises — 274

10 The 2 × K Table — 279

- 10.1 Introduction — 279
- 10.2 An Ordered Table — 280
- 10.3 An Unordered Table — 285
- 10.4 Test Statistics — 287
- 10.5 An Illustration — 290
- 10.6 Checking Linearity — 294
- 10.7 Other Sampling Designs — 295
- 10.8 Incidence Density Data — 299
- 10.9 An Inverse Sampling Design — 303
- 10.10 Additional Topics — 305
- 10.11 Extensions — 312
- 10.12 Derivation — 314
- 10.13 Relevant Literature — 315
- 10.14 Exercises — 316

11 Polynomial Algorithms: I — 323

- 11.1 Introduction — 323
- 11.2 Exhaustive Enumeration — 323
- 11.3 Monte-Carlo Simulation — 326
- 11.4 Recursive Multiplication — 329
- 11.5 Exponent Checks — 330
- 11.6 Applications — 334
- 11.7 The Fast Fourier Transform — 339
- 11.8 Relevant Literature — 344
- 11.9 Exercises — 344

CONTENTS

12 Polynomial Algorithms: II — 349
- 12.1 Introduction — 349
- 12.2 Bivariate Polynomials — 349
- 12.3 A Conditional Polynomial — 352
- 12.4 Backward Induction — 355
- 12.5 Conditional Values — 357
- 12.6 Applications — 360
- 12.7 Trivariate Polynomials — 362
- 12.8 An Extension — 365
- 12.9 Network Algorithms — 366
- 12.10 Power Computation — 368
- 12.11 Practical Implementation — 372
- 12.12 Relevant Literature — 372
- 12.13 Exercises — 373

13 Multinomial Models — 377
- 13.1 Introduction — 377
- 13.2 Compositions and Partitions — 377
- 13.3 A Single Multinomial — 380
- 13.4 Trinary Response Models — 388
- 13.5 Conditional Polynomials — 394
- 13.6 Several $3 \times K$ Tables — 400
- 13.7 $J \times K$ Tables — 402
- 13.8 Relevant Literature — 408
- 13.9 Exercises — 408

14 Matched and Dependent Data — 415
- 14.1 Introduction — 415
- 14.2 Matched Designs — 415
- 14.3 Paired Binary Outcomes — 424
- 14.4 Markov Chain Models — 432
- 14.5 Relevant Literature — 442
- 14.6 Exercises — 443

15 Reflections On Exactness — 449
- 15.1 Introduction — 449
- 15.2 Inexact Terminology — 449
- 15.3 Bayesians and Frequentists — 451
- 15.4 Design and Analysis — 455
- 15.5 Status Quo Exactness — 462
- 15.6 Practical Inexactness — 468
- 15.7 Formal Exactness — 471
- 15.8 In Praise of Exactness — 477
- 15.9 Relevant Literature — 481
- 15.10 Exercises — 482

References — 485

Index — 515

List of figures

2.1	$Pr[\mathcal{T} \geq 2; \phi]$ as a Function of ϕ.	33
2.2	A One-sided Binomial Evidence Function for $n = 10$, $t = 6$.	41
3.1	TST Evidence Function for $\mathcal{T} = 6$, and $n = 10$.	59
3.2	TST Mid-p Evidence Function for $\mathcal{T} = 6$.	61
3.3	Asymptotic Score Evidence Function.	67
5.1	Mid-p Evidence Function for HCG data.	142
6.1	Mid-p Evidence Function for Ganciclovir Data.	176
6.2	Mid-p Evidence Function for Arsenic Data.	178
7.1	An Exact Score Power Function.	200
7.2	An Ideal Shape for a Power Function.	215
10.1	Mid-p Evidence Function for Table 10.7 Data.	303
12.1	A Five Stage Network.	367

List of tables

1.1	Post Surgical Complications	3
1.2	ESR and Pulmonary Infection by Age in Kwashiorkor	4
1.3	Hypergeometric Coefficients	16
1.4	Generating Polynomials	17
1.5	Trinomial Sample Space and Coefficients	19
3.1	Test Statistics for $B(12; \pi_0 = 0.1)$	74
3.2	Exact and Asymptotic p-values	75
3.3	Analysis of Genetic Data	77
4.1	A Sample of Exponents and Coefficients	112
5.1	A 2×2 Table	123
5.2	Ophthalmic Data	124
5.3	Hypothetical Data	125
5.4	Sample Points and Coefficients	126
5.5	HCG Data	127
5.6	Artificial Data	129
5.7	Specification of the Distribution of $(\mathcal{T}_1, \mathcal{T}_2, \mathcal{T}_3)$	129
5.8	Aquatic Toxicity Data	130
5.9	Exposure and Gender	131
5.10	Case-Control Data	133
5.11	Inverse Sampling Data	134
5.12	Sample Points and Coefficients for HCG Data	142
5.13	Analysis of HCG Data	143
5.14	Sampling Design and p-values	144
5.15	Five Clinical Trials	154
5.16	Aspirin and Ulcer	155
5.17	CTS Data	155
5.18	Disease Incidence Data	155
6.1	The ith 2×2 Table	161
6.2	Three Clinical Trial Scenarios	162
6.3	CMV Prevention Trial	163
6.4	Bacteriuria and Catheterization	163
6.5	Gene and Environment Interaction Data	164
6.6	Mortality and Arsenic Exposure Among Smelter Workers	165

6.7	Success Probabilities	167
6.8	Conventional Analysis of Ganciclovir Data	172
6.9	Conditional Generating Polynomial for T_3	175
6.10	Mid-p Analysis of Ganciclovir Data	175
6.11	Cell Probability Estimates	176
6.12	Interaction Analysis of Arsenic Data	177
6.13	Interaction or No Interaction?	182
6.14	Antibiotics for the Common Cold	184
6.15	Race and Outcome of Trial for Murder	184
6.16	Bleeding and Compression Device	185
6.17	Bleeding and Aspirin	185
6.18	Five $2 \times 2 \times 2$ Data Sets	186
6.19	Five More $2 \times 2 \times 2$ Data Sets	186
6.20	Three Follow Up Design Data Sets	187
7.1	Hypothetical Data	191
7.2	Exact Unconditional Analysis	192
7.3	Method and p-values	193
7.4	Two Score Rejection Regions	199
7.5	Data for a Conditional Test	201
7.6	Actual Coverage for Three Methods	206
7.7	Comparing Fisher and Irwin Exact Tests	208
7.8	Data from Ten Clinical Trials	222
8.1	Stratified Clinical Trial Data	228
8.2	Exponents and Coefficients of $f(\phi)$	235
8.3	Analysis of the Data in Table 8.1	236
8.4	Hypothetical Clinical Data	240
8.5	Stratum-wise gps for Table 8.1	242
8.6	Analysis of the TOR Model	243
8.7	ESR and Infection by Age in Kwashiorkor	247
8.8	Cerebral Atrophy Data	248
8.9	HAART Data: A	249
8.10	HAART Data: B	249
8.11	HAART Data: C	250
8.12	HAART Data: D	250
8.13	Data for Comparative Analyses	251
9.1	Data from Four Clinical Trials	256
9.2	Exact Computation Results	261
9.3	Death by Region, Gender and Occupation	264
9.4	Analysis of the Data in Table 9.3	266
9.5	Hypothetical Meta-Analysis Data	275
9.6	Data for Comparative Analyses	275
9.7	Events by Study Factors	276

LIST OF TABLES

10.1	Notation for a $2 \times K$ Table	279
10.2	Hypothetical Dose Response Data	281
10.3	Dose Level Generating Polynomials	283
10.4	Outcome at Five Years	285
10.5	Tables with Same Margins as Table 10.4	290
10.6	Expected Values for Table 10.4	291
10.7	Cases by Genetic Marker and Environment	302
10.8	Toxicology Datasets	317
10.9	Clinical Trial Datasets	317
10.10	Liver Function Data	318
10.11	Gender and Spinal Disease	318
10.12	XX Gene Mutations in Lung Cancer	318
10.13	Ear Infection and Hours at Day Care	318
10.14	Deaths by Gender and Exposure	319
10.15	Inverse Sampling Trial Datasets	320
11.1	A Comparison of EE and RPM Algorithms ($K = 4$)	330
13.1	ESR and Pulmonary Infection in Kwashiorkor	388
13.2	Notation for a $3 \times K$ Table	389
13.3	A Comparison of EE and RPM + EC 13.02	396
13.4	Analyses of Table 13.1 Data	400
13.5	ESR and Pulmonary Infection by Age in Marasmus	401
13.6	Conditional Generating Polynomial for Example 13.4	403
13.7	Analyses of Table 13.5 Data	404
13.8	A 3×3 Table	407
13.9	ESR and Infection by Age in Marasmic Kwashiorkor	413
14.1	1:1 Case-Control Design	418
14.2	1:n Case-Control Design	419
14.3	1:n_j Case-Control Design	421
14.4	Number of Pairs by Response	424
14.5	Paired Outcomes with Two Binary Covariates	426
14.6	Paired Outcomes with One Binary Covariate	427
14.7	Two State Transition Probabilities	433
14.8	Transition Counts for mth Chain	434
14.9	Conditional GP for Telephone Data	438
14.10	Conditional GP for Example 14.5	441
14.11	1:2 Case-Control Design	443
14.12	A Paired Design with Two Binary Covariates	443
14.13	A 3:3 Case-Control Design	444
14.14	Eye Data	445
14.15	BSE Maternal Cohort Study	446
15.1	Study Type and Design	463
15.2	Analysis Type and Design	463

15.3	One- and Two-Tailed Exact Tests	468
15.4	A Matched Pairs Trial	472
15.5	Three Trials for Table 15.4	473
15.6	Promotions by Year	473
15.7	Promotions by Years of Service	474
15.8	Years to Promotions by Gender	475

Abbreviations

ACL	:	actual coverage level
ASL	:	actual significance level
CAM	:	Cochran–Armitage–Mantel
cdf	:	cumulative distribution function
CI	:	confidence interval
cmle	:	conditional maximum likelihood estimate
cmue	:	conditional median unbiased estimate
COR	:	common odds ratio
CS	:	conditional score
CT	:	combined tails
df	:	degrees of freedom
EC	:	exponent check(s)
FH	:	Freeman–Halton
gp	:	generating polynomial
HOR	:	heterogeneous odds ratio
iid	:	independent and identically distributed
iip	:	independent with identical parameters
lhs	:	left hand side
LR	:	likelihood ratio
max	:	maximum
MH	:	Mantel–Haenszel
min	:	minimum
mle	:	maximum likelihood estimate
mue	:	median unbiased estimate
rhs	:	right hand side
PBD	:	polynomial based distribution
RAM	:	random access memory
RCT	:	randomized controlled trial
RBG	:	Robins–Breslow–Greenland
RPM	:	recursive polynomial multiplication
SMR	:	standardized mortality ratio
TBD	:	Tarone–Breslow–Day
TOR	:	trend in odds ratios
TST	:	twice the smaller tail

Foreword

Researchers in biology, medicine, public health, psychology, sociology, law and economics regularly encounter variables that are discrete or categorical in nature. And there is no dearth of books - elementary to advanced - on the methods of analysis and interpretation of such data. These books mainly cover large sample methods. When the sample size is not large, or the data are otherwise sparse, their accuracy is suspect. In that event, methods not based on asymptotic theory, called exact methods, are desirable.

The origins of the exact method for analysis of discrete data lie in the early analysis of binomial and Poisson outcomes, and is related to the growth of nonparametric analysis of continuous data. Its emergence as a distinct branch of statistics, however, dates to the works of Sir Ronald A. Fisher, Frank Yates and James O. Irwin (Fisher 1934, 1935b; Irwin 1935; Yates 1934). All of them worked on the exact analysis of a 2×2 table. Despite the early start, the forms of discrete data for which exact analysis was feasible were, until 1980, rather limited. The analysis was mostly restricted to univariate data, an equiprobable multinomial, or a 2×2 table. For complex models, the exact analysis was formulated theoretically, and creative applications of the Monte–Carlo method were devised. Yet, it was not a practical option since the computational effort rose exponentially with sample size. In fact, more energy was expended in the periodic controversies about exact analysis than on expanding its scope.

A dramatic change occurred during the 1980s. Application of fast Fourier transform based techniques by Marcello Pagano, David Tritchler and their coworkers, on the one hand, and of network algorithms by Cyrus Mehta, Nitin Patel and their coworkers, on the other, were principally behind that turnaround. I had the fortune to be associated with the latter group, and worked on extending exact analysis of logistic models and the multivariate shift algorithm. Today, such efficient algorithms and enhanced computing power have made exact analysis eminently feasible for a vastly enlarged spectrum of discrete data problems. As such, it is often performed when the traditional large sample methods are in question. (The references relevant to this paragraph are noted later in the text.)

Yet, despite the plethora of original research and review papers on the subject, and the existence of computer software, no book with exact analysis of discrete data as its prime focus is currently available. Most books on discrete or categorical data analysis tend to have a small section on exact tests. The Fisher exact test for a 2×2 table is always covered. Recent books contain more material on the topic. The first book with the word "exact" in its title has one half of a chapter on exact tests for discrete data (Weerahandi 1995). Even when they cover more complex problems, the books essentially present exact methods as a set of recipes. None develops them from first principles, covers a broad class of models, gives a variety of worked examples, addresses the conceptual issues and also presents related computational algorithms, all within an integrated yet accessible framework.

This book begins the task of filling the void. My aim has been to present, in a unified but elementary and applications-oriented framework, the distributional theory, statistical methods and computational methods for exact conditional analysis of discrete data. To shape it into a sturdy and coherent edifice, I have relied on two key ideas, namely, that of a polynomial based distribution and an evidence function. Their roots lie in the pioneering work of Sir R. A. Fisher. His analysis of the odds ratio in a 2×2 table employed the conditional hypergeometric probability. For this purpose, he formulated the tail probability as a ratio of two polynomials in which the numerator was a segment of the denominator. The specific example in Fisher (1935b) was:

$$\mathsf{F}(\psi) = \frac{1 + 102\psi + 2992\psi^2}{1 + 102\psi + \cdots + 476\psi^{12}}$$

Fisher used this ratio of polynomials to assess a significance level for testing if the odds ratio was equal to one, and determine what we now call 99% and 95% upper confidence bounds for it. A two-sided version of the basic idea, which is relevant to continuous data and large sample analysis as well, was later elaborated in a generalized form by Birnbaum (1961). Allan Birnbaum named this entity a confidence curve; the Fisher polynomial ratio then is but a one-sided confidence curve.

The confidence curve, or in our terminology, the evidence function is, after a long hiatus, steadily showing up in the texts on statistics and epidemiology; a recent case in point is Rothman (2002), where it occupies a prominent position. Over the years, it has been given several different names, based on the aspect emphasized. I have chosen to call it an evidence function, a name that captures the overall spirit of what it stands for. No matter what the name, its primacy lies in that it embodies the three key tools of data analysis, namely, a *p*-value, a point estimate and a confidence interval within a single construct. Thereby, it serves a positive pedagogic purpose and allows a conceptually unified presentation of the results of data analysis.

The importance of the polynomial form for distributions lies in that not only the hypergeometric but other common discrete distributions like the binomial, Poisson, negative binomial, multinomial, product multinomial as well as many of the distributions derived from them are polynomial based distributions. The polynomial formulation is not rare; it is implicit in several classic works on discrete data analysis, e.g., Cox (1970) and Zelen (1971), and in the run of the mill old and new papers like Bennett and Nakamura (1964) and Emerson (1994), to name a few. But, apart from a specialized and rarely cited branch of research, its primacy was not noted, and, until recently, its computational utility was not appreciated.

This book utilizes the research published in the 1990s showing that the polynomial formulation produces an integrated framework for exact inference for discrete data, both in terms of theory and computation. It links the many algorithms in the field and, unlike the other formulations, it is based on a simple idea. In fact, the basic idea underlying it is not too distinct from the high school algebra method of multiplying a set of polynomials in a step by step fashion and selecting some terms from the product.

My research on polynomial multiplication algorithms for exact analysis began in 1988. On my own and with Stein E. Vollset, Isildinha Reis, Man Lai Tang and Timothy Johnson - my doctoral students at UCLA - I wrote a number of papers on exact analysis using such algorithms. The idea was also independently developed by David O. Martin and Harland Austin. Further, the pioneering ideas of Cyrus Mehta and Nitin Patel in a related algorithmic area are well reflected in the polynomial algorithms based literature. Two other papers, Baglivo, Pagano and Spino (1996) and van de Wiel, Di Bucchianico and van der Laan (1999), contain equivalent formulations of the same basic idea. (The other references relevant to this paragraph are noted later in the text.)

Thus far this material has been buried in specialized journals. It is time a wider audience of

FOREWORD

students, data analysts, applied researchers and statisticians has access to it. I hope this book will serve that purpose.

****

The first chapter reviews relevant discrete distributions, lays out the notation and defines key concepts. Apart from Chapter 4 and Chapter 7, the chapters which follow, up to and including Chapter 10, develop and illustrate exact conditional methods for various models for discrete data. Of these, Chapters 2, 3 & 5 cover simple one and two variable models. Chapters 6, 8, 9 & 10 deal with several 2×2 tables, and one and several $2 \times K$ tables. Chapters 4, 11 & 12, on the other hand, deal with computational techniques needed for implementing the exact method. Chapters 13 and 14 cover statistical and computational material in an integrated fashion, respectively dealing with multinomial data, and matched and dependent data. Chapter 7 is on assessing the tools of inference, and Chapter 15 deals with conceptual matters, and addresses the use and misuse of exact methods in practice. The relevant large sample methods are presented throughout the text.

Readers and teachers seeking a self-contained but elementary exposure to the field may study Chapters 1, 2, 3, 5, 6, parts of 7, 8, 9 and 10. Chapters 4, 11 and 12, and parts of Chapters 13 and 14 are of special importance to those who also want to explored the computational techniques underlying exact analysis.

This is an introductory work. A basic course in statistics, biostatistics or research methods (at the level say of Rosner (2000) or Daniel (2005)) is all that is needed to access most of its material. It developed from my notes for a course on discrete data analysis for masters and doctoral students in biostatistics at UCLA. I hope that graduate students, teachers and practitioners in statistics and biostatistics as well as quantitatively inclined researchers in fields like epidemiology, genetics, sociology, education, psychology and business studies will find it of value, both as a learning and teaching text, and a reference work.

I have avoided a cookbook approach. Instead, the analytic methods are developed in a step by step manner from first principles. Yet, elaborate theory is absent. Each chapter contains relevant worked examples and exercises. Some of them can be taken up as class research projects by students. Relevant material on computer implementation is also included.

This book is not related to any existing statistical software. Those who have access to software such as StatXact or SAS, which cover exact methods, can use them with the book. Many of the illustrative examples can be worked out on a calculator, or with simple programming in common software.

****

Over the years, the field of exact analysis has been fraught with much controversy. In essence, the debates were about the role of conditioning in, and definition of the frame of reference for, data analysis. They as such pertain to all forms of statistical analysis, including the traditional large sample methods. What is called an exact analysis can actually be an unconditional, a partly conditional or a fully conditional analysis. In discrete data analysis, moreover, there are distortions of the form not found in the analysis of continuous data. But these, in different ways no doubt, afflict both approximate and exact methods. Yet mainly because of the context within which the debates were aired, an impression has been created that such controversies are inextricably tied to the exact method. At times, some leading statisticians have not helped the matter by not sufficiently disentangling the separate strands of the arguments. To some, exact analysis is, as we illustrate in Chapter 15, synonymous with contentious analysis while some others swear by it!

The term "exact" also has had a variety of meanings in the statistical literature. We discuss these issues in the text. But for now, given the historical baggage attached to it, we state that to us the label "exact" surely does not imply that the method in question is the "correct" or "best" one in any absolute sense. It just refers to a method that avoids large sample approximations. In that spirit, we see and present exact methods as an integral part of the spectrum of data analytic techniques available today. After noting their advantages and limitations, we hold that the former often outweigh the latter. Exact analysis is, in our view, not just a valid, but often a better option, for many sparse data problems.

Acknowledgments

This book has benefitted in many ways from the participation of, and valuable suggestions from Dr. Stein E. Vollset, University of Bergen, Norway. He was associated with the initial stages of developing its material, and is a coauthor of Chapter 4. His wisdom, though, prevails in many parts of the work. I am very grateful to him for all his help. Dr. Elliot Landaw, Department of Biomathematics, UCLA and Dr. Roshan Bastani, UCLA School of Public Health, kindly allowed me access to their departmental facilities and resources. Karla Morales helped with the literature search and securing permissions for copyrighted material. My four doctoral students at UCLA (named above) played a key part in developing some of the ideas contained in this work. All of them also have my gratitude.

Dr. Mahmood F. Hameer generously provided data from his research; these have been used for illustrative purposes in several chapters. And Rob Calver, my editor at CRC Press, professionally expedited the various stages of production of this book.

And above all, without the sustained support of Farida, my wife, this book could not have seen the light of the day. Her love and patience are more than any one can ask for. To her, I only say: 'I love you.'

CHAPTER 1

Discrete Distributions

1.1 Introduction

When the variables of a **scientific model** are related through a chance mechanism, we call it a **statistical model.** This book covers statistical models for discrete variables. Its main focus is on the methods for analysis of discrete variable statistical models called exact conditional methods. This chapter prepares the groundwork. This includes introducing a unified framework for representing common discrete probability distributions, defining key concepts and describing the properties of these distributions. The specific aims here are:

- To state the properties of the binomial, Poisson, negative binomial, hypergeometric and related distributions.
- To introduce the multinomial distribution, a distribution that forms the basis of many statistical models for discrete data.
- To present the polynomial formulations for common univariate and multivariate discrete probability distributions.
- To introduce three essential ideas for exact analysis of discrete data, namely, the generating polynomial, conditioning and sufficient statistics.

We do not develop the probability theory for discrete random variables in a rigorous manner. The proofs for the results stated are also not all given. Many introductory probability theory texts contain such material. However, results that are not usually found in standard texts, but which are especially relevant for this book are accorded due elaboration.

1.2 Discrete Random Variables

Variability is an inescapable fact of nature and life. One child gets an earache frequently but another hardly ever suffers from the malady. The symptoms of a common cold may clear up in a few days, or may linger on for weeks. Will this child have an earache in the next six months? How long will a cold persist? Even a trained physician is hard placed to make the prediction since events like these do not have the level of certainty we associate with the statement: "A stone released in midair will fall to the ground." It is fair to regard whether or not a child will get an earache, or the number of days a cold will last as random or chance phenomena.

Randomness does not mean complete unpredictability or chaos. Usually, random phenomena incorporate some systematic components as well. When a large number of cases of the event in question are examined, a pattern often emerges. For example, boys show a higher tendency than girls to develop ear infections. Or, that more than 50% of the cases of the common cold tend to resolve spontaneously within a week or so.

To study processes that are random at a micro-level but which exhibit some degree of regularity when viewed on an aggregate scale, we use the idea of a **random variable**. A simple example of a random variable obtains from envisioning an event which either occurs or does not occur. Let π denote the probability of occurrence of the event. In that case,

$$0 \leq \pi \leq 1 \tag{1.1}$$

The probability of nonoccurrence of the event then is $1 - \pi$. We define an associated random variable, $\mathcal{Y}$, as follows. Let $\mathcal{Y} = 0$ if the event does not occur (called a failure), and $\mathcal{Y} = 1$ if it does (called a success). Then we write

$$P[\mathcal{Y} = 1] = \pi \quad \text{and} \quad P[\mathcal{Y} = 0] = 1 - \pi \tag{1.2}$$

A compact way of writing this is: For $y \in \{0, 1\}$,

$$P[\mathcal{Y} = y] = \pi^y (1 - \pi)^{1-y} \tag{1.3}$$

$\mathcal{Y}$ is called a **binary random variable**, and the random process is referred to as a **Bernoulli trial**. Examples include the outcome of a coin toss, cure or failure in treatment for a disease, the presence or absence of a genetic trait, exposure or nonexposure to a potential occupational carcinogen, and the presence or absence of an ear infection. Equation 1.3 is also a simple statistical model.

A random variable that assumes a finite, or at most, a countable number of values is called a **discrete random variable**. The number of days an episode of the common cold persists is an example of a discrete random variable.

Let the symbol $\mathcal{T}$ designate a discrete random variable. The set of values, realizations, or outcomes of $\mathcal{T}$, denoted by Ω, is the **sample space** or **support set** of $\mathcal{T}$. We assume throughout that Ω is a countable set of real numbers, or vectors with real elements, that is, it can be put into a one-to-one correspondence with a subset of the set of integers.

We let the function $f(t) = P[\mathcal{T} = t]$ represent the probability of an outcome $t \in \Omega$. This **probability function** has to satisfy the following basic properties:

Property 1.1: For any $t \in \Omega$, $P[\mathcal{T} = t] > 0$

Property 1.2: If $\Omega_1, \cdots, \Omega_k$ are mutually exclusive subsets of Ω, then

$$P[\mathcal{T} \in \Omega_1 \bigcup \cdots \bigcup \Omega_k] = \sum_{j=1}^{k} P[\mathcal{T} \in \Omega_j] \tag{1.4}$$

for $k = 1, \cdots, \infty$.

Property 1.3:

$$\sum_{t \in \Omega} P[\mathcal{T} = t] = 1 \tag{1.5}$$

As indicated by Property 1.1, throughout this book we consider events with strictly nonzero probabilities as the relevant subsets of Ω.

****

DISCRETE RANDOM VARIABLES

We consider two scientific investigations with discrete random variables.

Example 1.1: Kiviluoto et al. (1998) report a clinical trial of two surgical procedures, labeled LC and OC, for acute cholecystitis. One outcome of interest was the occurrence of major complications (including death) after surgery. The relevant data from this trial are shown in Table 1.1.

Table 1.1 *Post Surgical Complications*

Major Complications	Surgical Procedure	
	LC	OC
No	32	24
Yes	0	7
Total	32	31

Source: Kiviluoto et al. (1998), Table 3.
© 1998 by The Lancet Ltd.; Used with permission.

The allocation of patients to treatment was done using a random device. Table 1.1 relates two binary random variables, the treatment, and the onset of major complications after treatment. What conclusion can we draw about the complication rates of these treatments?

Example 1.2: Next consider Table 1.2 with data on the status of young children in the pediatrics ward of a hospital in Tanzania. All these children had Kwashiorkor. The main aim of the study was to evaluate the change in the erythrocyte sedimentation rate (ESR) level by type of infection in children with malnutrition. Other children in the study had marasmus and marasmic kwashiorkor, two other forms of malnutrition. The data for these children are given in Table 13.5 and Table 13.9, respectively, and are also summarized at other places in this text.

All the three variables in Table 1.2 are in a discretized form. None is a binary variable. This was a cross sectional type of study; at the time of examination, the age and infection status were known, and the ESR value was the random entity. The latter was converted into a discrete variable using common clinical cut points. After adjusting for the effect, if any, of age, does the ESR profile vary by infection status? - that was a key question of interest.

At this juncture, we emphasize that the above data examples and all the other data examples in this book are given for the sole purpose of illustrating and comparing statistical methods. **They are not used to draw substantive conclusions about the underlying biomedical or other substantive issues.** The latter has to be based on the full data set for the appropriate study.

These examples show us that discrete variables come in various forms. Some are **nominal** variables, also known as **nominal scale**, **qualitative**, or **categorical** variables. Their levels are devoid of any intrinsic order. A case in point is the variable 'type of infection' in Table 1.2 which has three unordered levels. As another example, the primary diagnosis of a hospitalized patient may be classified as cardiac, lung, liver, renal or other disease. For record keeping, we may attach numeric labels such as 1, 2, 3, 4 or 5 to these disease states. But they are arbitrary labels. When infection status is analyzed at two levels 'infection' or 'no infection,' it becomes a binary variable.

Table 1.2 *ESR and Pulmonary Infection by Age in Kwashiorkor*

	ESR Level			
	Age ≤ 12 Months			
	≤ 10	11–25	26–99	100+
No Infection	12	7	3	0
Pulmonary Tuberculosis	0	0	2	1
Other Pneumonia	4	0	0	0
	12 Months < Age ≤ 24 Months			
	≤ 10	11–25	26–99	100+
No Infection	11	1	0	0
Pulmonary Tuberculosis	0	0	1	1
Other Pneumonia	3	0	0	0
	Age > 24 Months			
	≤ 10	11–25	26–99	100+
No Infection	3	1	1	0
Pulmonary Tuberculosis	1	0	0	1
Other Pneumonia	0	0	1	0

Source: Hameer (1990); Used with permission.

A discrete variable whose levels have a built-in order is called an **ordinal** variable. The variables 'age group' and 'ESR level' in Table 1.2 are of this type. To take another case, the side effects of a medicinal drug may be depicted as: none, not life threatening and life threatening. The second level is more serious than the first, and the third is more serious than the second. Despite the ordering, no quantitative magnitude is attached to any level. Some ordinal variables come with a numeric score attached to each level. The number of side effects from a drug is a case in point. We call an ordered discrete variable which has a numeric score as a **scored** variable. The ESR categories in Table 1.2 (≤ 10, $11-25$, $26-90$ and $100+$) are generally viewed as normal, elevated, highly elevated and very highly elevated levels. They may respectively be assigned **natural** scores 0, 1, 2 and 3 in an analysis. Or, the scores may be the midpoints of the associated numeric range.

For the purpose of data analysis, the same underlying entity is at times deemed a nominal variable and at other times, an ordinal variable. The age of a subject, measured in years or months, as in Table 1.2, is often treated as a discrete variable. Then 'age group' may be used as a nominal variable or as a discrete variable with a numeric score.

Categorizing or scoring a variable may imply a loss of information. If not done with care, it has

PROBABILITY DISTRIBUTIONS

the potential to mislead. At times, it may simplify data analysis and interpretation, and at times, it yields valuable insights into nonlinear relationships. Good science requires that discretization or categorization be done at the stage of study design and not after the data are collected and summarized. Further, it should not be done in a way that deviates drastically from the underlying substantive meaning of the variable in question.

Table 1.1 and Table 1.2 are examples of **sparse data**. One or more of the cell counts in both are zero or small. Also, there is an imbalance in the counts at the levels of some variables. Such data occur even with large sample sizes. The methods given in this text are particularly suitable for sparse data.

1.3 Probability Distributions

The **cumulative distribution function (cdf)** of the random variable $\mathcal{T}$ is defined as

$$\mathsf{F}(t) = \mathrm{P}[\mathcal{T} \leq t] \tag{1.6}$$

The **mean** (or **expectation**) and the **variance** of $\mathcal{T}$ respectively are

$$\mu = \mathbf{E}[\mathcal{T}] = \sum_{t \in \Omega} t \mathrm{P}[\mathcal{T} = t] \tag{1.7}$$

$$\sigma^2 = \mathbf{E}[(\mathcal{T} - \mu)^2] = \sum_{t \in \Omega} (t - \mu)^2 \mathrm{P}[\mathcal{T} = t] \tag{1.8}$$

The mean is a measure of location and the variance indicates the extent of variability around the mean. The **median** is another measure of location. Conceptually, it is the value of the middle item of a population when all the items are arranged in an increasing order. In a population of discretized values, such a middle item may not be uniquely identifiable. For a discrete random variable taking two or more values, the median is defined as follows:

- Suppose there exists a $t_l \in \Omega$ such that $\mathrm{P}[\mathcal{T} \leq t_l] = 0.5$. Then $t_r \in \Omega$ is the value such that $\mathrm{P}[\mathcal{T} \geq t_r] = 0.5$. In this case, the median is not unique, and may be taken as t_l or t_r. In practice, we let $t_m = (t_l + t_r)/2$.
- Suppose there does not exist a $t \in \Omega$ such that $\mathrm{P}[\mathcal{T} \leq t] = 0.5$. Then the median is the unique value $t_m \in \Omega$ such that $\mathrm{P}[\mathcal{T} < t_m] < 0.5$ and $\mathrm{P}[\mathcal{T} \geq t_m] > 0.5$.

Other definitions also exist. The median, like the mean, is not necessarily a member of Ω. It is the preferred measure of location in distributions that are not symmetric. The mean has a key additive property. Suppose $\mathcal{T}_k$ is a random variable with mean μ_k, $k = 1, \cdots, K$. Then, for constants a_k, $k = 1, \cdots, K$,

$$\mathbf{E}[\sum_{k=1}^{K} a_k \mathcal{T}_k] = \sum_{k=1}^{K} a_k \mu_k \tag{1.9}$$

Unless otherwise specified, when we refer to a random variable from here on, we will mean a discrete random variable.

****

The **conditional probability** of the event Ω_1 given the event Ω_2 is defined by

$$P[\mathcal{T} \in \Omega_1 \mid \mathcal{T} \in \Omega_2] = \frac{P[\mathcal{T} \in \Omega_1 \cap \Omega_2]}{P[\mathcal{T} \in \Omega_2]} \tag{1.10}$$

provided $P[\mathcal{T} \in \Omega_2] > 0$. Any two events Ω_1 and Ω_2 are said to be **independent events** if the probability of any of them is not affected by whether the other has occurred or not. Formally, this means that

$$P[\mathcal{T} \in \Omega_1 \mid \mathcal{T} \in \Omega_2] = P[\mathcal{T} \in \Omega_1] \tag{1.11}$$

If Ω_1 and Ω_2 are independent events then it follows that

$$P[\mathcal{T} \in \Omega_1 \cap \mathcal{T} \in \Omega_2] = P[\mathcal{T} \in \Omega_1] P[\mathcal{T} \in \Omega_2] \tag{1.12}$$

In particular, if the random variables $\mathcal{T}_1$ and $\mathcal{T}_2$ are independent, then

$$P[\mathcal{T}_1 = t_1, \mathcal{T}_2 = t_2] = P[\mathcal{T}_1 = t_1] P[\mathcal{T}_2 = t_2] \tag{1.13}$$

for all t_1 and t_2.

If $\mathcal{T}_1$ and $\mathcal{T}_2$ are independent random variables with respective finite variances σ_1^2 and σ_2^2, and a_1, a_2 are some constants, then

$$\text{var}[a_1 \mathcal{T}_1 + a_2 \mathcal{T}_2] = a_1^2 \sigma_1^2 + a_2^2 \sigma_2^2 \tag{1.14}$$

where $\text{var}[\mathcal{T}]$ denotes the variance of $\mathcal{T}$.

1.4 Polynomial Based Distributions

A **polynomial based distribution**, denoted **PBD**, is defined as a probability distribution constructed from a polynomial. For example, consider the polynomial

$$f(\phi) = 2 + 7\phi^2 + 3\phi^5$$

with $\phi \geq 0$. Using the exponents of ϕ, we construct the sample space $\{0,2,5\}$. Then we define the random variable $\mathcal{T}$ on this sample space as

$$P[\mathcal{T} = 0; \phi] = \frac{2}{f(\phi)}$$
$$P[\mathcal{T} = 2; \phi] = \frac{7\phi^2}{f(\phi)}$$
$$P[\mathcal{T} = 5; \phi] = \frac{3\phi^5}{f(\phi)}$$

This satisfies all the required properties for a probability distribution. In general, a polynomial based distribution is defined as follows. Let Ω be a countable set and, for each $u \in \Omega$, let $c(u) > 0$ be an associated coefficient. With a parameter $\phi \geq 0$, we then construct the polynomial

$$f(\phi) = \sum_{u \in \Omega} c(u)\phi^u \qquad (1.15)$$

From this, we set up a discrete probability distribution for a random variable $\mathcal{T}$ as follows. For any $t \in \Omega$,

$$P[\mathcal{T} = t;\, \phi] = \frac{c(t)\phi^t}{f(\phi)} \qquad (1.16)$$

The generic polynomial $f(\phi)$ completely specifies this distribution. The set of its exponents constitutes the sample space, and for any point t in this space, the term with this exponent divided by the polynomial constitutes its probability.

Distribution (1.16) is expressed in an **exponential form** using the parameter $\beta = \ln(\phi)$ and the function

$$h(\beta) = f(\exp(\beta)) = \sum_{u \in \Omega} c(u)\exp(\beta u) \qquad (1.17)$$

where $\exp(\beta) = e^\beta$. Then

$$P[\mathcal{T} = t;\, \beta] = \frac{c(t)\exp(\beta t)}{h(\beta)} \qquad (1.18)$$

In the statistical literature, distribution (1.16) is called a **Power Series Distribution**. (1.18) is the **exponential form** of the power series distribution. In this book, we prefer to use the less technical sounding and more affable name, namely, Polynomial Based Distribution.

The polynomial $f(\phi)$ is called the **coefficient generating function**, the **series function**, or simply the **generating function** of the PBD. We shall refer to it as the **generating polynomial**, or **gp** of the PBD.

In probability theory, the term generating function refers to functions like the **probability generating function**, or the **moment generating function**, or the **characteristic function** (Gordon 1997). Such functions facilitate the study of properties of probability distributions. For example, the mean and variance are at times easier to determine by using such a function.

In the case of a PBD, the generating polynomial $f(\phi)$ is an equivalent replacement for the more elaborate generating functions. This is one of the many advantages of using the polynomial form. In fact, $f(\phi)$ is an unnormalized version of the probability generating function and also completely specifies it (Exercise 1.33). As we proceed in this text, it will become clear that for exact analysis of many discrete data models, it is simpler and more natural to deal with the generating polynomial than with the other generating functions.

For emphasis, we reiterate the basic property that the gp of a PBD completely and uniquely specifies the distribution of $\mathcal{T}$ in the following sense.

```
                The Generating Polynomial

    • The set of exponents of the generating polynomial of
      a PBD specifies the sample space, Ω.
    • The probability of a point in this space equals the
      term of this polynomial with that exponent divided
      by the whole polynomial.
```

The gp of a PBD can also be used to compute its mean and variance.

Theorem 1.1: Suppose $\mathcal{T}$ is a PBD variate with gp $f(\phi)$. If the mean, μ, and variance, σ^2, of $\mathcal{T}$ are finite, then

$$\mu = \phi \frac{d}{d\phi}[\ln f(\phi)] \tag{1.19}$$

$$= \frac{f_*(\phi)}{f(\phi)} = \frac{h'(\beta)}{h(\beta)} \tag{1.20}$$

$$\sigma^2 = \mu + \phi^2 \frac{d^2}{d^2\phi}[\ln f(\phi)] \tag{1.21}$$

$$= \frac{f_{**}(\phi)}{f(\phi)} - \left(\frac{f_*(\phi)}{f(\phi)}\right)^2 = \frac{h''(\beta)}{h(\beta)} - \left(\frac{h'(\beta)}{h(\beta)}\right)^2 \tag{1.22}$$

where

$$f_*(\phi) = \sum_{u \in \Omega} u c(u) \phi^u \quad \text{and} \quad f_{**}(\phi) = \sum_{u \in \Omega} u^2 c(u) \phi^u \tag{1.23}$$

further where $h'(\beta)$ and $h''(\beta)$ are the first two derivatives with respect to β of $h(\beta)$.

Proof: Consider the first portion rhs of the first equation.

$$\frac{d}{d\phi}[\ln f(\phi)] = \sum_{u \in \Omega} u c(u) \phi^{u-1} / f(\phi)$$

Multiplying by ϕ, this equals

$$\sum_{u \in \Omega} u \mathrm{P}[\mathcal{T} = u] = \mu$$

The other relations are proved similarly. □

In a majority of the applications we study, we need to combine a series of independent PBDs. The distribution of the sum of a set of random variables is called the **convolution** of these variables. It is not always easy to directly specify a convolved distribution. When the variables are independent, using the generating polynomial often provides an easier method.

Theorem 1.2: Let $\mathcal{T}_k$, $k = 1, 2$, be independent PBD variates with gp

$$f_k(\phi) = \sum_{u \in \Omega_k} c_k(u) \phi^u \tag{1.24}$$

Then $\mathcal{T} = \mathcal{T}_1 + \mathcal{T}_2$ is a PBD variate with probability function

$$\mathrm{P}[\mathcal{T} = t; \phi] = \frac{c(t)\phi^t}{f(\phi)} \tag{1.25}$$

POLYNOMIAL BASED DISTRIBUTIONS

where

$$c(t) = \sum_u c_1(u)c_2(t-u) \qquad (1.26)$$

and

$$f(\phi) = f_1(\phi)f_2(\phi) \qquad (1.27)$$

Further, the sample space of T is the set of exponents of the polynomial $f(\phi)$.

Proof: First note that

$$\begin{aligned} P[T=t;\phi] &= \sum_u P[T_1=u, T_2=t-u] \\ &= \sum_u P[T_1=u]P[T_2=t-u] \end{aligned}$$

The last step follows by the independence property. Substituting the probability formula, and after some rearrangement, we have

$$P[T=t;\phi] = \frac{\{\sum_u c_1(u)c_2(t-u)\}\,\phi^t}{f_1(\phi)f_2(\phi)}$$

The desired results are a direct consequence. □

Theorem 1.3: Suppose $T_1, \cdots, T_K$ are independent PBD variates with gps $f_1(\phi), \cdots, f_K(\phi)$, and sample spaces $\Omega_1, \cdots, \Omega_K$, respectively. Then $T = T_1 + \cdots + T_K$ has a PBD with gp $f(\phi)$ given by

$$f(\phi) = \prod_{k=1}^{K} f_k(\phi) \qquad (1.28)$$

Proof: Follows by induction from Theorem 1.2. □

In some applications we consider, $\Omega_k = \{l_k, l_k+1, \ldots, u_k\}$, where l_k and u_k are integers. In that case $\Omega = \{l, l+1, \ldots, u\}$ where $l = \Sigma_k l_k$ and $u = \Sigma_k u_k$.

The main message thereby is that the distribution of the sum of independent PBD variates with the same parameters is obtained as a PBD from the product of their generating polynomials.

****

The polynomial formulation is particularly useful because:

Central Observation

> Many of the discrete probability distributions commonly used in statistical analysis can be expressed in the form of a polynomial based distribution.

We demonstrate this observation below.

1.5 Binomial Distribution

Suppose n random patients from a homogeneous population receive a therapy for an acute disease. The outcome is cured ($\mathcal{Y} = 1$), or not cured ($\mathcal{Y} = 0$). Let $\mathcal{Y}_i$ denote the treatment result for the ith person, $i = 1, \ldots, n$. We make two assumptions: (i) All patients have the same chance, π, of being cured; (ii) Their outcomes are statistically independent of each other.

For this set of n independent and identically distributed (**iid**) Bernoulli trials, $\mathcal{T} = \Sigma_i \mathcal{Y}_i$ is the random total number of cures (or, in general, successes). The sample space of $\mathcal{T}$ is $\Omega = \{0, 1, \cdots, n\}$. Combinatorial arguments show that $\mathcal{T}$ has a **binomial distribution**, $B(n, \pi)$, given by

$$P[\mathcal{T} = t; \pi] = \binom{n}{t} \pi^t (1-\pi)^{n-t}, \qquad t \in \{0, 1, \ldots, n\} \tag{1.29}$$

where

$$\binom{n}{t} = \frac{n!}{t!(n-t)!} \tag{1.30}$$

Formulating the binomial parameter in terms of **odds**, i.e., the chance of success relative to the chance of failure, we have:

$$\phi = \pi/(1-\pi),$$

or, in terms of the logarithm of the odds (in short, **log-odds**):

$$\beta = \ln\{\pi/(1-\pi)\}$$

These transformations have the following properties:

- $0 \leq \pi \leq 1$ implies $0 \leq \phi \leq +\infty$, and $-\infty \leq \beta \leq +\infty$.
- These are one-to-one monotonic transformations with

$$\pi = \frac{\phi}{1+\phi} = \frac{\exp(\beta)}{1 + \exp(\beta)} \tag{1.31}$$

- Any inference on ϕ or β can be translated into one for π and vice versa.

With these transformations, the binomial probability becomes

$$P[\mathcal{T} = t; \phi] = \frac{c(t)\phi^t}{(1+\phi)^n} = \frac{c(t)\exp(\beta t)}{\{1 + \exp(\beta t)\}^n} \tag{1.32}$$

where $c(t) = n!/\{t!(n-t)!\}$. Hence, the binomial distribution is a polynomial based distribution with gp equal to

$$(1 + \phi)^n = \sum_{u=0}^{n} c(u)\phi^u \tag{1.33}$$

Example 1.3: In a study of mammography screening, Elmore et al. (1998) found that about

BINOMIAL DISTRIBUTION

one in two women tend to get a false positive report over a decade of annual screening for breast cancer. Then, for a random sample of twenty women, we need to compute the chance that at least 5 women will have a false positive report after a decade of mammography screening. First we compute the chance that at most 4 will have such an outcome. This is

$$2^{-20}\left(1 + 20 + \frac{20 \times 19}{2} + \frac{20 \times 19 \times 18}{3 \times 2} + \frac{20 \times 19 \times 18 \times 17}{4 \times 3 \times 2}\right)$$

which equals 0.006. Hence the probability that at least 5 women will have a false positive result in a decade of screening is about $1 - 0.006 = 0.994$.

****

For a binomial, we have that

$$\mu = n\pi = \frac{n\phi}{(1+\phi)} \quad \text{and} \quad \sigma^2 = n\pi(1-\pi) = \frac{n\phi}{(1+\phi)^2} \quad (1.34)$$

Suppose T_1 and T_2 are independent $B(n, \pi)$ and $B(m, \pi)$ variables. Then the gp of $T_1 + T_2$ is

$$(1+\phi)^n(1+\phi)^m = (1+\phi)^{(n+m)}$$

Thus $T_1 + T_2$ is a $B(n+m, \pi)$ variable.

****

For large n, the probability distribution of T is approximated by the **normal distribution**. The approximation holds well when π is not far from 0.5. Let $\Phi(x)$ denote the cumulative distribution function of the standard normal. From asymptotic theory, we know that

$$P[T \leq t] \approx \Phi\left(\frac{t - n\pi}{\sqrt{n\pi(1-\pi)}}\right) \quad (1.35)$$

A better approximation is provided with the use of the **continuity correction**. This gives

$$P[T \leq t] \approx \Phi\left(\frac{t + 0.5 - n\pi}{\sqrt{n\pi(1-\pi)}}\right) \quad (1.36)$$

We apply the normal approximation (without the continuity correction) to the breast cancer screening example given above. In this case, $n = 20$, $\pi = 0.5$ and $t = 5$. Then

$$z = \left(\frac{t - n\pi}{\sqrt{n\pi(1-\pi)}}\right) = -\sqrt{5}$$

From the standard normal distribution, this gives the desired probability as 0.988. This is close to the exact binomial probability, 0.994.

The assumption of a uniform cure rate (or identical trials) used to derive the binomial probability may not hold in practice. Alternatively, we may view the trials as a series of independent but nonidentically distributed Bernoulli trials. Suppose the chance of success in the ith trial is π_i ($i = 1, \cdots, n$), and π_i is modeled as a function of some features, called covariates, of the subject, or the study. For example, in a study where π_i stands for the chance the ith child will contract an ear infection, it is written as $\pi(x_i)$, where x_i indicates the gender of the ith child. Considerations of such additional variables or covariates often lead to the data being presented in the form of two or more 2×2 tables. Several chapters in this book deal with data in this format.

In some situations it may not be appropriate to assume that events are statistically independent. Modeling the total number of successes in such trials needs to account for the **extra binomial variation** or the interdependence between outcomes. There are many ways of doing it. Chapter 14 examines binary response models that relax the assumption of independence.

1.6 Poisson Distribution

The **Poisson distribution** is used for rare events, for example, for the number of accidents per day at a section of a roadway, or the number of new cases of a disease with a low incidence rate. Empirical studies indicate that it often models such processes with a reasonable accuracy.

Let Ω equal the set of nonnegative integers, $\{0, 1, 2, \ldots\}$. Then, for $t \in \Omega$, the Poisson probability is

$$P[\mathcal{T} = t; \lambda] = \lambda^t e^{-\lambda}/t! \quad \text{where} \quad \lambda > 0 \tag{1.37}$$

This can also be written as

$$P[\mathcal{T} = t; \lambda] = \frac{c(t)\lambda^t}{f(\lambda)} \tag{1.38}$$

where

$$c(t) = 1/t! \quad \text{and} \quad f(\lambda) = e^\lambda = \sum_{u=0}^{\infty} c(u)\lambda^u \tag{1.39}$$

Hence the Poisson is a PBD with $\Omega = \{0, 1, 2, \ldots, \}$, and $\phi = \lambda$.

****

Example 1.4: In a population based study of selected areas in Wisconsin, Nordstrom et al. (1998) estimated that the cases of newly diagnosed probable or definite carpal tunnel syndrome occur at the rate of 3.46 cases per 1000 person years. Assume that a random cohort of 100 subjects from the population without the condition is followed up for three years. We model this as a Poisson distribution with mean

$$\lambda = 3 \times 100 \times 0.00346 = 1.038$$

The chance that there at most two cases of carpal tunnel syndrome will occur in this sample during the study period is

POISSON DISTRIBUTION

$$e^{-1.038}\left(1 + 1.038 + \frac{1.038^2}{2}\right) = 0.913$$

****

For the Poisson variate T,

$$\mu = \sigma^2 = \lambda \qquad (1.40)$$

If T represents the number of events occurring within a unit period of time, then λ is the average rate of occurrence per unit time.

An important property of Poisson variables is additivity. That is, if T_i is Poisson distributed with parameter $\lambda_i, i = 1, \cdots, n$, and if these random variables are mutually independent, then ΣT_i is a Poisson variate with parameter $\Sigma \lambda_i$. This follows from the product of the n gps:

$$\prod_{i=1}^{n} \exp(\lambda_i) = \exp\left(\sum_{i=1}^{n} \lambda_i\right)$$

and recognizing that this is a gp of Poisson variable with mean $\Sigma \lambda_i$.

The binomial distribution arises from the Poisson as follows. Let T_1 and T_2 be independent Poisson variates with parameters λ_1 and λ_2. Consider the conditional distribution of T_1 when their sum is fixed:

$$P[T_1 = t \mid T_1 + T_2 = s] = \frac{P[T_1 = t, T_1 + T_2 = s]}{P[T_1 + T_2 = s]}$$

which is equal to

$$\frac{P[T_1 = t]P[T_2 = s - t]}{\sum_u P[T_1 = u]P[T_2 = s - u]}$$

Substituting the Poisson formula and simplifying, we get

$$P[T_1 = t \mid T_1 + T_2 = s] = \binom{s}{t}\pi^t(1-\pi)^{s-t} \qquad (1.41)$$

for $t = 0, 1, \ldots, s$, and where

$$\pi = \frac{\lambda_1}{\lambda_1 + \lambda_2} \qquad (1.42)$$

The conditional distribution of T_1 is thus a $B(s, \pi)$ distribution.

****

The Poisson distribution also arises as a limiting distribution to the Binomial. Let T have $B(n, \pi)$ distribution. Suppose n is large and π close to zero. Then

$$P[T = t] \approx \lambda^t e^{-\lambda}/t! \qquad (1.43)$$

where $\lambda = n\pi$. This property is known as the Poisson approximation to the binomial. The normal approximation to the Poisson is based on the standardized variate

$$z = \frac{(t-\lambda)}{\sqrt{\lambda}}$$

Normal approximations to the Poisson and binomial distributions may, however, not be adequate even at fairly large sample sizes, as succinctly demonstrated by Jollife (1995).

Poisson distributions are used to model disease occurrence over time or space. The number of cases of leukemia in the vicinity of a nuclear reactor has been modeled in terms of a Poisson distribution. The number of incident cases of HIV infection in a large population during a time period τ is another case. If λ is the rate of new HIV infection per unit time, the number of new cases during the period may be a Poisson variate with mean $\lambda\tau$.

For events distributed over time or space, the Poisson probability is derived under the following assumptions:

- The probability of an event in a small interval is proportional to the size of the interval.
- The probability of two or more events in a small time interval is negligible.
- Events in disjoint intervals are independent of one another.

The Poisson distribution is also used to model the rate λ as a function of covariates. In the case of incident HIV cases, for example, these may be urban or rural residence, illicit drug use, and gender.

1.7 Negative Binomial Distribution

Consider a series of iid Bernoulli trials, $\mathcal{Y}_1, \mathcal{Y}_2, \mathcal{Y}_3, \mathcal{Y}_4, \ldots$, with a common success probability π. For a fixed integer $r \geq 1$, let $\mathcal{T}$ denote the number of successes before the rth failure in the series $\mathcal{Y}_1, \mathcal{Y}_2, \mathcal{Y}_3, \mathcal{Y}_4, \ldots$. Then for $t = 0, 1, 2, \ldots$,

$$P[\mathcal{T} = t; \pi] = \binom{t+r-1}{r-1} \pi^t (1-\pi)^r \qquad (1.44)$$

This is called the **negative binomial distribution**. The special case when $r = 1$ is called the **geometric distribution**.

Now let $\phi = \pi$, $f(\phi) = (1-\phi)^{-r}$ and

$$c(t) = \binom{t+r-1}{r-1}$$

Then we can write

$$P[\mathcal{T} = t; \phi] = \frac{c(t)\phi^t}{f(\phi)} \qquad (1.45)$$

showing that the negative binomial is also a PBD on the sample space $\Omega = \{0, 1, 2, \ldots\}$. Note that here, $0 < \phi < 1$. In (1.45), we used the property that for ϕ in this interval,

HYPERGEOMETRIC DISTRIBUTION

$$(1 - \phi)^{-r} = \sum_{u=0}^{\infty} \binom{u + r - 1}{r - 1} \phi^u \qquad (1.46)$$

****

Example 1.5: Lazarou, Pomeranz and Corey (1998) reported a 6.7% incidence rate of serious or fatal adverse drug reaction among hospitalized patients in the USA. Suppose a hospital institutes a monitoring program for newly admitted patients. Assuming the 6.7% incidence rate, the chance that the first serious or fatal adverse drug reaction will occur in the tenth patient is

$$0.933^9 \times 0.067 = 0.035$$

****

For the negative binomial variable $\mathcal{T}_r$, we can show that

$$\mu = \frac{r\pi}{1 - \pi} \quad \text{and} \quad \sigma^2 = \frac{r\pi}{(1 - \pi)^2} \qquad (1.47)$$

1.8 Hypergeometric Distribution

Of a group of N individuals, assume n have a specific disease, and m do not. If we sample s persons at random without replacement from this group, the chance that t of them have the condition is

$$P[\mathcal{T} = t] = \binom{n}{t}\binom{m}{s-t}\binom{n+m}{s}^{-1} \qquad (1.48)$$

Let $l_1 = \max(0, s - m)$, $l_2 = \min(n, s)$. The sample space of $\mathcal{T}$ is $\Omega = \{l_1, l_1 + 1, \ldots, l_2\}$. This is the central **hypergeometric distribution** for which

$$\mu = \frac{ns}{(n + m)} \qquad (1.49)$$

$$\sigma^2 = \frac{nms(n + m - s)}{(n + m)^2(n + m - 1)} \qquad (1.50)$$

A more general version of this distribution arises from two binomial distributions. Let $\mathcal{A}$ be $B(n, \pi_1)$, and let $\mathcal{B}$ be $B(n, \pi_0)$, with $\mathcal{A}$ independent of $\mathcal{B}$. Consider the distribution of one of the variables if their sum is fixed, that is, $P[\mathcal{A} = t \mid \mathcal{A} + \mathcal{B} = s]$. By definition, this equals

$$\frac{P[\mathcal{A} = t]P[\mathcal{B} = s - t]}{P[\mathcal{A} + \mathcal{B} = s]} = \frac{P[\mathcal{A} = t]P[\mathcal{B} = s - t]}{\sum_u P[\mathcal{A} = u]P[\mathcal{B} = s - u]}$$

If we substitute the binomial expressions in the above, and let

$$c(t) = \binom{n}{t}\binom{m}{s-t} \quad \text{and} \quad \phi = \frac{\pi_1/(1-\pi_1)}{\pi_0/(1-\pi_0)} \tag{1.51}$$

then we get that

$$P[\mathcal{A} = t \mid \mathcal{A} + \mathcal{B} = s] = \frac{c(t)\phi^t}{\sum_{u \in \Omega} c(u)\phi^u} \tag{1.52}$$

where Ω is as defined above. (1.52) is called the noncentral hypergeometric distribution. The central hypergeometric is the special case when $\phi = 1$. The parameter ϕ is the **odds ratio**. We will have the occasion to consider it in depth later.

Now define

$$f(\phi) = \sum_{u \in \Omega} c(u)\phi^u \tag{1.53}$$

Then

$$P[\mathcal{A} = t \mid \mathcal{A} + \mathcal{B} = s] = \frac{c(t)\phi^t}{f(\phi)} \tag{1.54}$$

which obviously is a PBD.

****

Example 1.6: Suppose we randomly select 5 girls and 4 boys, all two years of age, from a population. They are monitored for six months for signs of ear disease. Let $\mathcal{A}$ denote the number of boys, and $\mathcal{B}$ denote the number of girls contracting an ear infection. Suppose a total of three children develop the infection. The conditional probability of $\mathcal{A}$, the number of boys with the infection, is obtained by first computing $c(t)$ for each t. The four possible values of t and the associated coefficients appear in Table 1.3

Table 1.3 *Hypergeometric Coefficients*

t	$c(t)$
0	10
1	40
2	30
3	4

With $f(\phi) = 10 + 40\phi + 30\phi^2 + 4\phi^3$, it follows that

$$P[\mathcal{A} = 0 \mid \mathcal{A} + \mathcal{B} = 3] = 10/f(\phi)$$
$$P[\mathcal{A} = 1 \mid \mathcal{A} + \mathcal{B} = 3] = 40\phi/f(\phi)$$
$$P[\mathcal{A} = 2 \mid \mathcal{A} + \mathcal{B} = 3] = 30\phi^2/f(\phi)$$
$$P[\mathcal{A} = 3 \mid \mathcal{A} + \mathcal{B} = 3] = 4\phi^3/f(\phi)$$

A GENERAL REPRESENTATION

****

For future reference, we list in Table 1.4 the gps of three commonly used univariate discrete distributions.

Table 1.4 Generating Polynomials

Distribution	ϕ	$f(\phi)$
Binomial	$\pi/(1-\pi)$	$(1+\phi)^n$
Poisson	λ	$\exp(\phi)$
Negative Binomial	π	$(1-\phi)^{-r}$

1.9 A General Representation

Now we define the multivariate polynomial based distribution. Consider a parameter vector, $\boldsymbol{\phi} = (\phi_1, \cdots, \phi_K)$, $\phi_k \geq 0$, and let $\boldsymbol{\mathcal{T}} = (\mathcal{T}_1, \cdots, \mathcal{T}_K)$ be a discrete random vector realizing values $\boldsymbol{t}$ in a countable K-dimensional set Ω. Define

$$\boldsymbol{\phi}^{\boldsymbol{t}} = \prod_{k=1}^{K} \phi^{t_k} \tag{1.55}$$

Note $\boldsymbol{\phi}^{\boldsymbol{t}}$ is **not** a vector. Let

$$f(\boldsymbol{\phi}) = \sum_{\boldsymbol{u} \in \Omega} c(\boldsymbol{u}) \boldsymbol{\phi}^{\boldsymbol{u}} \tag{1.56}$$

be a polynomial in K parameters. $\boldsymbol{\mathcal{T}}$ is said to have a multivariate polynomial based distribution with gp $f(\boldsymbol{\phi})$ if

$$\text{P}[\boldsymbol{\mathcal{T}} = \boldsymbol{t}] = \frac{c(\boldsymbol{t}) \boldsymbol{\phi}^{\boldsymbol{t}}}{f(\boldsymbol{\phi})} \tag{1.57}$$

where $c(\boldsymbol{t}) > 0$ for all $\boldsymbol{t} \in \Omega$.

To express this distribution in exponential form, we let $\phi_k = \exp(\beta_k)$ with $\boldsymbol{\beta} = (\beta_1, \cdots, \beta_K)$. Then

$$\text{P}[\boldsymbol{\mathcal{T}} = \boldsymbol{t}] = \frac{c(\boldsymbol{t}) \exp(\boldsymbol{\beta} \boldsymbol{t}')}{h(\boldsymbol{\beta})} \tag{1.58}$$

where

$$h(\boldsymbol{\beta}) = \sum_{\boldsymbol{u} \in \Omega} c(\boldsymbol{u}) \exp(\boldsymbol{\beta} \boldsymbol{u}') \tag{1.59}$$

The gp $f(\phi)$ has properties similar to those of the univariate gp $f(\phi)$. Some of these are listed below:

Property 1.4: $f(\phi)$ completely and uniquely specifies the multivariate PBD.

Property 1.5: $f(\phi)$ is an equivalent substitute for the probability generating function.

Property 1.6: The moments of a multivariate PBD are derived from the gp in a manner analogous to that for a univariate PBD.

Property 1.7: The sum of independent identically distributed multivariate PBD vectors is a multivariate PBD vector whose gp is a product of the gps of the vectors in the sum.

The proofs of these assertions are straightforward generalizations of respective proofs given for their univariate counterparts and are left to the exercises. Now let us consider an example.

1.10 The Multinomial Distribution

The multivariate distribution frequently applied in discrete data settings is the **multinomial distribution**. Consider a series of trials such that in each trial there are K possible outcome categories. An example of a three category case is: a child may either have no earache, have an ache in a single ear, or have the problem in both ears. Let the chance the kth outcome will materialize be π_k with $0 \le \pi_k \le 1$ and $\Sigma \pi_k = 1$. Let $\mathcal{T}_k$ denote the random number of times the kth category occurs in a series of n independent trials. The joint probability for these random variables is

$$P[\mathcal{T}_1 = t_1, \cdots, \mathcal{T}_K = t_K] = \frac{n!}{t_1! t_2! \cdots t_K!} \prod_k \pi_k^{t_k} \qquad (1.60)$$

We denote this distribution as $M(n; \pi_1, \cdots, \pi_K)$. The sample space Ω is the space of vectors $(t_1, \ldots, t_k)$ that satisfy

$$0 \le t_k \le n, \ k = 1, \cdots, K \quad \text{and} \quad \Sigma t_k = n$$

With the special case of $K = 3$, we illustrate a multivariate PBD. Suppose we perform n independent trials, and each trial has one of three outcomes labeled A, B and C. Let $(\mathcal{Y}_{1i}, \mathcal{Y}_{2i})$ be the outcome of the ith trial with $\mathcal{Y}_{1i} = 1$ if the ith trial results in "A", and $= 0$ otherwise; and $\mathcal{Y}_{2i} = 1$ if it results in "B", and $= 0$ otherwise. Let $\mathcal{T}_1, \mathcal{T}_2$ and $\mathcal{T}_3$ be the numbers of trials with outcomes A, B and C, respectively. Obviously,

$$\mathcal{T}_1 = \sum_{i=1}^{n} \mathcal{Y}_{1i}$$
$$\mathcal{T}_2 = \sum_{i=1}^{n} \mathcal{Y}_{2i}$$
$$\mathcal{T}_3 = n - (\mathcal{T}_1 + \mathcal{T}_2)$$

Further,

$$P[\mathcal{T}_1 = t_1, \mathcal{T}_2 = t_2] = c(t_1, t_2) \pi_1^{t_1} \pi_2^{t_2} \pi_3^{t_3}$$

THE MULTINOMIAL DISTRIBUTION

Table 1.5 *Trinomial Sample Space and Coefficients*

t_1	t_2	t_3	$c(t_1,t_2)$	t_1	t_2	t_3	$c(t_1,t_2)$
0	0	4	1	1	3	0	4
0	1	3	4	2	0	2	6
0	2	2	6	2	1	1	12
0	3	1	4	2	2	0	6
0	4	0	4	3	0	1	4
1	0	3	1	3	1	0	4
1	1	2	12	4	0	0	1
1	2	1	12				

where π_1, π_2 and π_3 are the chances of A, B and C, respectively, in any trial, with $0 \leq \pi_1, \pi_2, \pi_3 \leq 1$, and $\pi_1 + \pi_2 + \pi_3 = 1$; where $t_1 + t_2 + t_3 = n$; and further, where

$$c(t_1, t_2) = \frac{n!}{t_1! t_2! (n - t_1 - t_2)!} \qquad (1.61)$$

Let $\phi_1 = \pi_1/(1 - \pi_1 - \pi_2)$, and $\phi_2 = \pi_2/(1 - \pi_1 - \pi_2)$. With a slight rearrangement, we write this trinomial probability as

$$P[\mathcal{T}_1 = t_1, \mathcal{T}_2 = t_2] = \frac{c(t_1, t_2)\phi_1^{t_1}\phi_2^{t_2}}{f(\phi_1, \phi_2)} \qquad (1.62)$$

where

$$f(\phi_1, \phi_2) = (1 + \phi_1 + \phi_2)^n \qquad (1.63)$$

From basic algebra, we know that

$$f(\phi_1, \phi_2) = \sum_{(u,v) \in \Omega} c(u,v)\phi_1^{u_1}\phi_2^{u_2} \qquad (1.64)$$

where $\Omega = \{(u,v) : 0 \leq u, v \leq n \text{ and } 0 \leq u + v \leq n\}$. Hence, the trinomial distribution is a PBD. The general multinomial distribution is also shown to be a PBD in a similar way.

Example 1.7: Consider a concrete case with $n = 4$ and $K = 3$. A polynomial representation of this trinomial is in Table 1.5.

****

For the multinomial distribution, we can show that

$$\mu_k = \mathbf{E}[\mathcal{T}_k] = n\pi_k \qquad (1.65)$$

$$\sigma_k^2 = \text{var}[\mathcal{T}_k] = n\pi_k(1 - \pi_k) \qquad (1.66)$$

$$\sigma_{kj} = \mathbf{E}[(\mathcal{T}_k - \mu_k)(\mathcal{T}_j - \mu_j)] = -n\pi_k\pi_j, \quad k \neq j \qquad (1.67)$$

The following properties of a multinomial are useful in discrete data analysis.

Property 1.8: Aggregating a set of multinomial random variables will yield a smaller set of random variables which are also multinomial.

For example, suppose $(\mathcal{T}_1, \cdots, \mathcal{T}_5)$ is $M(n; \pi_1, \cdots, \pi_5)$. Define $\mathcal{S}_1, \mathcal{S}_2, \mathcal{S}_3$ as: $\mathcal{S}_1 = \mathcal{T}_1 + \mathcal{T}_2$, $\mathcal{S}_2 = \mathcal{T}_3 + \mathcal{T}_4$, and $\mathcal{S}_3 = \mathcal{T}_5$. Then $(\mathcal{S}_1, \mathcal{S}_2, \mathcal{S}_3)$ is $M(n; \pi_1 + \pi_2, \pi_3 + \pi_4, \pi_5)$.

Property 1.9: The distribution obtained by conditioning on distinct sums of multinomial outcomes is a product of multinomial distributions.

Suppose $(\mathcal{T}_1, \cdots, \mathcal{T}_5)$ is $M(n; \pi_1, \cdots, \pi_5)$. Consider, for example, the probability of $\mathcal{T}_1, \mathcal{T}_3, \mathcal{T}_4$ given $\mathcal{T}_1 + \mathcal{T}_2 = t$ (and thus $\mathcal{T}_3 + \mathcal{T}_4 + \mathcal{T}_5 = n - t$). This distribution is a product of $M(t; \pi_1^*, \pi_2^*)$ and $M(n - t; \pi_1^{**}, \pi_2^{**}, \pi_3^{**})$, where

$$\pi_1^* = \frac{\pi_1}{\pi_1 + \pi_2}$$
$$\pi_2^* = \frac{\pi_2}{\pi_1 + \pi_2}$$
$$\pi_1^{**} = \frac{\pi_3}{\pi_3 + \pi_4 + \pi_5}$$
$$\pi_2^{**} = \frac{\pi_4}{\pi_3 + \pi_4 + \pi_5}$$
$$\pi_3^{**} = \frac{\pi_5}{\pi_3 + \pi_4 + \pi_5}$$

Property 1.10: If $\mathcal{T}_k$, $k = 1, \ldots, K$, are independent Poisson(λ_k), then $P[\mathcal{T}_1 = t_1, \cdots, \mathcal{T}_K = t_K \mid \Sigma_k \mathcal{T}_k = s]$ is $M(s; \pi_1, \cdots, \pi_K)$ where $\pi_j = \lambda_j / \Sigma_k \lambda_k$.

The multinomial is the subtext for many distributions used to model and analyze discrete data. Properties 1.8, 1.9 and 1.10 are used to produce the conditional distribution or likelihood from which inference on the data is drawn.

The multinomial distribution is also used to model dependent discrete variables. For example, in the case of two Bernoulli variables, consider the occurrence or otherwise of cataract in the two eyes of an individual. These may not be independent events. One formulation of the possible dependence between them is as follows. Consider the variables $\mathcal{Y}_1$ and $\mathcal{Y}_0$ with respective observed value $y_1, y_0 \in \{0, 1\}$. Then let

$$P[\mathcal{Y}_1 = y_1, \mathcal{Y}_0 = y_0] = \frac{\theta_1^{y_1} \theta_0^{y_0} \theta_{10}^{y_1 y_0}}{1 + \theta_1 + \theta_0 + \theta_1 \theta_0 \theta_{10}} \quad (1.68)$$

where all the θ's are nonnegative parameters. Let $\pi_{ij} = P[\mathcal{Y}_1 = i, \mathcal{Y}_0 = j]$. This is a special case of a four outcome multinomial. In this distribution, if $\theta_{10} = 1$, then we get a product of two independent Bernoulli variables with success probabilities $\theta_j/(1 + \theta_j)$, $j = 1, 0$. We use models based on such a distribution in Chapter 14. This formulation can also be extended to more than two dependent binary variables.

1.11 The Negative Trinomial

Consider a sequence of independent trials with three outcomes, labeled $\{1, 2, 3\}$, in each trial. Let their respective probabilities of occurrence be π_1, π_2 and π_3 with $\pi_1 + \pi_2 + \pi_3 = 1$. We

SUFFICIENT STATISTICS

conduct the experiment until the total number of outcomes of type 3 reaches r. Let T_1, T_2 and T_3 respectively be the total number of outcomes of each type. The resulting probability distribution is a **negative trinomial distribution**. Using the arguments like those for the negative binomial, we can show that it has a polynomial form:

$$P[T_1 = t_1, T_2 = t_2] = \frac{c(t_1, t_2)\phi_1^{t_1}\phi_2^{t_2}}{f(\phi_1, \phi_2)} \qquad (1.69)$$

where $\phi_1 = \pi_1$, $\phi_2 = \pi_2$

$$c(t_1, t_2) = \binom{t_1 + t_2 + r - 1}{t_1 \quad t_2}$$

and

$$f(\phi_1, \phi_2) = (1 - \phi_1 - \phi_2)^{-r} \qquad (1.70)$$

Note that here $0 < \phi_1, \phi_2 < 1$.

This can be readily generalized to the case with K outcome categories.

****

The other cases of the multivariate PBD, which we will encounter later, include several forms of multivariate extensions of the hypergeometric and conditional distributions that are derived from multivariate discrete probability models.

1.12 Sufficient Statistics

Sufficiency is a key concept in conditional methods for data analysis. We introduce it through a sequence of n iid Bernoulli variables, $\{\mathcal{Y}_1, \cdots, \mathcal{Y}_n\}$. Set

$$\pi = \exp(\beta)/\{1 + \exp(\beta)\}$$

Then

$$P[\mathcal{Y}_1 = y_1, \ldots, \mathcal{Y}_n = y_n] = \frac{\exp(\beta \Sigma_i y_i)}{\{1 + \exp(\beta)\}^n}$$

Let $T = \Sigma_i \mathcal{Y}_i$, and consider the conditional probability

$$P[\mathcal{Y}_1 = y_1, \ldots, \mathcal{Y}_n = y_n \mid T = t]$$

$$= \frac{P[\mathcal{Y}_1 = y_1, \ldots, \mathcal{Y}_n = y_n]}{P[T = t]} = \frac{t!(n-t)!}{n!}$$

The conditional probability of the n iid Bernoullis given their sum does not contain the parameter β or π. In this sense, the knowledge of T has served to extricate the parameter from the distribution. Formally, we define:

Sufficient Statistic

A random variable $\mathcal{T}$ is sufficient for the parameter β if the conditional probability of the data given $\mathcal{T}$ does not contain the parameter.

Note $\mathcal{T}$ and β may be vectors. The **factorization theorem**, stated below without proof, often facilitates identification of sufficient statistics.

Theorem 1.4: Let $\mathcal{T}$ be a function of discrete random variable(s) $\mathcal{Y}$ (both may be vectors). $\mathcal{T}$ is sufficient for parameter β if and only if, for any y,

$$P[\mathcal{Y} = y; \beta] = g(t(y), \beta) q(y) \tag{1.71}$$

□

Two important results on sufficiency and PBD need to be stated.

Theorem 1.5: Let $\mathcal{T} = (\mathcal{T}_1, \cdots, \mathcal{T}_K)$ be multivariate PBD. Then $\mathcal{T}_j$ is sufficient for ϕ_j (or for β_j).

Proof: Use the definition or the factorization theorem. □

Theorem 1.6: Let $\mathcal{T} = (\mathcal{T}_1, \cdots, \mathcal{T}_K)$ have a multivariate PBD. The conditional distribution of a set of $\mathcal{T}_i$'s given some other $\mathcal{T}_j$'s is also multivariate PBD.

Proof: Let $\mathcal{T}_1$ and $\mathcal{T}_2$ be vectors of distinct elements of the vector $\mathcal{T}$ with ϕ_1 and ϕ_2, the corresponding parameter vectors. Write the joint generating polynomial as

$$f(\phi_1, \phi_2) = \sum_{(u_1, u_2) \in \Omega} c(u_1, u_2) \phi_1^{u_1} \phi_2^{u_2} \tag{1.72}$$

Then $P[\mathcal{T}_1 = t_1 \mid \mathcal{T}_2 = t_2]$ is

$$\frac{P[\mathcal{T}_1 = t_1, \mathcal{T}_2 = t_2]}{\sum_{u \in \Omega(., t_2)} P[\mathcal{T}_1 = u, \mathcal{T}_2 = t_2]} = \frac{c(t_1, t_2) \phi_1^{t_1}}{f(\phi_1, 1;., t_2)} \tag{1.73}$$

where

$$f(\phi_1, 1;., t_2) = \sum_{u \in \Omega(., t_2)} c(u, t_2) \phi_1^{u} \tag{1.74}$$

and where $\Omega(., t_2)$ is the set of values $\mathcal{T}_1$ from the vectors $(\mathcal{T}_1, \mathcal{T}_2) \in \Omega$ in which $\mathcal{T}_2 = t_2$. □

This conditional distribution excludes the parameters in ϕ_2. It is thus of use when the analysis only concerns the parameters in ϕ_1. For models with sufficient statistics, appropriate conditioning allows us to focus on the parameters of interest. It was observations along such lines that historically gave birth to the field of exact analysis of discrete data. In summary:

THE POLYNOMIAL FORM

- Sufficient statistics in a PBD are readily identifiable.
- The conditional distribution (1.73) is PBD.
- The conditional gp (1.74) has the same properties as the gp of an unconditional PBD.

The conditional PBDs we have seen thus far are: (i) The noncentral hypergeometric of §1.8, (ii) the binomial derived from conditioning on two Poisson distributions in §1.6 and (iii) the multinomial derived by conditioning on the sum of several Poisson variates noted in Property 1.10 of §1.10.

1.13 The Polynomial Form

We have shown many common discrete probability distributions can be expressed in a polynomial form. Using this form for exact (and even large sample) analysis of discrete distributions is additionally suggested for the following reasons. These reasons will become clearer as we proceed through the chapters of this text.

Why Use The Polynomial Form?

- Discrete data analysis often involves polynomial
  based distributions.
- Discrete distributions for which exact conditional
  methods have been developed so far mostly are those
  with the form of a univariate or a multivariate PBD.
- The common underlying distributional form allows
  us to construct a unified strategy for exact and
  asymptotic inference.
- The parameterizations that produce the polynomial
  forms are in accord with the common usage of odds
  ratios and log-odds ratios in conventional analysis
  of discrete data.
- The polynomial form promotes a unified development,
  and simple portrayal, of efficient computational
  algorithms for exact analysis of discrete data.

Computational algorithms for exact inference are too often described in unnecessarily intricate ways and appear to be very complex. A key aim of this book is to demonstrate that that is not the case.

Before we end this section, we give a general notation for polynomials, first invoked in Theorem 1.6 above, for later use. For example, consider a trivariate polynomial

$$f(\phi_1, \phi_2, \phi_3) = \sum_{(u_1, u_2, u_3) \in \Omega} c(u_1, u_2, u_3) \phi_1^{u_1} \phi_2^{u_2} \phi_3^{u_3} \qquad (1.75)$$

Suppose we want terms in this polynomial in which the exponents of ϕ_2 are equal to or greater than t_2, and the exponents of ϕ_3 are equal to t_3. We write this subpolynomial as $f(\phi_1, \phi_2, \phi_3; ., \geq t_2, t_3)$. That is

$$f(\phi_1, \phi_2, \phi_3; ., \geq t_2, t_3) = \phi_3^{t_3} \sum_{u_1} \sum_{u_2 \geq t_2} c(u_1, u_2, t_3) \phi_1^{u_1} \phi_2^{u_2} \qquad (1.76)$$

Next suppose we want terms in (1.75) in which the exponents of ϕ_2 are equal to t_2 and those of ϕ_3 are equal to t_3 when $\phi_2 = \phi_3 = 1$. We write this subpolynomial as $f(\phi_1, 1, 1;., t_2, t_3)$. That is

$$f(\phi_1, 1, 1;., t_2, t_3) = \sum_{u_1} c(u_1, t_2, t_3)\phi_1^{u_1} \qquad (1.77)$$

Such a notation is also useful for the sample space Ω. For example, $\Omega(., t_2, \geq t_3)$ represents that segment of Ω when the value along the second dimension is fixed at t_2 and the points along the third dimension are all greater than or equal to t_3.

This notation, which readily generalizes to higher dimensions and other forms of restrictions, is used later to represent conditional distributions derived from multivariate polynomial based distributions as well as their tail portions.

1.14 Relevant Literature

The material in this chapter is but a tiny portion from the vast field of probability and statistics. The references below, among many others, provide the broader background and further elaboration.

The main discrete distributions, the normal distribution and concepts like independence and conditioning are covered in many elementary books. For example, Thomas (1986) covers a wide ground in a readable manner. Gordon (1997) contains a particularly lucid account of discrete probability. Wild and George (2000) is an ideal elementary introduction to probability and statistical inference while Roussas (2003) clearly explains intermediate level material.

Many texts relate common continuous and discrete probability distributions. For instance, the cumulative binomial is linked to the F distribution and the Poisson, to the chisquare distribution. Many other asymptotic approximations to discrete distributions also exist; for example, the arcsine approximation of the binomial to the normal. For a comprehensive overview of discrete distributions, see Johnson and Kotz (1969) and Johnson, Kotz and Kemp (1992).

Books on stochastic processes generally derive the Poisson probability under mild assumptions. A readable work is Goodman (1988). Properties of the multinomial are discussed in Bishop, Fienberg and Holland (1975).

A sizeable theoretical literature on power series distributions, or what we call the PBD, exists. See Johnson and Kotz (1969) and Patil (1986). A multivariate PBD is a member of the discrete version of the multivariate exponential family of distributions (Patil 1985). A proof of a portion of Theorem 1.1 is in Patil (1986). Bivariate PBDs are well covered in Kocherlakota and Kocherlakota (1992). Pe'rez–Abreu (1991) has given a proof of the applicability of the Poisson approximation to power series distributions in general.

Distributions expressed in a polynomial form (or its exponential version) are found in many papers dealing with exact inference on discrete data. Cases in point: The exact distributions in the seminal paper, Zelen (1971), are in a polynomial form. So are several in the comprehensive review, Agresti (1992). The exact distributions for logistic regression models of discrete data in the influential text, Cox and Snell (1989), are in the exponential polynomial form. An explicit linkage between power series distributions and those used in the analysis of discrete data was made in Hirji (1997a).

The history of generating functions in exact inference on discrete data is a long one. See Cox (1958), Cox (1970) and Hirji, Mehta and Patel (1987). Two recent papers showing their use for

EXERCISES

discrete data analysis are Baglivo, Pagano and Spino (1996) and Hirji (1997a). van de Wiel, Di Bucchianico and van der Laan (1999) has a broad perspective on the subject and notes other relevant material.

For a rigorous approach to sufficiency, see Lehmann (1986) and Cox and Hinkley (1974). Application of sufficiency to discrete data models is covered in Cox and Snell (1989) and Bishop, Fienberg and Holland (1975). Pratt and Gibbons (1981) has an elementary proof of the factorization theorem.

1.15 Exercises

1.1. Construct a probability distribution based on each of the following polynomials: (i) $f_1(\phi) = 3\phi^3 + 11\phi^5 + 7\phi^7$; (ii) $f_2(\phi) = \phi^3 + 4\phi^4 + 5\phi^5 + 6\phi^6 + 7\phi^7$; and (iii) $f_3(\phi) = \phi^{-1} + 3 + 4\phi + 4\phi^2 + 3\phi^5 + \phi^9$. For each distribution compute $P[T = 5; \phi = 1.5]$ and $P[T < 5; \phi = 1.5]$.

1.2. With $f_1(\phi)$ and $f_2(\phi)$ defined above, let $f(\phi) = f_1(\phi) f_2(\phi)$ be the generating polynomial of a discrete PBD random variable T. Compute $P[T = 10; \phi = 1.5]$ and $P[T \geq 10; \phi = 1.5]$. Also compute the mean, median and variance of this distribution when $\phi = 1.5$.

1.3. A rare disease occurs in region A at an average rate of three cases a month, and in region B, at an average rate of two cases a month. Assume that cases in any one region occur independently of that in the other. Compute the probability that in a given month: (i) a total of at most three cases are reported from the two regions, (ii) at least two cases are reported in each region and (iii) no cases are reported in either region.

1.4. Kessler (1993) estimated that the reporting rate for serious adverse reactions to prescription drugs in the U.S. is about 1%. Using this as the underlying reporting rate, compute the probability that of the 100 serious reactions occurring in a given period at a medical facility, at most 5 will be reported. Use three methods for this: the exact binomial probability, the Poisson approximation to the binomial and the normal approximation to the binomial. Comment on the results.

1.5. For the geometric variate T with $\pi = 0.2$, what is (i) $P[T = 2]$ and (ii) $P[T > 4]$?

1.6. Show that the negative binomial variable arises as the sum of r independent geometric variables. Use this to derive its mean and variance.

1.7. Determine the mean, median and variance of the PBDs with generating polynomials

$$f(\phi) = (1+\phi)^2 (1 + 2\phi + \phi^2)^2$$

and

$$f(\phi) = (2 + 3\phi)(3 + 2\phi)^2$$

1.8. Let $\Omega = \{1, 2, \cdots, n\}$ and let $P[T = t] = 1/n$ for $t \in \Omega$. T is the discrete uniform random variable. Is T a PBD variate? Compute the mean, median and variance of T.

1.9. If T_1 and T_2 are discrete uniform variates on $\Omega_1 = \{1, 2, \cdots, n_1\}$ and $\Omega_2 = \{1, 2, \cdots, n_2\}$, what is the distribution of (i) $T_1 + T_2$ and (ii) $\min(T_1, T_2)$?

1.10. For a Poisson variate T, find the probability that (i) T is even and (ii) T is odd.

1.11. Prove the variance formulae in Theorem 1.1.

1.12. Derive the mean and variance of the binomial, Poisson, central hypergeometric and negative binomial distributions in two ways, directly and by using Theorem 1.1.

1.13. What is the median of $B(n, 0.5)$ when (i) n is even and (ii) n is odd?

1.14. Determine the medians of the Poisson distributions with $\lambda = 0.5, 1.0, 1.5$.

1.15. What is the median of the hypergeometric distribution when (i) $\phi = 1.0$, and (ii) $\phi = 0.5$ and $n = m$.

1.16. Determine the medians of the negative binomial distributions with $r = 5$ and $\pi = 0.25, 0.50, 0.75$.

1.17. How would you determine the median of the geometric distribution?

1.18. The kth factorial moment of $\mathcal{T}$ is

$$\mathbf{E}[\mathcal{T}(\mathcal{T}-1)\ldots(\mathcal{T}-k+1)]$$

Determine the kth factorial moment of a PBD in terms of its gp, $f(\phi)$. Use this to find the kth factorial moments of the binomial, Poisson and negative binomial distributions.

1.19. (i) If $\mathcal{T}$ is a $B(n, \pi)$ variate, then prove that

$$\mathbf{E}[(\mathcal{T} - n\pi)^3] = n\pi(1-\pi)(1-2\pi)$$

(ii) If $\mathcal{T}$ is a Poisson variate with parameter λ, then prove that

$$\mathbf{E}[(\mathcal{T} - \lambda)^3] = \lambda$$

1.20. Let $(\mathcal{T}_1, \mathcal{T}_2, \mathcal{T}_3)$ have $M(12; 0.5, 0.25, 0.25)$ distribution. What is the (i) marginal probability of $\mathcal{T}_1$ and (ii) conditional probability of $\mathcal{T}_1$ given $\mathcal{T}_2 = t_2$? Further, compute: (i) $P[\mathcal{T}_1 = 4, \mathcal{T}_2 = 3]$, (ii) $P[\mathcal{T}_1 \geq 3, \mathcal{T}_2 \geq 2]$, (iii) $P[\mathcal{T}_1 = 4]$, and (iv) $P[\mathcal{T}_1 \leq 4]$.

1.21. What is the generating polynomial for a multinomial variate with n trials and K outcome categories? For this distribution, show that

$$\sum_{\boldsymbol{t} \in \Omega} c(\boldsymbol{t}) = K^n$$

1.22. For the trinomial distribution of §1.10, show that the conditional distribution of $\mathcal{T}_1$ given $\mathcal{T}_2$ is a PBD.

1.23. Derive the mean, variance and covariance formulae for the multinomial distribution stated in §1.10.

1.24. Consider the model for two dependent Bernoulli variates in §1.10. What is the probability of having cataracts in both eyes given that at least one is affected?

1.25. Suppose $\mathcal{T}_1$ is a $B(n, \pi_1)$ and $\mathcal{T}_2$ is a $B(m, \pi_2)$ random variable. If they are independent, determine the distribution of $\mathcal{T}_1 + \mathcal{T}_2$. Is it a PBD?

1.26. Give a formal proof of Theorem 1.3.

1.27. Prove that the sum of iid multivariate PBD vectors is a multivariate PBD vector whose gp is a product of the gps of the vectors in the sum.

1.28. Prove the properties 1.8, 1.9 and 1.10 of the multinomial distribution stated in §1.10, and state them in a general form.

1.29. If $(\mathcal{T}_1, \mathcal{T}_2)$ has a bivariate PBD, then derive the covariance of $\mathcal{T}_1$ and $\mathcal{T}_2$ in terms of its generating function. Extend Theorem 1.1 to a multivariate PBD including the specification of the covariances in terms of the generating polynomial $f(\boldsymbol{\phi})$. Apply these results to the multinomial distribution and the negative trinomial distribution.

1.30. Suppose $\mathcal{T}_j, j = 1, \ldots, K$ is a binomial variate with success probability π and total number of trials equal to n_j. How would you obtain the distribution of $\Sigma_j \mathcal{T}_j^2$? Extend your result to K Poisson, negative binomial and general PBD distributions.

EXERCISES

1.31. Give a detailed proof that the negative trinomial of §1.11 is a PBD and clearly specify its sample space. Further: (i) Derive the mean and variance of T_1 and T_2 as well as the covariance between them. (ii) What is the distribution of T_1+T_2? (iii) What is the marginal distribution of T_1? (iv) Generalize this to the case with K outcome categories.

1.32. Consider an experiment with three outcome categories in each independent trial. It is continued until r outcomes of either type 2 or type 3 occur. Derive the joint distribution of (T_1, T_2). Is it a PBD?

1.33. Suppose $T_1, \cdots, T_K$ are independent but not necessarily identically distributed PBD variates with gps $f_1(\phi_1), \cdots, f_K(\phi_K)$, and sample spaces $\Omega_1, \cdots, \Omega_K$, respectively. Is the distribution of $T = T_1 + \cdots + T_K$ necessarily a PBD? Apply this to: (i) three independent binomial variates, (ii) three independent Poisson variates and (iii) two independent negative binomial variates.

1.34. For a nonnegative random variable T, and for $|\psi| < 1$, the probability generating function, $G_T(\psi)$, is defined by

$$G_T(\psi) = \mathbf{E}[\psi^T] = \sum_t \psi^t \mathrm{P}[T=t]$$

Suppose T is a PBD variate with gp $f(\phi)$. Then show that

$$G_T(\psi) = \frac{f(\psi\phi)}{f(\phi)}$$

Therefore, the probability generating function of a PBD is the ratio of two functions, each of which is obtained from the gp, $f(\phi)$. Extend this to the multivariate PBD.

1.35. Prove the Factorization Theorem for sufficient statistics (Theorem 1.4).

1.36. For a discrete random variable taking nonnegative integer values, T, show that

$$\mathbf{E}[T] = \sum_{j=0}^{\infty} \mathrm{P}[T > j]$$

Use this to find the mean of the geometric distribution.

1.37. Suppose the distribution of T given $\mathcal{N} = n$ is $\mathrm{B}(n, \pi)$, with the distribution of $\mathcal{N}$ being Poisson with mean λ. What is the unconditional distribution of T? Is it a PBD?

1.38. Suppose the distribution of T given $\mathcal{N} = n$ is $\mathrm{B}(n, \pi)$, with the distribution of $\mathcal{N}$ being a negative binomial with r, the maximal number of failures allowed, and success probability, π_*. What is the unconditional distribution of T? Is it a PBD?

1.39. Suppose the distribution of T given $\mathcal{R} = r$ is a negative binomial with r, the maximal number of failures allowed, and success probability, π. Further, the distribution of $\mathcal{R}$ is Poisson with mean λ. What is the unconditional distribution of T? Is it a PBD?

1.40. Consider K independent events, $A_1, \cdots, A_K$. Show that the probability that at least one of them will occur is

$$\mathrm{P}[\bigcup_{k=1}^{K} A_k] = 1 - \prod_{k=1}^{K}(1 - \mathrm{P}[A_k])$$

CHAPTER 2

One-Sided Univariate Analysis

2.1 Introduction

This chapter introduces exact methods for analyzing data from polynomial based distributions. The corresponding asymptotic methods are also described. Its specific aims are:

- To formulate one-sided and two-sided hypotheses setups for the parameter of a statistical model.
- To show how the tail area of a probability distribution can be used as a measure of evidence for a one-sided hypothesis.
- To formulate one-sided conventional exact, mid-*p* exact and asymptotic evidence functions for univariate PBDs.
- To define one-sided *p*-values and confidence intervals, and relate them to the one-sided evidence function.
- To define the size, significance level and critical region of a statistical test.
- To begin the discussion, to be continued in later chapters, of the linkage between study design and method of data analysis.

While the binomial, Poisson and negative binomial distributions are used to illustrate the ideas of this chapter, they apply broadly to other naturally univariate PBDs and to those univariate PBDs constructed by conditioning from a multivariate PBD. For now, we focus on one-sided methods. Though less frequently applied in practice, they form a useful device for explaining the logic of the tail area approach to statistical inference, and a foundation upon which the more common two-sided methods are formulated.

2.2 One Parameter Inference

The scientific aims of a study often require assessing the value of one or more of the parameters of a statistical model. We seek, for example, to evaluate the diagnostic utility of a neural network algorithm for subjects who have had a myocardial infarction. Its actual but unknown false diagnosis rate among the patients with an infarct is π. We want to check if this rate exceeds π_0, the maximally acceptable rate.

We begin with defining a key idea in study design. Suppose we have a large population of subjects whose characteristics are defined by the random variable $\mathcal{T}$ (it may be a vector). Suppose we sample n subjects from this population. Let $\mathcal{T}_j$ denote the characteristic of the jth subject in this sample. This sample is said to be **a random sample** from the parent population if (i) for any t, $\mathrm{P}[\mathcal{T}_j = t] = \mathrm{P}[\mathcal{T} = t]$ and (ii) $\mathcal{T}_j$, $j = 1, \ldots, n$ are mutually independent.

A random sample has the feature that probability statements we make about it apply to the population from which it was drawn as well. It is then said to protect the **generalizability** or **external validity** of the study.

In particular, for a random sample with binary variables, the expected mean value is the population proportion of the binary characteristic.

$$\mathbf{E}\left[\frac{1}{n}\sum_j \mathcal{T}_j\right] = \mathbf{E}[\mathcal{T}] = \pi$$

Example 2.1: Suppose $\pi_0 = 0.05$, or 5%. In one study, a particular neural network algorithm correctly identified 35 of the 36 patients who actually had a myocardial infarct. The observed error rate thereby was $1/36$ or 2.8% (Baxt 1991; Newman 1995). What do we conclude?

****

The objective of such a study may be stated as a choice between two competing hypotheses regarding the parameter of interest. They are the **null hypothesis**, denoted H_0, and the **alternative hypothesis**, denoted H_1. Several ways of framing such hypotheses exist. In a **one-sided setup,** the null and the alternative hypotheses represent a segment of the real line to the left or right of a specified value. For example, we write:

$$H_0 : \pi \leq \pi_0 \quad \text{versus} \quad H_1 : \pi > \pi_0 \qquad (2.1)$$

Alternatively, the directions of the one-sided null and alternative hypotheses may be reversed. That is:

$$H_0 : \pi \geq \pi_0 \quad \text{versus} \quad H_1 : \pi < \pi_0 \qquad (2.2)$$

One-sided hypotheses are also called directional hypotheses. At times, the one-sided hypothesis testing setup (for example, (2.2)) is, not quite accurately, written as testing a simple $H_0 : \pi = \pi_0$ versus $H_1 : \pi < \pi_0$.

****

Example 2.2: The gender of a newborn baby is a random entity. Some environmental factors may, however, alter the male to female ratio in the newborn babies. Irgens et al. (1997) looked at this ratio among the offsprings of workers who had been exposed to low frequency electromagnetic fields (LFEMF) on the job. Let us assume that the proportion of male offsprings in an unexposed but otherwise comparable reference population is known. Call it π_0, and let π be the unknown proportion of males in the LFEMF exposed population. Then we ask:

Does occupational exposure to LFEMF affect the offspring sex ratio? That is, is π different from π_0?

****

Such a query is formalized in a **two-sided hypotheses setup.** Here the null and alternative hypotheses are couched in terms of equivalence or nonequivalence to a given value.

$$H_0 : \pi = \pi_0 \quad \text{versus} \quad H_1 : \pi \neq \pi_0 \qquad (2.3)$$

Science does not deal with absolute certainties. Statistical analysis of data, in that spirit, also does not strive to prove or disprove an hypothesis with the certitude that a mathematical postulate is proved or disproved. Instead it constructs and applies measures of evidence. The measures indicate whether the data favor this or that hypothesis, and allow us to draw plausible conclusions about the relevant statistical model.

These statistical measures of evidence, however, are themselves subject to error. These errors basically come in two guises: (i) errors that intrinsically favor one or the other hypothesis, or tend to deviate from the truth in a fixed direction are called **systematic errors**; and (ii) errors that deviate from the truth in a random manner are called **random errors**. Reducing the first kind of error improves the **accuracy**, and reducing the second kind improves the **precision** of the analysis. We thus strive to construct measures of evidence that have high or best possible levels of accuracy and precision.

The *p*-value is one basic measure of evidence used in data analysis. It is a gauge of the level of evidence against the null hypothesis, and allows us to assess the degree to which the evidence suggests if it is false or not. In a two-sided setup, when the null hypothesis postulates the absence of a systematic effect, the *p*-value indicates whether the observed effect emanates from purely random fluctuations or not.

The *p*-value tells us something about the existence of an effect. We generally need an estimate of its magnitude as well. In the case at hand, we need to estimate π. This may be in the form of a **point estimate**, or a range of values within which the true parameter may be declared to lie. Such a range is called a **confidence interval**. Saying that the true value of the parameter lies within such an interval is also subject to systematic or random error.

The basic aim of statistical analysis then is to apply measures of evidence whose error rates are either known or can be otherwise controlled, and at the same time, are small or within acceptable limits.

Other than testing and estimating the parameters of a statistical model with reasonable accuracy, data analysis involves other activities like variable selection, prediction, checking the type of model used, detecting outliers in the data, adjusting for missing values, and so on, as well. These, however, are not the focus of the present work.

2.3 Tail Probability and Evidence

To develop one-sided methods, we take the case of a particular adverse effect that may be induced by a medicinal drug. Let π be the unknown underlying proportion at which it occurs in the patient population taking the drug. Suppose we pick, at random, n subjects from this population. T, the number of them exhibiting the side effect, is a $B(n, \pi)$ variable. Under the polynomial formulation, we write

$$P[T = t] = \frac{c(t)\phi^t}{f(\phi)} \qquad (2.4)$$

where

$$\phi = \frac{\pi}{1 - \pi}, \quad c(t) = \binom{n}{t},$$

and

$$f(\phi) = \sum_{u \in \Omega} c(u)\phi^u$$

and further where $\Omega = \{0, 1, \ldots, n\}$. The parameter of interest is ϕ (or, equivalently, π).

Example 2.3: It is hypothesized that in a population of eligible patients, not more than 10% would suffer the adverse effect if given the drug. Consider the one-sided setup:

$$H_0 : \pi \leq 0.1 \quad \text{or} \quad \phi \leq 1/9$$
$$\text{versus}$$
$$H_1 : \pi > 0.1 \quad \text{or} \quad \phi > 1/9$$

If the study has nine subjects ($n = 9$), the distribution of the number of subjects experiencing the effect is specified by the generating polynomial

$$\begin{aligned} f(\phi) &= (1 + \phi)^9 \\ &= 1 + 9\phi + 36\phi^2 + 84\phi^3 + 126\phi^4 + \\ &\quad 126\phi^5 + 84\phi^6 + 36\phi^7 + 9\phi^8 + \phi^9 \end{aligned}$$

Assume that two of the nine patients showed the effect, giving an observed proportion of 22%. Is this sufficient evidence to doubt the hypothesis that the true proportion does not exceed 10%? Or is it reflective of sampling variability? How do we gauge the level of evidence for or against H_0?

One entity that has the potential to be a measure of evidence in this context is the right tail probability of the distribution (2.4). Let us examine this possibility. In particular, the right tail probability for the B$(9, \pi)$ example with $t = 2$ is

$$P[T \geq 2; \phi] = 1 - P[T < 2; \phi] = 1 - \frac{1 + 9\phi}{(1 + \phi)^9}$$

Consider two specific values of π or ϕ, namely $\pi = 0.10$ ($\phi = 0.1111$) and $\pi = 0.40$ ($\phi = 0.6667$). For the former, $P[T \geq 2] = 0.225$ and for the latter, $P[T \geq 2] = 0.925$. Higher values of ϕ then lead to a larger mass in the right tail. This is also seen in a plot of this tail probability as a function of ϕ. Figure 2.1 shows that the right tail probability increases with ϕ (and thus with π). We formally prove this by taking the derivative of $P[T \geq 2; \phi]$ with respect to ϕ.

$$\frac{d}{d\phi} P[T \geq 2; \phi] = \frac{72\phi}{(1 + \phi)^8}$$

which is strictly positive for $\phi > 0$. $P[T \geq 2; \phi]$ is therefore a strictly increasing function of ϕ.

****

TAIL PROBABILITY AND EVIDENCE

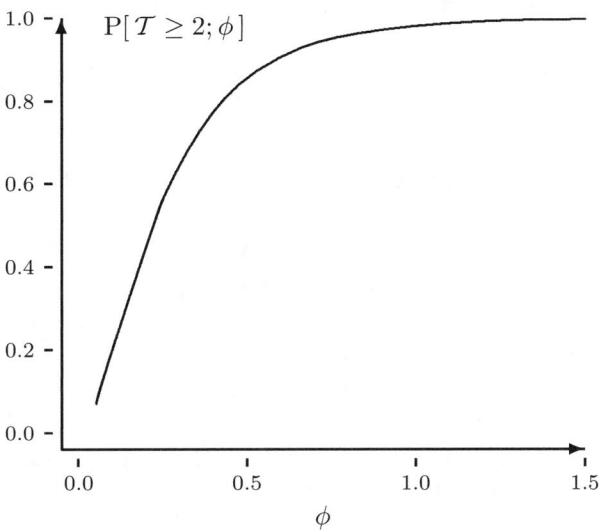

Figure 2.1 $Pr[\mathcal{T} \geq 2; \phi]$ as a Function of ϕ.

This result holds for a PBD in general. Recall that under the notation of §1.13, selected terms of the gp $f(\phi)$ are written as:

$$f(\phi;t) = c(t)\phi^t \tag{2.5}$$

$$f(\phi;\leq t) = \sum_{u \leq t} c(u)\phi^u \tag{2.6}$$

$$f(\phi;\geq t) = \sum_{u \geq t} c(u)\phi^u \tag{2.7}$$

Then we write the observed, and the left and right tail probabilities as

$$P[\mathcal{T} = t; \phi] = \frac{f(\phi;t)}{f(\phi)} \tag{2.8}$$

$$P[\mathcal{T} \leq t; \phi] = \frac{f(\phi;\leq t)}{f(\phi)} \tag{2.9}$$

$$P[\mathcal{T} \geq t; \phi] = \frac{f(\phi;\geq t)}{f(\phi)} \tag{2.10}$$

Theorem 2.1: Suppose $\mathcal{T}$ follows a univariate PBD. (i) If for some $t \in \Omega$, there exists $t_* \in \Omega$ such that $t_* < t$, then $P[\mathcal{T} \geq t; \phi]$ is a monotonically increasing function of ϕ. (ii) If for some $t \in \Omega$, there exists $t_{**} \in \Omega$ such that $t_{**} > t$, then $P[\mathcal{T} \leq t; \phi]$ is a monotonically decreasing function of ϕ.

Proof: With the above notation,

$$P[\mathcal{T} \geq t;\ \phi] = \frac{f(\phi;\geq t)}{f(\phi)}$$

Taking the derivative with respect to ϕ and simplifying, we get that the derivative of the right tail probability equals

$$\{f(\phi)\}^{-2} \sum_{u<t} \sum_{v \geq t} (v-u) c(u) c(v) \phi^{u+v-1}$$

which is positive for $\phi > 0$. Part (ii) is proved similarly. □.

Observation for a PBD

- At large values of ϕ, the probability mass is concentrated in the right tail.
- At small values of ϕ, it is concentrated in the left tail.
- Since the one-sided null and alternative hypotheses divide the parameter space into two parts, with smaller or larger values of ϕ, it is logical to regard these tail probabilities as measures of evidence in relation to them.

A measure of evidence derived from such reasoning is called a **tail area based measure of evidence**; those who adopt it follow the **Tail Area School** of inference. If an experiment yields $\mathcal{T} = t$ and if the probability mass in the right tail for some value of $\phi = \phi_0$ is "small," we then infer that the observation may be from a distribution with a larger actual value of ϕ. To be specific:

Key Result

For an observed t,

- The right tail area, $P[\mathcal{T} \geq t; \phi]$, as a function of ϕ, is a one-sided evidence function for the one-sided hypothesis setup (2.1).
- The left tail area, $P[\mathcal{T} \leq t; \phi]$, as a function of ϕ, is a one-sided evidence function for the one-sided hypothesis setup (2.2).

From here on, we use the notation $E(t;\phi)$ to depict an evidence function. (Note the difference with the symbol $\mathbf{E}[.]$ which denotes the expected value.) Evidence functions or the basic idea they embody are rarely referred to in explicit terms. Yet they, and especially the two-sided evidence function we define in Chapter 3, lie at the heart of most of the univariate data analysis methods employed in the applied literature. Further, when they have been explicitly proposed, it has been in a two-sided context. Also, in the literature, the evidence function is actually known by other names.

EXACT EVIDENCE FUNCTION

The Other Names

An evidence function is also called (i) a confidence curve, (ii) a confidence interval function, (iii) a p-value function or (iv) a confluence function.

2.4 Exact Evidence Function

The p-value is a commonly used index of evidence. For the hypothesis, $H_0 : \phi \leq \phi_0$, and an observed $\mathcal{T} = t$, the one-sided exact p-value is defined as

$$p(t; \phi_0) = \max_{\phi} \{P[\mathcal{T} \geq t; \phi] : H_0\} = \frac{f(\phi_0; \geq t)}{f(\phi_0)} \qquad (2.11)$$

This one-sided p-value is the maximum probability, over the null parameter values, of realizing a value for $\mathcal{T}$ as large as, or larger than that actually observed. For a PBD, it equals the value of the associated one-sided evidence function at the null, and is thus intrinsically linked to this function. The monotonicity of this function implies that:

p-Values and Evidence

- Small p-values do not support H_0
- Large p-values favor H_0

Example 2.3 (continued): For the drug adverse effect study ($n = 9, t = 2$), the p-value is

$$P[\mathcal{T} \geq 2; \phi_0 = 0.1111] = 0.2252$$

****

Under the one-sided setup $H_0 : \phi \geq \phi_0$ against the alternative $H_1 : \phi < \phi_0$, the left tail probability is an appropriate evidence function and is similarly used to compute a p-value. As other definitions of a p-value are also plausible, we denote the p-value in (2.11) as p_c, and call it the **conventional exact p-value**, or simply the **exact p-value**. The symbol p refers to a generic p-value without regard to any specific definition.

Several other types of evidence functions for one-sided setups also exist. Below we define two, namely, the one-sided mid-p and the one-sided asymptotic evidence functions.

2.5 Mid-p Evidence Function

The above definition of one-sided evidence function and related p-value assigned a specific meaning to the term "right tail of a probability distribution." This was taken as $P[\mathcal{T} \geq t]$. Alternatively it may be set as $P[\mathcal{T} > t]$ (Smith et al. 1979). The difference is immaterial for a continuous distribution. In the discrete case, it can be consequential. As a compromise, we average the two to produce a **mid-p evidence function** for the one-sided setup (2.1):

$$E_m(t;\phi) = P[\mathcal{T} \geq t;\ \phi] - 0.5P[\mathcal{T} = t;\ \phi] \qquad (2.12)$$

This function has the same monotonicity property as the conventional right tail in Theorem 2.1 (Exercise 2.13) and is thus a legitimate evidence function. A *p*-value based on it is called a **one-sided mid-p-value**, and is denoted as p_m. This is the value of the function at ϕ_0 and so is

$$p_m(t;\phi_0) = p_c(t;\phi_0) - 0.5\frac{f(\phi_0;t)}{f(\phi_0)} = \frac{f(\phi_0;\geq t) - 0.5f(\phi_0;t)}{f(\phi_0)} \qquad (2.13)$$

The mid-*p* approach puts half the probability of the realized observation in the right tail and the other half in the left tail. The total probability in the two tails now adds up to one; this was not the case with the previous definition of a tail probability.

Example 2.3 (continued): For the drug adverse effects study with $B(9, \pi)$, the mid-*p*-value for $H_0 : \phi \geq 1/9$ versus $H_1 : \phi < 1/9$ at $t = 2$ is $0.2252 - 0.1722/2 = 0.1371$. In this example, as generally, the mid-*p*-value is smaller than the conventional exact *p*-value and is thus more favorable towards the alternative hypothesis.

****

Example 2.4: Another application involves a Poisson model. Suppose, to examine the incidence of a disease in a population, a random sample of 550 persons was followed up. The follow up times among subjects were varied. The cases of the disease diagnosed in the sample numbered 4 with the total follow up time being to 1350 person years.

Let λ be the occurrence rate of the disease per person per year. For a given person time τ, the number of cases, $\mathcal{T}$, is then modeled as

$$P[\mathcal{T} = t; \phi] = \frac{c(t)\phi^t}{f(\phi)}$$

where $\phi = \lambda\tau$, $c(t) = 1/t!$, and $f(\phi) = \exp(\phi)$. Do the observed data support the hypothesis that the disease incidence rate does not exceed 100 cases per year per 100,000 people? In other words, we test $H_0 : \lambda \leq 0.001$ versus $H_1 : \lambda > 0.001$. Here, $\phi_0 = 1350 \times 0.001 = 1.35$. So the conventional exact one-sided *p*-value is

$$P[\mathcal{T} \geq 4; \phi_0 = 1.35] = 1 - P[\mathcal{T} \leq 3; \phi_0 = 1.35]$$

$$= 1 - \exp(-\phi_0)(1 + \phi_0 + \phi_0^2/2 + \phi_0^3/6) = 0.048.$$

Thus $p_c = 0.048$. The mid-*p*-value is

$$p_m = p_c - 0.5P[\mathcal{T} = 4; \phi_0 = 1.35] = 0.030$$

As expected, the mid-*p*-value makes a stronger case to doubt H_0 than the conventional *p*-value does.

2.6 Asymptotic Evidence Function

Another construction of an evidence function derives when the normal distribution is used to asymptotic a discrete probability. Let $\Phi(z)$ be the cdf of the standard normal distribution, and $\mathcal{T}$ be a PBD variable with mean and standard deviation equal to μ and σ, respectively. Suppose also that

$$P[\mathcal{T} \leq t;\, \phi] \approx \Phi(z)$$

where

$$z = \frac{t - \mu}{\sigma}$$

Using this, we define an **asymptotic evidence function** for the one-sided setup (2.1) as

$$\mathsf{E}_a(t;\phi) = 1 - \Phi(z) \qquad (2.14)$$

With this evidence function, an **asymptotic one-sided p-value**, denoted by p_a, is defined as

$$p_a(t;\phi_0) = 1 - \Phi\left(\frac{t - \mu_0}{\sigma_0}\right) \qquad (2.15)$$

where μ_0 and σ_0 are respectively the mean and standard deviation of $\mathcal{T}$ when $\phi = \phi_0$.

Example 2.3 (continued): For illustration, we return to the side effects data with $B(9;\pi)$. Here, $t = 2$, $\pi_0 = 0.1$, $\mu_0 = 0.9$ and $\sigma_0 = 0.9$. Thus,

$$z = \frac{t - \mu_0}{\sigma_0} = 1.2$$

and hence

$$p_a = 1 - \Phi(1.2) = 0.1151$$

****

Note that for this problem, p_a is smaller than both p_c and p_m. The asymptotic evidence function for the setup (2.2) is defined similarly.

2.7 Matters of Significance

In this section we deal with three practical matters relating to the *p*-value. These are (i) the interpretation or meaning of a *p*-value, (ii) use of fixed cut points or significance levels and (iii) the application of one-sided *p*-values.

We set the ball rolling above by defining the *p*-value as a tail probability. A subsequent version of the idea, the mid-*p*-value was, however, no longer a tail probability. But since it had the same monotonicity property, it was deemed a legitimate evidentiary entity. The asymptotic *p*-value was also **not a tail probability** for the sample space of the observations. But it too had the property we need.

In the applied literature, p-values are stated in terms of probabilities. For continuous data, a p-value can normally be linked to the probability of an event in the study sample space. A discrete data p-value may, on the other hand, often not correspond to the probability of an event in the sample space. When we use it, the following points should therefore be kept in mind:

- Primarily, a p-value is an index of evidence that takes on values between 0 and 1.
- It is derived from, looks like and shares some features of a probability. But it is not, by definition, a probability.
- Most importantly, **it is not the probability that the null hypothesis is false** (see below).
- A p-value is NOT identical to a significance level or the type I error rate.
- A p-value is a random variable computed from a (random) realization, namely, the study data.
- Studying the properties of this random variable can yield an insight into the errors associated with the particular method of generating p-values (see Chapter 7).

Clarity on these points will avoid the confusion regarding the use and interpretation of p-values that persists in parts of the literature.

****

If we state that small p-values give grounds to doubt the null hypothesis, the inevitable question is how small is small? If $p = 0.80$, the evidence against H_0 is considered weak. On the other hand, $p = 0.001$ does call H_0 into question (that is, if we set aside, for now, issues like sample size). Where do we draw the line? Data analysis often aims to provide a guide action or policy: Should a treatment be endorsed or not? Should a chemical be categorized as a carcinogen or not? A cut point for p-values may be of use in such cases. The p-value is then linked to a decision to reject or not reject the null.

Let α_* $(0 < \alpha_* < 1)$ denote a cut-off level for small p-values. It is variously known as the α_* **level**, **critical level**, **size** or **nominal size**, or **significance level** or **nominal significance level**. (The label "nominal" will be clarified below.) Two cut points, 0.05 or 0.01, are commonly used in practice.

- If a p-value falls below 0.05, it is said to be **statistically significant**.
- If it is below 0.01, it is called **statistically highly significant**.

In either case, the decision is to reject H_0. These levels (0.05 or 0.01) do not derive from an objective criterion. Rather, they come from a long standing convention that was in part fostered by the absence of computational facilities at a time when many statistical methods were being developed. Researchers had to use statistical tables which could only have a few cut points for relevant distributions. The fixed levels are now entrenched in practice even though that rationale for their use no longer exists.

The overuse and misuse of the 0.05 level is a serious problem that afflicts practical data analysis and reporting to this day. But such a fixed level has its positive aspects as well. The problems and advantages of a fixed level are:

- With a fixed level, a complex scientific decision is reduced to whether a single number falls above or below a specified value.

MATTERS OF SIGNIFICANCE

- This may oversimplify issues and heighten the potential for misuse of statistical methods.
- On the other hand, a common level for different studies and investigators does, in a sense, provide an objective standard of judgment.
- The use of a cut-off level as such is not identical to using the same fixed level under all conditions. Varied cut points can be employed in different studies.
- A fixed significance level may play a useful role even in complex situations provided it is embedded in a comprehensive data analytic, reporting and decision making framework.
- The idea of a fixed level provides the basis for formulating measures of evidence that address some of the shortcomings of using the *p*-value as a measure of evidence.

Under a fixed α_* level rule, we reject H_0 if $p \leq \alpha_*$; otherwise we do not. For each outcome t in the sample space Ω, this decision rule gives an indicator function, $\mathcal{I}(t; \alpha_*)$, such that

$$p(t) \leq \alpha_* \Rightarrow \mathcal{I}(t; \alpha_*) = 1 \Rightarrow \text{ reject } H_0 \qquad (2.16)$$
$$p(t) > \alpha_* \Rightarrow \mathcal{I}(t; \alpha) = 0 \Rightarrow \text{ do not reject } H_0 \qquad (2.17)$$

The function $\mathcal{I}(t; \alpha_*)$ is called a **test**. The subset of Ω for which $\mathcal{I}(t; \alpha_*) = 1$ is then the **rejection region** or **critical region** of the test. Such a region is also found by comparing the test statistic to a preset value. In that case, there is an implicit α_* level embodied in the test.

The probability of rejecting the null hypothesis when it is true is called the **type I error rate**, **actual α_* level**, **actual significance level** or **actual size** of the test. This is:

$$\alpha_{*a} = \sum_{t \in \Omega} \mathcal{I}(t; \alpha_*) P[\mathcal{T} = t; H_0] \qquad (2.18)$$

In continuous data, the actual significance level generally equals the nominal significance level; in discrete data problems, that is generally not the case. (This issue is taken up in Chapter 7.)

As noted, fixed significance levels are often used mechanically, and otherwise misused in practice. Nevertheless, a fixed α_* level provides a fulcrum for constructing statistical indices that remedy some of the shortcomings of a *p*-value. These include **confidence interval** and **power**. Conceptually, the idea of a fixed significance level is thus a vital one even if we strongly decry the overuse and misuse of *p*-values and the 0.05 level in practice.

****

Now we consider issues about the use of one-sided analysis and tests in practice. For this let us define a **two-sided analysis** as an analysis which also looks at the other tail (more precise definitions are given in Chapter 3). One aspect of the improper use of *p*-value then relates to one-sided *p*-values. This occurs when a one-sided value is reported just because the two-sided *p*-value is not small enough. In this case, the directionality of null and alternative hypotheses is specified after examining the data.

Partly in response to such abuse, statisticians have debated over the years when, where and if at all the one-sided *p*-value ought to be used. Before listing the arguments in that debate, we give two instances where a one-sided analysis is advocated, and, in our view, is justified.

Suppose we need to compare a new therapy with a standard one. The former may have more adverse effects. Its use is then advocated only if its efficacy level exceeds the standard by a given level, say, δ. In this case we test,

$$H_0 : \pi \leq \pi_0 + \delta \quad \text{versus} \quad H_1 : \pi > \pi_0 + \delta \tag{2.19}$$

A test for these hypotheses is named **a test for noninferiority.** Another instance where a one-sided test is deemed appropriate is when if the new action is found to be inferior to the standard; one would still advocate it, because it has some other proven benefits and hardly any risks. This is a rare but possible situation.

Situations where a one-sided test or confidence interval may be appropriate, and/or where one-sided tests were not used in a satisfactory manner are described in Andersen (1990), pages 234–236; Bland and Altman (1994); and Dunnet and Gent (1996). For a critique of the exclusive focus on two-sided tests, see Sackett (2004).

The choice of a one or two-sided analysis can have practical ramifications. Among statisticians, widely dissenting views on the utility of one-sided analysis exist. To simplify, one viewpoint holds that a one-sided analysis is appropriate when the aims of the study or what is to be expected from it fall on one side of a comparison. Preliminary evidence may indicate that the new therapy is unlikely to be worse than the standard. Or, in such a comparison, one should focus on the possible benefits of the new therapy. That it may harm needs a separate investigation. Others, however, point to the multitude of examples where actual results of a study completely contradicted prior expectations. The choice of the sidedness of a test should thus be based on all possibilities and not on what the investigators would like to get out of it.

In practice, a one-sided analysis is not common. In the LFEMF study, for instance, the authors state that prior to their study, preliminary data had indicated a low male offspring ratio for some occupations and an unchanged ratio in others. Several potentially causative exposures were noted. One among them was low frequency electromagnetic fields. Their aim was to obtain unbiased evidence as to how LFEMF affects the sex ratio. Notwithstanding what the initial data showed, if there is an effect, it may well be that exposure to LFEMF increases the male to female ratio. A one-sided analysis does seem not appropriate, and we prefer to analyze the LFEMF data with two-sided methods presented in Chapter 3.

There is at the moment a broad consensus about the use of one-sided methods, especially in the health, medical and psychological studies. In accordance with this consensus:

- Two-sided procedures (tests and confidence intervals) are the default norm in practical data analysis.
- There are a few instances when a one-sided procedure may be used.
- A clear justification is needed when a one-sided procedure is to be used.
- The justification for and the directionality of the one-sided procedure have to be declared at the design stage of the study. Otherwise, a one-sided analysis is to be viewed with suspicion.
- Making the decision to use a one-sided procedure during analysis, simply to attain "statistical significance," is not good science, bordering on scientific misconduct (Andersen 1990, pages 235–236).

For details on the above issues, see the literature cited in §2.11. Also, we further deal with controversies about fixed levels and one-sided procedures in Chapter 15. At this juncture, we note that in part these controversies take on more importance in a setting where p-values are the primary indices of evidence. But when they are used in conjuction with confidence intervals (see below), a more flexible and nuanced attitude towards these matters is permissible.

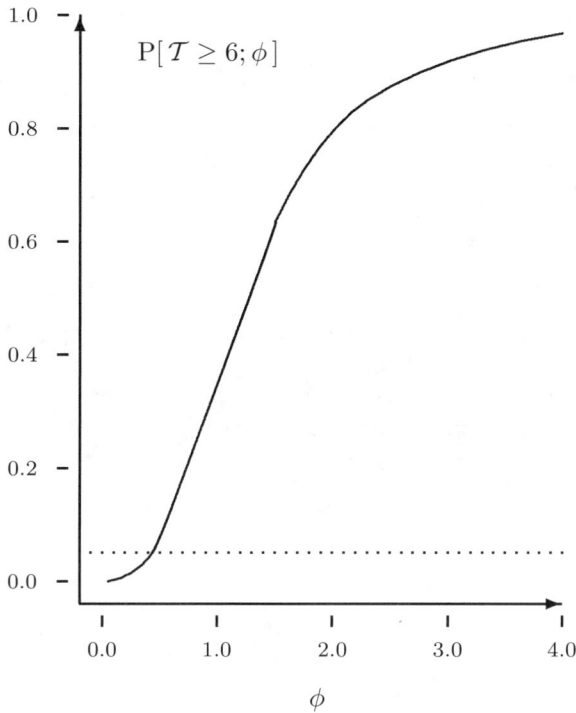

Figure 2.2 *A One-sided Binomial Evidence Function for* $n = 10$, $t = 6$.

2.8 Confidence Intervals

Consider a binomial trial to test $H_0 : \phi \leq 1.0$ versus $H_1 : \phi > 1.0$ (or $\pi \leq 0.5$ versus $\pi > 0.5$). If $n = 100$, and $t = 55$, then $p_c = \mathrm{P}[T \geq t; \phi = 1.0] = 0.1841$. On the other hand, if $n = 1000$, and $t = 550$, then $p_c = 0.0009$. The observed response rate in both cases is 55%. Two similar results in one respect give vastly dissimilar *p*-values, a disparity arising from the unequal sample sizes.

A *p*-value may fail to capture key aspects of the data. In a clinical trial, the *p*-value does not answer a prime research question: What is the extent of the difference between the therapies? It only gives an index that shows the extremeness of the observed outcome under treatment equivalence. In the medical parlance, we say a statistically significant difference may not be a clinically significant difference. In an epidemiologic study, the *p*-value does not measure the size of an exposure effect or the precision of the estimate of effect size as well. In a toxicology experiment, a *p*-value lacks information about the magnitude of a possible relationship between the dose of the chemical used and the probability of a toxic effect.

Further, when sample sizes are very large, even small, perhaps inconsequential effects may be associated with small *p*-values. The *p*-value has to be complemented with an assessment of the magnitudes of parameters of interest. One way to do this is with a confidence interval.

Example 2.5: To obtain a one-sided confidence interval, consider $H_0 : \phi \leq \phi_0$ for a study with

10 independent Bernoulli trials. The number of successes were 6. The conventional *p*-value is $p_c(6; \phi) = P[T \geq 6; \phi]$ when $\phi = \phi_0$. Figure 2.2 is a graph of $p_c(6; \phi)$ as a function of ϕ. It shows that if $\phi \leq 0.44$, the value of the evidence function is ≤ 0.05. At null values not smaller than 0.44, H_0 is rejected at 0.05 level. Thus 0.44 is the smallest value of ϕ consistent with the observed value of T if the 0.05 level of significance is adopted. For all values of $\phi = \phi_0$ larger than it, the one-sided null hypothesis is not rejected at 0.05 level.

We call $(0.44, \infty)$ a 0.95 level **upper confidence interval** for ϕ, and call 0.44 a 95% **lower confidence bound** or **lower confidence limit** for ϕ. In the plot of the evidence function, the horizontal line at 0.05 intersects the function at the lower 95% confidence limit.

****

In general, with an observation $T = t$, a $(1 - \alpha_*)100\%$ upper confidence region for ϕ comprises all values of ϕ_0 for which $H_0 : \phi \leq \phi_0$ is not rejected at a nominal level α_*. The monotonicity property ensures that this region is a connected interval. The smallest value in it, denoted $\phi_l(\alpha_*, t)$ or simply ϕ_l, is the lower confidence bound. It is also the solution of the equation

$$P[T \geq t; \phi] = \alpha_* \tag{2.20}$$

For a PBD, this becomes

$$f(\phi; \geq t) = \alpha_* f(\phi) \tag{2.21}$$

or

$$\sum_{u \geq t} c(u)\phi^u = \alpha_* \sum_{u \in \Omega} c(u)\phi^u \tag{2.22}$$

****

Example 2.5 (continued): For the binomial data with $n = 10$ and $t = 6$, equation (2.22) becomes

$$210\phi^6 + 120\phi^7 + 45\phi^8 + 10\phi^9 + \phi^{10} = \alpha_*(1 + \phi)^{10}$$

with the solution $\phi = 0.44$ when $\alpha_* = 0.05$.

****

An upper $(1 - \alpha_*)100\%$ confidence bound ϕ_u obtains by considering the null hypothesis $H_0 : \phi \geq \phi_0$. The equation to compute ϕ_u is

$$P[T \leq t; \phi] = \alpha_* \tag{2.23}$$

For a PBD, this becomes

$$f(\phi; \leq t) = \alpha_* f(\phi) \tag{2.24}$$

CONFIDENCE INTERVALS

or

$$\sum_{u \leq t} c(u)\phi^u = \alpha_* \sum_{u \in \Omega} c(u)\phi^u \qquad (2.25)$$

This is the method of constructing a confidence interval by **inverting an hypothesis test**. The parameter space is divided into two regions; that in which, given an outcome, an α_* level test would not be rejected and that in which it would be. In general, such regions need not be intervals. For the PBD, the monotonicity of the cdf ensures that a one-sided confidence region is a continuous interval.

Different tests give different confidence regions. A mid-p based confidence region is derived from the mid-p evidence function. We can show that one-sided lower and upper $(1 - \alpha_*)100\%$ level mid-p confidence limits for ϕ are respectively computed from

$$f(\phi; \geq t) - 0.5 f(\phi; t) = \alpha_* f(\phi) \qquad (2.26)$$
$$f(\phi; \leq t) - 0.5 f(\phi; t) = \alpha_* f(\phi) \qquad (2.27)$$

An asymptotic confidence region is derived from the asymptotic evidence function. One-sided lower and upper asymptotic $(1 - \alpha_*)100\%$ confidence limits for ϕ are, assuming the normal approximation, respectively computed from

$$1 - \Phi\left(\frac{t - \mu(\phi)}{\sigma(\phi)}\right) = \alpha_* \quad \text{and} \quad \Phi\left(\frac{t - \mu(\phi)}{\sigma(\phi)}\right) = \alpha_* \qquad (2.28)$$

In the binomial case, the symmetric nature of the normal distribution implies that both the lower and upper confidence limits, π_l and π_u, obtain from solving for π in

$$(t - n\pi)^2 = z_{\alpha_*}^2 n\pi(1 - \pi) \qquad (2.29)$$

where z_{α_*} is the α_*th percentile of the standard normal distribution. The solutions to this quadratic equation are

$$0.5\left(b \pm \sqrt{b^2 - 4c}\right) \qquad (2.30)$$

where

$$b = \frac{2t + z_{\alpha_*}^2}{n + z_{\alpha_*}^2} \quad \text{and} \quad c = \frac{t^2}{n(n + z_{\alpha_*}^2)} \qquad (2.31)$$

The smaller solution to this equation gives the one-sided lower confidence limit, and the larger gives the one-sided upper limit.

Example 2.3 (continued): For one-sided 95% bounds, $z_{.05} = 1.645$. With $n = 9$ and $t = 2$, the equation in question is

$$(2 - 10\pi)^2 = 1.645^2 \times 10\pi(1 - \pi)$$

with solutions $\pi_l = 0.0688$ and $\pi_u = 0.4582$.

****

With confidence limits for ϕ at hand, the limits for β, or in the binomial case, for π as well, are obtained by a simple transformation. This holds for the conventional exact, mid-p and asymptotic bounds. In each case,

$$\beta_l = \ln(\phi_l) \quad \text{and} \quad \pi_l = \frac{\phi_l}{1+\phi_l} \qquad (2.32)$$

$$\beta_u = \ln(\phi_u) \quad \text{and} \quad \pi_u = \frac{\phi_u}{1+\phi_u} \qquad (2.33)$$

We summarize the main points of what we have shown so far.

One-Sided Inference

- For inference on a parameter of a PBD, one-sided evidence functions are constructed from conventional exact, mid-p or asymptotic tail probabilities.
- One-sided p-values obtain by drawing vertical lines in a plot of this function.
- One-sided confidence intervals obtain by drawing horizontal lines in the plot.

2.9 Illustrative Examples

Now we give examples that have some specific features. First consider a study where the event of interest does not transpire, a form of sparse data. How are data with a **zero numerator** to be analyzed? Hanley and Lippman–Hand (1983) cite several examples of zero event occurrence in research on adverse medical events. Noting the varied meanings ascribed to the zero numerator in medical studies, they muse: "*If nothing goes wrong, is everything all right?*" Calculating an upper confidence bound for the underlying rate of occurrence when the observed proportion is 0% gives a clue.

Example 2.6: One study they cite followed up 14 boys who had undergone chemotherapy for leukemia. The event of concern was abnormal testicular function but none of the boys exhibited it during five and a half years of median follow up.

****

Let π be the true abnormality rate in the eligible population. The binomial left tail probability with no events is

$$(1+\phi)^{-n}$$

Exact, mid-p and asymptotic $(1-\alpha_*)100\%$ upper confidence bounds for ϕ (and π) therefore derive respectively from

ILLUSTRATIVE EXAMPLES

Exact: $\quad (1+\phi_u)^{-n} = \alpha_* \quad \Rightarrow \quad \phi_u = \alpha_*^{-n^{-1}} - 1$

Mid-p: $\quad 0.5(1+\phi_u)^{-n} = \alpha_* \quad \Rightarrow \quad \phi_u = (2\alpha_*)^{-n^{-1}} - 1$

Asymptotic: $\quad \sqrt{n\phi_u} = z_{\alpha_*} \quad \Rightarrow \quad \phi_u = n^{-1} z_{\alpha_*}^2$

The respective solutions to these equations in terms of π are

Exact: $\quad \pi_u = 1 - \alpha_*^{n^{-1}}$

Mid-p: $\quad \pi_u = 1 - (2\alpha_*)^{n^{-1}}$

Asymptotic: $\quad \pi_u = z_{\alpha_*}^2 (n + z_{\alpha_*}^2)^{-1}$

Example 2.6 (continued): With $n = 14$ and $t = 0$, and setting $\alpha = 0.05$ in the leukemia study, the exact approach gives $\pi_u = 0.1926$, the mid-p method gives $\pi_u = 0.1517$ and the asymptotic method yields $\pi_u = 0.1620$ as the 95% upper confidence bounds for π. We conclude with 95% confidence that the true rate of testicular abnormality could nevertheless be as high as somewhere from 15% to 19%.

****

Now we consider follow up data with zero counts.

Example 2.7: Levin et al. (1998) examined the causes of death among former workers of a plant producing asbestos pipe insulation materials. Death rates from different causes were compared to rates in the general population. Let τ denote the total person years under risk for the study cohort. The distribution of number of events in person years τ is assumed as Poisson with mean $\lambda\tau$. Table 1 of this paper gives 95% confidence limits only for causes for which at least one event was observed. No confidence interval is provided for malignant neoplasms of the stomach since no death from this cause of death occurred in this cohort.

****

In general, if no events are observed, then

$$P[\mathcal{T} \leq 0; \lambda] = \exp(-\lambda\tau)$$

Exact, mid-p and asymptotic upper confidence bounds for λ therefore obtain respectively from

Exact: $\quad \exp(-\lambda\tau) = \alpha_* \quad \Rightarrow \quad \lambda_u = -\ln(\alpha_*)/\tau$

Mid-p: $\quad 0.5\exp(-\lambda\tau) = \alpha_* \quad \Rightarrow \quad \lambda_u = -\ln(2\alpha_*)/\tau$

Asymptotic: $\quad (0 - \lambda\tau)/(\lambda\tau)^{0.5} = z_{1-\alpha_*} \quad \Rightarrow \quad \lambda_u = -z_{1-\alpha_*}^2/\tau$

Levin et al. (1998) do not state the data in terms of person years. Instead, for each cause of death, the expected number of deaths based on the standard population rate is given. If λ_s is the known standard rate for stomach neoplasms, the expected number of deaths from this cause in the cohort is $\lambda_s \tau$. The **standardized mortality ratio** or **SMR**, denoted by ν, is

$$\nu = \frac{\text{(Cohort Expected Number)}}{\text{(Population Expected Number)}} \times 100$$

$$= \frac{\lambda \tau}{\lambda_s \tau} \times 100 = \frac{\lambda}{\lambda_s} \times 100$$

The exact upper confidence bound for ν then is

$$\nu_u = \frac{\lambda_u \tau}{\lambda_s \tau} \times 100$$

$$= \frac{-\ln(\alpha)}{\text{(Population Expected Number)}} \times 100$$

The upper confidence bounds for the mid-p and asymptotic methods are computed similarly.

Example 2.7 (continued): For stomach neoplasms, the expected value was $\lambda_s \tau = 1.2$. With relevant substitutions, we get that for the stomach neoplasms,

$$\begin{aligned}
\text{Exact:} \quad & \nu_u = 272 \\
\text{Mid-}p\text{:} \quad & \nu_u = 209 \\
\text{asymptotic:} \quad & \nu_u = 246
\end{aligned}$$

Each of these confidence bounds generously allows for the possibility that the SMR may be greater than 100.

****

Example 2.1 (continued): We return to the study of the value of the neural network algorithm in myocardial infarction (Baxt 1991). 1 out of the 36 subjects (or 2.8%) with the ailment were missed by the algorithm. Upper 95% confidence bounds for the false diagnosis rate of the procedure under the exact, mid-p and asymptotic methods for ϕ (and π) derive respectively from

$$\begin{aligned}
\text{Exact:} \quad & 1 + 36\phi_u = 0.05(1 + \phi_u)^{36} \\
\text{Mid-}p\text{:} \quad & 1 + 18\phi_u = 0.05(1 + \phi_u)^{36} \\
\text{asymptotic:} \quad & 1 - 36\pi_u = 1.645\sqrt{36\pi_u(1 - \pi_u)}
\end{aligned}$$

Solutions to these are found graphically by plotting the three evidence functions and drawing horizontal lines at 0.05. This gives $\pi_u = 0.125$ for the exact, $\pi_u = 0.109$ for the mid-p, and $\pi_u = 0.115$ for the asymptotic method 95% confidence bound for π. We cannot rule out with 95% confidence the possibility that the error rate does not exceed the permissible rate 5%.

****

DESIGN AND ANALYSIS

When no events occur in a binomial experiment, $n\pi_u$ from the exact, mid-p and asymptotic methods approach fixed values for large n. From the formula for the asymptotic bound, it is clear that $n\pi_u$ tends to $z_{\alpha_*}^2$. When $\alpha_* = 0.05$, this equals $1.645^2 = 2.71$. For the 95% exact upper bound, $n\pi_u$ tends to 3.0 and for the mid-p bound, it approaches 2.30, Hence

$$\text{Exact:} \quad \pi_u \approx 3.00/n$$
$$\text{Mid-}p\text{:} \quad \pi_u \approx 2.30/n$$
$$\text{asymptotic:} \quad \pi_u \approx 2.71/n$$

These values are useful for a quick assessment of the confidence limits given in the literature (Hanley and Lippman–Hand 1983). Such checks were extended to small nonzero numerators by Newman (1995). They are also useful for computing the lower confidence bound for π when most or all of the outcomes of the trial were the event of interest.

2.10 Design and Analysis

A key question underpinning all forms of statistical analyses is: "Does how the data were collected matter?" In other words, does the study design affect, constrain or determine the choice of the data analytic method? If so, then how? These important questions have been at the center stage of controversy in the annals of statistics and science.

Let us consider a scenario with five study designs. Suppose we need to estimate the rate of liver toxicity among patients taking a drug for bronchial pneumonia. The drug is prescribed for hospitalized patients. A list of about a million patients who took the drug has been compiled. For each patient, it has his or her location and the hospital identification number. Other relevant information is to be extracted from the patient files. It has been decided that about 1,000 patient files will have to be examined for this study. Suppose that five investigators each independently carry out the study at this sample size. But each does it in his or her own way, as described below:

Study A: Investigator A does not have a plan; she just looks around nearby hospitals, manages to find 50 files of patients who had liver toxicity, and adds to that 950 files of patients who did not have liver toxicity.

Study B: Investigator B selects the first 1,000 patients from the list and locates their files. By a coincidence, in her sample as well, 50 patients had liver toxicity, and 950 patients did not have liver toxicity.

Study C: Investigator C selects the last 1,000 patients from the list and locates their files. It also happens that in his sample as well 50 patients had liver toxicity, and 950 patients did not have liver toxicity.

Study D: Investigator D selects his sample of 1,000 patients using random numbers and sampling with replacement. In this sample as well, 50 patients had liver toxicity, and 950 patients did not have liver toxicity.

Study E: Investigator E has a different plan. She also uses random numbers and sampling with replacement. However, each times she selects a patient at random, she obtains his or her file, and examines it for the presence of liver toxicity. She keeps on doing this until a total of 50 patients with liver toxicity has been attained. To get these 50 cases, she had to examine a total of 1,000 files. That is, in her sample as well there were 950 patients who did not have liver toxicity. Such a study is called **an inverse sampling** study or design.

So now we have the same set of numbers, the same estimate of liver toxicity, namely, 5%, but obtained in five different ways. Should we use all of them (in some combined fashion) to compute a lower confidence bound for the true rate of liver toxicity?

It is obvious that data from Study A are not worth analyzing; haphazard data collection is not science. We may also have qualms about using the numbers from Study B and Study C. The first 1,000 and the last 1,000 patients in the list may be from a particular area, or time period, and these patients may have characteristics that may predispose them to lower or higher rates of liver toxicity. These data may generate systematic errors. Therefore, we would only use the data from Study D and Study E in our combined assessment.

****

If we randomly select 5 healthy subjects, and then randomly select 5 subjects with a disease, we cannot use these numbers to estimate the prevalence rate of the disease. We need a random sample from the total population to get an admissible estimate.

From this example, and the above instance of five investigators, we can say that at one fundamental level then, how we collected the data, or the study design, does matter. It is not just the numbers we have but the process through which they were obtained that is also important.

At another level, the answer may not be that obvious. Thus, in Study D and Study E above, the designs were different but the results were the same. Are or should the confidence intervals for toxicity rate from the two studies be the same? If they are not, does that call into question the very approach to data analysis?

Let us return to the neural network study (Example 2.1). We note that the investigators had chosen a fixed number of subjects to be diagnosed by the neural network algorithm. Of these 36 patients, one was misdiagnosed.

Assuming a common underlying false diagnosis rate and independence between the subjects, we employed the binomial probability model to analyze the data. Whether the misdiagnosed subject was the second, the tenth, or some other order subject did not impact our analysis. In that regards, our analysis did ignore one aspect of how the data arose. But this is justified under the assumptions of the binomial model. Here, the distribution of the total number of subjects misdiagnosed depends only on the parameter of interest, π, and the total sample size. The order of successes and failures is immaterial. We also know that the total number of successes is sufficient for π.

Let us consider a different study design. Suppose the investigator had planned that the neural network algorithm would be applied until one subject was wrongly diagnosed. This is another case of the inverse sampling design. Suppose the same overall result was obtained, that is, 35 were correctly diagnosed and 1 was misdiagnosed. Of course, here the last subject was the one erroneously diagnosed. Also assume as before that subjects were diagnosed independently of one another, and a constant underlying misdiagnosis rate prevailed.

The probability model for the inverse sampling design is the **geometric distribution**, a special case of the negative binomial model. Let T denote the number of failures before a success in iid trials with rate π. Then

$$P[T = t; \pi] = (1-\pi)^t \pi, \quad t = 0, 1, 2, \ldots \tag{2.34}$$

We know that this is a PBD with $\phi = 1 - \pi$ and gp,

$$f(\phi) = (1-\phi)^{-1} = \sum_{u=0}^{\infty} \phi^u \qquad (2.35)$$

At $\mathcal{T} = t$, the one-sided $100(1-\alpha_*)\%$ lower bound for ϕ obtains from the right tail probability

$$P[\mathcal{T} \geq t; \phi] = \{f(\phi)\}^{-1} \sum_{u \geq t} \phi^u = \phi^t$$

The exact lower bound for ϕ then obtains from solving

$$\phi^t = \alpha_* \qquad (2.36)$$

Hence we get

$$\phi_l = \alpha_*^{t^{-1}} \quad \text{and} \quad \pi_u = 1 - \alpha_*^{t^{-1}} \qquad (2.37)$$

The asymptotic confidence bounds are obtained by using the normal approximation to the geometric distribution. The mean and variance of $\mathcal{T}$ in terms of π are

$$\mu = \frac{1-\pi}{\pi} \quad \text{and} \quad \sigma^2 = \frac{1-\pi}{\pi^2} \qquad (2.38)$$

An asymptotic $100(1-\alpha_*)\%$ level upper one-sided confidence bound for π is the larger of the two solutions for

$$\frac{(t-\mu)^2}{\sigma^2} = z_{\alpha_*}^2 \qquad (2.39)$$

which after simplification becomes

$$[\pi(t+1) - 1]^2 = (1-\pi)z_{\alpha_*}^2$$

Applying these to the neural network data ($t = 35$ and $\alpha_* = 0.05$), we get

Exact: $\quad \pi_u = 0.082$

Asymptotic: $\quad \pi_u = 0.072$

The exact and asymptotic upper confidence bounds for π for these data under the binomial sampling scheme were 0.125 and 0.115, respectively. These are about one and a half times larger than those obtained from the geometric sampling scheme. In general, the geometric upper confidence limit is always smaller than the binomial upper confidence limit (George and Elston 1993).

The sample space of the negative binomial design, even if we fix the total number of observations, differs distinctly from that of the binomial design. In the former, the event with $t+1$ failures is, for example, not a permissible event while the latter scheme allows it. According to the above noted authors:

Information and Design

The fact that there were no occurrences in the first $n-1$ trials gives more information compared to one occurrence in n trials, and it is possible to obtain a shorter confidence interval by basing the analysis on the geometric rather than the binomial distribution. George and Elston (1993).

This comparative analysis leads us to an important observation:

A Word of Caution

When analyzing data, not taking into account how they were collected can lead to an inaccurate or erroneous conclusion.

That is what we are led to if we adopt, as we have done, the tail area based approach to statistical inference. But many statisticians do not agree with this observation or the line of reasoning given above. Statisticians of the **likelihood school** regard the likelihood as a more basic measure of evidence (see Chapters 3 and 15, and Barnard, Jenkins and Winsten 1962). Those adopting the **Bayesian framework** analyze data in terms of prior and posterior probabilities (see Chapter 15). Both produce identical results from the binomial and negative binomial designs. Also, to some, the idea of a sample size fixed by design is problematic. To complicate the issue further, later we will find classes of designs for which one can, for certain kinds of analysis and even under a tail area based analysis, ignore how the data arose. The issues thereby are complex; what has been presented above is but the start of a discussion that will be pursued further.

2.11 Relevant Literature

The theory and application of hypothesis tests and confidence intervals are covered in many statistical books and monographs. The primary work is Lehmann (1986). Other references relating to exact methods are noted in several chapters in this books. Since evidence functions, or confidence interval functions, and *p*-value functions have mostly been described in a two-sided context, we note the literature related to them in Chapter 3.

2.11.1 The mid-p

For a general introduction to the *p*-value, see Moyé (2000) and Gibbons and Pratt (1975).

The idea of the mid-*p* was formulated around the year 1950. Lancaster (1949, 1952, 1961) are the primary works. Seneta and Phipps (2001), though, trace its roots to earlier works. A remarkable thing is that it has been endorsed by statisticians coming from very different philosophical and methodological perspectives. See Stone (1969), Lancaster (1969), Miettinen (1974), Pratt and Gibbons (1981), Anscombe (1981), Plackett (1984), Miettinen (1985), Franck (1986), Rothman (1986b), Guess et al. (1987), Nurminen and Mutan (1987), Williams (1988a, 1988b), Barnard (1989, 1990), Upton (1992), Routledge (1992, 1994), Kulkarni, Tripathi and Michalek (1998), Pierce and Peters (1999), Hwang and Yang (2001) and Seneta and Phipps (2001) for the vast diversity of views on, and rationale for, it.

Empirical properties of mid-p-values and mid-p confidence intervals have been studied in many different settings. These include binomial data (Vollset 1993; Newcombe 1998a, 1998b; Reiczigel 2003); Poisson person time data models (Cohen and Yang 1994; Martin and Austin 1996; Kulkarni, Tripathi and Michalek 1998); a 2×2 table (Miettinen 1974; Haber 1986b; Hirji, Tan and Elashoff 1991; Davis 1993; O'Brien 1994; Lydersen and Laake 2003); several 2×2 tables (Mehta and Walsh 1992; Kim and Agresti 1995); an ordered $R \times C$ table (Agresti, Lang and Mehta 1993); the power divergence goodness-of-fit tests (Tang 1998); comparison of two correlated proportions (May and Johnson 1997; Tango 1998); matched case-control data (Vollset, Hirji and Afifi 1991; Hirji 1991) and general permutation tests (Routledge 1997; Potter 2005). The overall message from this body of work is that the mid-p based approach is either among or not too distinct from the preferred methods. However, some authors like May and Johnson (1997), Tango (1998) and Reiczigel (2003) are not inclined towards it.

Berry and Armitage (1995) reviewed the concept of mid-p; see also Sahai and Khurshid (1995). Mid-p methods now appear in textbooks on statistics, epidemiology and biostatistics. In addition to those listed in the previous paragraph, they include Miettinen (1985), Ahlbom (1993), Agresti (1996), Sahai and Khurshid (1996) and Rothman and Greenland (1998). It has also been promoted in relation to confidence interval functions. Sullivan and Foster (1990) promote the mid-p method in a similar spirit as Cohen and Yang (1994), that is, it is among the *"sensible tools for the applied statistician."* Agresti and Min (2001) include mid-p confidence interval among the recommended methods for a variety of discrete data problems.

Seneta, Berry and Macaskill (1999) construct an adjusted version of the one-sided mid-p-value to bring its distribution closer to the distribution of p-values in continuous data. Chen (2002) extended the mid-p idea to exact unconditional inference for a 2×2 table.

2.11.2 Inverse Sampling

Haldane (1945a, 1945b) and Finney (1947, 1949) are some of the early works on the inverse sampling design. The construction of confidence limits for geometric sampling was tackled by Clemans (1959) and extended by George and Elston (1993). This paper applies the geometric distribution to DNA testing. With a focus on exact confidence limits, Liu (1995b) extended the work of George and Elston (1993) to the negative binomial. For summaries of the properties of the negative binomial, see Williamson and Bretherton (1963), pages 7–15, and Bishop, Fienberg and Holland (1975), pages 452–456. They also outline the history of the negative binomial model. We note papers on the inverse sampling design for multivariate settings in later chapters.

2.11.3 One-Sided Procedures

A comprehensive debate on one-sided tests in the *Journal of Biopharmaceutical Statistics* presents the varied perspectives on the issue. The relevant papers are Dubey (1991), Fisher (1991), Koch (1991), Overall (1991) and Peace (1991). Dunnett and Gent (1996) summarize the points made there and give alternatives. Bland and Altman (1994) is a recommended short, well illustrated paper. Other relevant papers and books are Andersen (1990), Collet (1991), Gatswirth (1998a) and Pocock (1983). Collet (1991) discourages application of one-sided methods for discrete data. The current consensus on the issue is enshrined in The Standards of Reporting Trials Group (1994) and its subsequent updates.

2.11.4 Poisson Models and Data

For exact analysis of univariate person time data and the SMR, see Bailar and Ederer (1964), Mulder (1983) and Liddell (1984). For other applications of the Poisson distribution, see Schinazi (2000) and Bishop, Fienberg and Holland (1975). More related references are in Chapter 10.

2.12 Exercises

2.1. An observation drawn from B(15; π) gives $T = 8$. Compute p-values for $\pi \leq 0.3$ versus $\pi > 0.3$. Plot the exact, mid-p and asymptotic evidence functions and obtain the appropriate one-sided confidence bounds for π.

2.2. An observation drawn from a Poisson distribution gives $T = 7$. Compute the p-value for $\lambda \leq 4$ versus $\lambda > 4$. Plot the exact, mid-p and asymptotic evidence functions and obtain the appropriate one-sided confidence bounds for λ.

2.3. Consider a univariate PBD specified by the gp

$$f(\phi) = 1 + 10\phi + 10\phi^2 + 24\phi^3 + 6\phi^4 + 6\phi^5 + 24\phi^6 + 10\phi^7 + 10\phi^8 + \phi^9$$

Note: At $\phi = 1$, this is a bimodal distribution. If the observed value is $T = 2$, plot the exact and mid-p evidence functions to test $H_0 : \phi \leq 0.8$ versus $H_1 : \phi > 0.8$. Are these monotonic functions? Use them to compute one-sided exact and mid-p-values, and corresponding 90% lower confidence bounds for ϕ. What do you conclude?

2.4. Suppose we draw a value from the univariate PBD specified by the exponents and coefficients of the polynomial given below.

u	c(u)	u	c(u)
0	41	5	4
1	10	6	4
2	10	7	10
3	4	8	10
4	4	9	41

Note: At $\phi = 1$, this is a U-shaped distribution. If the observed value is $T = 3$, plot the exact and mid-p evidence functions to test $H_0 : \phi \geq 0.8$ versus $H_1 : \phi < 0.8$. Are they monotonic functions? Use them to compute one-sided exact and mid-p-values, and corresponding 80% upper confidence bounds for ϕ. What do you conclude?

2.5. Let the sample space $\Omega = \{0, 1, 2, 3, 4, 5, 6, 7, 8, 9\}$ and the coefficients $c(t) = 1$ for all $t \in \Omega$ represent a PBD. When $\phi = 1$, this is the uniform discrete distribution. We test $H_0 : \phi \leq 1.0$ versus $H_1 : \phi > 1.0$. The observed value is $T = 9$. Plot exact and mid-p evidence functions for these data. Are these functions monotonic? Use them to compute one-sided exact and mid-p-values, and corresponding 95% lower confidence bounds for ϕ. What do you conclude?

2.6. In a binomial problem with $n = 32$, we need to test $\pi \leq 0.25$ versus $\pi > 0.25$. Determine the exact, mid and asymptotic p-values if $T = 13$ is observed. Compute the corresponding 95% lower confidence bounds for π.

EXERCISES

2.7. Clayton and Hills (1993) consider a study of genetic marker halotypes among sibling pairs who may share both halotypes, one halotype or no halotype. Ignoring the pairs with one halotype shared, they consider the odds of sharing both halotypes in relation to sharing no halotypes (ϕ). For the genetic model they consider, values of ϕ less than one are not permissible. From their data, 16 pairs shared both halotypes and 3 pairs shared none (page 97). Using the setup $\phi = 1$ versus $\phi > 1$, analyze the data using exact, mid-p and asymptotic methods.

2.8. Among subjects exposed to bis(chloromethyl) ether on the job for at least five years, 5 cases of lung cancer were detected during a total of 702 person years under risk follow up (Lemen et al. 1976). The expected number of cases was 0.54. Let ν denote the SMR. Test the hypothesis $\nu \leq 100$ versus $\nu > 100$, and compute a 95% lower confidence bound for ν. Use the plots of exact, mid-p and asymptotic evidence functions for this purpose. Do the three approaches lead to a similar conclusion?

2.9. Waxweiler et al. (1976) reported a total of 35 cancer cases among workers occupationally exposed to vinyl chloride in a sample for which the expected number of cases was 23.5. Compute one-sided asymptotic 90% and 95% upper confidence bounds for the SMR.

2.10. Infante (1976) detected 73 congenital malformations from 4,109 births in three communities with polyvinyl chloride production facilities. The expected number of malformations in general was 40.75. The observed and expected numbers of subjects with malformations of the eye were 0 and 0.58, respectively, and the corresponding numbers for the central nervous system were 17 and 5.62. Analyze these data using a one-sided approach and state your conclusions.

2.11. Formally show that the evidence functions used in the first five problems above are monotonic functions of ϕ.

2.12. Prove for a Poisson distribution that the conventional left tail probability and its mid-p version are both a monotonically increasing function of ϕ.

2.13. Prove that for a PBD, the one-sided left tail mid-p evidence function is a monotonically increasing function of ϕ.

2.14. In a binomial problem with $n = 30$, we need to test $\pi \leq 0.5$ versus $\pi > 0.5$. Determine the critical regions for the exact, mid-p and asymptotic 0.05 nominal level tests and compute their actual significance levels. Comment on your results.

2.15. In a Poisson data problem, we need to test $\lambda \leq 1.0$ versus $\lambda > 1.0$. Find the critical regions for the exact, mid-p and asymptotic 0.05 nominal level tests and compute their actual significance levels. Comment on your results.

2.16. In a geometric data problem, we need to test $\pi \leq 0.5$ versus $\pi > 0.5$. Determine the critical regions for the exact, mid-p and asymptotic 0.05 nominal level tests and compute their actual significance levels. Comment on your results.

2.17. In a negative binomial data problem with $r = 5$, we need to test $\pi \leq 0.5$ versus $\pi > 0.5$. Determine the critical regions for the exact, mid-p and asymptotic 0.05 nominal level tests and compute their actual significance levels. Comment on your results.

2.18. Noting that $z = \exp\{\ln(z)\}$, apply the expansion

$$\exp(z) = 1 + z + \frac{z^2}{2!} + \frac{z^3}{3!} + \ldots$$

to the problem of computing $(1 - \alpha_*)100\%$ upper confidence bounds when zero events are observed in a binomial setting. Use this to show that $n\pi_u$, with π_u determined by exact and mid-p methods, approach fixed values. What are these fixed values?

2.19. Consider the computation of upper 95% confidence bounds for the binomial parameter when $t = 1, 2, 3, 4$. For each of these values of t, compute exact and mid-p bounds for π_u when $n = 10, 30, 50, 100, 500$. Tabulate the values of $n\pi_u$. What do you conclude? Compare your results with Newman (1995) and account for any differences.

2.20. Compute the mid-p 95% upper confidence bound for π, under a geometric sampling design, for the neural network data of Example 2.1. How does it compare with the mid-p upper confidence bound under the binomial design? Repeat this comparison for other hypothetical data.

2.21. Suppose we observe 3 successes in 30 iid Bernoulli trials. Analyze these data using a one-sided approach assuming (i) the binomial sampling design, and (ii) the negative binomial sampling design. Compare your findings in terms of exact, mid-p and asymptotic indices of inference.

2.22. For any given number of trials n, and number of successes $t = 1$, prove that the upper exact binomial confidence limit for π is always larger than the upper exact geometric confidence limit (George and Elston 1993). Using either theoretical arguments or empirical comparisons, determine how the two lower confidence limits relate to each other.

Do these results extend to the comparison of the binomial and negative binomial confidence limits when $t > 1$? Do they extend to mid-p and asymptotic confidence limits?

CHAPTER 3

Two-Sided Univariate Analysis

3.1 Introduction

This chapter extends the one-sided data analysis methods of Chapter 2 to two-sided methods. The focus remains on the parameter of a univariate polynomial based distribution (PBD). The specific aims now are:

- To define two-sided evidence functions and formulate the twice the smallest tail (TST) exact, TST mid-p and asymptotic two-sided evidence functions for the parameter of a univariate PBD.
- To define three indices of statistical inference, namely, the two-sided p-value, two-sided confidence interval and point estimate, related to the evidence function, and illustrate their use.
- To introduce the median unbiased estimate, the likelihood, maximum likelihood estimate and test statistic, and define the idea of equivalent statistics.
- To describe the score, likelihood ratio, Wald, probability, distance from the center and the combined tails methods for constructing exact and, where appropriate, large sample two-sided evidence functions for discrete data.
- To formulate an evidence function based reporting of the results of discrete data analysis.
- To describe point and interval estimation at the boundary of the sample space.

3.2 Two-Sided Inference

We give two real data examples to introduce the issues.

Example 3.1: We reconsider the study of the sex ratio among children of subjects occupationally exposed to low frequency electromagnetic fields used in Example 2.2. Set π and π_0 as the proportions of males among the offsprings in the exposed and reference populations. π_0 is known. We argued in Chapter 2 that the appropriate hypothesis setup here is:

$$H_0 : \pi = \pi_0 \quad \text{versus} \quad H_1 : \pi \neq \pi_0$$

Among 277 offsprings of the mothers employed at Norwegian smelter works, 125, or 45%, were male. In a comparable unexposed reference Norwegian population, this rate was 51% (Irgens et al. 1997). We thus test $H_0 : \pi = 0.51$ versus $H_1 : \pi \neq 0.51$ and compute an estimate for π.

Example 3.2: Let π be the proportion of women diagnosed with breast cancer before age 35 years who have a *BRCA1* gene mutation. This mutation has been postulated to increase the risk of breast cancer. In a population based sample, 12 of the 193 women diagnosed with breast cancer in this age at diagnosis group had such a germ line gene mutation (Malone et al. 1998). From these data, we need lower and upper confidence bounds for π.

****

The two-sided hypotheses for a proportion is also formulated in terms of odds, $\phi = \pi/(1-\pi)$, or of log-odds, $\beta = \ln\{\pi/(1-\pi)\}$.

$$H_0 : \phi = \phi_0 \quad \text{versus} \quad H_1 : \phi \neq \phi_0 \tag{3.1}$$

$$H_0 : \beta = \beta_0 \quad \text{versus} \quad H_1 : \beta \neq \beta_0 \tag{3.2}$$

As noted in Chapter 2, two-sided hypotheses are the norm in practical data analysis. In a clinical trial comparing two therapies, the null hypothesis is that they have equivalent efficacies. Under the alternative, we allow that the new treatment may be better or worse than the standard treatment. This is justified by historical and recent experience with new therapies. In an epidemiologic study of the role of a lifestyle factor in a disease, the neutral *a priori* null hypothesis states that it has no role in the disease process. Alternatively, it may either cause, promote or protect from the disease.

The null value often encountered is $\phi_0 = 1$, or $\beta_0 = 0$. Formulating the issue in terms of β divides the alternative parameter space into two symmetric parts, namely, $\beta > 0$ or $\beta < 0$. This, to an extent, enhances interpretation and is often preferred. For large sample analysis, statistical models are, for a variety of reasons, usually formulated under this type of parameterization.

Throughout this text, we formulate and analyze the PBD in an interchangeable manner in terms of either β or ϕ. We go from one form to the other with the understanding that they are equivalent ways of looking at the same problem. To restate the PBD,

$$P[\mathcal{T} = t; \beta \text{ or } \phi] = \frac{c(t)\exp(\beta t)}{h(\beta)} = \frac{c(t)\phi^t}{f(\phi)} \tag{3.3}$$

where

$$h(\beta) = \sum_{u \in \Omega} c(u)\exp(\beta u) = f(\phi) = \sum_{u \in \Omega} c(u)\phi^u \tag{3.4}$$

When $\beta = 0$, or $\phi = 1$,

$$P[\mathcal{T} = t; \beta = 0] = \frac{c(t)}{\sum_u c(u)} \tag{3.5}$$

Upon observing $\mathcal{T} = t$, we need to assess if the evidence does, or does not favor the null, and want a confidence interval (CI) for β. Let $E(t;\beta)$ be the two-sided evidence function that we plan to use for this purpose. Then the features of one-sided evidence functions developed in Chapter 2 lead us to propose the following features for its two-sided version:

Two-Sided Evidence Function

- $E(t;\beta)$ is a continuous function of β taking values between 0 and 1.
- The right portion of $E(t;\beta)$ decreases monotonically as β increases, with zero as the positive infinity asymptote.
- The left portion of $E(t;\beta)$ decreases monotonically as β decreases with zero as the negative infinity asymptote.

A function that, starting from a maximal value at a central point, decreases monotonically in either direction is called a **bimonotonic decreasing function**. Beyond some point, the larger the value of β, the smaller the evidence for it. Below some point, the smaller the value of β, the smaller the evidence for it. Then, large values of the function support H_0 and small values support H_1.

For a PBD, a large β is linked with low probability in the left tail, and a small β value with low probability in its right tail. It is reasonable to combine, in some way, the probability in the two tails to construct a two-sided evidence function. Consider a function defined from the smaller of the right and left tails.

$$E(t;\beta) = 2\min\{P[\mathcal{T} \geq t; \beta], P[\mathcal{T} \leq t; \beta]\} \qquad (3.6)$$

Since the value of (3.6) can exceed 1.0, we set it equal to 1.0 when it does.

Example 3.3: Take the binomial trial with $n = 10$ and $t = 6$. The graph of $E(6;\beta)$, shown in Figure 3.1, has the shape of an inverted V, the middle part flattened, and with asymptotes that approach zero in each direction. $E(t;\beta)$ is hence a valid **two-sided evidence function**. As noted before, it is also known as a *p*-value function, or a confidence interval function.

****

As in the one-sided situation, a two-sided *p*-value is obtained as a vertical intercept from the two-sided evidence function, and a two-sided CI, from the interval obtained from its horizontal intercepts. A two-sided evidence function moreover yields another useful quantity, namely a **point estimate**, or an **estimate**. This may be regarded as the 'most plausible' value of the parameter based on the data.

Two-Sided Inference

- The two-sided p-value for H_0 is the value of the evidence function at the null.

$$p(t; \beta_0) = \mathsf{E}(t; \beta_0) \qquad (3.7)$$

- The two-sided $100(1 - \alpha_*)\%$ CI for β is the set of values of β where $\mathsf{E}(t; \beta) \geq \alpha_*$. It comprises the set of β_0 at which $H_0 : \beta = \beta_0$ is *not rejected* at level α_*. The two-sided lower and upper confidence limits, respectively denoted $\beta_l(\alpha_*, t)$ and $\beta_u(\alpha_*, t)$ or just β_l and β_u, obtain from the equation

$$\mathsf{E}(t; \beta) = \alpha_* \qquad (3.8)$$

- The value of β contained in all possible two-sided CIs for β is a point estimate for β. In other words,

$$\hat{\beta}(t) = \bigcap_{0 \leq \alpha_* \leq 1} (\beta_l(\alpha_*, t), \beta_u(\alpha_*, t)) \qquad (3.9)$$

At times, this may not be a unique value.

Example 3.3 (continued): Suppose we need a 90% CI for β from binomial data with $n = 10$ and $t = 6$. Drawing a horizontal line at 0.1 in Figure 3.1, the values of β for which the function is greater than 0.1 lie in the interval $(-0.830, 1.734)$. This is a 90% exact CI for β. Since this function has a flat apex, it does not yield a unique point estimate. However, if the left and right portions were not truncated at 1.0, they would meet at a point which is the **median unbiased estimate** (see §3.3 below).

****

The continuous, bimonotonic decreasing nature of the evidence function ensures that statistical indices derived from it have the features that are generally regarded as desirable features. In particular:

- For any $1 > \alpha_{*1} > \alpha_{*2} > 0$, the two-sided test significant at level α_{*2} is also significant at level α_{*1}.
- For any $1 > \alpha_{*1} > \alpha_{*2} > 0$, the $100(1 - \alpha_{*1})\%$ confidence region is smaller than and is contained in the $100(1 - \alpha_{*2})\%$ confidence region. This is the property of **nestedness**.
- For any $1 > \alpha_* > 0$, the $100(1 - \alpha_*)\%$ confidence region is a single interval. This is the property of **connectedness**.

Many different methods of computing estimates, two-sided p-values and confidence intervals exist. Some derive from evidence functions and some do not. We do not focus on the latter in this text. There are also many ways of constructing two-sided evidence functions, especially when the underlying distribution is not unimodal or symmetric. Not all of these functions satisfy all of the above noted properties. The varied evidence functions derive from varied conceptualizations of a tail probability, and of how to combine the two tails. We describe some common ways constructing two-sided evidence functions below.

TWICE THE SMALLER TAIL METHOD

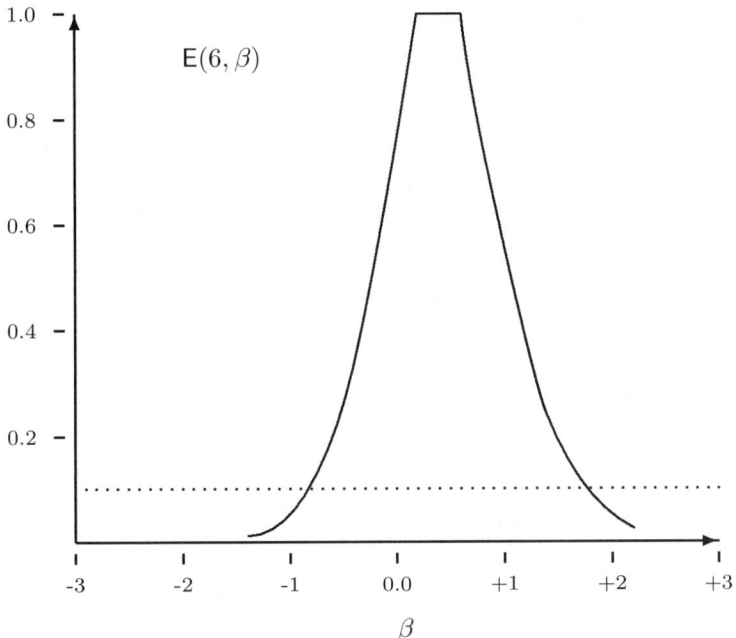

Figure 3.1 *TST Evidence Function for* $T = 6$, *and* $n = 10$.

3.3 Twice the Smaller Tail Method

The function (3.5), set to 1.0 when it exceeds 1.0, is the TST conventional evidence function. The TST conventional exact *p*-value is its value at the null. Thus, for the PBD,

$$p_c(t; \phi_0) = \frac{2 \min \{f(\phi_0; \geq t), f(\phi_0; \leq t)\}}{f(\phi_0)} \qquad (3.10)$$

When $\phi_0 = 1$, or $\beta_0 = 0$, this becomes

$$p_c(t; \phi_0) = \frac{2 \min \left\{\sum_{u \geq t} c(u), \sum_{u \leq t} c(u)\right\}}{\sum_{u \in \Omega} c(u)} \qquad (3.11)$$

Note, p_c is set equal to 1.0 when these formulas yield a value greater than 1.0.

The TST $(1-\alpha_*)100\%$ lower and upper confidence limits for ϕ obtain by drawing the horizontal line at α_* in the graph of the function. Respectively denoted ϕ_l and ϕ_u, these limits obtain from solving the equations

$$f(\phi_l; \geq t) = 0.5\alpha_* f(\phi_l) \quad \text{and} \quad f(\phi_u; \leq t) = 0.5\alpha_* f(\phi_u) \qquad (3.12)$$

This, the most common way of computing an exact two-sided p-value and CI for discrete data, is also equivalent to combining two one-sided $\alpha_*/2$ level tests.

****

The TST method counts the observed point probability twice. A mid-p version in a two-sided setting assigns one half of this probability to the left tail and the other half to the right tail. The mid-p TST evidence function, $E_m(t;\beta)$, is defined as

$$2\min\{P[\mathcal{T} > t;\phi], P[\mathcal{T} < t;\phi]\} + P[\mathcal{T} = t;\phi] \tag{3.13}$$

For a PBD, the two-sided mid-p-value equals

$$p_m(t;\phi_0) = \frac{2\min\{f(\phi_0;>t), f(\phi_0;<t)\} + f(\phi_0;t)}{f(\phi_0)} \tag{3.14}$$

Since the observed point probability is not counted twice, we have that $0 \leq p_m \leq 1$. The TST mid-p-value is also equal to

$$p_m(t;\phi_0) = p_c(t;\phi_0) - \frac{f(\phi_0;t)}{f(\phi_0)} \tag{3.15}$$

The mid-p $(1-\alpha_*)100\%$ lower and upper confidence limits obtain from solving the respective equations

$$f(\phi_l;>t) + 0.5f(\phi_l;t) = 0.5\alpha_* f(\phi_l) \tag{3.16}$$

and

$$f(\phi_u;<t) + 0.5f(\phi_u;t) = 0.5\alpha_* f(\phi_u) \tag{3.17}$$

The parameter value that makes the observed point the median of the distribution assigns equal probabilities to its right and left tails. Equating the tails of the TST mid-p evidence function, we get

$$f(\phi;>t) = f(\phi;<t) \tag{3.18}$$

The solution to (3.18) is the **median unbiased estimate** or **mue** for ϕ and β, denoted respectively $\hat{\phi}_{\text{mue}}$ and $\hat{\beta}_{\text{mue}}$. It also maximizes the TST mid-p evidence function and is contained in all TST mid-p confidence intervals. To restate in terms of β, it is the solution of

$$\sum_{u<t} c(u)\exp(\beta u) = \sum_{u>t} c(u)\exp(\beta u) \tag{3.19}$$

As noted earlier, the same estimate obtains from equating the two portions of the conventional exact TST evidence function. So, even though the 0% CI from a conventional TST evidence function is not necessarily a single point, we report the mue as the point estimate associated with the exact TST method.

EXAMPLES

3.4 Examples

Example 3.4: First consider a disease for which two diagnostic procedures (A or B) are used. Are they used equally often? In a random sample of 12 patients, 4 were subjected to procedure A and 8 to B. Let π denote the true rate at which A is employed; so $1 - \pi$ is the rate at which B is employed. We have $B(12, \pi)$ problem with $t = 4$. Under the null, $\pi = 0.5$ and $\beta = 0$, which is to be tested against $\beta \neq 0$. The TST conventional p-value is

$$p_c = 2\{c(0) + c(1) + c(2) + c(3) + c(4)\}/2^{12}$$

where $c(t) = 12!/\{t!(12-t)!\}$. Hence $p_c = 0.388$. The mid-p-value is

$$p_m = p_c - c(4)/2^{12} = 0.388 - 0.121 = 0.267$$

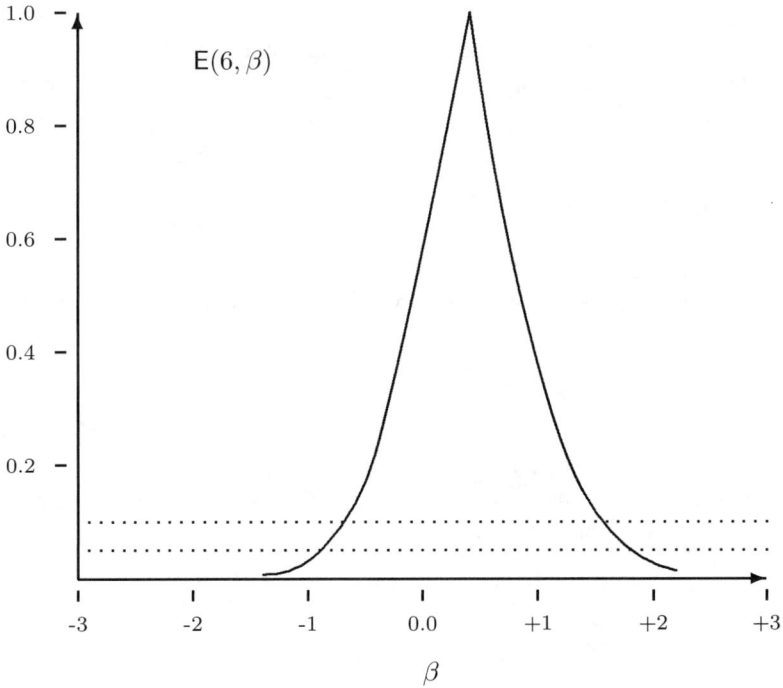

Figure 3.2 *TST Mid-p Evidence Function for $T = 6$.*

Example 3.3 (continued): For an example of a two-sided mid-p confidence interval we return to the binomial data with $n = 10$ and $t = 6$. Figure 3.2 is a graph of mid-p $E(t; \beta)$ as a function of β when $t = 6$. The 95% mid-p confidence limits are obtained by drawing the horizontal line at 0.05. These are -0.890 and 1.796 which are also the solutions to the two equations

$$\frac{120e^{7\beta} + 45e^{8\beta} + 10e^{9\beta} + e^{10\beta} + 105e^{6\beta}}{(1 + e^{\beta})^{10}} = 0.025$$

$$\frac{1 + 10e^{\beta} + 45e^{2\beta} + 120e^{3\beta} + 210e^{4\beta} + 252e^{5\beta} + 105e^{6\beta}}{(1 + e^{\beta})^{10}} = 0.025$$

Consider TST mid-p 95% and 90% CIs for β for binomial data with $n = 10$ and $t = 6$. The horizontal lines at 0.05 and 0.10 in Figure 3.2 show that the latter is included in the former. Similarly, the 80% interval is contained in the 90% interval.

Our estimate of β, the mue, is the value contained in all the possible confidence intervals. For the above data, we get the mue from Figure 3.2. The apex of the function is at $\hat{\beta}_{\text{mue}} = 0.393$. Therefore,

$$\hat{\phi}_{\text{mue}} = \exp(\hat{\beta}_{\text{mue}}) = 1.481 \quad \text{and} \quad \hat{\pi}_{\text{mue}} = \frac{\hat{\phi}_{\text{mue}}}{1 + \hat{\phi}_{\text{mue}}} = 0.597$$

****

Having been introduced to statistics via the normal distribution, students assume that the point estimate is always the mid-point of a confidence interval. That, however, is not necessarily so. A more general way of conceptualizing an estimate is as the value obtained by progressively decreasing the levels of confidence of the confidence intervals.

3.5 The Likelihood Function

The **likelihood** is the probability $P[\mathcal{T} = t; \beta]$ regarded as a function of the parameter β. Its logarithm is called the **log-likelihood function**. For the exponential form of the PBD, the log-likelihood is

$$\ell(\beta) = \beta t - \ln\{h(\beta)\} \qquad (3.20)$$

Additive factors in the log-likelihood that do not depend on β are ignored. The **maximum likelihood estimate** is the parameter value at which the likelihood is maximized. Denoted $\hat{\beta}$, and abbreviated the **mle**, it is computed by solving

$$\frac{\partial \ell}{\partial \beta} = 0 \qquad (3.21)$$

and checking at the solution that

$$\frac{\partial^2 \ell}{\partial \beta^2} < 0 \qquad (3.22)$$

Let $h'(\beta)$ and $h''(\beta)$ respectively denote (as in Theorem 1.1) the first and second derivatives of $h(\beta)$ with respect to β. Then

THE LIKELIHOOD FUNCTION

$$h'(\beta) = \sum_{u \in \Omega} uc(u)\exp(\beta u) \tag{3.23}$$

$$h''(\beta) = \sum_{u \in \Omega} u^2 c(u)\exp(\beta u) \tag{3.24}$$

Hence we get

$$\frac{\partial \ell}{\partial \beta} = t - \frac{h'(\beta)}{h(\beta)} = t - E[\mathcal{T};\beta] \tag{3.25}$$

Setting this to zero and solving for β translates to solving

$$E[\mathcal{T};\beta] = t \quad \text{or} \quad \mu(\beta) = t \tag{3.26}$$

which is equivalent to

$$\sum_{u \in \Omega} (u-t)c(u)\exp(\beta u) = 0 \tag{3.27}$$

Differentiating the log-likelihood twice we get

$$\frac{\partial^2 \ell}{\partial \beta^2} = -\frac{h(\beta)h''(\beta) - (h'(\beta))^2}{(h(\beta))^2} \tag{3.28}$$

Substituting the expressions for $h(\beta)$, $h'(\beta)$ and $h''(\beta)$ in this and simplifying gives

$$\frac{\partial^2 \ell}{\partial \beta^2} = -(E[\mathcal{T}^2;\beta] + \{E[\mathcal{T};\beta]\}^2) = -\sigma^2 \tag{3.29}$$

where σ^2 is the variance of $\mathcal{T}$. Since $\sigma^2 > 0$, we see that requirement (3.22) is satisfied. The mle of ϕ then is $\exp(\hat{\beta})$.

****

For binomial data:

$$\frac{\partial \ell}{\partial \beta} = t - n\ln\left(\frac{\exp(\beta)}{1+\exp(\beta)}\right)$$

Putting this equal to zero and solving, we have that

$$\hat{\beta} = \ln\left(\frac{t}{n-t}\right) \quad \text{and} \quad \hat{\pi} = \frac{t}{n}$$

For the Poisson distribution with parameter λ, $\ell = -\lambda + t\ln\lambda$. Thus

$$\frac{\partial \ell}{\partial \lambda} = -1 + \frac{t}{\lambda} \quad \rightarrow \quad \hat{\lambda} = t$$

In both these cases, $\ell'' < 0$, except at the boundary points of the sample space.

3.6 The Score Method

This section deals with the score based method of constructing evidence functions. First we define a key concept.

> Test Statistic
>
> A test statistic is a quantity that depends only on observable data and known parameters, and which can be used to draw inference about one or more unknown parameters of a statistical model.

Test statistics are not just used to construct statistical tests but are also used to construct evidence functions, and thereby derive confidence intervals. So they may as well be called **evidentiary statistics**.

A test statistic is a random variable dependent on the random data variables and known parameters. We denoted a general test statistic as $\mathcal{W}$. In the one-sided setting in Chapter 2, we implicitly used the statistic $\mathcal{W} = \mathcal{T}$. In a two-sided setting, many test statistics are available. The tail areas of their distributions give evidence in relation to H_0 versus H_1. Often the statistics are nonnegative. At times, large values provide evidence against H_0; and at times, small values do not support H_0; and at times, both small and large values have this feature.

The three test statistics often used in large sample discrete data analysis are the score, likelihood ratio and Wald statistics. They are also used for exact and mid-p inference. The **score statistic** for β is defined by

$$\mathcal{W} = \left(\frac{\partial \ell}{\partial \beta}\right)^2 \left(-\mathbf{E}\left[\frac{\partial^2 \ell}{\partial \beta^2}\right]\right)^{-1} \tag{3.30}$$

Relations (3.25) and (3.29) show that for a PBD

$$\mathcal{W} = (\mathcal{T} - \mu)^2 \sigma^{-2} \tag{3.31}$$

The following properties, stated here without proof, hold for the univariate (unconditional or conditional) PBDs considered in this book:

- A one parameter score statistic approximately has a chisquare distribution with one **degree of freedom (df)** when the data counts or the sample size is sufficiently large.
- The score statistic does not depend on whether the model is parameterized in the regular form with ϕ, or in the exponential form with $\exp(\beta)$.
- Score statistics often have a common form that is written as

$$\sum_{\text{all cells}} \frac{(\text{Observed} - \text{Expected})^2}{\text{Expected}}$$

Such statistics, known as **chisquare statistics,** are regularly found in the analysis of discrete data.

THE SCORE METHOD

****

The binomial, Poisson and negative binomial data score statistics **under specified parameter values** are:

$$\textbf{Binomial:} \qquad \mathcal{W} = \frac{(\mathcal{T} - n\pi)^2}{n\pi(1-\pi)}$$

$$\textbf{Poisson:} \qquad \mathcal{W} = \frac{(\mathcal{T} - \lambda)^2}{\lambda}$$

$$\textbf{Negative Binomial:} \qquad \mathcal{W} = \frac{[\mathcal{T}(1-\pi) + r\pi]^2}{r\pi}$$

The Poisson score statistic clearly has the chisquare form. That the binomial score statistic has this form emerges from rewriting it as

$$\mathcal{W} = \frac{(\mathcal{T} - n\pi)^2}{n\pi} + \frac{([n - \mathcal{T}] - n[1 - \pi])^2}{n(1-\pi)}$$

****

At $\mathcal{T} = t$, the observed value of the score statistic is

$$w = (t - \mu)^2 \sigma^{-2}$$

The two-sided **score statistic based evidence function** is

$$\mathrm{E}(t;\beta) = \mathrm{P}[\mathcal{W} \geq w; \beta] \qquad (3.32)$$

- Using the exact distribution of the score statistic gives the exact score evidence function, and using an approximation for the tail probability in (3.32) gives an asymptotic score evidence function.
- The two-sided conventional exact score p-value is the probability, under the null, of all points with score statistic value greater than or equal to its value at the observed point. This is obtained by putting $\beta = \beta_0$ in (3.32).
- Confidence intervals from these functions are obtained in the usual manner with appropriate horizontal intercepts.
- An exact score evidence function may violate the bimonotonicity property. The violations usually are minor in nature. The asymptotic score evidence function is, on the other hand, a continuous bimonotonic function.

The point estimate associated with the score evidence function, if it exists, is the value of β which makes its observed value equal to zero, or that which makes the right hand side of the (3.32) equal to one. It thus corresponds to a 0% confidence interval for β and is the value of β at which

$$\mathbf{E}[\mathcal{T};\beta] = t \qquad (3.33)$$

The mle

For a PBD, the mle is contained in all score based conventional exact and large sample score confidence intervals. It also makes the observed point the mean of the distribution and maximizes the likelihood.

In many commonly used univariate distributions, the mle, unlike the mue, is explicitly specified.

An estimator is called **mean unbiased**, or simply **unbiased** if its expected value equals the parameter value. For binomial data, the mle of π is an unbiased estimator of π. But the mle of $\phi = \pi/(1-\pi)$ is not an unbiased estimator of ϕ. Another problem with using mean bias to evaluate estimators in discrete data is that more often than not, the finite sample expected values of the mle of ϕ and β are not finite.

To emphasize: the three items usually produced in a traditional data analysis, namely a two-sided p-value, a confidence interval and a point estimate, emanate naturally from the two-sided evidence function.

3.7 Additional Illustrations

Example 3.1 (continued): We analyze data from the low frequency electromagnetic fields study with the asymptotic score method. 125, or 45%, of the 277 births to mothers employed at a Norwegian smelter works were male (Irgens et al. 1997). Assume the data have the $B(n, \pi)$ distribution, where π is the true proportion of males births among the exposed group. Using the reference population value of $\pi_0 = 51\%$, we test $H_0 : \pi = 0.51$ versus $H_1 : \pi \neq 0.51$ by constructing the asymptotic score evidence function.

- The observed score statistic value is

$$w = \frac{(t - n\pi)^2}{n\pi(1-\pi)} = \frac{(125 - 277\pi)^2}{277\pi(1-\pi)}$$

- Using the one df **chisquare approximation**, we plot the two-sided evidence function

$$\mathsf{E}(w;\pi) = \mathsf{P}[\,\mathcal{W} \geq w;\,\pi\,]$$

as a function of π. We derive the three quantities of interest from this plot, which is given in Figure 3.3.

The p-value: At $\pi_0 = 0.51$, $w = 3.82$. The vertical line at $\pi = \pi_0 = 0.51$ reaches the evidence function at 0.0505, the corresponding p-value.

Confidence Interval: The horizontal line $\alpha_* = 0.05$ intersects this function at two points, giving the 95% confidence interval for π as $(0.394, 0.510)$.

In general, the lower and upper limits of the $(1 - \alpha_*)100\%$ CI for π is obtained by solving the equation

$$\frac{(t - n\pi)^2}{n\pi(1-\pi)} = \chi^2_{1-\alpha_*} \qquad (3.34)$$

where $\chi^2_{1-\alpha_*}$ is the value such that the area to the right of it under the chisquare curve with one df is α_*. So these 95% asymptotic limits obtain from the equation

ADDITIONAL ILLUSTRATIONS

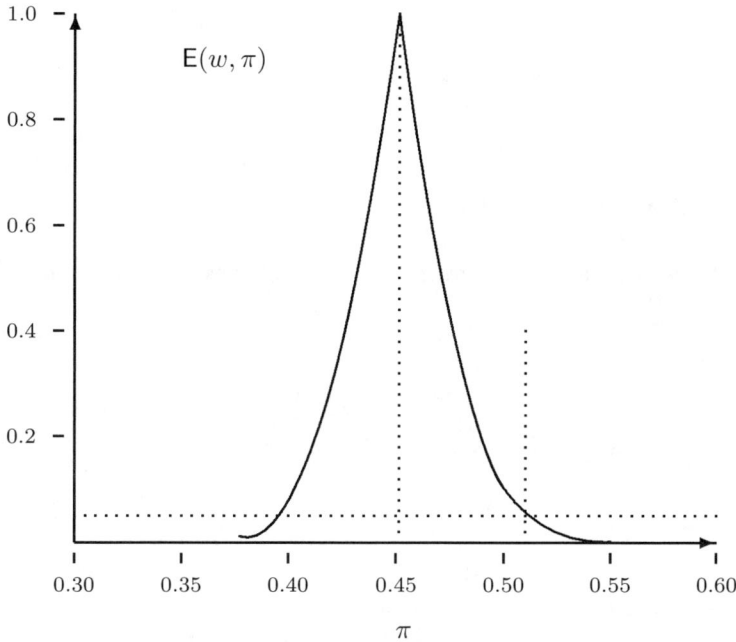

Figure 3.3 *Asymptotic Score Evidence Function.*

$$(t - n\pi)^2 = 3.84n\pi(1 - \pi) \tag{3.35}$$

Point Estimate: The value of π at which the left and right portions of the evidence curve meet is t/n, which is also the solution to the above equation for $\alpha = 1$ and the mle for π. For the Irgens et al. (1997) data, the estimate of π from the evidence function is 0.451, which also equals $125/277$.

****

Example 3.5: We revisit Example 2.4. There the number of cases T in the period τ was modeled as a Poisson variate with mean $\phi = \lambda\tau$. With $t = 4$ as the observed count, the mue for ϕ obtains by solving

$$P[T \geq 4; \phi] = P[T \leq 4; \phi]$$

which becomes

$$e^\phi = 2 + 2\phi + \phi^2 + \phi^3/3 + \phi^4/24$$

giving the solution $\phi = 4.1611$. With $\tau = 1350$, the mue for λ is

$$\hat{\lambda}_{\text{mue}} = \frac{\hat{\phi}_{\text{mue}}}{1350} = 0.00308 \quad \text{cases per person year}$$

The mle for ϕ is $t\tau = 4\tau$. Thus

$$\hat{\lambda}_{mle} = 4/\tau = 4/1350 = 0.00296 \quad \text{cases per person year}$$

Hence, over a year, with the mle, we expect about 300 cases in a population of size 100,000 over a year while the mue leads us to anticipate about 310 cases.

3.8 Likelihood Ratio and Wald Methods

For observed $\mathcal{T} = t$, the **likelihood ratio statistic**, or the **LR statistic**, denoted by $\mathcal{Q}$, is defined as

$$\mathcal{Q} = 2\{\ell(\hat{\beta}) - \ell(\beta)\} \tag{3.36}$$

where $\hat{\beta}$ is the mle of β. For a univariate PBD, it becomes

$$\mathcal{Q} = 2\{(\hat{\beta} - \beta)\mathcal{T} - \ln[h(\hat{\beta})] + \ln[h(\beta)]\} \tag{3.37}$$

- For the univariate PBDs we consider, the LR statistic follows approximately a chisquare distribution with one df when the cell counts in the data or the sample size are sufficiently large.
- Like the score statistic, the LR statistic for a PBD does not depend on whether the model is parameterized in the regular form or in the exponential form.
- The LR statistics often has form:

$$2 \sum_{\text{all counts}} (\text{Observed}) \ln \frac{(\text{Observed})}{(\text{Expected})}$$

The LR statistic can also be used for exact analysis. Write

$$\mathcal{Q} = 2\{(\hat{\beta}(\mathcal{T}) - \beta)\mathcal{T} - \ln[h(\hat{\beta}(\mathcal{T}))] + \ln[h(\beta)]\} \tag{3.38}$$

with the observed value

$$q = 2\{(\hat{\beta}(t) - \beta)t - \ln[h(\hat{\beta}(t))] + \ln[h(\beta)]\} \tag{3.39}$$

The two-sided **likelihood ratio based evidence function** is

$$E(t; \beta) = P[\mathcal{Q} \geq q; \beta] \tag{3.40}$$

- The exact LR evidence function may not be bimonotonic. The asymptotic LR evidence function, however, is a continuous bimonotonic function.
- Two-sided conventional exact and asymptotic LR based *p*-values and confidence intervals are computed from these functions in the usual manner.

LIKELIHOOD RATIO AND WALD METHODS

The LR statistics for the binomial and Poisson distributions are

Binomial: $$Q = 2t\ln\left\{\frac{t}{n\pi}\right\} + 2(n-t)\ln\left\{\frac{(n-t)}{n(1-\pi)}\right\}$$

Poisson: $$Q = 2\left\{\lambda - t + t\ln\frac{t}{\lambda}\right\}$$

Note the LR statistic for the negative binomial is the same as the LR statistic for the binomial model.

****

Another common statistic used in discrete data analysis is the Wald statistic. Denoted Z, it is defined as

$$Z = \frac{(\hat{\beta} - \beta)^2}{\text{var}[\hat{\beta}]} \quad (3.41)$$

where $\hat{\beta}$ is the mle of β and $\text{var}[\hat{\beta}]$ is the large sample variance of the mle, $\hat{\beta}$. For the PBDs found in this text, we state here without proof that in large samples this is given by

$$\text{var}[\hat{\beta}] = \left(-\mathbf{E}\left[\frac{\partial^2 \ell}{\partial \beta^2}\right]\right)^{-1} \quad (3.42)$$

From (3.29), we have that

$$\text{var}[\hat{\beta}] = \sigma^{-2} \quad (3.43)$$

That is, the large sample variance of the mle of β is the inverse of the exact variance of T.

- The one parameter Wald statistic in univariate PBDs we consider generally has an approximately one df chisquare distribution in large samples.
- Unlike the score and LR statistics, this statistic depends on how the distribution is parameterized. Under the Wald method, inference on ϕ thus does not directly translate into inference on β.
- The Wald statistic often used in practice is the empirical form of (3.41) in which an estimate of the variance, $\text{var}[\hat{\beta}]$, instead of the variance itself, is used. The empirical Wald statistic is thereby

$$Z_e = \frac{(\hat{\beta}(T) - \beta)^2}{\widehat{\text{var}}[\hat{\beta}(T)]} \quad (3.44)$$

For the binomial distribution, $B(n; \pi)$, the Wald statistic in terms of π is

$$Z_* = \frac{(\hat{\pi} - \pi)^2}{\text{var}[\hat{\pi}]} = \frac{(T - n\pi)^2}{n\pi(1-\pi)}$$

In terms of β, the log-odds, it is

$$Z_{**} = \frac{(\hat{\beta} - \beta)^2}{\text{var}[\hat{\beta}]}$$

$$= \left(\ln\frac{\hat{\pi}}{1-\hat{\pi}} - \ln\frac{\pi}{1-\pi}\right)^2 E\left[-\frac{\partial^2 \ell}{\partial \beta^2}\right]$$

$$= \left(\ln\frac{T}{n-T} - \ln\frac{\pi}{1-\pi}\right)^2 \frac{n\pi(1-\pi)}{n}$$

Further, the binomial empirical Wald statistic in terms of π is

$$Z_{e*} = \frac{(\hat{\pi} - \pi)^2}{\widehat{\text{var}}[\hat{\pi}]} = \frac{(T - n\pi)^2 n}{T(n-T)}$$

For the binomial, the Wald statistic in terms of π is identical to the score statistic. A similar property also holds for the Poisson case. But that is not always the case.

****

Using the parameter β, the two-sided regular **Wald statistic based evidence function** is

$$E(t;\beta) = P[\mathcal{Z} \geq z;\beta] \qquad (3.45)$$

The regular and empirical Wald statistics are used to construct their own versions of two-sided asymptotic and exact evidence functions. In that case, we have a total of **eight** Wald evidence functions, that is, four exact and asymptotic, regular and empirical functions in terms of ϕ and corresponding **four** in terms of β. At least two of these are, however, equivalent.

One advantage of the empirical Wald statistic is that it often gives an explicit formula for the large sample confidence intervals for β or ϕ. In terms of the former, the 95% confidence interval is

$$\hat{\beta}(t) \pm 1.96\sqrt{\widehat{\text{var}}[\hat{\beta}(t)]} \qquad (3.46)$$

The empirical Wald statistic is commonly produced by statistical software packages, and is often used in large sample analysis. However, its forms usually employed in practice have shown poor, and in fact erratic, performance in a variety of theoretical and empirical studies. We therefore do not favor the use of the Wald method.

3.9 Three More Methods

We note three additional methods of formulating test statistics and evidence functions. These are used in some specific cases and are important from a theoretical viewpoint. They are (i) the **probability based (PB) method**, (ii) the **distance from the center (DC) method** and (iii) the **combined tails (CT) method**.

THREE MORE METHODS

3.9.1 The Probability Based Method

In a **unimodal** distribution, ranking the sample points according to their probability and aggregating the points whose probability does not exceed a specified value generates a set of points that lie either in the right or left tail of the distribution. We may thus regard the probability, $f(t;\beta) = P[\mathcal{T} = t; \beta]$ itself as a test statistic. The two-sided **probability based evidence function** using this statistic is then

$$E(t;\beta) = P[f(\mathcal{T};\beta) \leq f(t;\beta)] \quad (3.47)$$

For a PBD, this becomes

$$E(t;\beta) = P[c(\mathcal{T})\exp(\beta\mathcal{T}) \leq c(t)\exp(\beta t)] \quad (3.48)$$

A two-sided probability based p-value is the null probability of all points whose null probability does not exceed the null probability of the attained value t. At $\beta = \beta_0$,

$$p_c(t;\beta_0) = P[c(\mathcal{T})\exp(\beta_0 \mathcal{T}) \leq c(t)\exp(\beta_0 t); \beta_0] \quad (3.49)$$

Under the usual null hypothesis $\beta_0 = 0$, this simplifies to

$$p_c(t;0) = P[c(\mathcal{T}) \leq c(t); \beta = 0] \quad (3.50)$$

Confidence regions, called **Stern** confidence regions, are obtained from the associated evidence function in the usual way. Many common discrete distributions, like the binomial, Poisson and hypergeometric, are **unimodal.** Other univariate conditional distributions like those found in some dose-response studies may, however, be multimodal. In that case, the probability method associates central observations with those in the tails.

The probability method is often used to compute exact p-values for a 2×2 table. Two-sided probability based p-values in hypergeometric or binomial distributions are often not too distinct from, and at times smaller than, those from other methods. A multivariate version of the probability method is applied in assessing association in unordered $r \times c$ contingency tables. The sole exact test for the homogeneity of odds ratios in several 2×2 tables available until recently was the Zelen test, which is based on the probability statistic (see Chapter 10).

3.9.2 Distance from the Center Method

The distance from a central point in the distribution can also be used to construct an evidence function. Let μ be a central point of the sample space Ω. This may be the mean or the median. For $\mathcal{T} \in \Omega$, consider the distance statistic

$$\mathcal{D} = \|\mathcal{T} - \mu\| \quad (3.51)$$

with observed value equal to

$$d = \|t - \mu\| \quad (3.52)$$

where $\|x\|$ denotes the absolute value of x. The two-sided **distance from the center evidence function** is

$$\mathsf{E}(t;\beta) = \mathsf{P}[\mathcal{D} \geq d;\ \beta] \qquad (3.53)$$

The two-sided exact *p*-value then is the probability, under the null, of all points which are at least as far from the center as the observed point. It is obtained by setting $\beta = \beta_0$ in the above. Note that

$$\frac{(\mathcal{T}-\mu)^2}{\sigma^2} \geq \frac{(t-\mu)^2}{\sigma^2} \quad \text{if and only if} \quad \|\mathcal{T}-\mu\| \geq \|t-\mu\|$$

When μ is the mean of $\mathcal{T}$, the left hand side of the first relation above is the score statistic. Hence, for univariate inference based on an exact univariate PBD, the **distance from the mean method** gives the same evidence as the exact score method. In particular, the two methods yield identical *p*-values and confidence intervals. The mle is also contained in all distance from the mean evidence functions. In this text, we use score based exact inference for a univariate PBD with the understanding that it is equivalent to the exact distance from the mean based inference. The score statistic has a widely used large sample counterpart and generalizes readily to multivariate settings.

3.9.3 The Combined Tails Method

Another method for assessing two-sided evidence for discrete data computes the smaller of the probability in the two tails. Let

$$\delta(t;\beta) = \min\{\mathsf{P}[\mathcal{T}\geq t;\beta], \mathsf{P}[\mathcal{T}\leq t;\beta]\} \qquad (3.54)$$

The two-sided **combined tails (CT) evidence function** is

$$\mathsf{E}(t;\beta) = \mathsf{P}[\delta(\mathcal{T};\beta) \leq \delta(t;\beta);\ \beta] \qquad (3.55)$$

According to this, a sample point is as or more extreme than the observed point if the probability in the smaller of its tail does not exceed the probability in the smaller of the observed tails. The set of extreme points is the set of points in the observed smaller tail and points in the other tail whose tail probability does not exceed the observed smaller tail probability. This combines the two tails of the distribution to construct the evidence.

- If $\mathsf{P}[\mathcal{T} \geq t;\beta] < \mathsf{P}[\mathcal{T} \leq t;\beta]$, let $t_* < t$ be the maximum $u \in \Omega$ such that $\mathsf{P}[\mathcal{T} \leq u;\beta] \leq \mathsf{P}[\mathcal{T} \geq t;\beta]$. Then,

$$\mathsf{E}(t;\beta) = \mathsf{P}[\mathcal{T}\leq t_*;\beta] + \mathsf{P}[\mathcal{T}\geq t;\beta] \qquad (3.56)$$

- If $\mathsf{P}[\mathcal{T} \geq t;\beta] > \mathsf{P}[\mathcal{T} \leq t;\beta]$, let $t_{**} > t$ be the minimum $u \in \Omega$ such that $\mathsf{P}[\mathcal{T} \geq u;\beta] \leq \mathsf{P}[\mathcal{T} \leq t;\beta]$. Then,

$$\mathsf{E}(t;\beta) = \mathsf{P}[\mathcal{T}\leq t;\beta] + \mathsf{P}[\mathcal{T}\geq t_{**};\beta] \qquad (3.57)$$

- If $\mathsf{P}[\mathcal{T} \geq t;\beta] = \mathsf{P}[\mathcal{T} \leq t;\beta]$, then

$$\mathsf{E}(t;\beta) = 1 \qquad (3.58)$$

COMPARATIVE COMPUTATIONS

The null p-value equals

$$p_c(t; \beta_0) = \mathrm{E}(t; \beta_0) \quad (3.59)$$

Confidence intervals are obtained as horizontal intercepts of this function. Unlike the TST method, the CT method does not count the observed probability twice. For a symmetric distribution, the CT p-value is, except possibly for a central point, equal to the TST p-value. For an asymmetric distribution, it generally is equal to or is smaller than the latter.

Theorem 3.1: Suppose T has a univariate PBD. Then for any observed point in the sample space, the CT p-value can never exceed the TST p-value.

Proof: Evident from considering each of the three possibilities stated in (3.56), (3.57) and (3.58). □.

This result also extends to confidence intervals (see Exercise 3.20).

****

A crucial point here is that the exact score, likelihood ratio, Wald, probability and distance from center evidence function each has an associated exact mid-p version. This obtains by subtracting half the probability of the observed statistic p-value from the probability of the respective conventional statistic based value. For a statistic for which the right tail area gives the evidence function, its mid-p version is

$$\mathrm{E}_m(t; \beta) = \mathrm{P}[\mathcal{W} \geq w; \beta] - 0.5\mathrm{P}[\mathcal{W} = w; \beta] \quad (3.60)$$

Unlike the TST mid-p-value and other exact and asymptotic p-values, the mid-p given by (3.60) cannot equal one. Estimates and confidence intervals from such functions are obtained in the usual manner. Such a mid-p is used in Lydersen and Laake (2003). Haber (1986b) used a mid-p-value based on the distance from the mean statistic.

3.10 Comparative Computations

Example 3.6: A weight loss program aims to produce sustained weight reduction among its participants. Assume that around 10% of overweight subjects experience long term weight loss simply when regularly followed up. (This is called the **Hawthorne effect**.) Let π denote the true proportion of sustained weight reduction among the subjects in the program. A study designed to evaluate it tests the hypothesis $\pi = 0.1$ against the alternative $\pi \neq 0.1$. The alternative includes the possibility of a positive effect over and above the Hawthorne effect, and a negative effect which reduces of the proportion of program persons reaching that goal. In one study, 12 randomly picked obese subjects are recruited into the program, and of them, 3 demonstrate long term weight loss. What do we conclude?

We use these data to compare the different types of p-values. We test $H_0 : \beta = \beta_0$ versus $H_1 : \beta \neq \beta_0$ where $\beta_0 = \ln(0.1/0.9) = -0.220$. The observed $T = 3$. The first two columns of Table 3.1 show the sample space and coefficients of the PBD. The null probability of each sample point is in the third column.

Table 3.1 *Test Statistics for* $B(12; \pi_0 = 0.1)$

	t	$c(t)$	$P[\mathcal{T}=t;\beta_0]$	$\mathcal{W}$	$\mathcal{Q}$	$\mathcal{Z}$
	0	1	0.2824	1.33	2.53	$+\infty$
	1	12	0.3766	0.04	0.04	0.04
	2	66	0.2301	0.59	0.50	0.38
$\Rightarrow$	3	220	0.0852	3.00	2.22	1.44
	4	495	0.0213	7.26	4.83	2.94
	5	792	0.0038	13.37	8.20	4.95
	6	924	0.0005	21.33	12.26	7.68
	7	792	< 0.0001	31.14	16.99	11.53
	8	495	< 0.0001	42.81	22.41	17.34
	9	220	< 0.0001	56.33	28.58	27.04
	10	66	< 0.0001	71.70	35.66	46.46
	11	12	< 0.0001	88.93	43.98	104.77
	12	1	< 0.0001	108.00	55.26	$+\infty$

Note: $\Rightarrow$ indicates the observed sample point.

TST Method: $P[\mathcal{T} \geq 3; \beta_0] = 0.1108$, $P[\mathcal{T} \leq 3; \beta_0] = 0.9743$ and $P[\mathcal{T} = 3; \beta_0] = 0.0852$. Then

$$p_c = 2\min\{0.1108, 0.9743\} = 0.2216$$
$$p_m = 0.2216 - 0.0852 = 0.1364$$

Probability Based Method: The null probability of the observed sample point is 0.0852. Then

$$p_c = P[\pi(\mathcal{T};\beta_0) \leq 0.0852] = 0.1108 \quad \text{and} \quad p_m = 0.0682$$

CT Method: $P[\mathcal{T} \geq 3; \beta_0] = 0.1108$, $P[\mathcal{T} \leq 3; \beta_0] = 0.9743$. With the observed right tail the smaller one, even the most extreme point in the left tail has higher null probability than the total null right tail probability. Therefore,

$$p_c = 0.1108$$

Score Method: For the binomial: $\mu_0 = 12 \times 0.1 = 1.2$ and $\sigma_0^2 = 12 \times 0.1 \times 0.9 = 1.08$. These give the score statistic values given in the fourth column of Table 3.1. At $\mathcal{T} = 3$,

$$\mathcal{W} = (\mathcal{T} - 1.2)^2/1.08 = 3.00$$

There are nine points in the sample space with a score statistic value equal to or greater than 3.00. Thus

$$p_c = P[\mathcal{W} \geq 3; \beta_0] = 0.1108 \quad \text{and} \quad p_m = 0.0682$$

THE ABC OF REPORTING

At $w = 3.00$, the chisquare distribution with one df gives the asymptotic score *p*-value equal to 0.0833.

Likelihood Ratio Method: First we need the mle for β: $\hat{\pi} = 3/12 = 0.25$, so $\hat{\beta} = \ln(0.25/0.75) = -1.10$. The fifth column of Table 3.1 gives the values of the LR statistic under $\beta = \beta_0$ at each sample point. At $T = 3$, $Q = 2.22$. There are ten sample points with an LR statistic value equal to or greater than 2.22. Thus

$$p_c = P[W \geq 2.22; \beta_0] = 0.3932 \quad \text{and} \quad p_m = 0.3506$$

Using the chisquare distribution with one df, the asymptotic *p*-value is equal to 0.1362.

Wald Method: The regular Wald statistic for binomial data is the same as the score statistic. Further, the exact and asymptotic *p*-values with this method are identical to those from the respective score methods.

Empirical Wald Method: The observed value of the empirical Wald statistic is 1.44. From Table 3.1, this gives an exact *p*-value equal to 0.392 and a mid-*p*-value equal to 0.3506. The asymptotic *p*-value for this statistic is 0.230.

Table 3.2 *Exact and Asymptotic p-values*

Method	Exact	Asymptotic	Mid-*p*
TST	0.2216	-	0.1364
Probability	0.1108	-	0.0682
CT	0.1108	-	-
Score	0.1108	0.0833	0.0682
LR	0.3932	0.1362	0.3506
Wald	0.1108	0.0833	0.0682
E-Wald	0.3932	0.2301	0.3506

Note: E-Wald is Empirical Wald.

A summary of the results, in Table 3.2, shows that where available, the mid-*p* is smaller than the respective conventional exact *p*-value. The LR and empirical Wald exact *p*-values are the largest of the *p*-values. The score, probability and regular Wald mid-*p*-values are the smallest. The wide range of the values here is striking; different methods then lead to different conclusions.

The relative behavior of these statistics cannot be gleaned from a single example. What this example indicates is that an analyst needs a guide for selection of the appropriate method(s) for different situations.

3.11 The ABC of Reporting

As noted in Chapter 2, the *p*-value remains the most frequent, and often the sole, index for reporting a data analysis. It has major limitations and a strong potential for misuse. This is an ongoing cause for concern. Recently, the situation has begun to improve somewhat. Major health, medical and social science journals now have guidelines requiring more complete reporting of the results. For the primary aim of the study, at least, many journals need a combination of the p-value, point estimate and confidence interval.

The current opinion about good reporting practice among the applied statistical community is summarized as follows.

A Reporting Guide

- The evidence from a research study cannot be judged on the basis of one, or a series of p-values alone. It is even less helpful to just state if they are below 0.05 or not. In a medical study, for example, statistical significance is not the same thing as clinical significance.
- A confidence interval combined with a point estimate give a better picture of the magnitude, precision and actual significance of the effect in question.
- Nevertheless, as a confidence interval is based on a fixed significance level, complementing it with the p-value provides a better assessment of the variability in the data.

A few statisticians have also stressed reporting the confidence interval (evidence) function. A plot of this function has a complete picture of the evidence relating to the parameter in question. It remedies the exclusive dependence on p-values yet does not dispense with them. It gives the confidence intervals, yet avoids the arbitrariness associated with reporting a 95% or a 90% interval. And it includes more. Its location indicates the general magnitude of the parameter, and its shape indicates the precision of and the logic behind the evidence. If it is narrow, then confidence intervals at all levels are narrow; If it is wide, then the opposite holds. The apex of the function is at the point estimate. Evidence functions include all the essential items suggested for good reporting of data analysis and more.

Sullivan and Foster (1990) present real illustrations of the use of the confidence interval (evidence) function in data analysis of a single study and a meta-analysis. Their examples include identifying effect modification, confounding, between study variability as well as comparing different statistical methods. Rothman (2002) makes a strong case for the confidence interval function, provides relevant examples and discusses the interpretation and relevance of the function.

Despite their positive features, evidence or confidence interval functions are never reported. When several parameters are to be analyzed, the interpretation of their confidence functions may be an unwieldy task. The weight of convention is, in our view, the main reason behind its absence in the applied literature.

One other aspect of reporting needs more scrutiny than it has been given. This concerns the use of one statistic for computing a p-value and quite another for a confidence interval and, perhaps, yet another one for an estimate. For example, a probability method based exact p-value is given with a confidence interval based on inverting the exact TST test, or even the asymptotic Wald test. An asymptotic score p-value is combined with a confidence interval derived from the empirical Wald method. Or, the estimate given with a TST 95% exact confidence interval is the (conditional) mle.

When estimates, p-values and intervals for the same parameter, be they asymptotic or exact,

THE ABC OF REPORTING

but derived from disparate underlying statistics are reported, the evidentiary consistency is lost. At times, contradictory results may ensue. That is, the *p*-value may be less than 0.05 while the 95% confidence interval includes the null value. The analysis based on evidence functions precludes such an eventuality. Therefore, even when the entire evidence function is not reported, the evidence is conveyed in a consistent manner by **reporting indices deriving from a common evidence function**.

A Unified Report

A unified report on a parameter presents a p-value, a confidence interval and an estimate computed from the same evidence function.

Example 3.2 (continued): In the *BRCA1* gene mutation study, π is the mutation probability with $t = 12$ and $n = 193$. We test $\pi = \pi_0 = 0.1$ against a two-sided alternative, compute a confidence interval and give an estimate, and report them in a unified way. The three items in each row of Table 3.3 represent an example of such a report. Each row here is derived from the same evidence function.

Table 3.3 *Analysis of Genetic Data*

Statistic	p-value	Estimate	90% CI
Exact TST	0.0895	0.0629	$(0.0324, 0.1061)$
Mid-p TST	0.0690	0.0629	$(0.0339, 0.1029)$
Asymptotic Score	0.0797	0.0622	$(0.0359, 0.1055)$

Note: The p-values are for $\pi_0 = 0.1$.

****

In an actual report, one generally gives only one of such a row. A TST mid-*p*-value goes with a TST mid-*p* CI and the mue; a conventional TST exact *p*-value goes with a TST conventional exact CI and the mue, and so on. The choice of the evidence function to use, namely, score, TST, mid-*p* TST or another, is done at the stage of study design and is a part of the data analysis plan. Such a decision should be based on considerations of the sample size, type of parameter to be analyzed and knowledge on the performance of the indices based on the respective evidence functions. In some cases, we may report more than one type of evidentiary index. This happens when their relative behavior has not been studied adequately, and when different inferences may emanate from the indices. If the score statistic based analysis gives a distinctive result from the TST mid-*p* analysis, we may want to make a note of that in the report.

A final note to this section: The manner of usage of terms such as exact, mid-*p* or even asymptotic at times conveys the impression that there is a unique method of computing an exact, mid-*p* or asymptotic *p*-value and confidence interval. Even some papers by prominent statisticians give that impression. This needs to be rectified. In data analysis reports, the specific method used should be stated clearly. When giving an exact *p*-value, for example, we should state whether the reported entity is a TST, probability or score statistic based value, or perhaps another form of the exact value.

3.12 Additional Comments

This section covers additional material on estimates, p-values and confidence intervals, and gives a general suggestion about their use.

For sparse or small sample discrete data, one way of computing point and interval estimates uses the **empirical logit** or, simply, **logit** approach. In the binomial case, for example, the empirical logit estimate is

$$\hat{\beta} = \ln\left(\frac{t+0.5}{n-t+0.5}\right) \qquad (3.61)$$

Another logit approach for the binomial adds the value 2 to the number of successes and 4 to the overall total. Then

$$\hat{\beta} = \ln\left(\frac{t+2}{n-t+2}\right) \qquad (3.62)$$

For general discrete data problems, the empirical logit approach entails adding 0.5 (or some other small value) to tabulated data points when zero counts are present, or even when they are not. It gives a finite mle for β when there are zero cell values and gives a finite estimate of its variance. The empirical Wald method employs such adjustments to obtain estimates, confidence intervals and p-values.

These adjustments have a long history. In some univariate cases, they perform well when the data are sparse data. Agresti and Coull (1998), Agresti and Caffo (2000) and Agresti (1999) argue for their use for the single binomial, the difference of two binomials and the analysis of the odds ratio in a 2×2 table. This approach is reminiscent of the multitude of continuity corrections that have been devised to improve the performance of large sample tests in sparse data.

In our view, the use of such adjustments as a general principle for discrete data analysis is to be discouraged. It amounts to tinkering with this and that index, with the extent of adjustment dependent on the problem at hand. For general discrete data, the value added to each count is arbitrary; the inference from adding 0.25 may differ markedly from that obtained when adding 0.5. Such adjustments also break up the underlying evidentiary unity between the p-value, the confidence interval and the point estimate.

****

The median unbiased estimate can be defined in more than one way. Another version of the mue obtains from solving

$$P[\mathcal{T} \geq t; \beta] = 0.5 \quad \text{and} \quad P[\mathcal{T} \leq t; \beta] = 0.5 \qquad (3.63)$$

and taking the mean of the two solutions. At the boundary, the estimate is the single finite solution to one of the two equations. This method always provides a finite estimate. This is not equivalent to (3.18). (As we will see below, the TST method based mue for β is not finite at the boundary.) Some empirical studies have shown that such an mue is more accurate than the mle. This approach, though, gives estimates for ϕ that are not directly related to estimates for β. Moreover, it is not directly linked to an evidence function.

According to Lehmann (1986), the mue is a value that is "*as likely to underestimate as to*

ADDITIONAL COMMENTS

overestimate" the true value. This is different from the definition we use. In a PBD, the two definitions give results that are generally not too far apart from each other.

****

A general version of a $(1 - \alpha_*)100\%$ two-sided confidence region for a parameter selects, either explicitly or implicitly, α_{*1} and α_{*2} that are such that $\alpha_{*1}, \alpha_{*2} \geq 0$, and $\alpha_{*1} + \alpha_{*2} = \alpha_*$, and solves

$$P[\mathcal{T} \geq t; \beta] = \alpha_{*1} \quad \text{and} \quad P[\mathcal{T} \leq t; \beta] = \alpha_{*2} \qquad (3.64)$$

In the TST method, we have $\alpha_{*1} = \alpha_{*2} = \alpha_*/2$. Some of the methods we have described, namely the exact score, likelihood ratio, Wald and probability methods are cases of this general approach. One problem with this approach, as with the Stern regions, is that the resulting confidence region may be composed of several disjoint intervals. This deficiency can be remedied by forming a single interval from the smallest and the largest value from the range covered by these intervals.

For specific distributions like the binomial and Poisson, some authors have developed ways of constructing shorter confidence intervals. Such intervals may, however, violate the key property of nestedness. Yet another method, not directly based on the inversion of a test, and which always yields finite two-sided intervals for discrete data is in Hirji, Mehta and Patel (1988a).

****

The key question then is: Which of the many evidence functions (and the associated *p*-values, estimates and confidence intervals) should we use and when? Which should we employ especially when the sample size is small or the data are sparse? Many empirical and theoretical studies have been done to answer such questions. And they have spawned many recommendations, not all of them consistent with one another. And these pertain to specific indices and conditions. Comparative studies to evaluate the overall methods of constructing evidence as such have hardly begun.

The answers, moreover, depend on the criteria used to evaluate statistical tests, confidence intervals and estimates. So far, with the exception of the actual significance level defined in Chapter 2, we have not specified such criteria. In Chapter 7, we cover the issue in greater depth and define two principal evaluative criteria, **the exact power of a test** and **the actual coverage level of a confidence interval** that can be applied for this purpose.

The research done thus far has not produced a broad consensus applicable to general discrete data problems. **Exact unconditional methods,** also briefly introduced in Chapter 7, and various alternatives for model parameterization, for example, have not been studied in any depth.

Despite such limitations, we have formulated a set of preliminary recommendations based on a synthesis of the research to date. These apply to the analysis of a single parameter in a conditional and an unconditional PBD. For some specific types of problems, though, exceptions to these suggestions can also be given.

- Statistical indices based on the mid-*p* approach are recommended for small samples and/or sparse data. This means reporting the combination of a median unbiased estimate, a TST mid-*p*-value and a TST mid-*p* confidence interval.

- When guaranteed bounds on type I error rates or coverage rates of confidence intervals are needed, the exact TST method or the exact CT method can be used. The latter provides smaller *p*-values and shorter intervals than the former. Its empirical behavior, however, needs to be studied for a wider variety of settings.
- The asymptotic (conditional) score method has good properties in small and large samples. Its performance is often close to, and at times better than, the mid-*p*. This approach entails reporting the mle, the score asymptotic *p*-value and the asymptotic score test based confidence interval.
- For the main parameter(s) of the study, presenting and discussing the features of the complete evidence function is advisable.

Usage of the probability method for a univariate analysis is, in our view, problematic because the associated Stern confidence regions are rarely used. Reiczigel (2003) advocates a broader usage of such regions and gives a computer intensive method, equivalent to the computation of the related evidence function, for their determination. For a comparative discussion of the TST and probability methods, see §7.6.

3.13 At the Boundary

Let t_* and t_{**} respectively be the minimum and maximum values of $t \in \Omega$ (one or both may not be finite). Such a point is called a **boundary point** or an **extreme point** of Ω. For example, the boundary points of $B(n; \pi)$ are 0 and n.

The probability of a boundary point may be low when the sample size is large but in small samples, it may not be negligible. The chance of $T = 10$ or $T = 0$ in $B(10, 0.9)$ is thus 0.29. The determination of estimates and confidence limits at the boundary needs special attention because evidence functions behave differently at such a point. Consider the mle for $B(n; \pi)$. Suppose, we observe $T = 0$. Then

$$\hat{\pi}_{mle} = 0 \quad \text{and} \quad \hat{\phi}_{mle} = 0 \quad \text{and} \quad \hat{\beta}_{mle} = -\infty$$

If we observe $T = n$ then

$$\hat{\pi}_{mle} = 1 \quad \text{and} \quad \hat{\phi}_{mle} = +\infty \quad \text{and} \quad \hat{\beta}_{mle} = +\infty$$

In a general univariate PBD, the mle here is the parameter value that makes the observed point the mean of T. The mean, a weighted average of the sample points, cannot equal a boundary value if all points in the discrete sample space have finite relative weights. If we observe $T = t_*$ then

$$\hat{\phi}_{mle} = 0 \quad \text{and} \quad \hat{\beta}_{mle} = -\infty$$

And if we observe $T = t_{**}$ then

$$\hat{\phi}_{mle} = +\infty \quad \text{and} \quad \hat{\beta}_{mle} = +\infty$$

Consider now the two-sided TST evidence function at $T = t_*$. Since $P[T \geq t_*; \beta] = 1$ for all β,

AT THE BOUNDARY

$$E(t_*; \beta) = 2P[T \leq t_*; \beta] \tag{3.65}$$

At $T = t_*$, the TST two-sided evidence function then has one monotonically decreasing segment, just like a one-sided evidence function. By drawing horizontal lines, we observe that the lower $(1 - \alpha_*)100\%$ confidence limit for β is $-\infty$. The upper limit is the solution of

$$2c(t_*)\exp(\beta t_*) = \alpha_* h(\beta) \tag{3.66}$$

In this case, the mue for β is also $-\infty$.

If the observed point is t_{**}, then

$$E(t_{**}; \beta) = 2P[T \geq t_{**}; \beta] \tag{3.67}$$

Then, the upper $(1 - \alpha_*)100\%$ confidence limit for β is $+\infty$. The mue is also $+\infty$. The lower limit is obtained from

$$2c(t_{**})\exp(\beta t_{**}) = \alpha_* h(\beta). \tag{3.68}$$

The mid-*p* version of the TST two-sided evidence function gives similar results at the boundary. At t_*, the lower limit and mue are $-\infty$, and at t_{**}, the upper limit and mue are $+\infty$. The equations to compute the upper limit at t_* and the lower limit at t_{**} are respectively

$$c(t_*)\exp(\beta t_*) = \alpha_* h(\beta) \tag{3.69}$$

and

$$c(t_{**})\exp(\beta t_{**}) = \alpha_* h(\beta). \tag{3.70}$$

When the parameter in question is β, one of the TST confidence limits and the mue are nonfinite at an extreme point. When it is ϕ, the negative infinity limits or estimates are replaced by zero, and positive infinity limits and estimates remain the same.

Consider the score statistic based exact evidence function. We saw that the mle for β is $-\infty$ at t_*, and $+\infty$ at t_{**}. Two-sided $(1 - \alpha_*)100\%$ lower and upper exact confidence limits using this statistic respectively obtain from solving for β in

$$P[T \geq t; \beta] + P[T \leq 2\mu(\beta) - t; \beta] = \alpha_* \tag{3.71}$$

$$P[T \leq t; \beta] + P[T \geq 2t - \mu(\beta); \beta] = \alpha_* \tag{3.72}$$

At t_*, $P[T \geq t_*; \beta] = 1$ for all β. Hence there is no finite lower limit here. At t_{**}, $P[T \leq t_{**}; \beta] = 1$ for all β. Hence there is no finite upper limit here.

For the large sample score evidence function with the chisquare approximation, the asymptotic confidence limits obtain from solving

$$(t - \mu)^2 = \chi^2_{1-\alpha_*} \sigma^2 \tag{3.73}$$

Here also at t_*, there is no finite lower confidence limit, and at t_{**}, there is no finite upper confidence limit. The situation with the LR statistic based confidence interval is more complex but essentially similar. In the previous section, we noted the empirical logit estimators, adjusted Wald method and adjusted median unbiased estimation for dealing with zero count data. Other methods of dealing with boundary observations are also available in the literature.

3.14 Equivalent Statistics

The concept of **equivalent statistics** is a helpful one for exact conditional analysis of discrete data.

> Equivalent Statistics
>
> For a specified model, a given sampling design and a particular form of analysis of a prescribed set of parameters, statistic $\mathcal{A}$ is said to be equivalent to statistic $\mathcal{B}$ if both yield identical inference for these parameters.

Equivalent statistics have identical evidence functions. The p-values, confidence intervals and estimates deriving from equivalent statistics are thus the same. To give a simple case: $\mathcal{W}$ and $2\mathcal{W}$ both yield the same exact and asymptotic p-values since $P[\mathcal{W} \geq w]$ if and only if $P[2\mathcal{W} \geq 2w]$.

For a PBD, the univariate score statistic and the distance from mean statistic give identical exact inference on its parameter (see §3.9.2). For this reason, we mostly deal with the former in this text. Note that two statistics that are equivalent for exact analysis are not necessarily so for asymptotic analysis.

Using an equivalent statistic often simplifies the computational effort for exact analysis of complex models. We will use it in the course of exact analysis of several 2×2 tables, and a $2 \times K$ table.

Our formulation of equivalent statistics is a generalized version of the definition given by Pratt and Gibbons (1981), page 22. It is, however, distinct from the idea of equivalent tests proposed by Krauth (1988), page 34.

3.15 Relevant Literature

The evidence function, under a variety of other names, has been promoted by several statisticians including Folks (1981) and Miettinen (1985). A rigorous formulation was given by Birnbaum (1961). An extensive debate on the utility of p-values and confidence intervals in epidemiology also featured this function, see Poole (1987a, 1987b, 1987c) and Borenstein (1994). See also Smith and Bates (1992, 1993). Its use in epidemiology and clinical trials was well laid out in Sullivan and Foster (1988, 1990) and Borenstein (1994). Foster and Sullivan (1987) gave a program for it. A comprehensive epidemiology text, Rothman and Greenland (1998), highlights the use of the function while Hirji and Vollset (1994c) and Blaker (2000) overviewed its features. In our view, the best elementary and well illustrated introduction to the topic is given by Rothman (2002).

Gibbons and Pratt (1975), though dated somewhat, remains a good reference on the methods

EXERCISES

of two-sided *p*-value computation for discrete data. Cox and Snell (1989) use the TST method for a two-sided *p*-value. The TST method is also known as the **double tail or doubling** method. Pratt and Gibbons (1981) call the probability method, the minimum likelihood method; a recent update is in Reiczigel (2003). The two-sided CT method for computing *p*-value also has a long history. See, for example, in Cox and Hinkley (1974), Gibbons and Pratt (1975) and Upton (1982); Blaker (2000) calls it the **acceptability method** while Newman (2001) refers to it as the **cumulative method.** The last but one paper used this method for CIs. For general discussions of exact and other two-sided *p*-values and CIs for discrete data, see Pratt and Gibbons (1981), Santner and Duffy (1989), Hirji and Vollset (1994c), Blaker (2000), Agresti and Min (2001), Lydersen and Laake (2003) and Hirji, Tan and Elashoff (1991). Many other papers on this issue also exist.

There is an extensive literature on the analysis of binomial and Poisson parameters. Some of the early papers are Clopper and Pearson (1934), Garwood (1936), Stevens (1950), Sterne (1954), Crow (1956), Crow and Gardner (1959). Other notable papers are Gosh (1979) and Blyth and Still (1983). For a broad view of the literature on these topics, consult the recent papers: Vollset (1993), Cohen and Yang (1994), Edwardes (1998), Newcombe (1998a), Blaker (2000), Agresti and Min (2001), Kabaila and Byrne (2001) and Reiczigel (2003). The references in these papers give a good snapshot of the extensive research in this field. Pratt and Gibbons (1981) and Santner and Duffy (1989) describe methods for computing two-sided confidence intervals for the binomial and Poisson parameters. See also Casella and Berger (1999).

Recent results on the adjusted Wald method are in Agresti and Coull (1998), Agresti (1999) and Agresti and Caffo (2000). The former paper notes that the asymptotic score confidence interval for the binomial was derived by Wilson (1927).

The diversity of views on which of the methods noted in this chapter to use, when to use them and how to judge them is truly a wide one. To appreciate the divide, see Edwardes (1998), Blaker (2000), Agresti and Min (2001) and Reiczigel (2003). More references on this topic are in Chapter 7.

Formal derivations of the large sample properties of the mle and the score, LR and Wald statistics are in many standard books including Rao (1973), Cox and Hinkley (1974), Bishop, Fienberg and Holland (1975) and Santner and Duffy (1989). The mue has garnered scant attention; the foundational paper is Birnbaum (1964); a good though dated review is in Read (1985). See Hirji, Tsiatis and Mehta (1989) and Salvan and Hirji (1991) for comparative studies of the mue. See also related references in the last noted paper.

3.16 Exercises

3.1. Lemen et al. (1976) conducted a study to assess the mutagenic and tumorigenic potential of bis(chloromethyl)ether (BCME). Among 44 workers with five or more years of exposure to BCME, 29 showed an abnormal sputum cytology. Construct an asymptotic score evidence function for π, the underlying abnormality rate in this population, and compute a 95% two-sided score confidence interval for it. What is the mle of π? Also plot a TST mid-*p* evidence function and compute the corresponding estimate and CI. How do the results of the two methods compare? Also analyze these data using the CT and probability methods and plot their evidence functions.

3.2. Plot the score and TST confidence interval functions for the *BRCA1* gene mutation data of Example 3.2. Analyze them using the probability and CT methods as well, and plot their evidence functions.

3.3. Let π denote the proportion of subjects with a family history of ovarian cancer among breast cancer cases with a family history of breast cancer. Malone et al. (1998) found that of the 208 subjects with breast cancer and family history of the disease, 22 also had a family history of ovarian cancer. Use the evidence function approach to analyze the data, and write a short report. Use several evidence functions for comparative purposes.

3.4. Compute asymptotic p-values for the data of Example 3.1 using the LR and Wald methods. Do these differ from the score method? How does the empirical Wald 95% CI compare with the score interval? Compare these intervals with the adjusted Wald method in which a count of two is added to the number of successes and four to the overall total.

3.5. Waxweiler et al. (1976) studied 31 deaths from malignant neoplasms among a cohort of workers exposed for more than 15 years to vinyl chloride. The expected value was 16.9. Assuming a Poisson model, plot a score, TST and mid-p TST evidence functions for the SMR. Obtain the mle, mue and related 95% CIs. Repeat the exercise when attention is restricted to Brain and CNS cancers. For these, the observed number was 3 and the expected value was 0.6. Compare the three methods used in the two analyses and also apply the CT method to these data.

3.6. In a study of side effects of nonsteroidal antiinflammatory drugs, Katz et al. (1992) looked at the disclosure of potential side effects of the drugs to patients being prescribed these drugs. Of the 46 first time patient encounters studied, 6 patients were informed about the possibility of gastrointestinal bleeding. Let π denote the disclosure rate. Analyze these data using three two-sided methods and write a short report. What concerns would you have about generalizing your results?

3.7. In a group of 595 subjects with chronic hepatitis B or chronic hepatitis C, 27 subjects subsequently acquired a hepatitis virus A superinfection (Vento et al. 1998). Analyze these data, giving plot(s) of the evidence function(s). Write a short report.

3.8. Wu et al. (1997) report the effect on lung function resulting from consumption of a weight loss fad vegetable *Sauropus androgynus*. 18 of the 100 subjects in whom lung function tests were done showed obstructive ventilatory impairment. Analyze these data using two-sided methods and write a short report.

3.9. Tillmann et al. (1997) reported on six subjects who had ingested a stimulant popularly known as Ecstasy. All six had developed fever and malaise. Analyze these data and write a short report.

3.10. Derive the score, likelihood ratio, Wald and the various empirical Wald statistics for negative binomial and geometric distributions. Write the equations for computing two-sided large sample confidence intervals based on these methods. Which methods give explicit solutions? Write the equation to compute the mue for π.

3.11. Show that the score and LR statistics for a PBD are not affected by posing the problem in terms of ϕ or β. Does that also hold for the Wald statistic?

3.12. Plot the score, LR, Wald and empirical Wald exact and asymptotic evidence functions for the following: (i) binomial with $n = 15$ and $t = 3$, (ii) Poisson with $t = 3$, (iii) binomial with $n = 12$ and $t = 7$, (iv) geometric with $t = 5$, (v) negative binomial with $r = 3$ and $t = 4$ and (vi) Poisson with $t = 7$. Are these functions strictly bimonotonic?

3.13. Repeat the previous exercise with the corresponding mid-p based evidence functions.

3.14. Repeat the exercise with the exact and mid-p probability based and CT evidence functions.

3.15. Show that the mle also derives from exact and asymptotic LR, regular Wald and empirical Wald evidence functions.

3.16. Formally prove the formulas for upper and lower confidence limits for the parameter of a PBD at the boundary points of the sample space given in §3.13.

EXERCISES

3.17. For the binomial problem, formulate the eighteen possible Wald evidence functions: Three methods (exact, mid-p or asymptotic) by type (regular or empirical) under three parameter formulations (π, ϕ or β). Which of them are equivalent, and which give explicit solutions for the confidence limits? Determine these limits.

3.18. Repeat the above exercise for the Poisson model, negative binomial and noncentral hypergeometric distributions and generalize, where possible, to the case of a general PBD.

3.19. Derive two-sided exact, mid-p and asymptotic methods of analysis for the parameter of a negative binomial distribution. Derive the score, LR and Wald statistics for this problem. How do they relate to the corresponding statistics for the binomial distribution?

3.20. Show that for a PBD, confidence intervals from the CT method are not longer than and are contained in those from the TST method (Blaker 2000). How is the CT method p-value related to the TST mid-p, and probability method p-value when the distribution is symmetric and not symmetric? How would you formulate a mid-p version of the CT method?

3.21. Consider an estimator derived from the probability method evidence function. With $\mathcal{T} = t$, consider the value(s) of β which makes the observed point the **mode** of the distribution. Investigate the properties of this estimator for the binomial, Poisson and negative binomial distributions.

3.22. Let $\mathcal{T}$ be a nonnegative PBD variate with finite mean μ. Then the Markov Inequality states that for $d > 0$,

$$P[\mathcal{T} \geq d] \leq \frac{\mu}{d} \qquad (3.74)$$

Give a proof of this inequality for a PBD. Can it provide useful upper bounds for exact p-values? For this purpose, consider the binomial, Poisson, hypergeometric and negative binomial distributions.

3.23. Let $\mathcal{T}$ be a PBD variate with finite mean μ and standard deviation σ. Then Chebychev's Inequality states that for $k > 0$,

$$P[\|\mathcal{T} - \mu\| \geq k\sigma] \leq \frac{1}{k^2} \qquad (3.75)$$

Give a proof of this inequality for a PBD. Can it provide useful upper bounds for exact score based p-values? Use the binomial, Poisson, hypergeometric and negative binomial distributions for this purpose.

3.24. Consider binomial data with $n = 10, 15, 20, 40$ and the problem of testing $\pi = 0.5$ versus $\pi \neq 0.5$. Find the actual significance levels of nominal level 0.05 and 0.10 TST and TST mid-p; exact, asymptotic and mid-p score; exact, asymptotic and mid-p LR; exact, asymptotic and mid-p Wald; exact, asymptotic and mid-p empirical and adjusted Wald; exact and exact and mid-p probability; exact and mid-p CT tests.

3.25. Consider geometric data and the problem of testing $\pi = 0.5$ versus $\pi \neq 0.5$. Find the actual significance levels of nominal level 0.05 and 0.10 TST and TST mid-p; exact, asymptotic and mid-p score; exact, asymptotic and mid-p LR; exact, asymptotic and mid-p Wald; exact, asymptotic and mid-p empirical and adjusted Wald; exact and exact and mid-p probability; exact and mid-p CT tests.

3.26. Consider Poisson data and the problem of testing $\lambda = 1.0$ versus $\pi \neq 1.0$. Find the actual significance levels of nominal level 0.05 and 0.10 TST and TST mid-p; exact, asymptotic and mid-p score; exact, asymptotic and mid-p LR; exact, asymptotic and mid-p Wald; exact, asymptotic and mid-p empirical and adjusted Wald; exact and exact and mid-p probability; exact and mid-p CT tests.

3.27. Define an evidence function based on the distance from the median statistic, and investigate its properties. Apply it, using real and simulated data, to the binomial, Poisson, negative binomial and hypergeometric distributions.

CHAPTER 4

Computing Fundamentals

Karim Hirji and Stein Vollset

4.1 Introduction

Exact analysis of discrete data is a computationally intensive exercise. The key role of modern computing power in making it accessible to the data analyst is not in doubt. Nonetheless, without efficient algorithms and sound implementation, it would still remain infeasible, especially for multivariate data problems. This chapter starts the presentation of computational issues and algorithms relating to exact conditional analysis of discrete data. The goal is to present methods for computing the probability distributions and tools of inference described in the previous chapters, and to lay a foundation upon which efficient algorithms for exact analysis of more complex data will later be constructed. Its specific aims are to present:

- Algorithms for computing the factorial, binomial, Poisson, hypergeometric and negative binomial coefficients, distributions and their tail areas.
- Techniques for the computer representation, storage and efficient evaluation of a polynomial and its derivatives.
- Equations arising in the computation of exact and large sample confidence limits and estimates for a univariate PBD.
- Iterative numerical methods for finding the roots of such equations.

Both conceptual and practical issues are covered. The details of computer programming are given only if deemed necessary. Apart from a few definitions, algorithmic theory is mostly avoided.

4.2 Computing Principles

A good numeric algorithm provides accurate results without too many arithmetic operations. Among such operations, addition or subtraction are preferred over multiplication or division, which in turn are preferred over taking logarithms or exponentiation. In addition, simplicity is a key feature. The practical significance of any numerical method has to be assessed in the context of the modern computing power. When actual differences between the algorithms are negligible, we would favor the conceptually and practically simpler algorithms.

Algorithms that retain their efficiency and accuracy in a diversity of problems and computing environments have a particular appeal. While a specific method may be the best for a specific problem, a robust method yields sufficiently accurate results in a generally efficient manner for a broad spectrum of situations encountered in practice. It also controls the errors arising from the varied ranges of real and integer valued numbers on the software and hardware used. This is a critical concern in exact analysis in which very large or very small values arise on a regular basis. In particular, the factorial, binomial, negative binomial, hypergeometric or multinomial coefficients can be huge numbers while the probabilities associated with them can be truly minute.

Another crucial, yet often neglected, characteristic is multiplicity of computation. An algorithm that is efficient for a one time effort is not necessarily so when that task has to be repeated several times. In the latter case, it may be more efficient to compute and store intermediate quantities, and apply what with respect to a one time effort may be the less efficient alternative.

The five main features we seek in computational algorithms in general, and those used for exact analysis of discrete data in particular, are as follows.

An Ideal Algorithm

A good computational algorithm is (i) accurate; (ii) efficient; (iii) simple; (iv) robust; and (v) suitable for multiple applications.

Two concepts, **recursion** and **iteration**, are central to the construction of an efficient algorithm. A recursion divides the overall effort into a specified number of stages. Starting from preliminary inputs into the initial stage, a stage by stage computation brings us to the final stage. The number of stages is fixed, and the termination of the process always produces what the answer need.

In an iteration, on the other hand, we start with a rough estimate of the quantity we need, and, in a step-wise fashion, improve its accuracy. The number of steps is not fixed, depending in part on the accuracy desired. Further, the process may at times not converge to the result we seek.

Both recursive and iterative techniques are used in to construct efficient algorithms for exact analysis of discrete data. The former is generally used to obtain exact distributions and its tail areas, while the latter is applied to these distributions to compute the various indices of inference. In special cases, the two techniques are used in a combined manner.

4.3 Combinatorial Coefficients

Virtually all discrete distributions we encounter in this book have the factorial quantity $n!$. Suppose we need to compute it for a positive integer n.

Algorithm 4.F.1: The obvious approach to get $n!$ is to set $0! = 1! = 1$, and implement the relation:

$$k! = k(k-1)! \qquad (4.1)$$

for $k = 2, \ldots, n$. The task is completed in $n-1$ steps, and needs $n-1$ multiplications. As a by product, it also yields the values of $k!$, $k = 0, 1, 2, \ldots, n-1$. This relation is a simple example of a **recursive relation**.

Implementing (4.1) with integer arithmetic has to be done with caution because the result may exceed the largest integer available, typically $2^{31} - 1$. Double precision real operation is a more robust alternative. Many environments permit real numbers up to around 10^{308}; a few give much larger numbers. As 64 bit computing becomes more common, the range will increase. Yet, even with double precision, the loss of accuracy for the applications we consider is usually not a major problem.

COMBINATORIAL COEFFICIENTS

When the desired factorials exceed the largest real number available, we need the logarithmic scale.

Algorithm 4.F.2: With $\ln(0!) = \ln(1!) = 0$, recursion (4.1) on the log-scale becomes

$$\ln(k!) = \ln(k) + \ln\{(k-1)!\} \tag{4.2}$$

(4.2) entails $n - 1$ calls to the logarithmic function. The log-scale, if used too often, markedly reduces efficiency. As a one time exercise to store the factorials up to a given number, it generally is not a critical impediment. This scale, though, brings forth a problem of its own. When such numbers are added repeatedly, the potential for numeric overflow and loss of accuracy is high. We deal with this issue in the next section.

Algorithm 4.F.3: When having more than three digits of accuracy is not critical, the approximations called **Stirling's formulae** are avaliable. These are:

$$n! \approx (2n\pi)^{0.5} \, n^n \, e^{-n} \tag{4.3}$$

or

$$n! \approx (2n\pi)^{0.5} \, n^n \, e^{[-n + 1/(12n)]} \tag{4.4}$$

While (4.3) underestimates $n!$, (4.4) overestimates it. For small n, both may mislead. Their impact on accuracy, if used in exact analysis of discrete data, has not been evaluated. In a univariate setting, the loss of accuracy is not likely to be large for *p*-values, but may be dramatic for confidence limits.

Algorithm 4.F.4: When complete accuracy is desired even for very large n, the factorial is represented as a vector of its prime factors. For example,

$$10! = 2^8 \, 3^4 \, 5^2 \, 7^1 \quad \text{and} \quad 8! = 2^7 \, 3^2 \, 5^1 \, 7^1$$

The former is written as (8,4,2,1) and the latter is (7,2,1,1). In this case, multiplying and dividing factorials becomes an exercise in adding and subtracting the elements of the prime factor vectors. So,

$$10!/8! = (1, 2, 1, 0) = 90$$

This method needs an efficient algorithm for identifications of the prime factors of a number. Recourse to high precision integer arithmetic may be needed for evaluating the final result, and for adding or subtracting the factorials represented in this manner.

****

Consider now the binomial coefficient. Suppose we need to compute

$$\binom{n}{m}$$

for some nonnegative integers n and m with $m \leq n$. We consider six methods for this task.

Algorithm 4.B.1: Use (4.1) to compute $n!$. In the process, also store the intermediate results $m!$ and $(n-m)!$. Evaluate the coefficient as $n!/\{m!(n-m)!\}$. This method, which needs n multiplications and one division, is not robust because the factorial may overflow even when the binomial coefficient is well within the allowable numeric range.

Algorithm 4.B.2: Compute $\ln(n!)$ with (4.2) and also store the intermediate results $\ln(m!)$ and $\ln[(n-m)!]$. Evaluate the binomial coefficient as

$$\exp\{\ln(n!) - \ln(m!) - \ln[(n-m)!]\} \tag{4.5}$$

This is less efficient than Algorithm 4.B.1 as it computes $n-1$ logarithms and does one exponentiation. But it is more robust. Other robust but more efficient alternatives also exist.

Algorithm 4.B.3: Set

$$\binom{n}{0} = 1$$

and let

$$q = \min\{m, n-m\}$$

Then use the recursion

$$\binom{n}{k} = \frac{n-k+1}{k}\binom{n}{k-1} \tag{4.6}$$

for $k = 1, \ldots, q$. This needs q multiplications and q divisions. If q is small relative to $n/2$, Algorithm 4.B.3 needs less effort than Algorithm 4.B.1. If q nearly equals $n/2$, then using stored factorials via (4.1) entails an equivalent effort. To enhance robustness, we compute (4.6) on the log-scale for n larger than a given value, or for certain (n, k) combinations.

Algorithm 4.B.4: Use the **Pascal triangle**, which involves additions only. Set, for all $r = 1, \cdots, n-1$

$$\binom{r}{0} = \binom{r}{r} = 1$$

Then for each $r = 2, \cdots, n-1$, we implement

$$\binom{r}{k} = \binom{r-1}{k-1} + \binom{r-1}{k} \tag{4.7}$$

for all $k = 1, \cdots, r/2$, if r is even, and for all $k = 1, \cdots, (r-1)/2$, if r is odd. For each r, the other coefficients are obtained by symmetry. And finally we get,

COMBINATORIAL COEFFICIENTS

$$\binom{n}{m} = \binom{n-1}{m-1} + \binom{n-1}{m}$$

This is a bivariate recursion and produces binomial coefficients for $r < n$ as well. Though no multiplications are done, the number of additions performed are a quadratic function of n.

Example 4.1: In a data example in the Chapter 3, we needed

$$\binom{277}{t}$$

for all $t \leq 125$. Now $277! \approx 10^{558}$. As the largest double precision real number on many compilers is not more than 10^{308}, computing factorials directly will produce a numeric overflow. The largest binomial coefficient with $n = 277 \approx 10^{82}$. Using (4.5), (4.6) or (4.7) gives the desired results without such a problem.

Algorithm 4.B.5: Express the factorial $n!$, $m!$ and $(n-m)!$ in terms of prime factors, cancel common factors from the numerator and denominator and multiply the remaining factors, using high precision integer arithmetic if necessary.

Algorithm 4.B.6: Use the Sterling formulae for the factorial to get the binomial coefficient.

****

The above algorithms, and the principles they embody, are of help for computing hypergeometric, negative binomial, inverse sampling, and multinomial coefficients. Care is needed in their computation as their values may be very large even at sample sizes that are not that large. Consider the hypergeometric case. We may compute these coefficients using stored factorials, stored binomials or a recursion.

Algorithm 4.H.1: In case of the latter, we first let, for fixed n and m,

$$c(u, s) = \binom{n}{u}\binom{m}{s-u} \qquad (4.8)$$

for $l_1 \leq t \leq l_2$ where $l_1 = \max(0, s - m)$ and $l_2 = \min(n, s)$. Then, after computing $c(l_1, s)$, we apply the recursion

$$c(u, s) = \frac{(n - u + 1)(s - t + 1)}{u(m - s + t)} c(u - 1, s) \qquad (4.9)$$

for $u = l_1 + 1, \ldots, l_2$.

Algorithm 4.H.2: If necessary, recursion (4.9) may be done on the log-scale as

$$\ln[c(u, s)] = \ln\left[\frac{(n - u + 1)(s - t + 1)}{u(m - s + t)}\right] + \ln[c(u - 1, s)] \qquad (4.10)$$

for $u = l_1 + 1, \ldots, l_2$.

The negative binomial or some multinomial coefficients are also computed using stored quantities or an appropriate recursion.

****

Which of these methods of computing combinatorial coefficients do we use in practice? The first but often ignored query is: For the problem at hand, does it make a difference? Suppose for a number of values of π, we need a histogram for the binomial probabilities $B(12, \pi)$. How we compute the 13 binomial coefficients has a negligible impact on the total time taken, even on a personal computer. What may be noticeable is the efficiency of drawing the graph on the screen or the printer.

On the other hand, in some cases how such coefficients are computed has a major impact on efficiency and accuracy. Take the case where we need to multiply many pairs of coefficients from a broad set of binomial, hypergeometric, negative binomial or multinomial coefficients. How a coefficient is computed each time it is needed can, in this case, makes a difference. Instead of computing them from scratch as needed, we may get them from intermediate quantities that have been determined at the outset.

Until recently the random access memory (RAM) available was a limiting factor for such an option. Today, a generous amount of RAM exists even on personal computers. When we repeatedly need such coefficients, we may use compactly stored factorial or binomial coefficients.

Algorithm 4.C.1: (i) Determine the largest possible integer n for which the expression $n!$ will appear in the computations. (ii) Compute and store $k!$, $k = 2, \ldots, n$ using (4.1) or (4.2). (iii) Use the stored factorials to compute the coefficients as needed.

Algorithm 4.C.2: (i) Determine the largest possible integer n for which a binomial expression of the form

$$\binom{n}{m}$$

will appear in the computations. (ii) For $k = 2, \cdots, n$, compute and store

$$\binom{k}{a}$$

for $a = 2, \cdots, n$ using (4.7). (With very large coefficients, use the log-scale on expression (4.6).) (iii) Use the stored binomial coefficients as needed, either as such, or to compute the hypergeometric or negative binomial coefficients on a repeated basis.

Note that the quantity $c(l_1, s)$ in Algorithm 4.H.1 and Algorithm 4.H.2 may be computed using stored factorials. However, it can often be set to 1 without affecting the final result. When we need multinomial coefficients on a repeated basis, using stored factorials, possibly on the log-scale, is the recommended alternative.

****

POLYNOMIAL STORAGE AND EVALUATION

The key point is that using a recursion each time a coefficient is needed can consume more computing time even if the recursion is an efficient one. Computing the binomial coefficient with prime factorization has a limited role in exact analysis of discrete data. For example, if we need the set of binomial coefficients with $n = 120$ and $m \leq 35$, expression (4.6) which generates the coefficients in a single recursive pass works well under most computing environments.

4.4 Polynomial Storage and Evaluation

Now we return to the general PBD where the random variable $\mathcal{T}$ on sample space Ω has probability

$$P[\mathcal{T} = t] = \frac{c(t)\phi^t}{f(\phi)} \qquad (4.11)$$

where $c(u) > 0$ for all relevant u, and

$$f(\phi) = \sum_{u \in \Omega} c(u) \phi^u \qquad (4.12)$$

First consider the storage of this distribution on a computer. Typically, for a specified ϕ, we store, at each point in Ω, its probability. Exact and large sample analysis tend to require repeated evaluation of the distribution at many values of ϕ. As the gp (4.12) gives a complete representation of the PBD, we then store (i) the set Ω, and (ii) for each $u \in \Omega$, the value of $c(u)$. The applications we consider are such that either Ω or the subset of Ω over which the calculations are done is a finite set. Storing items (i) and (ii) thereby corresponds to how such a polynomial is stored in a computer.

In the hypergeometric example of Table 1.3, the polynomial in question was $f(\phi) = 10 + 40\phi + 30\phi^2 + 4\phi^3$. The tabular representation sufficed to fully specify the generating polynomial. In the case of a multivariate PBD, similar considerations apply. The gp for the multinomial distribution of §1.10 is represented as an array corresponding to Table 1.5.

In practical terms, two basic options for storing a generic polynomial on a computer exist. Either we use a conventional array, say, $C(:)$, and set its value at location u equal to $c(u)$. Or, we have a linear array with two values at any location, namely, an exponent or vector of exponents, $\boldsymbol{u}$, and the associated coefficient value, $c(\boldsymbol{u})$. The linear array saves memory and processing time when dealing with sparse polynomials or multivariate polynomials having noncontiguous exponent values. Consider, for example, the bivariate polynomial

$$f(\theta, \phi) = 1 + \theta^2 \phi + 5\theta^3 \phi^8 + 9\theta^7 \phi^{10}$$

To store this polynomial in a conventional manner, we declare an array of dimension $C(0 : 7, 0 : 10)$, and set the values $C(0,0) = 1$, $C(2,1) = 1$, $C(3,8) = 5$, $C(7,10) = 9$, and $C(I, J) = 0$ for all other I, J. This method needs 88 units of storage, and each time we process the polynomial, we cycle through the 88 array units. On the other hand, we may store the values in a linear array as

0	2	3	-1	7	-1	-1	...
0	1	8	0	10	0	0	...
1	1	5	0	9	0	0	...

Each column in the array is a possible term of the polynomial. An entry in the first row is an exponent of θ, and that in the second row, an exponent of ϕ. The third row contains the coefficient of the term with these exponents. Empty columns are indicated by setting the first or last row element equal to an infeasible value, like the value -1 used here. The number of columns is less than 88. If the coefficient $c(u, v)$ is not integer valued, it is kept in a separate array but at the same linear address.

The linear array storage method needs a function that uniquely maps a record (a polynomial term) to a given location. Suppose we need to store a nonnegative integer valued record (set of exponents), $(u_1, \cdots, u_K)$, and its coefficient. Let Q be an integer greater than the total number of records. We then use a $K \times Q$ record array, and a parallel $1 \times Q$ coefficient array. Also let $\mathrm{l}(u_1, \cdots, u_K)$ be the (preliminary) value of the address (location) function for this record. Then we proceed as follows:

- If the location given by $\mathrm{l}(u_1, \cdots, u_K)$ is empty, store this record here, and place the coefficient $c(u_1, \cdots, u_K)$ at the same location in the parallel array.
- If the location given by $\mathrm{l}(u_1, \cdots, u_K)$ is not empty, but the existent record at this place has the same values as $(u_1, \cdots, u_K)$, then just increase the parallel coefficient value by the amount $c(u_1, \cdots, u_K)$.
- If the location given by $\mathrm{l}(u_1, \cdots, u_K)$ is not empty, and the existent record at this place has at least one element that is not the same as the corresponding element of $(u_1, \cdots, u_K)$, then we look at the succeeding records one by one, and perform the above two checks until we find either an empty location or an identical record.
- If the end of the linear array is reached without detecting a viable location, then we continue this process from the first record in the array onwards.
- If still no viable location is found, then the array is full and more memory is needed.

The two key ingredients needed to make this process efficient are (i) an adequate amount of memory, and (ii) a location function that results in a minimum or small number of collisions and searches. Various choices for a location function are available. The use of a **hash function** often serves the purpose well. Other techniques are also available; for details, see the references in §4.8.

A hash function that is generally adequate is defined as follows. Suppose all the elements of $(u_1, \cdots, u_K)$ are nonnegative integers and M is the largest prime number that does not exceed Q. Select $m_* \geq \max\{u_1, \cdots, u_K\}$. Then we use the hash function

$$\mathrm{l}(u_1, \cdots, u_K) = \left(\sum_{k=1}^{K} u_k m_*^{(k-1)} \right) \mod M \qquad (4.13)$$

This function needs to be evaluated in such a way that avoids possible integer overflow. For elaboration, see Exercise 4.32, and Hirji, Mehta and Patel (1987).

Another variation of the linear method is to store one component of the record using a conventional address method (that is, its value gives its location), and to use a hash function or

POLYNOMIAL STORAGE AND EVALUATION

another method to store records having a common first component value. This method is useful when the first component assumes a small number of contiguous integer values.

The linear storage method can also be used to compactly store a bivariate array. For example, we can use it to store the binomial coefficients by placing the records in the following order:

$$\binom{1}{0}, \binom{2}{0}, \binom{2}{1}, \binom{3}{0}, \binom{3}{1}, \binom{3}{2}, \binom{4}{0}, \binom{4}{1}, \binom{4}{2}, \binom{5}{0}, \binom{5}{1}, \binom{5}{2}, \cdots$$

This placement also takes advantage of the symmetry of the coefficients. A general function that can be used for placing and retrieving records in this way is easily specified (Exercise 4.2).

****

When a PBD is stored as values of the coefficient at each point in the sample space, the term "computing the distribution" at a specific parameter value is understood to mean evaluating the gp, and one of its components, at that value. This also applies to computing the right or left tail probability of the PBD, which is a ratio of two polynomials. To plot one-sided or two-sided evidence functions, or to compute exact confidence limits and point estimates, such polynomials then need to be evaluated at a series of parameter values.

The computation of the tail areas of particular distributions like the Poisson or the binomial distribution is done by specifically tailored recursive schemes. These are given in the next section. For now we examine the more general situation. Consider then a polynomial of degree n

$$f(\phi) = \sum_{u=0}^{n} c(u) \phi^u \tag{4.14}$$

A direct evaluation of $f(\phi)$ at a given ϕ requires $(2n - 1)$ multiplications and n additions.

Algorithm 4.EP.1: On a computer, we set $f_0 = c(0)$, $e_0 = 1.0$, and for $j = 1, \cdots, n$, implement

$$e_j = e_{j-1}\phi \quad \text{and} \quad f_j = f_{j-1} + c(j-1)e_j \tag{4.15}$$

At the nth step, we get $f(\phi) = f_n$.

When the evaluation has to be done for many values of ϕ, a better method is desirable. **Horner's scheme,** based on writing the polynomial in a nested form, reduces the number of multiplications by a half. For example, a polynomial of degree 4 is written as

$$\begin{aligned} f(\phi) &= c(0) + c(1)\phi + c(2)\phi^2 + c(3)\phi^3 + c(4)\phi^4 \\ &= c(0) + [\,c(1) + [\,c(2) + [\,c(3) + c(4)\phi\,]\phi\,]\phi\,]\phi \end{aligned}$$

Evaluation starts in the innermost bracket and proceeds outwards.

Algorithm 4.EP.2: For the general case, we set $f_0 = c(n)$, and for $j = 1, \ldots, n$, implement

$$f_j = f_{j-1}\phi + c(n-j) \tag{4.16}$$

At the nth step, we get $f(\phi) = f_n$.

This method requires n multiplications and n additions. An added advantage is that, with minor augmentation, it also yields the first two derivatives of the polynomial. These are needed, for example, in some of the iterative methods for solving equations described in §4.7, and to compute the mean and variance of the PBD.

Algorithm 4.EP.3: Consider the first derivative. We write

$$f'(\phi) = \sum_{u=1}^{n} uc(u)\phi^{u-1} = \sum_{u=0}^{n-1} b(u)\phi^u \qquad (4.17)$$

where $b(u) = f_{n-u-1}$. For $f'(\phi)$, set $f'_1 = f_0$, and implement, for $j = 2, \ldots, n$,

$$f'_j = f'_{j-1}\phi + f_{j-1} \qquad (4.18)$$

The required derivative is given by f'_n. To compute the second derivative, we use the recursion

$$f''_j = f''_{j-1}\phi + f'_{j-1} \qquad (4.19)$$

Since the f_j's are obtained in the implementation of Horner's scheme (4.16), we combine the two recursions to produce the values of the polynomial and the first (and even the second derivative) in a single recursive pass. Horner's scheme is not that efficient for polynomials with noncontiguous coefficients that are found in a number of discrete data applications.

****

When the values of the coefficients, powers of ϕ or their products are so large that numeric overflow may ensue, the direct Horner's scheme does not provide much of an advantage. To evaluate a polynomial whose coefficients are maintained on a log-scale, we need consider the problem of adding numbers on this scale. Suppose we know $a^\diamond = \ln(a)$ and $b^\diamond = \ln(b)$, and we need to compute

$$d^\diamond = \ln(a + b)$$

Exponentiating $a^\diamond$ and $b^\diamond$ may risk a numerical overflow. To reduce that danger, assume without loss of generality that $a^\diamond < b^\diamond$. Following Mehta and Patel (1983), we then compute

$$d^\diamond = b^\diamond + \ln\left[1 + e^{(a^\diamond - b^\diamond)}\right] \qquad (4.20)$$

This approach has two advantages. First, the danger of overflow is reduced. Second, instead of two exponentiations, only one is done. Possible loss of accuracy due to numeric underflow is, however, a concern. Our experience with real and simulated data and other papers indicate that if the computations are done in double precision, the final results of exact conditional analysis are not markedly affected.

Let $c^\diamond(u) = \ln[c(u)]$ and suppose we need $\ln\{f(\phi)\}$ for a specific ϕ. Let $\phi^\diamond = \ln(\phi)$ and $d^\diamond(0) = c^\diamond(u)$. Then we first implement, for $j = 1, \cdots, n$

COMPUTING DISTRIBUTIONS

$$\phi_j^\diamond = \phi_{j-1}^\diamond + \phi^\diamond \quad \text{and} \quad d^\diamond(j) = c^\diamond(j-1) + \phi_j^\diamond \qquad (4.21)$$

The next task is to get the logarithmic sum

$$\ln\{f(\phi)\} = \sum_{j=0}^{n} \exp[d^\diamond(j)] \qquad (4.22)$$

This may be obtained by using procedure (4.20) n times. Or, we may extend that method by adding three or more log-scale numbers at a time. For example, with $a^\diamond < c^\diamond$ and $b^\diamond < c^\diamond$, we use

$$d^\diamond = c^\diamond + \ln\left[1 + e^{(a^\diamond - c^\diamond)} + e^{(b^\diamond - c^\diamond)}\right] \qquad (4.23)$$

(4.20) involves more logarithmic operations than (4.23). But the latter may produce a greater underflow error. In any case, to reduce the decrease in efficiency, the log-scale should be invoked under a **dynamic sensing** approach (see §4.5).

Other more elaborate methods of polynomial evaluation, such as the **fast Fourier transform**, are covered in Chapter 11.

4.5 Computing Distributions

Discrete distributions such as the Poisson or the binomial are at times computed using a normal approximation. For some cases, this is sufficiently accurate. But it can have serious limitations at even large sample sizes. This happens when the Poisson mean is small, or the binomial success probability is close to 0.0 or 1.0 (Joliffe 1995; Kennedy and Gentle 1980). These are the situations where exact analysis is indicated.

Now we deal with computation of the point and tail probabilities of the binomial, Poisson, hypergeometric and negative binomial distributions. We focus on computing them at a given value of the parameter of the distribution. We continue using the notation for a discrete random variable $\mathcal{T}$:

$$f(t;\phi) = P[\mathcal{T} = t; \phi] \quad \text{and} \quad F(t;\phi) = P[\mathcal{T} \leq t; \phi] \qquad (4.24)$$

The general approach of the previous section, of course, also applies here.

****

First, consider the Poisson distribution with $\lambda > 0$ and

$$f(t;\lambda) = \frac{e^{-\lambda}\lambda^t}{t!} \qquad (4.25)$$

Algorithm 4.PP.1: Compute (4.25) directly with initial separate computation of $\exp(-\lambda)$, λ^t and $t!$. This method may produce a numeric overflow or underflow if t is large, or the value of λ is too high or too low. The danger is reduced by using logarithms:

$$\ln[f(t;\lambda)] = -\lambda + t\ln(\lambda) - \ln(t!) \qquad (4.26)$$

and exponentiating the result. Either of these options has a special appeal if the factorials or log-factorials have been computed and stored earlier.

Algorithm 4.PP.2: Alternatively, we use a recursion. On the natural scale, we set $f(0;\lambda) = \exp(-\lambda)$, and for $u = 1, \cdots, t$, implement

$$f(u;\lambda) = \frac{\lambda}{u} f(u-1;\lambda) \qquad (4.27)$$

while on the log-scale, we set $\ln[f(0;\lambda)] = -\lambda$, and, for $u = 1, \cdots, t$, implement

$$\ln[f(u;\lambda)] = \ln(\lambda) - \ln(u) + \ln[f(u-1;\lambda)] \qquad (4.28)$$

The recursive approach obviates the need to compute factorials or powers of λ. Compared to (4.25), (4.27) enhances accuracy by extending the domain over which actual scale computations are done.

For the Poisson left tail probability, we set $F(0;\lambda) = \exp(-\lambda)$, and in conjunction with (4.27), we implement, for $u = 1, \cdots, t$,

$$F(u;\lambda) = F(u-1;\lambda) + f(u;\lambda) \qquad (4.29)$$

Or, on the logarithmic scale, with $F(0;\lambda) = \exp(-\lambda)$ and together with (4.26) or (4.28), we implement, for $u = 1, \cdots, t$,

$$F(u;\lambda) = F(u-1;\lambda) + \exp\{\ln[f(u;\lambda)]\} \qquad (4.30)$$

Or, we maintain all terms in (4.30) on the log-scale until the final step in the recursion. The process is made more robust by assessing the potential for underflow within an initial set of terms and adjusting the computation accordingly. Distinct sets of recursions for $\lambda > 1$ and $\lambda < 1$ also enhance robustness (Tietjen 1994).

Algorithm 4.PP.3: The Poisson tail area can also be evaluated recursively with a Horner type of scheme. Let $F_{*0}(t;\lambda) = 1$, and implement, for $j = 1, \cdots, t$,

$$F_{*j}(t;\lambda) = 1 + \left(\frac{\lambda}{t-j+1}\right) F_{*,j-1}(t;\lambda) \qquad (4.31)$$

At the end, we get

$$F(u;\lambda) = \exp\{\ln[F_{*t}(t;\lambda)] - \lambda\} \qquad (4.32)$$

This method gives the tail probability without yielding intermediate point probabilities as a by product.

Algorithm 4.PP.4: Finally, another way of computing Poisson tail areas is based on its relation

COMPUTING DISTRIBUTIONS

to the cumulative chisquare probability. This latter is available in many software packages and published programs (Ling 1992; Tietjen 1994).

****

Now consider the binomial probability with

$$f_n(t;\phi) = \binom{n}{t}\pi^t(1-\pi)^{(n-t)} = \binom{n}{t}\phi^t(1+\phi)^{-n} \qquad (4.33)$$

Several methods for computing point and tail probabilities for some t and given ϕ are avaliable.

Algorithm 4.BP.1: Compute individual terms directly and add them for the tail probability. This may generate numeric underflow or overflow. Log-scale computations reduce the danger. This method has appeal when either the log-factorials or the logarithms of binomial coefficients have been stored.

Algorithm 4.BP.2: A recursive alternative reduces the underflow or overflow risk. Let $f_{*n}(0;\phi) = F_{*n}(0;\phi) = 1$. Then, for $u = 1, \ldots, t$, we implement

$$f_{*n}(u;\phi) = \frac{(n-u+1)\phi}{u} f_{*n}(u-1;\phi) \qquad (4.34)$$

$$F_{*n}(u;\phi) = F_{*n}(u-1;\phi) + f_{*n}(u;\phi) \qquad (4.35)$$

These recursions may also be implemented on the log-scale. Then

$$\ln[f_{*n}(u;\phi)] = \ln(\phi) + \ln\left[\frac{(n-u+1)}{u}\right] + \ln[f_{*n}(u-1;\phi)] \qquad (4.36)$$

$$F_{*n}(u;\phi) = F_{*n}(u-1;\phi) + \exp\{\ln[f_{*n}(u;\phi)]\} \qquad (4.37)$$

Under either of the above options, at the end we get

$$f(u;\phi) = \exp\{\ln[f_{*n}(u;\phi)] - n\ln[1+\phi]\} \qquad (4.38)$$

$$F(u;\phi) = \exp\{\ln[F_{*n}(u;\phi)] - n\ln[1+\phi]\} \qquad (4.39)$$

The order of the recursions may be reversed when n is large and $1-\pi$ is small. Otherwise, when n is large and π small, we start in the usual order with $u = 0$.

Algorithm 4.BP.3: A bivariate Pascal triangle-like method may also be used here. Set, for all $r = 1, \cdots, n$,

$$f_{*r}(0;\phi) = 1 \quad \text{and} \quad f_{*r}(r;\phi) = \phi^r$$

Then for each $r = 1, \cdots, n-1$, we implement

$$f_{*,r+1}(\phi;u) = \phi f_{*r}(\phi;u-1) + f_{*r}(\phi;u) \tag{4.40}$$

for $u = 1, \cdots, r$. The final result is divided by the gp $(1+\phi)^n$. The log-scale may be used at one or all of the stages here as well. This method is particularly useful if binomial probabilities for smaller values of n are also needed. The tail probability is obtained by adding the point probabilities.

Algorithm 4.BP.4: Another method to get the tail area is to employ stored coefficients and Horner's scheme to evaluate the left portion of the gp, $f(\phi;\leq t)$, and divide it with $f(\phi) = (1+\phi)^n$.

Algorithm 4.BP.5: Finally, the exact relation between the cumulative binomial probability and the F distribution may also be used. If $\mathcal{T}$ has a $B(n,\pi)$ distribution, then

$$P[\mathcal{T} \leq t] = F_F((t+1)(1-\pi)/(\pi(n-t)); 2(n-t), 2(t+1)) \tag{4.41}$$

where $F_F(x;\mu,\nu)$ is the cumulative distribution function of the F distribution with μ and ν degrees of freedom. Exact TST lower and upper confidence limits for the binomial parameter also emerge from this relation.

****

Now consider the noncentral hypergeometric distribution. Using the notation given for (4.8), let

$$f_s(u;\phi) = \frac{c(u,s)\phi^{(u-l_1)}}{f_s(\phi)} \tag{4.42}$$

where

$$f_s(\phi) = \sum_{v=0}^{l_2-l_1} c(v+l_1, s)\phi^v \tag{4.43}$$

is an adjusted version of the hypergeometric gp.

This gp and its terms satisfy the following often used recursions:

$$(s+1)f_{s+1}(\phi) = [(n-s)\phi + (m-s)]f_s(\phi) + (n+m-s+1)\phi f_{s-1}(\phi) \tag{4.44}$$

for $s \geq 2$. Further, with

$$f_s(\phi;0) = c(l_1, s)$$

we have

$$f_s(\phi;u) = \frac{(n-u+1)(s-u+1)\phi}{u(m-s+u)} f_s(\phi;u-1) \tag{4.45}$$

COMPUTING DISTRIBUTIONS

for $u = 1, \cdots, l_2 - l_1$.

Then we consider the following algorithms for the hypergeometric.

Algorithm 4.HP.1: Compute and store the binomial coefficients. Then use Horner's scheme to evaluate $f_s(\phi)$, and the left part, $f_s(\phi; \leq u)$, if needed. These quantities are applied to compute the point and tail probabilities. The log-scale is used when the possibility of an overflow or underflow is detected.

This algorithm is useful when the overall computation is done in several stages, and at each stage, hypergeometric probabilities for different values n, m and s, and all corresponding possible values of u are needed. Examples of this arise in the exact analysis of several 2×2 and $2 \times K$ tables.

Algorithm 4.HP.2: Alternatively, we first set $c(l_1, s) = 1$, and then apply, for $u = 1, \cdots, l_2 - l_1$, the recursion (4.45) and accumulate

$$f_s(\phi; \leq u) = f_s(\phi; u-1) + f_s(\phi; u) \tag{4.46}$$

And at the end, we get

$$\mathsf{f}_s(u; \phi) = \frac{f_s(\phi; u - l_1)}{f_s(\phi)} \tag{4.47}$$

$$\mathsf{F}_s(u; \phi) = \frac{f_s(\phi; \leq u - l_1)}{f_s(\phi)} \tag{4.48}$$

Later parts of this process may be performed on the logarithmic scale. We also may compute $f_1(\phi)$ and $f_2(\phi)$ and then use (4.44) to get $f_s(\phi)$.

Algorithm 4.HP.3: **Granville and Schifflers (1993)** developed a series of recursions to implement (4.44) and (4.45) in an efficient and robust manner. Their method gives hypergeometric point and tail probabilities as well as the associated mle and confidence limits in a rapid and accurate fashion for cases with large values of s, n and m. This obviates the need for the logarithmic scale.

Algorithm 4.HP.4: A number of efficient and robust methods for the central hypergeometric distribution ($\phi = 1$) also exist. Some develop an efficient way of implementing Algorithm 4.HP.2, another employs the prime factor formulation of the factorials and another uses Chebychev polynomials. Alvo and Cabilio (2000) and the references therein provide the details.

Algorithm 4.HMV.1: Recursive methods for directly computing the mean and variance of the hypergeometric have also been proposed. They are useful for computation of the conditional mle and asymptotic inference conditional score based inference in a 2×2 table (Chapter 5). The following have been shown to be numerically stable, robust and fast:

Let $\mu(n, s, N; \phi)$ be the mean of (4.42). Then

$$\mu(n, s, N; \phi) = \frac{ns\phi}{N + (1 - \phi)[\mu(n-1, s-1, N-1; \phi) - n - s + 1]} \tag{4.49}$$

Implementing this starts with $\mu(0, s-n, m; \phi) = 0$ and computing $\mu(1, s-n+1, m+1; \phi)$, and so on. Also, $\sigma^2(n, s, N; \phi)$, the variance of (4.42), is given by

$$\mu(n, s, N; \phi)[1 + \mu(n-1, s-1, N-1; \phi) - \mu(n, s, N; \phi)] \qquad (4.50)$$

and also by

$$\frac{\mu(N-n-s+\mu) - \phi(n-\mu)(s-\mu)}{\phi - 1} \qquad (4.51)$$

where, in (4.51), $\mu = \mu(n, s, N; \phi)$. See Harkness (1965), Liao (1992) and Satten and Kupper (1990) for proofs and further details on these relations.

Algorithm 4.HMV.2: When the factorials or binomial coefficients have been stored, we compute the gp (4.43) and its first and second derivatives in a single pass with Horner's method and then apply Theorem 1.2.

Algorithm 4.HMV.1 is more robust than algorithm 4.HMV.2 but the latter is potentially more efficient for multiple applications. Using the latter on a log-scale may reduce its efficiency somewhat but will enhance its robustness. Comparisons of these alternatives in the context of conditional analysis for discrete data have yet to be done.

****

Similar methods for computing point and tail probabilities of the negative binomial distribution also exist. If $\mathcal{T}$ denotes the number of successes till the rth failure, then

$$f(u; \pi) = \binom{u+r-1}{r-1} \pi^u (1-\pi)^r \qquad (4.52)$$

As elsewhere, direct computation is the initial option. Alternative methods take the recursion below as a starting point.

$$f(u; \pi) = \frac{(u+r-1)\pi}{u} f(u-1; \pi) \qquad (4.53)$$

We leave the details as an exercise.

****

Given these options to compute the point and tail probabilities, which one does one use? Our recommendations are as follows.

- If the factorials or binomial coefficients, or their logarithms have been stored, and multiple analyses are to be done, direct computations are advisable.
- Otherwise, recursive methods under double precision are suggested. For the binomial some favor Algorithm 4.BP.2 while others consider Algorithm 4.BP.3 as the more robust alternative.

ROOTS OF EQUATIONS

- The use of the Horner's scheme has not been evaluated in this context. It has a good potential when the derivatives are also needed.
- The method of Granville and Schifflers (1993) is advisable for the hypergeometric though it needs to assessed in the multiple applications context.
- Ling (1992) argues for the universal use of relations like (4.41) to compute discrete distributions. This is because many software programs already provide the F and chisquare probabilities. We do not consider it an efficient option.

We also strongly recommend the dynamic application of a **scale sensing test**. For this test, we start with rough estimates of the largest and smallest number that can arise in the next round of computation. The estimates are obtained on the logarithmic scale. If testing these numbers indicates a potential for overflow or underflow, all computations are converted to the log-scale. Otherwise, they are continued on the natural scale. When the computation is to be done in several stages, the test is performed at the beginning of each stage. This test enhances efficiency and controls the overflow and underflow errors. Reliable estimates of the largest or the smallest possible numbers are usually easy to obtain.

Multiplying the initial set of numbers either by a very large or a very small number also delays the conversion to the logarithmic scale. The former is done when the quantities being computed are decreasing from stage to stage (often when $\phi < 1$), and the latter is done in the reverse situation.

4.6 Roots of Equations

So far we have relied on a graphical method to obtain confidence intervals and point estimates from the evidence functions. But it has a limited accuracy. More precise values are obtained directly by solving the nonlinear equations derived from these functions. In a few cases, the equations admit explicit solutions. In general, iterative algorithms are required. A variety of iterative techniques, covered in books on numerical methods, are avaliable for this purpose. We overview the methods of interest, point to special problems found in implementing them for the types of equations we deal with, and give an example.

Though the basic principles used to solve one variable equations are straightforward, robust and efficient execution requires a degree of care. Special attention has to be paid to the behavior of the functions at extreme arguments. Also, some functions have multiple roots in close proximity to one another, and may also exhibit discontinuities. A poor initial guess may lead to erratic iterative behavior. The potential for numerical overflow or underflow has to be kept in mind. This issue also relates to the choice of the scale on which the computations are done.

4.6.1 The Main Equations

We first list the equations that arise in computing point estimates and confidence intervals for discrete data. They are of relevance to the problems in most chapters of this text. We first give equations for the TST and asymptotic score methods as these are most often used.

First consider the equations relating to estimation. These equations and other are stated in terms of ϕ (and not β) to emphasize their polynomial form. In that case, only solutions with $\phi \geq 0$ are relevant. For the TST and score (and some other) methods, estimates and confidence limits for β are just the logarithms of the corresponding estimates and limits for ϕ.

For a univariate PBD with the gp (4.12), recall from Theorem 1.1 that with

$$f_*(\phi) = \sum_{u \in \Omega} uc(u)\phi^u$$

$$f_{**}(\phi) = \sum_{u \in \Omega} u^2 c(u)\phi^u$$

the equations to compute the mle and the mue of ϕ are respectively

$$tf(\phi) - f_*(\phi) = 0 \qquad (4.54)$$

$$f(\phi;<t) - f(\phi;>t) = 0 \qquad (4.55)$$

The $(1-\alpha_*)100\%$ lower and upper confidence limits for ϕ using the TST conservative and mid-p methods obtain respectively from the equations:

$$\delta c(t)\phi^t + \frac{\alpha_*}{2} f(\phi;<t) - \left(1 - \frac{\alpha_*}{2}\right) f(\phi;\geq t) = 0 \qquad (4.56)$$

$$\delta c(t)\phi^t + \left(1 - \frac{\alpha_*}{2}\right) f(\phi;\leq t) - \frac{\alpha_*}{2} f(\phi;>t) = 0 \qquad (4.57)$$

For conservative exact limits, we set $\delta = 0$ in (4.56) and (4.57), and for mid-p limits, we set $\delta = -0.5$. If in large samples the score statistic is an approximately chisquare variate with one degree of freedom, the equation for computing approximate score confidence limits is

$$[tf(\phi) - f_*(\phi)]^2 - z_{\alpha_*}^2 [f(\phi)f_{**}(\phi) - (f_*(\phi))^2] = 0 \qquad (4.58)$$

We note key aspects of the above equations.

- All these equations are simple polynomial equations. In many applications, the exponents are nonnegative integers.
- These equations yield unique estimates and unique upper and lower confidence limits.
- When the observed point is a boundary point, the equations given in §3.14 should be used.

4.6.2 Other Equations

The equations for the exact score, LR and probability method confidence limits share a similar form which is

$$\sum_{u \in \Lambda(t;\phi)} c(u)\phi^u - \alpha_* f(\phi) = 0 \qquad (4.59)$$

The lower and upper limits are respectively the smallest and largest solutions of this equation. The set $\Lambda(t;\phi)$ depends on the method. For the score method, it is

$$\Lambda(t;\phi) = \{u : [uf(\phi) - f_*(\phi)]^2 \geq [tf(\phi) - f_*(\phi)]^2\} \qquad (4.60)$$

ROOTS OF EQUATIONS

For the LR method, it is

$$\Lambda(t;\phi) = \left\{ u : f(\hat{\phi}(u))\left(\frac{\hat{\phi}(u)}{\phi}\right)^u \geq f(\hat{\phi}(t))\left(\frac{\hat{\phi}(t)}{\phi}\right)^t \right\} \tag{4.61}$$

where $\hat{\phi}(u)$ is the mle of ϕ when $\mathcal{T} = u$. For the probability method, the set Λ is

$$\Lambda(t;\phi) = \left\{ u : c(u)\phi^u \leq c(t)\phi^t \right\} \tag{4.62}$$

The equation for exact Wald method is stated in terms of β because it may not yield the same inference when stated in terms of ϕ. The equation for the latter may easily be derived by the reader. The equation for the Wald method then is

$$\sum_{u \in \Lambda(t;\beta)} c(u)\exp(\beta u) - \alpha_* \sum_{u \in \Omega} c(u)\exp(\beta u) = 0 \tag{4.63}$$

where, for the regular Wald method,

$$\Lambda(t;\beta) = \left\{ u : (\hat{\beta}(u) - \beta)^2 \geq (\hat{\beta}(t) - \beta)^2 \right\} \tag{4.64}$$

while for the empirical Wald method it is

$$\Lambda(t;\beta) = \left\{ u : \frac{(\hat{\beta}(u) - \beta)^2}{\hat{\sigma}^2(\hat{\beta};u)} \geq \frac{(\hat{\beta}(t) - \beta)^2}{\hat{\sigma}^2(\hat{\beta};t)} \right\} \tag{4.65}$$

where

$$\hat{\sigma}^2(\hat{\beta};u) = \frac{(h(\hat{\beta}(u)))^2}{h(\hat{\beta}(u))h''(\hat{\beta}(u)) - (h'(\hat{\beta}(u)))^2} \tag{4.66}$$

In these expressions, $\sigma^2(\hat{\beta})$ denotes the asymptotic variance of $\hat{\beta}$, and $\hat{\sigma}^2(\hat{\beta};u)$ is an estimate for it given $\mathcal{T} = u$. As noted in §3.8, in a PBD, $\sigma^2(\hat{\beta})$ is the inverse of the exact variance of $\mathcal{T}$.

For the problems where the large sample LR and Wald statistics are approximately chisquare variates with one degree of freedom, the respective equations for computing approximate LR, regular Wald and empirical Wald confidence limits are

$$\phi^t f(\phi) - (\hat{\phi}(t))^t f(\hat{\phi}(t)) \exp\left(-0.5 z_{\alpha_*}^2\right) = 0 \tag{4.67}$$

$$(\hat{\beta}(t) - \beta)^2 [h(\beta)h''(\beta) - (h'(\beta))^2] - z_{\alpha_*}^2 (h(\beta))^2 = 0 \tag{4.68}$$

$$(\hat{\beta}(t) - \beta)^2 - z_{\alpha_*}^2 \hat{\sigma}^2(\hat{\beta}(t)) = 0 \tag{4.69}$$

And, finally, as it is in a class of its own, we list the equation obtaining exact confidence limits for the combined tail probabilities (CT) or the acceptability method last. This is:

$$f(\phi;\leq\nu_1(t,t_*;\phi)) + f(\phi;\geq\nu_2(t,t_{**};\phi)) = \alpha_* f(\phi) \quad (4.70)$$

where

- If $f(\phi;\geq t) < f(\phi;\leq t)$, then $\nu_1(t,t_*;\phi) = t_*$ and $\nu_2(t,t_{**}) = t$, and $t_* < t$ is the maximum $u \in \Omega$ such that $f(\phi;\leq u) \leq f(\phi;\geq t)$.
- If $f(\phi;\geq t) > f(\phi;\leq t)$, then $\nu_1(t,t_*;\phi) = t$ and $\nu_2(t,t_{**}) = t_{**}$, and $t_{**} > t$ is the minimum $u \in \Omega$ such that $f(\phi;\geq u) \leq f(\phi;\leq t)$.
- If $f(\phi;\geq t) = f(\phi;\leq t)$, then $\nu_1(t,t_*;\phi) = t$ and $\nu_2(t,t_{**}) = t$.

****

We note key aspects of the above equations.

- The equation for asymptotic empirical Wald confidence limits, namely expression (4.69), has explicit solutions.
- Equation (4.67) for asymptotic LR confidence limits is a polynomial equation with nonnegative exponents in general.
- Expression (4.68) for the asymptotic regular Wald method is a mixed polynomial and exponential expression.
- These expressions involve continuous and differentiable functions of ϕ or β, and so generally yield unique estimates, or unique confidence limits.
- Expression (4.59) which pertains to exact score, LR, and the probability method confidence limits, and equation (4.63) for the exact Wald methods are NOT simple polynomial expressions. The complicating feature is that the range of summation in the set Λ involves the parameter value. These functions are potentially discontinuous and may not have a derivative some points in the parameter space. Solving these equations requires particular care, and may need random or systematic searches for the largest and smallest of the solutions.
- Though equation (4.70) for the CT method appears like a simple polynomial expression, it actually is not so, and stands in a class of its own.
- Boundary points need special attention.

4.7 Iterative Methods

Suppose we need to solve the equation

$$\phi^2 - 4\phi + 1 = 0$$

Instead of using the usual quadratic formula, we rewrite this equation as

$$\phi = 4 - \frac{1}{\phi}$$

We then write this as an iteration by considering the rhs as an updated version of the lhs, and set

ITERATIVE METHODS

$$\phi_i = 4 - \frac{1}{\phi_{i-1}}$$

Beginning with a guess ϕ_0, we implement this relation for $i = 1, 2, 3, \cdots$. Starting with $\phi_0 = 1$ and implementing this, it is easy to verify that the process does converge to one of the solutions of the quadratic. The number of steps, or the stopping point, depends on the desired accuracy.

This is a simple example of an iterative approach. In general, it takes the existence of a real solution as its basis. The starting value may also affect whether it converges or not. When there are multiple solutions, different starting values may converge to the different solutions. If the accuracy increases by a fixed level at each step, the **rate of convergence** of the iteration is said to be **linear**. If it doubles at each step, it is **quadratic**.

Where it can be so formulated, this approach works well for complicated equations as well. For example, consider, for $k = 1, \ldots, K$, let $t_k, n_k, a_k, b_k \geq 0$ and where $t_k \leq n_k$, the following equation

$$\sum_{k=1}^{K} t_k = \sum_{k=1}^{K} \frac{n_k a_k \phi}{a_k \phi + b_k} \tag{4.71}$$

Such an equation arises in the analysis of follow up studies, and similar forms arise in other discrete data problems. We need to solve it for ϕ. Then we rewrite it as

$$\phi = \left\{ \sum_{k=1}^{K} \frac{t_k a_1 k}{a_k \phi + b_k} \right\} \left\{ \sum_{k=1}^{K} \frac{(n_k - t_k) a_k}{a_k \phi + b_k} \right\}^{-1} \tag{4.72}$$

and apply the iteration

$$\phi_i = \left\{ \sum_{k=1}^{K} \frac{t_k a_k}{a_k \phi_{i-1} + b_k} \right\} \left\{ \sum_{k=1}^{K} \frac{(n_k - t_k) a_k}{a_k \phi_{i-1} + n_k} \right\}^{-1} \tag{4.73}$$

Starting with $\phi = 1$, this iteration usually converges rapidly, giving an accurate solution within four or five steps (Clayton 1982; Newman 2001, pages 219–220).

Such specific iterations are, however, not always available. So now we consider iterative methods for the general case. Suppose we need the real root(s) of the equation

$$g(\phi) = 0 \tag{4.74}$$

at $\epsilon > 0$ level of accuracy. Several iterative methods to determine the root(s) exist. The stopping rule used for all methods we give is: Continue until the desired level of accuracy is attained, or the maximum number of iterations, denoted I_m, is reached.

All methods need an initial guess. This may be a single point, or an interval with or without the root. The current point or interval is then improved in terms of closeness to the root until the convergence criteria are met. Some methods are guaranteed to improve the solution at each iteration whereas others may diverge in unfavorable conditions. First consider two commonly used methods, namely the bisection and Newton's methods.

4.7.1 Bisection Method

The bisection method is a simple and reasonably efficient method for most equations we deal with. Though slower than other methods, it always produces a solution if started with an interval containing the root. At the $(i-1)$th iteration let, without loss of generality, (ϕ_{i-1}, ϕ_{i-2}) denote the root bracketing interval. Therefore, $g(\phi_{i-1})g(\phi_{i-2}) < 0$, or that the sign of $g(\phi)$ changes sign at least once between ϕ_{i-1} and ϕ_{i-2}. Assume $g(\phi)$ is continuous over this interval. The subsequent steps are:

- Set $\phi_i = 0.5(\phi_{i-1} + \phi_{i-2})$ as the current estimate of the root.
- If $\mid \phi_{i-1} - \phi_{i-2} \mid \leq \epsilon$, or if the number of iterations exceeds I_m, then stop.
- If $g(\phi_i)g(\phi_{i-1}) < 0$, set (ϕ_{i-1}, ϕ_i) as the bracketing current interval. If $g(\phi_i)g(\phi_{i-2}) < 0$, set (ϕ_i, ϕ_{i-2}) as the current interval. Go to the initial step.

At each iteration, the midpoint of the interval is the estimate of the root. It deviates by at most $0.5 \mid \phi_{i-1} - \phi_{i-2} \mid$ from the root.

At the outset, we evaluate the function at three points. At each further iteration, that is done once, at the midpoint of the current interval. The maximum error or tolerance is reduced to a half its previous size by each iteration, that is, the rate of convergence for the bisection method is linear.

4.7.2 Newton's Method

Provided that the function is smooth in the neighborhood of the root and an adequate starting value is given, Newton's (or the Newton–Raphson) method is faster than bisection. The first order Taylor approximation at an initial guess, ϕ_0, is used as a linear approximation of the function. The crossing of the approximating line and the ϕ-axis provides an updated guess of the root. This is repeated until desired accuracy is attained. The current estimate, ϕ_i, is obtained from the previous by

$$\phi_i = \phi_{i-1} - \frac{g(\phi_{i-1})}{g'(\phi_{i-1})} \tag{4.75}$$

In contrast to the bisection method, Newton's method is not guaranteed to succeed in its search for the root. In the neighborhood of the root, its rate of convergence often is quadratic. This makes it the method of choice when evaluation of the first derivative is possible at a moderate computational cost, and the danger of divergence is minimal.

For computation of the mle for a PBD, the equation can be written as

$$g(\phi) = t - \frac{f_*(\phi)}{f(\phi)} = t - \mu(\phi) = 0 \tag{4.76}$$

The Newton's recursion for this equation simplifies to

$$\phi_i = \phi_{i-1} + \frac{\phi_{i-1}(t - \mu(\phi_{i-1}))}{\sigma^2(\phi_{i-1})} \tag{4.77}$$

where μ and σ^2 are the mean and variance of $\mathcal{T}$. This version is useful when the mean and

ITERATIVE METHODS

variance of T can be computed rapidly for any ϕ. In this case, the potential for floating point overflow or underflow associated with computing the generating polynomial is reduced.

Otherwise, we use the version

$$g(\phi) = tf(\phi) - f_*(\phi) = 0 \qquad (4.78)$$

In this version, the use of an integrated version of the Horner's scheme to compute, for a given ϕ, the gp $f(\phi)$ and its two derivatives in a single pass is useful in the implementation of the Newton's method.

Some modifications of these two methods have rates of convergence that are in between the linear rate of bisection and the quadratic rate of Newton–Raphson methods. Two of these are described below.

4.7.3 False Position Method

As in Newton's method, the method of false position approximates the function with a line. Like the bisection method, it needs an initial specification of an interval with the root. The approximating line is the line drawn between the two most recent function evaluations bracketing the root. The crossing of the approximating line with the ϕ-axis is the updated guess. With (ϕ_{i-1}, ϕ_{i-2}) as the current root bracketing interval, we have

$$\phi_i = \frac{\phi_{i-1} g(\phi_{i-2}) - \phi_{i-2} g(\phi_{i-1})}{g(\phi_{i-2}) - g(\phi_{i-1})} \qquad (4.79)$$

Note that brackets are kept on the root throughout the iterations. The method is often faster than bisection, but the number of iterations is unpredictable and under pathological conditions it may converge more slowly than bisection. Unlike the Newton's method, it does not need the computation of a derivative.

4.7.4 Secant Method

The secant method also uses a linear approximation of the function. It requires initial specification of an interval, but unlike the bisection and false position methods, it need not bracket the root. Without loss of generality, let (ϕ_{i-1}, ϕ_{i-2}) denote the current interval. Then we use

$$g'_{i-1}(\phi_{i-1}) \approx \frac{g(\phi_{i-2}) - g(\phi_{i-1})}{\phi_{i-2} - \phi_{i-1}} \qquad (4.80)$$

in (4.75) to get an updated value. This method tends to be faster than the false position method. Its primary advantage over Newton's method is that instead of obtaining the derivative analytically, it numerically approximates the derivative.

4.7.5 Hybrid Methods

It is possible to combine the sureness of bisection with faster methods in a way that the guarantee of convergence is upheld at substantial savings in computational effort. Two such combinations are the Newton-bisection and secant-bisection methods. Both Newton's method and the secant method require good initial guesses. If the function is too steep or nonmonotone these methods may fail except when started with a good approximation to the root. Convergence may be secured by taking a few simple measures. One is to require initial specification of an interval containing the root. The bisection method is used if the faster method produces an update outside the root bracketing interval. The rate of convergence for different methods may also be monitored at each iteration so that the bisection method can take over whenever it will provide a better guess than the higher order methods.

Another recommended one dimensional root solver is Brent's method (Or the Van Wijngaarden Dekker Brent method). It combines the bisection method with a second order approximation of the function. Brent's method is guaranteed to find the root, usually converging with rates not far from the secant or Newton's methods. An initial specification of an interval containing the root is required. An updated guess obtains from three points by quadratic interpolation. The bisection method takes over if the update is outside the bracketing interval or if monitoring of convergence shows that the bisection update would be closer to the root.

For a detailed specification of these methods, see the references in §4.8.

4.7.6 Recommendations

A detailed study of the relative efficiency and accuracy of the iterative methods for the types of equations given in §4.6 has not been done. Our experience suggests the following: For equations (4.54), (4.55), (4.56), (4.57) and (4.58) the secant-bisection method provides a good performance. Note that for some of these equations, we need two roots, one for the lower limit and one for the upper limit.

For equations of the form (4.59) which pertain to exact score, LR, Wald and probability method confidence limits, the bisection or false position methods provide a safe option. The possibilities of multiple roots for the lower or the upper confidence limit, and unpredictable behavior needs to kept in mind. For these equations, plotting the corresponding two-sided evidence function is a good check. We may also perform iterative interpolation of the evidence function or random search to deal with such problems.

Only the bisection method is applicable for computing the CIs based on the CT method. This may equivalently be formulated in the form of a systematic plotting of the evidence function.

When the observed value of T is at the boundary of Ω, computing the estimates and confidence limits under any of these methods should allow for nonfinite solutions as noted in §3.13.

For the hypergeometric distribution, methods that combine recursion with iteration are also available. For example, the algorithm of Liao (1992), as noted in Algorithm 4.HMV.1, is used to recursively compute the conditional mle for a given ϕ. This is used to iteratively solve the equation

$$\mu(n, s, N; \phi) = t \qquad (4.81)$$

If we use the bisection method here, then at each iteration, we only need to compute the conditional mean for the updated value of ϕ. For Newton's method, both the conditional mean and variance are needed. These recursions also apply for iteratively computing $(1 - \alpha_*)100\%$ confidence limits for ϕ from inverting the asymptotic conditional score test. These are the two solutions to the equation

$$\frac{(t - \mu(n, s, N; \phi))^2}{\sigma^2(n, s, N; \phi)} = \chi^2_{1-\alpha_*} \qquad (4.82)$$

ITERATIVE METHODS

Granville and Schifflers (1993) used, in conjuction with their modified recursions for the hypergeometric, the bisection method for computing the mle and exact confidence limits for the hypergeometric. For the conditional mle and asymptotic score interval, the algorithm of Liao (1992), and for some forms of exact analysis, the approach of Granville and Schifflers (1993) deserve a close scrutiny.

4.7.7 Starting Values

The secant and Newton methods are quite sensitive to the quality of the initial guess. An initial guess may be obtained from a method where related explicit values are available. For example, in the binomial or Poisson case, the computation of the mue can begin in the vicinity of the mle. Confidence limit computation may use the empirical adjusted Wald limits, where possible. The starting value in general depends on the problem at hand. For example in the case of interval estimation of the common odds ratio in several 2×2 tables, the interval of Robins–Breslow–Greenland is a suitable starting point (see chapter 8). For other cases, an adjusted empirical logit type of estimate is a good starting point to compute the mle or the mue.

4.7.8 Computational Scale

In discrete data analysis, often the coefficients $c(u)$ are quite large. A typical problem (for details, see Chapter 8) relates to a series of 2×2 tables. Consider a case with all cell entries in a table equal to 10. The gp of a single table has coefficients shown in Table 4.1. If there are ten such tables in total, the overall coefficients are obtained from the product of ten such polynomials. Some of these will be small and some may be larger than the largest real number available on the compiler. Such cases call for the logarithmic scale for storing the coefficients.

Exponentiation and taking logarithms are computationally expensive and significantly increase the computing time needed for root solving. Some compilers allow extended real precision where representation of numbers from 10^{-4951} to 10^{+4932} is possible. This permits exclusive use of the actual scale for a wide range of problems.

Where resorting to the logarithmic scale is unavoidable, care needs to be exercised to ensure minimal loss of accuracy as explained in §4.4. The methods of polynomial evaluations, also in this section, are relevant for all the iterative schemes.

4.7.9 An Example

Consider computation of the roots of a polynomial that is the product of ten identical polynomials, each having the coefficients and exponents given in Table 4.1, to illustrate the use of the root solvers. Suppose the product polynomial is the gp of a PBD for which we need 95% exact confidence limits for ϕ. Under a variety of environments, with the bisection method on the logarithmic scale as the reference, the secant method reduced the computing time by a factor of between 3 to 5. Avoiding the logarithmic scale on a compiler supporting large exponents reduced computing time with another factor of between 4 to 8. In our experience Brent's method, which we have not discussed, is generally a little slower than the secant method.

Table 4.1 *A Sample of Exponents and Coefficients*

t	$c(t)$
0	1
1	400
2	36100
3	1299600
4	23474025
5	240374016
6	1502337600
7	6009350400
8	15868440900
9	28210561600
10	34134779536
11	28210561600
12	15868440900
13	6009350400
14	1502337600
15	240374016
16	23474025
17	1299600
18	36100
19	400
20	1

4.8 Relevant Literature

The concept of a recursion, and the combinatorics and computation of binomial and other coefficients are covered in many books; for example, Anderson (1989) and Sedgewick (1988). The computation of discrete distributions in general is discussed in Kennedy and Gentle (1980), Korfhage (1984), Ling (1992) and Tietjen (1998).

Many textbooks describe how to use the F distribution to compute the binomial tail probability and the chisquare distribution to compute Poisson probability; see, for instance, Santner and Duffy (1989). Ling (1992) makes a case for the universal use of such relations, and decries the recourse to tabulated values.

Tietjen (1994, 1998) give the state of the art approaches for computing binomial and Poisson tail probabilities. The references in Alvo and Cabilio (2000) illustrate the varied approaches to computing the central hypergeometric probability. This paper gives a new method based on Chebychev polynomials. Harkness (1965), Liao (1992) and Satten and Kupper (1990) give methods and programs for recursive computation of the mean and variance of the hypergeometric. Granville and Schifflers (1993) is a good reference in this regard. For recursive computation of negative binomial probabilities, see Williamson and Bretherton (1963).

Recursions (4.49), (4.50) and (4.51) are in Liao (1992) who based them on Harkness (1965). Another method is in Satten and Kupper (1990). For proofs, see these papers. Liao (1992) also gave a FORTRAN program. Beginning with the work of Jerome Cornfield in the 1950s, many papers have tackled computational issues for asymptotic and exact conditional inference for a 2×2 table (Chapter 5). These are related to the topics we have covered in the present chapter. Thomas (1971) and Baptista and Pike (1977) gave programs to compute exact confidence limits for the odds ratio in a 2×2 table. The former uses the bisection method. See the next chapter for further references.

Basic and more elaborate methods of evaluating polynomials and their derivatives as well as assessments of their efficiency are in many texts on algorithms and numerical methods. See, for example, Sedgewick (1988) and Kronsjö (1987).

Mehta and Patel (1983) suggested use of (4.20) for log-scale computation. Consult this paper for a discussion of the magnitude of the error involved and related references.

Readable accounts of iterative methods for finding roots of equations appear in Burden and Faires (1993) and Kronsjö (1987). Works directly relating to statistical applications are Kennedy and Gentle (1980), and Rustagi (1994). Kronsjö (1987) also covers some advanced methods. These books also deal with efficient evaluation of polynomials. Borse (1985) and Press et al. (1992) give FORTRAN codes for root solving methods. A list of other sources for computer codes is in Burden and Faires (1993).

4.9 Exercises

4.1. Determine the computer RAM required to store the log-factorial values, $\log(n!), n = 0, 1, \cdots, 1000$, in single or double precision.

4.2. Determine the computer RAM required if all binomial coefficients for $n = 0, 1, \cdots, 1000$ are stored in single or double precision. Compare two storage methods: a conventional bivariate array and the linear array given in §4.4. Derive the address and retrieval function to be used for the latter method.

4.3. Write computer programs for Algorithms 4.F.1, 4.F.2 and 4.F.3 for the factorials and compare them in terms of accuracy, efficiency and robustness for values of n up to 1000, under single precision and double precision arithmetic.

4.4. Write computer programs for Algorithms 4.B.1, 4.B.2, 4.B.3, 4.B.4 and 4.B.6 for the binomial coefficients. Compare them in terms of accuracy, efficiency and robustness for values of n up to 1000, under single precision and double precision arithmetic.

4.5. Detail the recursive and other approaches for computing the hypergeometric and negative binomial coefficients. Write computer programs for them and compare them in terms of accuracy, efficiency and robustness for values of n up to 1000 under single precision and double precision arithmetic.

4.6. How do these comparisons relate to Algorithm 4.C.1 and Algorithm 4.C.2?

4.7. Write a computer program for Horner's scheme (4.16) for evaluating a polynomial. Implement a robust version that reduces the danger of numeric overflow.

4.8. Write a computer program based on an integrated Horner's scheme to compute a polynomial and its first two derivatives. Implement a robust version of the scheme to reduce the danger of numeric overflow or underflow.

4.9. Compare method (4.20) with method (4.23) for adding a series of numbers on the log-scale.

4.10. Give the details of and implement on a computer Algorithms 4.PP.1, 4.PP.2, 4.PP.3 and 4.PP.4 for the Poisson distribution. Compare them in terms of accuracy, efficiency and robustness for values of t up to 1000 and for variety of λ under single precision and double precision arithmetic. Assess the memory requirements in these methods. Which method is more suited to multiple applications?

Compare your programs in terms of accuracy and speed with those of Tietjen (1998). Can accuracy be improved by separate consideration of the cases with $\lambda < 1$ and $\lambda > 1$?

4.11. Give the details of, and implement on a computer, Algorithms 4.BP.1, 4.BP.2, 4.BP.3, 4.BP.4 and 4.BP.5 for the binomial distribution. Compare them in terms of accuracy, efficiency and robustness for values of n up to 1000 and for a broad range of π, under single precision and double precision arithmetic. Assess the memory requirements of these methods. Can accuracy be improved by separate consideration of the cases with small $\pi < 1$ and π close to 1? Compare your programs in terms of accuracy and speed with those of Tietjen (1998). Which method is more suited to multiple applications?

4.12. Suppose T is a B(31, 0.25) random variable. Compute: $P[T = 5]$, $P[T > 3]$ and $P[T \leq 8]$.

4.13. Suppose T is a B(120, 0.40) variate. Compute $P[115 \leq T \leq 130]$ exactly and using the normal approximation.

4.14. Suppose T is a B(190, 0.10) variate. Compute $P[T \geq 41]$ exactly and through the Poisson approximation.

4.15. Give the details of and implement on a computer Algorithms 4.HP.1, 4.HP.2, 4.HP.3 and 4.HP.4 for the hypergeometric distribution. Compare them in terms of accuracy, efficiency and robustness for values of n and m up to 1000 and broad range of ϕ under single precision and double precision arithmetic. Assess the memory requirements in these methods. Which method is more suited to multiple applications?

4.16. Give the details of, and implement on a computer, Algorithms 4.HMV.1 and 4.HMV.2 for computing the mean and variance of the hypergeometric distribution. Compare them in terms of accuracy, efficiency and robustness for values of n and m up to 1000 and broad range of ϕ under single precision and double precision arithmetic. Apply them to compute relevant estimate and confidence limits and assess the memory requirements in these methods. Compare this method with that of Granville and Schifflers (1993).

4.17. Give a proof of the following expressions for the hypergeometric: (4.44), (4.45), (4.49), (4.50) and (4.51), as well of the recursions given in Granville and Schifflers (1993).

4.18. Write a computer program to compute the medians of the Poisson, binomial, negative binomial and noncentral hypergeometric distributions. Extend this to the case of a general PBD.

4.19. Implement and compare three approaches to computation of the negative binomial point and tail probabilities, namely: (i) a direct term by term method, (ii) an improved recursive approach, and (iii) a method based on the use of logarithms at all stages. Compare these methods in terms of accuracy and speed. See Williamson and Bretherton (1963) and the references in Ling (1992).

4.20. For selected values of n and t, compute and plot the cumulative binomial probability π as a function of ϕ. Do this for $\phi = 0.01 \times j$, $j = 1, 2, \cdots, 300$. Compare the various approaches available for this purpose.

4.21. Repeat the previous exercise, modified as appropriate, for the Poisson, hypergeometric and negative binomial tail probabilities.

4.22. State the exact relations between the TST lower and upper confidence limits for the binomial parameter and the F distribution tail probabilities. Implement this method on the computer. Extend this to the Poisson distribution.

4.23. Write and implement an efficient computer program to plot one-sided and two-sided exact, mid-p and approximate evidence functions for a PBD. Use both actual scale and log-scale options for robust implementation.

EXERCISES

4.24. Empirically compare the naive approach, Horner's scheme and the method of preprocessing coefficients (Kronsjö 1987) for repeated evaluation of the polynomials derived from binomial and hypergeometric generating functions.

4.25. Write a computer program to compute mle and mue for a PBD. Implement actual and log-scale options as well as an efficient method of polynomial evaluation in the program.

4.26. Write a robust and efficient computer program to compute exact, mid-*p* and asymptotic confidence intervals for the parameter of a PBD.

4.27. Write computer programs for and compare the bisection, Newton, false position, secant, Newton-bisection and secant-bisection methods for computing an mue for a PBD and, specifically, for binomial and Poisson data. Which do you recommend?

4.28. Write computer programs for and compare the bisection, Newton, false position, secant, Newton-bisection and secant-bisection methods for computing confidence limits (exact, mid-*p* and approximate) for a PBD. Apply these to variety of simulated data. Which do you recommend?

4.29. Give detailed derivations of all the equations given in §4.6.

4.30. Write computer programs for and compare the bisection, Newton, false position, secant, Newton-bisection and secant-bisection methods for computing confidence limits for the exact TST, mid-*p* and asymptotic score, LR, probability methods for a PBD. Which do you recommend?

4.31. Write a computer program for computing estimates and confidence limits from the acceptability (CT) method for a PBD. Can a method other than the bisection method be used here? If so, give the details. Which do you recommend?

4.32. For a prime number M and nonnegative integers a and b, prove the following:

$$(ab) \mod M = [(a \mod M)(b \mod M)] \mod M$$

and

$$(a+b) \mod M = [(a \mod M) + (b \mod M)] \mod M$$

Use these expressions to derive a method for computing the hash function (4.13) that reduces the danger of numeric overflow (see Hirji, Mehta and Patel (1987) for details).

CHAPTER 5

Elements of Conditional Analysis

5.1 Introduction

This chapter introduces conditional methods for the analysis of discrete data. To provide a comparative framework, relevant unconditional large sample methods are also described. Models for the 2×2 table of counts, and two and four group Poisson data serve as the raw material for constructing these methods. The specific aims here are:

- To define and contrast the unconditional and conditional methods of data analysis.
- To relate the concepts of ancillarity and sufficiency to the use of conditional methods of inference on discrete data.
- To present five independent observation designs for a 2×2 table of counts. They include three prospective designs (one margin fixed, overall total fixed and nothing fixed designs), one retrospective design with one margin fixed, and the inverse sampling design.
- To formulate statistical models for these designs, and portray their probability distributions as polynomial based distributions.
- To present unconditional large sample, and conditional exact and large sample methods for analyzing key parameters from these designs.
- To develop unconditional and conditional analysis of a two group person time design.
- To further explore the relationship between study design and data analysis method.

5.2 Design and Analysis

Example 5.1: Assume we have obtained two lists of preschool attendees in a district. One lists four year old boys and the other, four year old girls. We randomly sample five children from each list, and for each child sampled, we find out whether or not he or she had an acute earache (acute otitis media or AOM) in the previous year. Our aim is to compare the AOM rates among the boys and girls of that age. We observe that two boys and one girl had the condition.

We have ten binary random variables corresponding to the disease status of the study children (0 = no AOM, 1 = AOM). If we consider each child as a distinct entity, we have 32 possible outcomes for each group, or $32 \times 32 = 1024$ outcomes for the combined sample.

Now, for each group, consider the total number of children with AOM. Assuming independent events between the participants, we get a two binomial model with the total in each group sufficient for its outcome probability. We summarize the 1024 possibilities by considering the total number who had AOM. In this case, there are 6 outcomes $\{0, 1, 2, 3, 4, 5\}$ per group, or $6 \times 6 = 36$ outcomes overall. For example, the latter includes the possibility that no boy or girl had AOM, or that four boys and two girls had it. This way of looking at the situation does not exclude any underlying outcome. Using sufficient statistics captures, in a sense, information about the parameters and allows us to view the situation in a more compact way. Hardly any statistician will raise an objection against this step.

Now suppose we go even further and restrict our attention to, as observed in our samples, only the set of outcomes in which a total of three children had the problem. This can arise in four different ways: (0 Boys, 3 Girls), (1 Boy, 2 Girls), (2 Boys, 1 Girl) and (3 Boys, 0 Girls). This includes the observed outcome, namely, (2 Boys, 1 Girl). In this case, the other outcomes are excluded from consideration.

Our aim is to compare the AOM rate between boys and girls. If we accept the last step, our analysis will be restricted to the four possibilities in which total number of AOM cases is equal to what was observed in the sample at hand. Or, should our analysis take into account all the 36 possible outcomes under the study design? Different answers give a different method of analysis for the data, and may lead to a different conclusion.

Analysis Methods

```
An unconditional method of analysis is based on the
complete statistical frame or outcome space of the
study design. In contrast, a conditional method is
based on a subset of outcomes from the complete outcome
space.
```

Conditional analysis fixes items observed in the course of the study. Is it a valid approach? How do we justify treating as constant what from the viewpoint of the design or *a priori* are unknown, random entities?

****

To further illustrate the issues, we consider another series of examples. They derive from a hypothetical study done by a nationwide network of child health care centers. The focus again is on children afflicted with AOM. A large list of such cases is available.

Example 5.2: We consider the proportion of children with AOM who are younger than two years. Denote it as π. A random sample of $n = 300$ children diagnosed with AOM is drawn. Let $\mathcal{T}$ be the number among them younger than two years. Assume that the data on road accidents in the same areas are available as well. The associated variable here is $\mathcal{S}$ and the parameter, λ. While a purist can argue that all phenomena are linked to each other in some way, we generally regard the two phenomena, ear infections and road accidents, as unrelated phenomena, and the associated random variables, statistically independent. Hence, we write

$$P[\mathcal{T} = t, \mathcal{S} = s; \pi, \lambda] = P[\mathcal{T} = t; \pi]P[\mathcal{S} = s; \lambda] \tag{5.1}$$

For the analysis of the AOM data, $\mathcal{S}$ is not deemed a relevant entity, and λ has no bearing on π, the **parameter of interest.** In relation to π, we call the latter a **nuisance parameter.** $\mathcal{S}$ is then an **unrelated statistic**, and for the purpose at hand, we ignore it, or what here is the same thing, treat it as fixed. Our analysis will thereby condition on the variable $\mathcal{S}$.

****

To appreciate the next scenario, we need another concept. A statistic $\mathcal{S}$ is called an **ancillary statistic** for ϕ, a parameter of interest, if we can write

DESIGN AND ANALYSIS

$$P[\mathcal{T}=t, \mathcal{S}=s; \phi, \theta] = P[\mathcal{T}=t \mid \mathcal{S}=s; \phi]P[\mathcal{S}=s; \theta] \qquad (5.2)$$

with the proviso that θ is also not related to ϕ. The two key points in (5.2) are: (i) $\mathcal{S}$ is sufficient for θ; conditioning on it removes the nuisance parameter; and (ii) the marginal distribution of $\mathcal{S}$ does not contain the parameter of interest.

Example 5.3: Let π_1 be the proportion of cases of AOM in whom both the ears are affected (bilateral AOM). The study design adopted for estimation of π_1 is as follows: A bent coin is tossed. If a head (x = 1) is seen, $n_1 = 200$ AOM cases will be randomly sampled; if not (x = 0), $n_0 = 20$ will be sampled. Either way, the sampling is to be done with replacement.

The chance of a head, π_*, is not known. For the realized x, let n_x be the number sampled, and $\mathcal{T}$, the number of cases with bilateral AOM. Then we have

$$P[\mathcal{T}=t, \mathcal{X}=x; \phi, \pi_*] = \frac{c(n_x, t)\phi^t}{fn_x(\phi)} \pi_*^{n_x}(1-\pi_*)^{n(1-x)} \qquad (5.3)$$

where, as usual,

$$c(n_x, t) = \binom{n_x}{t}; \quad \phi = \frac{\pi_1}{1-\pi_1}; \quad fn_x(\phi) = (1+\phi)^{n_x} \qquad (5.4)$$

The design sample space is the set of values of (x, u) that satisfy $x = 0, 1$ and $0 \leq u \leq n_x$. In the above study, there are $201 + 21 = 222$ possible outcomes. ϕ is the parameter of interest while π_* is a nuisance parameter. If we condition on the outcome of the coin toss, we get the binomial probability:

$$P[\mathcal{T}=t \mid \mathcal{X}=x; \phi] = \frac{c(n_x, t)\phi^t}{fn_x(\phi)} \qquad (5.5)$$

$\mathcal{X}$ is hence ancillary for ϕ. Now suppose we toss the coin, observe $x = 0$, sample 20 children with AOM, and find that 6 among them had bilateral AOM. To draw inference about ϕ, should we use the conditional distribution with $n_0 = 20$, or in some way, use the design (unconditional) distribution (5.3)?

****

In general, if $\mathcal{S}$ is ancillary for ϕ, is fixing it appropriate for an analysis of ϕ, even though it was not fixed by design? This analysis would take as its starting point the conditional distribution

$$P[\mathcal{T}=t \mid \mathcal{S}=s; \phi] \qquad (5.6)$$

To analyze the bilateral AOM data, most statisticians would use the conditional binomial distribution (5.5). It would be argued that for estimating ϕ, how n_x was determined is not relevant. This is a reasonable point of view. Yet, unlike in the first example, the value of π_* and the outcome of $\mathcal{S}$ do affect the inference on ϕ. In particular, the sample size impacts the width of the confidence interval for ϕ. While observing 6 out of 20 cases yields the same mle for ϕ as observing 60 out of 200, the precision of the estimate is higher in the latter case. Further, it affects the method we would use to analyze the data as well; we may be inclined to adopt a small sample method in the former while the approximate large sample method will suffice for the latter situation.

The ideas of sufficiency and ancillarity were developed by R. A. Fisher, the father of modern statistics (Fisher 1935b). While they are beneficial in developing methods of data analysis, they have also generated controversy. In part, this is because of the persistent lack of clarity about the idea of ancillarity. Its seeds were laid by Fisher himself. The quote below illustrates his terminological vagueness on the issue.

Ancillary Statistics

> ... ancillary statistics .... tell us nothing about the value of the parameter, but, instead tell us how good an estimate we have made of it. ...
> Their function is, in fact, analogous to the part which the *size* of our sample is always expected to play, in telling us *what reliance* to place on the result.
> (Fisher 1935b).

An ancillary statistic first tells us 'nothing' about the parameter but then also tells us something about it! To avoid such a confusion, we must specify that an ancillary statistic is not identical to an unrelated statistic. To declare, in an absolute sense, that it contains 'no information' about the parameter of interest is misleading.

For now, we set this conceptual tangle aside. What is important is that the ideas of ancillarity and sufficiency allow us to examine possibilities for alternative methods of data analysis in which quantities not fixed by design are treated as fixed, and which may facilitate the handling of nuisance parameters.

****

Example 5.4: Consider cases of multidrug resistant AOM. They are few in number but are very serious. Let T_0 and T_1 respectively be the number of such cases seen during three summer and three winter months. We assume T_i is Poisson distributed with mean λ_i, and T_0 and T_1 are independent. Suppose seven such cases were seen during the previous winter, and two during the past summer. We need to test

$$H_0 : \lambda_0 = \lambda_1 \quad \text{versus} \quad H_1 : \lambda_0 \neq \lambda_1 \qquad (5.7)$$

and to estimate the magnitude of the disparity between the mean rates. The disparity between the rates may be gauged by one of two indices:

$$\textbf{Rate Ratio:} \quad \phi = \frac{\lambda_1}{\lambda_0} \quad \text{or} \quad \textbf{Rate Difference:} \quad \delta = \lambda_1 - \lambda_0 \qquad (5.8)$$

ϕ is also called the **relative rate**, or **relative risk**. δ is also the **risk difference**. The former employs the multiplicative scale while the latter is based on the additive scale.

Consider first the rate ratio. Writing $\lambda_0 = \lambda$ and $\lambda_1 = \phi\lambda$, the equivalent problem is to test

$$H_0 : \phi = 1 \quad \text{versus} \quad H_1 : \phi \neq 1 \qquad (5.9)$$

DESIGN AND ANALYSIS

Then ϕ is the parameter of interest, and λ, the nuisance parameter. The unconditional distribution for the data is

$$P[\mathcal{T}_0 = t_0, \mathcal{T}_1 = t_1; \lambda, \phi] = \frac{1}{t_0!\, t_1!} \exp[-\lambda(1+\phi)] \lambda^{(t_0+t_1)} \phi^{t_1} \qquad (5.10)$$

Here $\mathcal{S} = \mathcal{T}_0 + \mathcal{T}_1$ is sufficient for λ, and the conditional distribution of $\mathcal{T}_1$ given $\mathcal{S} = n$ is the binomial $B(n; \pi_3)$, where

$$\pi_3 = \frac{\lambda_1}{\lambda_0 + \lambda_1} = \frac{\phi}{1+\phi} \qquad (5.11)$$

After a little algebra, we get that for $t = 0, 1, \cdots, n$,

$$P[\mathcal{T}_1 = t \mid \mathcal{S} = n; \phi] = \frac{c(t)\phi^t}{f(\phi)} \qquad (5.12)$$

where $c(t) = n!/\{t!(n-t)!\}$ and $f(\phi) = (1+\phi)^n$.

Conditioning then gives a univariate PBD containing only the parameter of interest. This may be analyzed with the methods of Chapters 2 and 3. For inference on ϕ, should we employ the conditional distribution (5.12), or the unconditional distribution (5.10)? To answer this, consider the marginal distribution of $\mathcal{S}$:

$$P[\mathcal{T}_0 + \mathcal{T}_1 = n; \lambda, \phi] = \frac{1}{n!} \exp[-\lambda(1+\phi)] \times [\lambda(1+\phi)]^n \qquad (5.13)$$

Unlike the bilateral AOM study example, here the distribution of $\mathcal{S}$ is not independent of the parameter of interest. $\mathcal{S}$ is not an ancillary statistic for the parameter of interest, though it is sufficient for the nuisance parameter. Under such a situation, often found in discrete data analysis, the use and justification of conditional analysis is even more contentious. And, even when considered appropriate, the reasons advanced for adopting it can be very different.

****

Example 5.5: Now consider the problem of multidrug resistant AOM in terms of the rate difference. Let $\lambda_0 = \lambda$ and $\lambda_1 = \lambda + \delta$. The problem now is to test

$$H_0 : \delta = 0 \quad \text{versus} \quad H_1 : \delta \neq 0 \qquad (5.14)$$

and to estimate δ. δ is the parameter of interest, and λ is then the nuisance parameter. The unconditional distribution now is

$$P[\mathcal{T}_0 = t_0, \mathcal{T}_1 = t_1; \lambda, \delta] = \frac{1}{t_0!\, t_1!} \exp[-2\lambda - \delta] \lambda^{t_0} (\lambda+\delta)^{t_1} \qquad (5.15)$$

The rate difference model differs from the rate ratio model in that there is no known way of removing the nuisance parameter from (5.15) by conditioning. To get a CI for δ, essentially we only have one option, namely to use the unconditional distribution (5.15). This is a situation where we cannot find ancillary or sufficient statistics to simplify our analytic task.

5.3 Modes of Inference

Before developing the details of unconditional and conditional analysis methods for discrete data models, we summarize the options available for analyzing such models in general terms.

A common problem confronting statistical analysis is to assess the values of one or more parameters of interest when nuisance parameters are present. The values of the latter are generally unknown. As such, they pose a distinct conundrum because the distribution of a relevant test statistic, or the formula for a CI may include an unknown entity. Several ways of dealing with such parameters have been devised. The common options for the tail area mode of inference are categorized as follows.

- **Asymptotic Unconditional Analysis**: This method applies large sample theory to the unconditional likelihood. In general, the test statistics used are such that their large sample distributions do not include the nuisance parameters.
- **Exact Unconditional Analysis**: This method employs the exact unconditional distribution for the statistical model under the study design. It takes care of the nuisance parameters by optimizing the indices of inference over the space of these parameters.
- **Exact Conditional Analysis**: This method conditions on the sufficient statistics for the nuisance parameters, if they exist, to obtain a distribution containing only the parameters of interest. Inference is performed with reference to this exact distribution.
- **Asymptotic Conditional Analysis**: This method also conditions on the sufficient statistics for the nuisance parameters. Then it applies large sample theory to the conditional likelihood to analyze the parameters of interest.

Data analysis methods that do not neatly fit into one of these categories also exist. They include those that use the marginal distribution, the marginal likelihood or the partial likelihood. Some deploy a partly conditional or semiconditional approach. Methods that fall outside the tail area mode of inference such as the Bayesian method, the empirical Bayes method or methods that employ the likelihood itself as an index of evidence are also advocated.

****

The asymptotic unconditional approach is the most frequently used in practice. Though it is remarkably versatile, its accuracy is in question under small to moderate sample sizes, or when the data are sparse.

Exact unconditional analysis, at least in conceptual terms, has been around since the middle of the 20th century. In principle, it is a broadly applicable method. Yet, until recently, its use for real data analysis was unheard of because (i) even for simple models and designs, it needs complex computations and (ii) in those problems where a conditional alternative exists, the preliminary studies had shown that it may be more conservative. In recent years, better methods of computation have emerged. Also recent empirical studies favor the method. It has thus gradually begun to gain ground, though for now its use is restricted to relatively simple models. More efficient algorithms and empirical studies may make it a practical option for a broader range of problems in the future. For now, that is not the case. We briefly explain the method of exact unconditional analysis in Chapter 7.

The conditional approach is available only for the models and designs where sufficient statistics for the nuisance parameters exist. This is the case for many discrete data models applied in practice. This holds for the models we study which for the most part are PBDs. A major gain then is that, with appropriate conditioning, the methods of Chapters 2 and 3 apply broadly to

THE 2 × 2 TABLE   123

the analysis of individual parameters, irrespective of the model or design from which the data emerged. Application of efficient computational algorithms in the last two decades of the 20th century vastly expanded the situations for which such an analysis is feasible. Conditional exact and large sample methods are thereby in extensive use today.

Conditional analysis, however, is not a viable option for a large class of models where sufficient statistics do not exist. The additive rates model given above is one example.

Conditional analysis, moreover, begs a basic question: How do we justify treating as fixed quantities that are, in the context of the study design, random quantities? And particularly when they are not ancillary? Do the concepts of ancillarity and sufficiency embody some fundamental principles that serve to provide the complete justification for a conditional analysis? Or do we still need to compare a conditional approach against the unconditional approach? An analysis of such key issues in any depth lies beyond the scope of the present text. But since they pertain to the methods we develop, we address them and related controversies in a limited fashion in other parts of this text, particularly in Chapter 7 and Chapter 15.

For now we attend to developing the conditional approach. We do this upon noting its main advantage, that it allows us to easily factor out nuisance parameters, and so to simplify and unify our data analytic methods.

5.4 The 2 × 2 Table

Table 5.1 *A 2 × 2 Table*

Row Variable	Column Variable		
	0	1	Total
0	d	c	r
1	b	a	s
Total	m	n	N

The 2 × 2 contingency table is the commonest form of discrete data found in applied research. It has four internal cells, each with a numeric count. We write a generic 2 × 2 table in Table 5.1 with $a + b = s$, $m + n = N$, etc. The sums of the row and column cells are the row and column margins, respectively. N is the overall total. Parts of the table may be fixed prior to data collection. The extent to which this happens depends on the design, or the sampling scheme. What remains is the random portion gathered during the study. When seen as random variables, we respectively denote these cell counts as $\mathcal{A}$, $\mathcal{B}$, $\mathcal{C}$ and $\mathcal{D}$.

We consider five study designs that generate data in this form. For each, we present a model, formulate its probability distribution as PBD and then give exact and large sample methods to analyze the model.

5.5 The One Margin Fixed Design

Example 5.6: Prummel et al. (1993) report a randomized trial of prednisone and radiation therapy for Graves' ophthalmopathy (an autoimmune eye disease). Twenty eight patients were recruited for each treatment arm. Using relevant clinical criteria, the outcome was assessed as success or failure. The main results appear in Table 5.2: a 50% success rate in the prednisone treated and a 46% rate in the irradiated patients were observed.

Table 5.2 *Ophthalmic Data*

Outcome	Treatment 0 Prednisone	Treatment 1 Radiotherapy
Failure	14	15
Success	14	13
Total	28	28

Source: Prummel et al. (1993).
© 1993 by The Lancet Ltd.; Used with permission.

Let the population success rates of prednisone and radiotherapy be π_0 and π_1, respectively. The aim of the trial was to compare these rates. The column totals m and n are fixed by design. We assume that patient outcomes are independent, and that within each treatment group, the response probabilities are homogeneous. This is called the independent observations **one (column) margin fixed design**. The two independent binomials distributions model is normally deemed the appropriate statistical model here. Because of random allocation, this is, strictly speaking, not the design distribution. In Chapter 6, we justify using it on the basis of ancillarity. So for now, assume that that is the case.

With $\mathcal{A} + \mathcal{C} = n$ and $\mathcal{B} + \mathcal{D} = m$, we then have

$$P[\mathcal{A} = a, \mathcal{B} = b; \pi_0, \pi_1] = \binom{m}{b}\binom{n}{a}\pi_0^b(1-\pi_0)^{m-b}\pi_1^a(1-\pi_1)^{n-a} \quad (5.16)$$

Since parameterization in terms of odds and odds ratio converts this distribution into a polynomial form, we define

$$\alpha = \ln(\theta) = \ln\left\{\frac{\pi_0}{1-\pi_0}\right\} \quad (5.17)$$

$$\beta = \ln(\phi) = \ln\left\{\frac{\pi_1(1-\pi_0)}{\pi_0(1-\pi_1)}\right\} \quad (5.18)$$

Let $x = 0, 1$ be a binary indicator for treatment. With $\pi(x) = \pi_x$, the above is equivalent to

$$\pi(x) = \frac{\exp(\alpha + \beta x)}{1 + \exp(\alpha + \beta x)} = \frac{\theta\phi^x}{1 + \theta\phi^x} \quad (5.19)$$

$$\ln\left(\frac{\pi(x)}{1-\pi(x)}\right) = \alpha + \beta x = \theta\phi^x \quad (5.20)$$

Note then:

THE ONE MARGIN FIXED DESIGN

- ϕ is the **odds ratio** or the **cross product ratio**.
- β is the corresponding **log-odds ratio**.
- $\beta = 0$ (or, $\phi = 1$) if and only if $\pi_0 = \pi_1$.
- The one-to-one relations between the pairs (π_0, π_1), (θ, ϕ) and (α, β) make these ways of expressing the model equivalent to one another.

Now let

$$S = \mathcal{A} + \mathcal{B} \quad \text{and} \quad \mathcal{T} = \mathcal{A} \tag{5.21}$$

(i) $\mathcal{S}$ is the total number of successes, and (ii) $\mathcal{T}$ is the number of successes in the treatment $x = 1$. Substituting for π_0 and π_1 in terms θ and ϕ in (5.21) and simplifying, it emerges that

$$P[\mathcal{S} = s; \mathcal{T} = t; \theta, \phi] = \frac{c(s,t)\theta^s \phi^t}{f(\theta, \phi)} \tag{5.22}$$

where

$$f(\theta, \phi) = (1+\theta)^m (1+\theta\phi)^n \tag{5.23}$$

$$= \sum_{(u,v) \in \Omega} c(u,v) \theta^u \phi^v \tag{5.24}$$

$$c(u,v) = \binom{m}{u-v}\binom{n}{v} \tag{5.25}$$

$$\Omega = \{(u,v) : 0 \leq u \leq m+n;\ \max(0, u-m) \leq v \leq \min(u,n)\} \tag{5.26}$$

$(\mathcal{S}, \mathcal{T})$ thus has a bivariate PBD in which $\mathcal{S}$ and $\mathcal{T}$ are sufficient for θ and ϕ, respectively.

Table 5.3 *Hypothetical Data*

	$x = 0$	$x = 1$
Not Cured	1	2
Cured	2	1
Total	3	3

Example 5.7: Table 5.3 shows an hypothetical trial with $m = n = 3$. The exponents and coefficients of the gp $f(\theta, \phi)$ are in Table 5.4. These obtain by enumerating all 2×2 tables with both the column sums equal to 3, or equivalently, by multiplying the polynomials

$$(1+\theta)^3 (1+\theta\phi)^3$$

Some remarks about this study design are in order.

Table 5.4 *Sample Points and Coefficients*

(s,t)	$c(s,t)$	(s,t)	$c(s,t)$	(s,t)	$c(s,t)$
(0,0)	1	(3,0)	1	(4,1)	3
(1,0)	3	(3,1)	9	(4,2)	9
(1,1)	3	(3,2)	9	(4,3)	3
(2,0)	3	(3,3)	1	(5,2)	3
(2,1)	9			(5,3)	3
(2,2)	3			(6,3)	1

Note: These values are for $m = n = 3$.

- Model (5.20) is a **logistic model** where the parameter of interest is β, or, equivalently, ϕ. The greater the disparity between the cure rates, the greater the absolute magnitude of β.
- The one margin fixed design is also called a **comparative trial**. A test for $\beta = 0$, or $\pi_0 = \pi_1$, is called a **test for homogeneity**.
- Two other models for the comparative trial are also used. The **relative or multiplicative risk model** is

$$\pi_1 = \phi \pi_0 \qquad (5.27)$$

while the **risk difference**, or **additive risk model** is

$$\pi_1 = \pi_0 + \delta \qquad (5.28)$$

The distributions for these two models are not of a polynomial form. Hence we cannot use conditional methods to analyze them. This situation is similar to the rate difference Poisson model in §5.2.

- At times the two treatments are compared in terms of a clinically specified minimum level of difference. The treatments are considered equivalent if the difference falls below this level. A statistical test for this is hence called a **test for equivalence**. See Dunnet and Gent (1996) for elaboration.
- In a clinical trial, factors other than treatment, known as **prognostic factors**, may influence the outcome. These need to be taken into account when comparing treatments. Model (5.20) does not directly do that. Under random allocation to treatment, the groups tend to be comparable in terms of known and unknown prognostic factors. Models which explicitly allow for prognostic factors are considered in later chapters.
- Another important requirement in clinical trials is to control other sources of **bias** or **systematic errors**. The lack of a uniform support services and monitoring scheme for all subjects, and the influence of prior expectations on outcome evaluation are examples of such sources of bias. A **double blind** design helps counter these biases.
- Selecting the study sample in a representative or random manner is important for clinical trials. It promotes the generalizability of the study results to the appropriate target population.

5.6 The Overall Total Fixed Design

Example 5.8: Fernandez et al. (1993) report on the use of methotrexate to treat ectopic pregnancy in 100 consecutive patients. The outcome was presented at two levels of the hormone

THE OVERALL TOTAL FIXED DESIGN

HCG. The total number of patients was set by the investigators. The number of subjects at each HCG level was a random variable, not known til the end of the recruitment. This is the independent observations **overall total fixed design** with two binary factors. The results are in Table 5.5.

Table 5.5 *HCG Data*

Outcome	HCG Level	
	≤ 5000	> 5000
Failure	7	10
Success	65	18

Source: Fernandez et al. (1993).
Reprinted by permission from the
American Society for Reproductive Medicine.

The success rate at HCG level ≤ 5000 was 90% and at HCG > 5000, it was 64%. Is there an association between HCG level and outcome? How do we assess the difference in the cure rates at the two levels of HCG?

To formulate the questions of interest, let $\mathcal{Y}$ denote the random indicator for the row factor, and $\mathcal{X}$, the random indicator for the column factor. If the level of $\mathcal{X}$ affects the probability of $\mathcal{Y}$, and vice versa, the two factors are associated with one another. For $x, y \in \{0, 1\}$, let

$$P[\mathcal{Y} = y, \mathcal{X} = x] = \pi(y, x) \tag{5.29}$$

Then

$$P[\mathcal{Y} = y] = \pi(y, +) = \pi(y, 0) + \pi(y, 1) \tag{5.30}$$

$$P[\mathcal{X} = x] = \pi(+, x) = \pi(0, x) + \pi(1, x) \tag{5.31}$$

If $\mathcal{Y}$ and $\mathcal{X}$ are independent factors, then

$$\pi(y, x) = \pi(y, +)\pi(+, x) \tag{5.32}$$

for all x, y. Selecting $\pi(0, 0)$ as the reference probability, we write the probabilities in terms of odds ratios or log-odds ratios as

$$\ln\left(\frac{\pi(y, x)}{\pi(0, 0)}\right) = \beta_0 + \beta_1 y + \beta_2 x + \beta_3 yx \tag{5.33}$$

or

$$\frac{\pi(y, x)}{\pi(0, 0)} = \phi_0 \phi_1^y \phi_2^x \phi_3^{yx} \tag{5.34}$$

Note the similarity of this model to that for paired binary data in §1.9. Applying the relation $\pi(0,0) + \pi(0,1) + \pi(1,0) + \pi(1,1) = 1$ to this shows that $\beta_0 = 0$ and $\phi_0 = 1$. Further

$$\pi(0,0) = \frac{1}{1 + \phi_1 + \phi_2 + \phi_1\phi_2\phi_3} \tag{5.35}$$

$$\pi(0,1) = \frac{\phi_1}{1 + \phi_1 + \phi_2 + \phi_1\phi_2\phi_3} \tag{5.36}$$

$$\pi(1,0) = \frac{\phi_2}{1 + \phi_1 + \phi_2 + \phi_1\phi_2\phi_3} \tag{5.37}$$

$$\pi(1,1) = \frac{\phi_1\phi_2\phi_3}{1 + \phi_1 + \phi_2 + \phi_1\phi_2\phi_3} \tag{5.38}$$

These relations can also be given in terms of β's. We can further express the β's or ϕ's in terms of the π's. In particular, we note that

$$\beta_3 = \ln(\phi_3) = \ln\left(\frac{\pi(0,0)\pi(1,1)}{\pi(1,0)\pi(0,1)}\right) \tag{5.39}$$

The independence relation (5.32) holds for all x and y if and only if $\phi_3 = 1$ or $\beta_3 = 0$. ϕ_3 or β_3 measure, on the logit scale, the magnitude of the dependence between these factors.

Suppose N patients from an eligible population are selected at random. The probability distribution for this design is a four outcome multinomial with probabilities $\pi(y,x)$, $y,x \in \{0,1\}$. With

$$\mathcal{A} + \mathcal{B} + \mathcal{C} + \mathcal{D} = N \tag{5.40}$$

we have

$$P[\mathcal{A} = a, \mathcal{B} = b, \mathcal{C} = c, \mathcal{D} = d] \tag{5.41}$$

equal to

$$\left(\frac{N!}{a!\,b!\,c!\,d!}\right)\pi(0,0)^d \pi(0,1)^c \pi(1,0)^b \pi(1,1)^a \tag{5.42}$$

Define

$$\mathcal{T}_1 = \mathcal{C} + \mathcal{A}, \quad \mathcal{T}_2 = \mathcal{B} + \mathcal{A} \quad \text{and} \quad \mathcal{T}_3 = \mathcal{A} \tag{5.43}$$

$$\mathcal{T} = (\mathcal{T}_1, \mathcal{T}_2, \mathcal{T}_3)', \quad \boldsymbol{\beta} = (\beta_1, \beta_2, \beta_3) \quad \text{and} \quad \boldsymbol{\phi} = (\phi_1, \phi_2, \phi_3) \tag{5.44}$$

and

$$c(\boldsymbol{t}) = \left(\frac{n!}{(N - t_1 - t_2 + t_3)!\,(t_1 - t_3)!\,(t_2 - t_3)!\,t_3!}\right) \tag{5.45}$$

Substituting the expressions for $\pi(0,0), \pi(0,1), \pi(1,0)$ and $\pi(1,1)$ in terms of ϕ's, and using the vector notation of Chapter 1, we get

THE OVERALL TOTAL FIXED DESIGN

$$P[\mathcal{T} = t] = \frac{c(t)\phi^t}{f(\phi)} = \frac{c(t)\exp(\beta't)}{h(\beta)} \qquad (5.46)$$

where

$$f(\phi) = (1 + \phi_1 + \phi_2 + \phi_1\phi_2\phi_3)^N = \sum_{u \in \Omega} c(u)\phi^u \qquad (5.47)$$

and

$$h(\beta) = \left(1 + e^{\beta_1} + e^{\beta_2} + e^{(\beta_1+\beta_2+\beta_3)}\right)^N = \sum_{u \in \Omega} c(u)\exp(\beta'u) \qquad (5.48)$$

and further where $\Omega = \{t : N \geq t_1 + t_2 - t_3 \geq 0; N \geq t_1 \geq t_3; N \geq t_2 \geq t_3; N \geq t_3 \geq 0\}$. This is a three variable PBD.

Example 5.9: For illustration, we choose an artificial study with $N = 4$.

Table 5.6 *Artificial Data*

	$\mathcal{X} = 0$	$\mathcal{X} = 1$
$\mathcal{Y} = 0$	0	2
$\mathcal{Y} = 1$	2	0

The sets Ω, and $c(t)$ for $t \in \Omega$, for the distribution of $\mathcal{T} = (\mathcal{T}_1, \mathcal{T}_2, \mathcal{T}_3)$ appear in Table 5.7.

Table 5.7 *Specification of the Distribution of* $(\mathcal{T}_1, \mathcal{T}_2, \mathcal{T}_3)$

(t_1, t_2, t_3)	$c(t_1, t_2, t_3)$	(t_1, t_2, t_3)	$c(t_1, t_2, t_3)$	(t_1, t_2, t_3)	$c(t_1, t_2, t_3)$
(0,0,0)	1	(2,0,0)	6	(3,0,0)	4
(0,1,0)	4	(2,1,0)	12	(3,1,0)	4
(0,2,0)	6	(2,1,1)	12	(3,1,1)	12
(0,3,0)	4	(2,2,0)	6	(3,2,1)	12
(0,4,0)	1	(2,2,1)	24	(3,2,2)	12
(1,0,0)	4	(2,2,2)	6	(3,3,2)	12
(1,1,0)	12	(2,3,1)	12	(3,3,3)	4
(1,1,1)	4	(2,3,2)	12	(3,4,3)	4
(1,2,0)	12	(2,4,2)	6	(4,0,0)	1
(1,2,1)	12			(4,1,1)	4
(1,3,0)	4			(4,2,2)	6
(1,3,1)	12			(4,3,3)	4
(1,4,1)	4			(4,4,4)	1

The total fixed design is sometimes called the 2×2 **independence trial** or the 2×2 **multinomial trial**. Note that samples picked in an ad hoc manner compromise the **external validity** of such a study.

5.7 The Nothing Fixed Design

Example 5.10: Piegorsch and Bailer (1997), page 422, give data from a study of the reproductive toxicity of the pesticide azinphosmethyl in an aquatic environment. The number of male and female offspring of the organism *Amphiascus tenuiremis* that survived to adulthood after a 26-day maturation period were determined. The original data include a control group and two other settings at two levels of exposure for the pesticide. Each condition was replicated in three separate but similar tanks. We give these data in a combined form below.

Table 5.8 *Aquatic Toxicity Data*

Group	Number of Offspring	
	Female	Male
Control	363	103
Exposed	376	182

Source: Piegorsch and Bailer (1997), Table 9.2.
© CRC Press, Boca Raton, Florida.
Reprinted with permission.

All cell counts, and thus the margins and overall total, were not known at the start; they are random variables. This is the independent observations **nothing fixed design**. Let $\mathcal{Y}$ be the exposure indicator and $\mathcal{X}$, the gender indicator. How do the counts of offspring surviving to adulthood vary by gender and exposure? Do the data provide evidence that one or both of the factors affect the mean cell counts?

With $x, y \in \{0, 1\}$, let $\lambda(y, x)$ be the mean rate of viable offspring of sex x at exposure level y. The **Poisson log-linear model** assumes that the number in cell (y, x) derives from a Poisson distribution with mean $\lambda(y, x)$, where

$$\ln(\lambda(y, x)) = \mu + \mu_1 y + \mu_2 x + \mu_{12} yx \qquad (5.49)$$

Here, μ is the overall effect parameter; μ_1 is the main effect of variable y; μ_2, the main effect of factor x; and μ_{12}, the effect of an **interaction** between y and x. For the pesticide study, the absence of a pesticide exposure effect implies $\mu_1 = \mu_{12} = 0$. If the exposure effect is the same for male and female offspring, then $\mu_{12} = 0$. In this case, we say that there is no interaction between these two variables. Inverting this model, we note that

$$\mu = \ln(\lambda(0,0)) \quad \text{and} \quad \mu_1 = \ln\left(\frac{\lambda(1,0)}{\lambda(0,0)}\right) \qquad (5.50)$$

$$\mu_2 = \ln\left(\frac{\lambda(0,1)}{\lambda(0,0)}\right) \quad \text{and} \quad \mu_{12} = \ln\left(\frac{\lambda(1,1)\lambda(0,0)}{\lambda(1,0)\lambda(0,1)}\right) \qquad (5.51)$$

The interaction term, μ_{12}, is the logarithm of a cross product ratio of cell means. Assuming the cell values are independent, and with $\mathcal{T}_{00} = \mathcal{D}$, $\mathcal{T}_{01} = \mathcal{C}$, $\mathcal{T}_{10} = \mathcal{B}$ and $\mathcal{T}_{11} = \mathcal{A}$, we have

$$P[\mathcal{T}_{yx} = t_{yx}; y, x \in \{0,1\}] = \prod_{y,x \in \{0,1\}} \frac{e^{-\lambda(y,x)} \lambda(y,x)^{t_{yx}}}{t_{yx}!} \qquad (5.52)$$

THE NOTHING FIXED DESIGN

The Poisson distribution is a PBD, and a product of independent PBDs also a PBD. Hence, the joint distribution of the $\mathcal{T}_{yx}$'s is a multivariate PBD. Its sample space is a four dimensional space of vectors whose elements are the nonnegative integers.

In some studies, each cell of the table is linked with the duration of observation, area of observation or a similar variable. For example, in a study of occupational cancer, a group of workers of similar age are followed up. For a worker, if the malignancy of interest is diagnosed, the time to diagnosis is noted; otherwise the duration of follow up is noted. Further, the family history of the disease and work place exposure to the agent under study are determined as well. The results for each of the four levels occupation and history combination may be given in a 2×2 table.

Example 5.11: Table 5.9 shows data from a study of work exposure to phenoxy herbicides and other chemicals. Do exposure and gender affect the event occurrence? That is the question.

Table 5.9 *Exposure and Gender*

Exposure	Gender	
	Male	Female
No	(482; 8773)	(85; 1974)
Yes	(549; 13,634)	(13; 303)

Note: Cell values are (case count;person years).
Source: Hooiveld et al. (1998), Table 1.
Used with permission of Oxford University Press.

With $y, x \in \{0, 1\}$, let $\mathcal{T}_{yx}$ be the number of cases, and $\tau(y, x)$ be the total follow up time, and $\lambda(y, x)$ be the rate of disease occurrence per unit person time in cell (y, x). The expected value in the (y, x) cell is

$$\lambda_*(y, x) = \lambda(y, x)\tau(y, x) \tag{5.53}$$

We assume homogeneity of rates over time, an exponential distribution of time to the occurrence of the event of interest, and no or little loss to follow up. Then we may regard $\tau(y, x)$ as fixed entities, and posit independent Poisson variability for each cell to get

$$P[\mathcal{T}_{yx} = t_{yx}; y, x \in \{0, 1\}] = \prod_{y, x \in \{0, 1\}} \frac{e^{-\lambda_*(y, x)} \lambda_*(y, x)^{t_{yx}}}{t_{yx}!} \tag{5.54}$$

For a justification for this model, see §1.6, and Breslow and Day (1987), pages 131–135. See also Chapter 15. Under this model, we can then show that

$$\ln(\lambda_*(y, x)) = \mu_{*0} + \mu_{*1}y + \mu_{*2}x + \mu_{*12}yx \tag{5.55}$$

with the μ^*'s related to the μ's by

$$\mu_{*1} = \delta_1 + \mu_1; \qquad (5.56)$$
$$\mu_{*2} = \delta_2 + \mu_2; \qquad (5.57)$$
$$\mu_{*12} = \delta_{12} + \mu_{12} \qquad (5.58)$$

and further where

$$\delta_1 = \ln\left(\frac{\tau(1,0)}{\tau(0,0)}\right) \quad \text{and} \quad \delta_2 = \ln\left(\frac{\tau(1,1)}{\tau(0,1)}\right) \qquad (5.59)$$

$$\delta_{12} = \ln\left(\frac{\tau(1,1)\tau(0,0)}{\tau(0,1)\tau(1,0)}\right) \qquad (5.60)$$

These relations facilitate the analysis of person time data using methods developed for pure count Poisson data. For example, the second term in the rhs of the expression for μ_{*12} depends only on follow up durations. For assessing the interaction between the variables, we analyze the data as if they were simple count data, and then offset the estimate and the CI by this constant term. Testing the hypothesis of no interaction effect is done by appropriately adjusting the null value. This is a straightforward task for unconditional and conditional analysis (see below).

5.8 A Retrospective Design

In the three designs given thus far, the data were gathered in a **prospective** manner. That is, a set of conditions are observed, then the outcome(s) of interest noted. For practical or ethical reasons, such a design is at times infeasible or deemed unacceptable. For example, the effect of maternal smoking on the health of the baby is not studied by a clinical trial in which some mothers are encouraged to smoke and others are not. We instead sample healthy and unhealthy newborns and inquire into the smoking behavior of the mother during pregnancy. Subjects in such a **retrospective case-control study** are selected on the basis of the outcome; the level of exposure to the factor(s) under investigation is determined later. A number of different case-control designs are used in epidemiological investigations.

Example 5.12: We look at a basic type of retrospective study design. An occupational exposure is thought to cause a rare lung disease that takes many years to manifest. The investigators sample subjects with proven cases of the disease and a control sample from the same background group but without the disease. The level of exposure to the suspect agent is assessed with a questionnaire administered in the same fashion by trained personnel. The case and control groups are similar in age. Hypothetical data from a study with 20 cases and 50 controls are shown in Table 5.10.

Let $\mathcal{X}$ be the binary factor of interest, and $\mathcal{Y}$, the disease status. Let $\pi_*(y)$ be the probability that a subject has a positive history of smoking given that his disease status is y. A logit model for this design is

$$\ln\left(\frac{\pi_*(y)}{1-\pi_*(y)}\right) = \gamma + \delta y \qquad (5.61)$$

This model, however, does not reflect the main aim of the study, which is to analyze $\pi(x)$, the

Table 5.10 *Case-Control Data*

	Exposed		
Disease	No	Yes	Total
No	12	38	50
Yes	2	18	20
Total	14	54	70

Note: Hypothetical data.

probability that a subject with a history of smoking indicated by x will develop lung cancer. The actual model of interest is

$$\ln\left(\frac{\pi(x)}{1-\pi(x)}\right) = \alpha + \beta x \qquad (5.62)$$

The parameter of interest is β but the study data relate to δ. Fortunately, this disparity can be resolved if the following conditions hold:

- The case sample is a random sample from the population of subjects with the disease.
- The control sample is an independent random sample from the population of subjects at risk from but not having the disease.
- The cases and controls were selected independently of the exposure factor.
- The populations from which the cases and controls were selected are homogeneous in terms of other factors that affect disease occurrence.

Under these conditions, we show in Chapter 6 that

$$\beta = \delta \qquad (5.63)$$

This implies that we may analyze the data from a case-control study as if they were from a prospective study provided the analysis concerns the parameter β. The parameter α is not estimable from such a study. This result, moreover, holds for the odds ratio model only, and does not extend to either the relative or the additive risk models.

The following aspects of retrospective studies are important:

- Sampling cases and controls has to be done with a great deal of care to avoid bias. The determination of exposure status has to be done accurately and in an equivalent manner among all subjects.
- The problem of **recall bias** is a major concern. Missing, unreliable or unreliably recorded information reduces the precision of the estimates.
- Influence of other factors on disease status needs to be controlled for in either the data analysis, the study design or both.

5.9 The Inverse Sampling Design

Finally, we consider a design in which instead of fixing the total number in each group, we fix the number of subjects with a particular outcome. Suppose a rare, acute and potentially lethal form of liver damage may be induced by two drugs used on patients hospitalized with a myocardial infarct. Let π_x be the probability of **not experiencing** this effect for drug x, where $x = 0$, or $x = 1$.

To compare the drugs, we use the **inverse sampling** design. Suppose lists of the patients prescribed each drug in the past five years are available from area hospitals. For each drug, we randomly select a patient from its list, and obtain the hospital chart to determine if he or she did or did not experience liver damage. For each drug, the process is continued until a fixed number of patients with liver damage are found. Suppose the fixed number for drug $x = 0$ is d, and for $x = 1$, it is c. The chart searches are done independently. Let b and a respectively denote the numbers in the two groups who did not have liver damage.

Though the data are shown as before in the form of Table 5.1, the two internal cells (and so one part of one margin) are now set by design.

Example 5.13: The data from a hypothetical inverse sampling study with $c = d = 5$ are shown in Table 5.11.

Table 5.11 *Inverse Sampling Data*

Liver Damage	Drug		
	$x = 0$	$x = 1$	Total
Yes	5	5	10
No	53	312	365
Total	58	317	375

Note: Hypothetical data.

Let $\mathcal{A}$ and $\mathcal{B}$ be the random number of subjects sampled who did not have the adverse reaction to obtain c and d subjects with an adverse reaction from $x = 1$ and $x = 0$, respectively. Their probability distribution is a product of two negative binomial distributions:

$$P[\mathcal{A} = a, \mathcal{B} = b; \pi_0, \pi_1] =$$

$$\binom{a + c - 1}{a}\binom{b + d - 1}{b}\pi_1^a(1 - \pi_1)^c \pi_0^b(1 - \pi_0)^d \quad (5.64)$$

for $a, b = 0, 1, 2, \ldots$. Now let $\pi_0 = \theta$ and $\pi_1 = \theta\phi$. Here ϕ is the relative risk of not having the event. Then, with $\mathcal{S} = \mathcal{A} + \mathcal{B}$ and $\mathcal{T} = \mathcal{A}$, we can write

$$P[\mathcal{S} = s, \mathcal{T} = t; \theta, \phi] = \frac{c(s, t)\theta^s \phi^t}{f(\theta, \phi)} \quad (5.65)$$

where

UNCONDITIONAL ANALYSIS

$$c(s,t) = \binom{t+c-1}{t}\binom{s-t+d-1}{s-t} \tag{5.66}$$

$$f(\theta,\phi) = (1-\theta\phi)^{-c}(1-\theta)^{-d} = \sum_{u=0}^{\infty}\sum_{v=0}^{u} c(u,v)\theta^u \phi^v \tag{5.67}$$

In the product binomial design, reparameterization in terms of odds produced a PBD. In the inverse sampling design, on the other hand, a formulation in terms of the relative risk of not having the event produces a bivariate PBD.

5.10 Unconditional Analysis

Now we develop conventional large sample unconditional methods for the analysis of the designs given above. These extend the univariate likelihood based methods for a PBD given in Chapter 3. Derivations of the statistics we use are given later. These statistics are also used for exact conditional analysis.

Traditional analysis of a 2×2 table is based on the unconditional likelihood. We first look at the one margin fixed prospective design. The logarithm of the unconditional likelihood is

$$\ell = a\ln(\pi_1) + c\ln(1-\pi_1) + b\ln(\pi_0) + d\ln(1-\pi_0) \tag{5.68}$$

Stated in terms of α and β in model (5.20), it is

$$\ell = s\alpha + t\beta - m\ln(1+e^\alpha) - n\ln(1+e^{\alpha+\beta}) \tag{5.69}$$

The partial derivatives of the log-likelihood are

$$\frac{\partial \ell}{\partial \alpha} = s - \frac{me^\alpha}{1+e^\alpha} - \frac{ne^{\alpha+\beta}}{1+e^{\alpha+\beta}} = s - m\pi_0 - n\pi_1 \tag{5.70}$$

$$\frac{\partial \ell}{\partial \beta} = t - \frac{ne^{\alpha+\beta}}{1+e^{\alpha+\beta}} = t - n\pi_1 \tag{5.71}$$

Setting the derivatives to zero and solving gives

$$\hat{\alpha} = \ln\left[\frac{b}{m-b}\right] \quad \text{and} \quad \hat{\beta} = \ln\left[\frac{a(m-b)}{b(n-a)}\right] \tag{5.72}$$

Or, in terms of π's, the unconditional mle's are

$$\hat{\pi}_0 = \frac{s-t}{m} \quad \text{and} \quad \hat{\pi}_1 = \frac{t}{n} \tag{5.73}$$

****

We observe that the mle for a key parameter in each of the five designs for a 2×2 table we have considered is the same.

Theorem 5.1: The unconditional mle for β in the prospective one margin fixed design, for β_3 in the overall total fixed design, for $\mu_{1,2}$ in the nothing fixed design, and also for the log-odds ratio in the inverse sampling and case-control designs are identical and equal to $\ln(ad/bc)$.

Proof: The proof for the prospective one margin fixed design is given above; assuming the validity of the estimation of the odds ratio in the case-control design, the same proof applies. Now consider the log-likelihood for the total fixed design in terms of $\pi(i,j)$'s. This is

$$\ell = a\ln[\pi(1,1)] + b\ln[\pi(1,0)] + c\ln[\pi(0,1)] + d\ln[\pi(0,0)] \tag{5.74}$$

Note, the above probabilities are constrained to sum to 1. Taking first derivatives and solving, we get

$$\hat{\pi}(0,0) = d/n \quad \text{and} \quad \hat{\pi}(0,1) = c/N$$

$$\hat{\pi}(1,0) = b/n \quad \text{and} \quad \hat{\pi}(1,1) = a/N$$

We get the desired mle for β_3 by substituting these estimates in expression (5.39). The proofs for the other designs are similar and left to the reader. For the case-control design, we also need to question the validity of an unconditional analysis, and invoke the results showing its validity for the estimation of the odds ratio in a conditional context. □

****

Returning to the prospective one margin fixed design, we note that the mles given by (5.72) and (5.73) are the **unrestricted mles**. In contrast, the **restricted mle** of α for an arbitrary β obtains by solving the equation

$$\frac{\partial \ell}{\partial \alpha} = s - \frac{me^\alpha}{1+e^\alpha} - \frac{ne^{\alpha+\beta}}{1+e^{\alpha+\beta}} = 0 \tag{5.75}$$

Denote the restricted mle by $\hat{\alpha}_r(\beta)$. This is obtained by solving the following quadratic equation (stated in terms of θ and ϕ):

$$(m+n-s)\phi\theta^2 + [m-s+(n-s)\phi]\theta - s = 0 \tag{5.76}$$

In particular, when $\beta = 0$,

$$\frac{\partial \ell}{\partial \alpha} = 0 \rightarrow \frac{e^\alpha}{1+e^\alpha} = s/N \rightarrow \hat{\alpha}_r = \ln\left(\frac{s}{N-s}\right) \tag{5.77}$$

The restricted mle of α is required for the construction of some test statistics for β. Restricted mles for the nuisance parameters in the other designs and models are defined similarly though their computation may be more involved.

****

The three commonly used test statistics for inference on a 2×2 table are as follows. (Derivations

UNCONDITIONAL ANALYSIS

of the statistics are in §5.14.) Again, focusing on the one margin fixed prospective design, the unconditional score statistic for β is

$$\mathcal{W}_u(\beta) = (t - n\pi_1)^2 \left\{ \frac{1}{m\pi_0(1-\pi_0)} + \frac{1}{n\pi_1(1-\pi_1)} \right\} \tag{5.78}$$

when π_0 and π_1 are evaluated at the restricted mle α obtained by solving (5.67). This is the **Wilson score statistic**. At $\beta = 0$, the statistic is obtained by substituting $\pi_0 = \pi_1 = \mathcal{S}/(m+n)$ in (5.78). After simplification, this becomes

$$\mathcal{W}_u = \frac{[\mathcal{A}(m-\mathcal{B}) - \mathcal{B}(n-\mathcal{A})]^2 N}{mn\mathcal{S}(N-\mathcal{S})} \tag{5.79}$$

This is the well known chisquare statistic for a 2×2 table.

****

The unconditional LR statistic for β is

$$0.5\mathcal{Q}_u(\beta) = \mathcal{S}(\hat{\alpha} - \hat{\alpha}_r) + \mathcal{T}(\hat{\beta} - \beta) + m\ln\left(\frac{1+e^{\hat{\alpha}_r}}{1+e^{\hat{\alpha}}}\right) + n\ln\left(\frac{1+e^{\hat{\alpha}_r+\beta}}{1+e^{\hat{\alpha}+\hat{\beta}}}\right) \tag{5.80}$$

When $\beta = 0$, this becomes

$$\mathcal{Q}_u/2 = \mathcal{A}\ln\left(\frac{\mathcal{A}N}{\mathcal{S}n}\right) + (n-\mathcal{A})\ln\left(\frac{(n-\mathcal{A})N}{(N-\mathcal{S})n}\right) + \\ \mathcal{B}\ln\left(\frac{\mathcal{B}N}{\mathcal{S}m}\right) + (m-\mathcal{B})\ln\left(\frac{(m-\mathcal{B})N}{(N-\mathcal{S})m}\right) \tag{5.81}$$

Note that in the above formula, we set $0\ln(0) = 0$.

****

Finally, the Wald statistic for β is

$$\mathcal{Z}_u(\beta) = \frac{(\hat{\beta} - \beta)^2}{\sigma^2(\hat{\beta})} \tag{5.82}$$

where $\sigma^2(\hat{\beta})$ is the large sample variance of $\hat{\beta}$ given by

$$\sigma^2(\hat{\beta}) = \left\{ \frac{1}{m\pi_0(1-\pi_0)} + \frac{1}{n\pi_1(1-\pi_1)} \right\} \tag{5.83}$$

The Wald statistic used in practice estimates this variance by replacing the parameters with their unconditional mles. This approximation is

$$\hat{\sigma}^2(\hat{\beta}) = \left\{ \frac{1}{b} + \frac{1}{m-b} + \frac{1}{a} + \frac{1}{n-a} \right\} \tag{5.84}$$

****

Large sample unconditional analysis of 2×2 tables from three of the five designs we have considered is considerably simplified by the following result.

Theorem 5.2: The unconditional score, LR and Wald statistics for the parameter β_3 in the total fixed design and for μ_{12} in the nothing fixed design at the null value 0 are the same as the corresponding statistics for $\beta = 0$ in the one margin fixed design.

Proof: A part of the proof appears in §5.14 below. The remainder is left to the reader. □

The analyses for a 2×2 table from these designs is facilitated by the following additional results:

- The statistics $\mathcal{W}_u$, $\mathcal{Q}_u$ and $\mathcal{Z}_u$ follow an approximately chisquare distribution with one df when the sample size is large. This is used to compute asymptotic p-values for various hypotheses for the parameters β, β_3 or μ_{12}.
- We construct unconditional large sample evidence functions for the above parameters in the usual way. For example, if w is the observed value of the unconditional score statistic for analyzing the parameter β in the one margin fixed prospective design, then the two-sided **unconditional score statistic based evidence function** is

$$\mathrm{E}(t; \beta) = \mathrm{P}[\mathcal{W} \geq w; \beta] \tag{5.85}$$

For the Wald statistic, this is a straightforward task. At null values other than zero, the statistics from the different designs are not necessarily equivalent and may not yield identical CIs for related parameters from the different designs.
- The null expected value for a cell is the product of the two marginal totals divided by the overall total. The null unconditional score and LR statistics then takes the general forms for the score and LR statistics given in §3.6 and §3.8 respectively.
- The Wald statistic depends on how the model is parameterized. The form of the statistic above applies when the model is analyzed in terms of log-odds.

In sum, as long as the focus is on testing whether the parameters β, β_3 or μ_{12} are zero, the same analysis method applies. For some forms of large sample unconditional analysis, we can ignore how the data actually arose. This is a useful observation for application. But its limitation should also be borne in mind; for example, even for these parameters, CIs under the Wilson score method are not necessarily equivalent. (This is issue will be pursued in Chapter 15.)

****

Inverse sampling data, however, differ in some respect from the above scheme. Here, the unconditional likelihood is identical to that from the product binomial design. It thus yields the same unconditional mles, and the same unconditional LR statistic to compare π_1 and π_0 as that given earlier. That however is not the case with score, or the various forms of the Wald statistic. The null score statistic to test $\pi_1 = \pi_0$ for the inverse sampling data, for example, is

$$\mathcal{W}_u = \frac{(\mathcal{A}d - \mathcal{B}c)^2 (c+d)}{N(\mathcal{A} + \mathcal{B})cd} \tag{5.86}$$

which is distinct from (5.79). We leave it to the reader to show the differences in relation to the Wald statistics.

5.11 Conditional Analysis

For conditional analysis, we eliminate nuisance parameters by fixing the values of their sufficient statistics. As the distributions in the five designs are PBDs when reparametrized appropriately, such statistics exist and are readily identified. Conditional analysis is available if, for example, we need to analyze the parameter β in model (5.19) or (5.62), β_3 in (5.33), μ_{12} in (5.49) or ϕ in (5.65). This is facilitated by the fact that the conditional distribution for the relevant sufficient statistic for the first four of these parameters and designs is the same.

Theorem 5.3: With reference to Table 5.1, let

$$\mathcal{T} = \mathcal{A} \quad \text{and} \quad t = a$$

Further, let BMF stand for "Both Margins Fixed" in the 2×2 table. Then, for the prospective and retrospective one margin fixed, the overall total fixed and the nothing fixed designs,

$$P[\mathcal{T} = t \mid \text{BMF}] = \frac{c(t)\phi^t}{f(\phi)} \tag{5.87}$$

where

$$f(\phi) = \sum_{v \in \Omega} c(v)\phi^v, \quad c(t) = \binom{n}{t}\binom{N-n}{s-t} \tag{5.88}$$

and

$$\Omega = \{v : \max(0, n+s-N) \leq v \leq \min(s, n)\} \tag{5.89}$$

and further where

$$\begin{aligned}
\phi &= \exp(\beta) &&: \text{for the one margin fixed designs} \\
\phi &= \exp(\beta_3) &&: \text{for the total fixed design} \\
\phi &= \exp(\mu_{12}) &&: \text{for the nothing fixed design}
\end{aligned}$$

Proof: (i) As the hypergeometric distribution (5.87) arises from conditioning on the total number of successes in a two independent binomials model (see §1.8), this proves the result for the one margin fixed design. (ii) For the total fixed design, Property 2 of the multinomial distribution in §1.10 shows that if we condition on one (row or column) margin for the data from this design, we get a product binomial model. Upon further conditioning on the remaining margin, we get the required hypergeometric distribution. (iii) For the nothing fixed design, invoking Property 3 of §1.10 means that if we first condition on the overall total here, we get a multinomial. Conditioning further on the row and then the column margin, we get the hypergeometric distribution. The detailed write up of this proof is left to the reader. □

This result has interesting implications:

- In the four models (5.19), (5.62), (5.33) or (5.49), associated designs and the noted parameters, conditioning on both margins gives a common univariate distribution that includes only the parameter of interest.

- This distribution is a univariate PBD. So we may apply methods of Chapters 2 and 3 to analyze it. As for the two group Poisson data, this is now a conditional analysis.
- However, the entities being fixed in each design are NOT ancillary statistics for the parameter of interest.

****

Example 5.7 (continued): To illustrate conditional analysis, consider the hypothetical data in Table 5.3 in which $n = m = s = 3$, and $\mathcal{T} = 1$. Extracting the relevant information from the unconditional sample space in Table 5.4, we have

$$f(\phi) = 1 + 9\phi + 9\phi^2 + \phi^3$$

The conditional mue and mle for ϕ obtain respectively from

$$1 = 9\phi^2 + \phi^3$$

and

$$9 + 18\phi + 3\phi^2 = 1 + 9\phi + 9\phi^2 + \phi^3$$

Solving these, we get that

$$\hat{\phi}_{\text{mue}} = 0.327 \quad \text{and} \quad \hat{\phi}_{\text{mle}} = 0.322$$

or

$$\hat{\beta}_{\text{mue}} = -1.116 \quad \text{and} \quad \hat{\beta}_{\text{mle}} = -1.133$$

Exact TST conventional and mid-p-values for testing $\phi = 1$ versus $\phi \neq 1$ then are 1.0 and 0.55, respectively. With $\mathcal{T} = 1$ as the observed value, the TST mid-p evidence function is

$$\mathsf{E}(1; \phi) = \frac{2\min\{1, (9\phi^2 + \phi^3)\} + 9\phi}{1 + 9\phi + 9\phi^2 + \phi^3}$$

The mue is at the apex of the evidence function (not shown). We get the 90% TST mid-p CI for ϕ by drawing a horizontal line at 0.10. This interval is $(0.01, 5.73)$ with $(-4.42, 1.75)$, the corresponding CI for β, the log-odds ratio.

For a 2×2 table of counts, the exact TST test is called the **Fisher exact test** while the exact probability based test is the **Irwin exact test**. For a comparative perspective, see Chapter 7.

****

The test statistics for analyzing a univariate PBD constructed in Chapter 3 also apply in the analysis of the conditional univariate hypergeometric distribution (5.87). In the present context, they are called **conditional** statistics. The conditional score, LR and Wald statistics for conditional analyses of a 2×2 table then are, respectively:

$$\mathcal{W} = (\mathcal{T} - \mu)^2 \sigma^{-2} \tag{5.90}$$

$$\mathcal{Q} = 2\{(\hat{\beta}_c - \beta)\mathcal{T} - \ln[h(\hat{\beta}_c)] + \ln[h(\beta)]\} \tag{5.91}$$

$$\mathcal{Z} = (\hat{\beta}_c - \beta)^2 \sigma^2 \tag{5.92}$$

CONDITIONAL ANALYSIS

where μ and σ^2 are the conditional mean and variance of the $\mathcal{T}$ and $\hat{\beta}_c$ is the conditional mle of β. For $\beta = 0$, we use the mean and variance of the central hypergeometric distribution noted in §1.7. Using these we get that the conditional score statistic at $\beta = 0$ is

$$\mathcal{W} = \frac{[\mathcal{T}(m-b) - b(n-\mathcal{T})]^2(N-1)}{mns(N-s)} \quad (5.93)$$

$\mathcal{W}$ is called the **scaled chisquare statistic** because of its relation to the unconditional score (chisquare) statistic. It is also called the **Mantel–Haenszel statistic**. The formulae for $\mathcal{Q}$ and $\mathcal{Z}$ at $\beta = 0$ are determined similarly. The probability and CT statistics of §3.9 also apply here. Exact (conditional) evidence functions from all of them are constructed in the usual way.

****

When the sample size is large, these three conditional statistics follow a chisquare distribution with one df. This is used to construct asymptotic (conditional) evidence functions, p-values, and CIs for β. If w is the observed value of the CS statistic then the two-sided **conditional score statistic based evidence function** is

$$\mathsf{E}(w; \beta) = \mathrm{P}[\mathcal{W} \geq w \mid \mathcal{S} = s; \beta] \quad (5.94)$$

The large sample $(1-\alpha_*)100\%$ conditional confidence limits for β are then the solutions of the equation

$$(t-\mu)^2 = \sigma^2 \chi^2_{1-\alpha_*} \quad (5.95)$$

The point estimate corresponding to this evidence function is the **conditional maximum likelihood estimate** or **cmle**. In terms of ϕ, it is also the solution of

$$\sum_{u \in \Omega} c(u)\phi^u = tf(\phi) \quad (5.96)$$

Large sample analyses with the conditional LR and conditional Wald statistics are done along similar lines.

If a boundary point (a zero or maximal cell value) is observed, then we report a lower or an upper confidence bound. Dewar and Armstrong (1992) give formulae for approximate confidence limits when one cell is zero.

****

Example 5.8 (continued): For illustration, we analyze the HCG data in Table 5.5. Table 5.12 gives the conditional sample space and the hypergeometric coefficients for the data.

Figure 5.1 shows the TST mid-p evidence function for these data. The asymptotic conditional score (CS) evidence function is almost identical, and is not shown. p-values, estimates and 90% CIs for β are also directly computed using the numeric methods of Chapter 4. The results, shown in Table 5.13, indicate a close agreement between the three methods. All support the hypothesis that low HCG levels improve the cure rate. The evidence functions are located, for the most part, to the left of 0.0. They, and the CIs are somewhat wide, indicating that the precision in estimation of the magnitude of β is not high.

Table 5.12 *Sample Points and Coefficients for HCG Data*

	t	$c(83, t)$	t	$c(83, t)$
	11	$0.214742E + 08$	20	$0.264540E + 18$
	12	$0.219037E + 10$	21	$0.634896E + 18$
	13	$0.957022E + 11$	22	$0.113862E + 19$
	14	$0.239255E + 13$	23	$0.150990E + 19$
	15	$0.385201E + 14$	24	$0.145183E + 19$
	16	$0.425647E + 15$	25	$0.978948E + 18$
	17	$0.335510E + 16$	26	$0.436762E + 18$
$\Rightarrow$	18	$0.193318E + 17$	27	$0.115257E + 18$
	19	$0.826688E + 17$	28	$0.135596E + 17$

Note: $\Rightarrow$ indicates the observed value of $\mathcal{T}$.

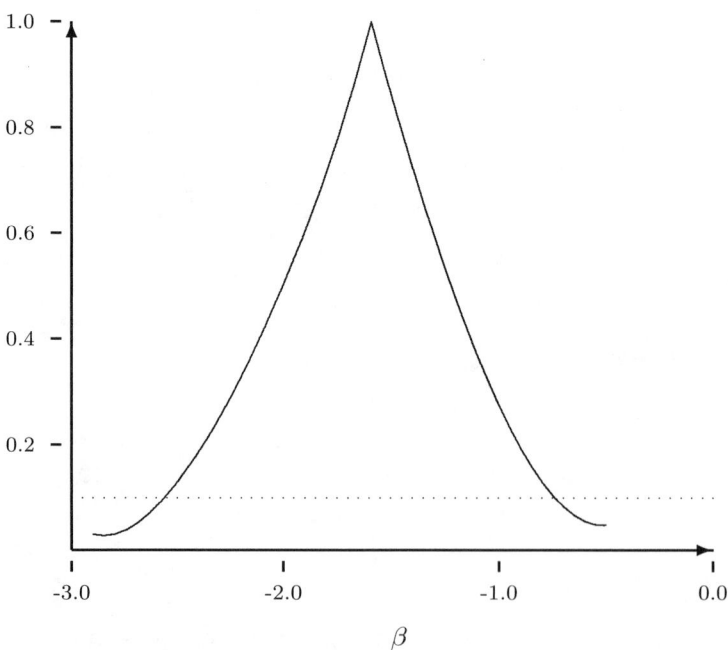

Figure 5.1 *Mid-p Evidence Function for HCG data.*

****

Now consider conditional analysis for the inverse sampling design under the model (5.65). As before, let $\mathcal{T} = \mathcal{A}$, and $\mathcal{S} = \mathcal{A} + \mathcal{B}$. Then, with $\pi_1 = \pi_0 \phi$, we can show that

CONDITIONAL ANALYSIS

Table 5.13 *Analysis of HCG Data*

Statistic	p-value	Estimate	90% CI
TST Exact	0.0070	−1.61	(−2.71, −0.58)
TST mid-p	0.0041	−1.61	(−2.58, −0.69)
CS Asymptotic	0.0020	−1.62	(−2.52, −0.72)

Note: These results relate to β.

$$P[\mathcal{T} = t \mid \mathcal{S} = s] = \frac{c(t)\phi^t}{f(\phi)} \qquad (5.97)$$

where

$$\phi = \frac{\pi_1}{\pi_0} \qquad (5.98)$$

$$f(\phi) = \sum_{v \in \Omega} c(v)\phi^v \qquad (5.99)$$

$$c(t) = \binom{t+c-1}{t}\binom{s-t+d-1}{s-t} \qquad (5.100)$$

and further where

$$\Omega = \{v : 0 \leq v \leq s\} \qquad (5.101)$$

ϕ is the **relative risk** of nonoccurrence of the event in question. The following remarks for this conditional distribution are in order.

- Conditioning on the total number of successes, $\mathcal{S}$, in this design gives a distribution which contains only the parameter ϕ.
- Distribution (5.97) is a PBD but it is NOT a hypergeometric distribution.
- Unlike the one margin fixed design, the parameter ϕ here is NOT the odds ratio. Rather, it is the relative risk of not contracting the adverse effect.

The conditional mean and variance of distribution (5.97) at $\phi = 1$ are respectively

$$\mu = \frac{cs}{c+d} \quad \text{and} \quad \sigma^2 = \frac{cdsN}{(c+d)^2(c+d+1)} \qquad (5.102)$$

Hence the conditional score statistic for testing $\phi = 1$ is

$$\mathcal{W} = \frac{[\mathcal{A}d - (s-\mathcal{T})c]^2(c+d+1)}{cdsN} \qquad (5.103)$$

This is different from the null conditional score statistic for the other four designs given earlier.

The conditional LR and Wald statistics are also different. Conditional analysis (exact or asymptotic) of data from the inverse sampling design is different from that for data from the other four 2×2 designs. The parameter of interest is distinct, the conditional distribution is different and the relevant statistics are also not the same.

****

Example 5.14: Consider a data set with $a = 14$, $b = 10$, $c = 1$ and $d = 5$. We analyze these data in two ways. First, assume that they are from a regular design (i.e., one of the other four designs for a 2×2 table given above). Then, assume that they arose from an inverse sampling design. Focusing on the *p*-values, the results of the two forms of analyses appear in Table 5.14.

Table 5.14 *Sampling Design and p-values*

	Statistic		*p*-value	
Type	Form	Name	Regular	Inverse
Unconditional	Asymptotic	Score	0.0678	0.0143
Unconditional	Asymptotic	LR	0.0583	0.0583
Conditional	Asymptotic	Score	0.0725	0.0081
Conditional	Exact	TST	0.1686	0.0506
Conditional	Mid-*p*	TST	0.0927	0.0421

Note: $a = 14$, $b = 10$, $c = 1$, $d = 5$.

The unconditional analyses in Table 5.14 relate to the odds ratio formulation for both designs. The LR *p*-values are the same but the asymptotic score values are markedly apart. Conditional analyses under inverse sampling relate to the relative risk while for the regular designs, it concerns the odds ratio. The former has smaller *p*-values compared to their counterparts for the other designs. A big difference is in terms of the conditional sample spaces. Under the former, there are seven 2×2 permissible conditional tables. The inverse sampling design, on the other hand, has thirty permissible tables in the conditional sample space.

The inverse sampling design requires a well defined sampling frame and the feasibility of evaluating each unit in an unbiased manner before selecting the next one. These conditions are hard to meet in biomedical investigations, though there are other settings where it can be usefully employed. More papers on analytical methods for it have appeared recently, and empirical studies to evaluate the statistical indices for it are emerging as well. These may facilitate its greater adoption in practice.

Especially for rare conditions, and even otherwise, usually the interest is on the relative risk of the occurrence of the event of interest. As this is not avaliable with conditional analysis of the inverse sampling design, as formulated above, we may prefer an unconditional method of analysis for this purpose.

5.12 Comparing Two Rates

This section furthers illustrate unconditional and conditional analysis using two group person time data.

COMPARING TWO RATES

Example 5.15: In a clinical trial for testing a drug for reducing the risk of breast cancer among women at high risk for the disease, 6707 women were allocated to the Tamoxifen group and 6681, to the placebo group (Fisher et al. 1998). The treatment group had 89 cases of invasive breast cancer with 26,247 person years of follow up, and in the placebo group, with 26,154 person years of follow up, 175 such cases were observed. At the outset, the two randomized groups were comparable in terms of other risk factors for the condition. What do we conclude?

With $i = 0$ indexing the control group, and $i = 1$, the study group, let the true rate per unit person time of the incidence of invasive cancer in group i be λ_i. Then we test

$$H_0 : \lambda_0 = \lambda_1 \quad \text{versus} \quad H_1 : \lambda_0 \neq \lambda_1$$

and estimate the magnitude of the disparity between the true rates. The disparity between the rates is usually gauged by one of two indices:

$$\phi = \frac{\lambda_1}{\lambda_0} \quad \text{or} \quad \delta = \lambda_1 - \lambda_0 \quad (5.104)$$

As stated before, ϕ is the rate ratio, and δ is the rate difference.

Now, let t_i and τ_i respectively be the total number of incident cases and total person time of observation for group i. We assume T_i is Poisson distributed with mean $\lambda_i \tau_i$, and that T_0 and T_1 are independent variates. In terms of the rate ratio, we write $\lambda_0 = \lambda$ and $\lambda_1 = \phi\lambda$. So, the equivalent problem is to test

$$H_0 : \phi = 1 \quad \text{versus} \quad H_1 : \phi \neq 1$$

and to estimate ϕ. The joint distribution of the variables, namely, $P[T_0 = t_0, T_1 = t_1; \lambda, \phi]$, is

$$\frac{\exp[-\lambda(\tau_0 + \phi\tau_1)](\lambda\tau_0)^{(t_0+t_1)}(\phi\tau_1)^{t_1}}{t_0!\ t_1!} \quad (5.105)$$

To a constant, the logarithm of the unconditional likelihood is

$$\ell_u(\lambda, \phi) = -\lambda(\tau_0 + \phi\tau_1) + (t_0 + t_1)\log(\lambda) + t_1 \log(\phi) \quad (5.106)$$

The unconditional mle for ϕ is

$$\hat{\phi}_u = \frac{t_1/\tau_1}{t_0/\tau_0} \quad (5.107)$$

and the unconditional score statistic for ϕ is

$$W_u = \frac{(T_1 - T_0\tau\phi)^2}{(T_1 + T_0)\phi\tau} \quad (5.108)$$

This has an approximate chisquare distribution with one df.

****

The conditional method notes that $T_0 + T_1$ is sufficient for λ and fixes its value at $n = t_0 + t_1$. Then the conditional distribution of T_1 is the binomial distribution $B(n; \pi)$, where

$$\pi = \frac{\lambda_1 \tau_1}{\lambda_0 \tau_0 + \lambda_1 \tau_1} = \frac{\phi \tau}{1 + \phi \tau} \tag{5.109}$$

and further where $\tau = \tau_1/\tau_0$. After a little algebra, we get that for $t = 0, 1 \cdots, n$,

$$P[T_1 = t \mid T_0 + T_1 = n] = \frac{c(t) \phi_*^t}{(1 + \phi_*)^n} \tag{5.110}$$

where $c(t) = n!/\{t!(n-t)!\}$, and $\phi_* = \phi \tau$. This is a univariate PBD. Then the cmle for ϕ is

$$\hat{\phi}_c = \frac{t_1/\tau_1}{t_0/\tau_0} \tag{5.111}$$

and the CS statistic is

$$W_c = \left\{T_1 - \frac{n\phi\tau}{1+\phi\tau}\right\}^2 \left\{\frac{n\phi\tau}{(1+\phi\tau)^2}\right\}^{-1} = \frac{[T_1 - (n-T_1)\tau\phi]^2}{n\phi\tau} \tag{5.112}$$

****

Example 5.15 (continued): Applying this to the Tamoxifen study, we have

$$\tau_0 = 26,154; \quad t_0 = 175; \quad \tau_1 = 26,247; \quad t_1 = 89$$

Then

$$\tau = \frac{26,154}{26,247} = 1.004; \quad \text{and} \quad n = 89 + 175 = 264$$

The cmle for ϕ is

$$\hat{\phi}_c = \frac{89 \times 26154}{175 \times 26247} = 0.507$$

At $\phi = 1$, the value of the CS statistic is

$$W_c = \frac{(89 - 132.26)^2}{66.00} = 28.35$$

Using the one df chisquare distribution, the asymptotic p-value is less than 0.0001. Approximate 95% confidence limits for ϕ obtain from solving the equation

$$[t_1 - (n - t_1)\phi_*]^2 = (1.96)^2 n\phi_*$$

which in the Tamoxifen study becomes

$$\phi_*^2 - 1.0502\phi_* + 0.2586 = 0$$

Solving and dividing by $\tau = 1.004$, the asymptotic CS 95% CI for ϕ is $(0.392, 0.653)$. In contrast to this asymptotic analysis, we could have done an exact conditional analysis; given the numbers, the two results would have been (and are) almost identical.

****

Now consider the issue in terms of the rate difference. Let $\lambda_0 = \lambda$ and $\lambda_1 = \lambda + \delta$. The problem is then to test

$$H_0 : \delta = 0 \quad \text{versus} \quad H_1 : \delta \neq 0$$

and to estimate δ. The distribution of interest, $P[\mathcal{T}_0 = t_0, \mathcal{T}_1 = t_1; \lambda, \delta]$, is

$$\frac{\exp[-\lambda(\tau_0 + \tau_1) - \delta\tau_1](\lambda\tau_0)^{t_0}[(\lambda + \delta)\tau_1]^{t_1}}{t_0!\, t_1!} \qquad (5.113)$$

To a constant, the unconditional likelihood is

$$l_u(\lambda, \delta) = -\lambda(\tau_0 + \tau_1) - \delta\tau_1 + t_0 \log(\lambda) + t_1 \log(\lambda + \delta) \qquad (5.114)$$

The unconditional mle of δ from this is

$$\hat{\delta}_u = \frac{t_1}{\tau_1} - \frac{t_0}{\tau_0} \qquad (5.115)$$

The unconditional score statistic for testing $\delta = 0$ is the same as that for testing $\phi = 1$ in the rate ratio model, namely,

$$W_u = \frac{(\mathcal{T}_1 - \mathcal{T}_0\tau)^2}{(\mathcal{T}_1 + \mathcal{T}_0)\tau} \qquad (5.116)$$

In the rate difference model, as noted previously, we cannot remove the nuisance parameter with conditioning. The general score unconditional statistics for the two models are also not equivalent.

5.13 Points to Ponder

When we analyze ϕ, a parameter of interest, in the presence of θ, a nuisance parameter, four types of situations, noted in §5.2, can occur. A statistic we plan to ignore, or fix in the analysis of ϕ, may be (i) unrelated to ϕ; (ii) ancillary to ϕ and sufficient for θ; (iii) sufficient for θ but not ancillary for ϕ; or (iv) not sufficient for θ, and thus not an ancillary statistic for ϕ.

Conditional analysis is generally seen as a reasonable approach in the first two situations. In the last case, it is not a viable option. The third situation is, thereby, the one fraught with controversy. And, it is also the one most of the discrete data models used in practice fall into. In the five designs for the 2×2 table, and the Poisson designs above, the statistic(s) sufficient for the nuisance parameter(s) is (are) not ancillary for the parameter of interest. Conditional analysis then needs a clear empirical and conceptual justification. In the literature, arguments supporting such an analysis but given in terms of the first two situations are at times uncritically extended to the third situation.

Historically the 2×2 table has been a favored vehicle for airing controversies on statistical inference. They have debated the validity of conditional analysis, the relation of design to data analysis, the merits of exact methods and tail areas for inference, the properties of various indices of inference, and the methods of evaluating such indices. We discuss these issues more fully in Chapter 7 and Chapter 15. For now, two points are apropos: (i) Mostly, when conditioning is discussed, it is done in relation to exact tests. Yet, it as much pertains to large sample tests. Arguments for or against exact methods are thus not necessarily arguments for or against conditioning; and (ii) An exclusive focus on the 2×2 table may mask the broader nature of the issues at hand.

Our study of the models and design in this chapter informs us that there are some problems in which conditional and unconditional analysis give identical results and there are other problems in which that is not so.

To give some examples:

- The conditional and unconditional mles of the relative risk for the two group person time data design are identical. But for the two binomials design, the conditional and unconditional mles of the odds ratio are not always the same.
- The conditional and unconditional score statistics for testing if the relative risk equals one are identical to each other for the case of two group Poisson data. But for the two binomials design, the conditional and unconditional score statistic for testing if the odds ratio equals one differ by a factor based on the total sample size.
- For the nothing fixed design, total fixed design and one margin fixed design, the unconditional score and LR statistics for testing whether the logit parameter embodying the association between the variable equal the usual null value are respectively equivalent.
- Full marginal conditioning on logit models for these three designs gives the same univariate PBD. But this does not extend to other models like the additive or relative risk model.
- For the inverse sampling design, the foregoing observation holds only for the relative risk model, and not for the logit model.

Such findings, at the minimum, tell us that the issue of the relevance of the study design to the data analytic technique and of conditioning are complex matters that should not be reduced to simple pronouncements. A careful elucidation for each class of designs, the models in question and the form of data analysis being considered needs to be undertaken.

****

Another complicating aspect for discrete data analysis is the variety of unconditional and conditional exact and asymptotic sample p-values, confidence intervals and evidence functions that are available. For example, Upton (1982) noted twenty two methods of computing p-values for a 2×2 table. From papers published since then, one may essentially double this number! Which one should an applied statistician use? And when?

The features and properties of evidence functions have to be evaluated when we need to select the methods of inference. Functions that give indices of inference (estimate, p-value and confidence interval) with good performance over a broad range of settings, and also have adequate properties for the problem at hand, are preferred. We deal with what is meant by 'good performance' in Chapter 7. For now, we note that for the analysis of a logit model parameter, the mid-p and conditional score methods perform well in many settings. Accordingly, in accordance with the reporting guide of Chapter 3, we recommend the following sets of items for the analysis of a parameter in such models for 2×2 tables, person time data and other higher dimensional discrete data.

- The conditional mle, the asymptotic conditional score *p*-value and asymptotic conditional score confidence limits.

- The conditional mue, the conditional mid-*p*-value and conditional mid-*p* confidence limits.

- The conditional mue, conditional exact TST *p*-value and exact TST conditional confidence limits, when guaranteed actual coverage and significance levels are needed.

Recent studies indicate that the conditional combined tails (CT) method may be a better alternative to the exact TST method. But more studies are needed before this can be firmly recommended.

The unconditional mle is the most frequently reported estimate for the odds ratio in a 2×2 table. Its advantage over the conditional mle (or the conditional mue) is that the latter can be zero or infinite while the former is a finite number.

****

Yates (1934) gave an adjustment to the chisquare test for a 2×2 table to give asymptotic *p*-values closer to that obtained from exact tests. Such an adjustment is known as a **continuity correction**. Many corrections have been proposed and a lively debate about their use has taken place. In this age of high speed computing, we concur with Agresti (1996) that "*[t]here is no longer any reason to use this [or similar] approximation*" (page 43). Sahai and Khurshid (1996), page 22, discourage their use for CIs, and give other references on this topic. An applied statistician today, in our view, may regard such corrections as interesting historic curiosities, though Upton (1992) has a different opinion.

Many adjusted estimates for the odds ratio are also available. But we do not recommend them. They perform well in selected settings, but make a marginal difference, and encourage an ad hoc approach to inference. Not only are they not in line with the integrated evidence function based approach to data analysis but, for a practitioner, they mostly add to the existing confusion. Their use is not in line with the sentiments expressed above about continuity corrections.

5.14 Derivation of Test Statistics

This section outlines the derivation of the unconditional score, LR and Wald statistics for the prospective product binomial design. They can be extended to other designs for count data and for the two group person time data readily. These derivations will be of use in constructing test statistics for more complex forms of discrete data.

Consider model (5.19) for the one margin fixed design. (5.69) gives its likelihood. The derivatives of the likelihood are in (5.70) and (5.71); the unrestricted mles in (5.71) and the restricted mle for α are obtained from (5.75) and (5.76). The second partial derivatives of the log-likelihood are

$$\frac{\partial^2 \ell}{\partial \alpha^2} = -\frac{me^\alpha}{(1+e^\alpha)^2} - \frac{ne^{\alpha+\beta}}{(1+e^{\alpha+\beta})^2} \tag{5.117}$$

$$= -m\pi_0(1-\pi_0) - n\pi_1(1-\pi_1) \tag{5.118}$$

$$\frac{\partial^2 \ell}{\partial \beta^2} = -\frac{ne^{\alpha+\beta}}{(1+e^{\alpha+\beta})^2} = -n\pi_1(1-\pi_1) \tag{5.119}$$

$$\frac{\partial^2 \ell}{\partial \alpha \partial \beta} = -\frac{ne^{\alpha+\beta}}{(1+e^{\alpha+\beta})^2} = -n\pi_1(1-\pi_1) \tag{5.120}$$

The **information matrix** is defined as

$$\mathbf{I} = -\mathbf{E}\begin{bmatrix} \dfrac{\partial^2 \ell}{\partial \alpha^2} & \dfrac{\partial^2 \ell}{\partial \alpha \partial \beta} \\ \dfrac{\partial^2 \ell}{\partial \alpha \partial \beta} & \dfrac{\partial^2 \ell}{\partial \beta^2} \end{bmatrix} \tag{5.121}$$

$$= \begin{bmatrix} m\pi_0(1-\pi_0) & 0 \\ 0 & 0 \end{bmatrix} + n\pi_1(1-\pi_1)\begin{bmatrix} 1 & 1 \\ 1 & 1 \end{bmatrix} \tag{5.122}$$

The inverse of the information matrix is called the **variance covariance matrix**. This is

$$\mathbf{\Sigma} = \mathbf{I}^{-1} = \begin{bmatrix} \sigma_{00}^2 & \sigma_{01}^2 \\ \sigma_{10}^2 & \sigma_{11}^2 \end{bmatrix} \tag{5.123}$$

where

$$\sigma_{00}^2 = -\sigma_{01}^2 = -\sigma_{10}^2 = \frac{1}{m\pi_0(1-\pi_0)} \tag{5.124}$$

and

$$\sigma_{11}^2 = \left\{\frac{1}{m\pi_0(1-\pi_0)} + \frac{1}{n\pi_1(1-\pi_1)}\right\} \tag{5.125}$$

The score vector is the vector of first partial derivatives. Thus

$$\mathbf{D} = \begin{pmatrix} \dfrac{\partial \ell}{\partial \alpha} \\ \dfrac{\partial \ell}{\partial \beta} \end{pmatrix} = \begin{pmatrix} s - m\pi_0 - n\pi_1 \\ t - n\pi_1 \end{pmatrix} \tag{5.126}$$

The unconditional score statistic is defined by the relation

$$\mathcal{W}_u = \mathbf{D}'\mathbf{\Sigma}\mathbf{D} \tag{5.127}$$

computed at the restricted mle for α. Under this restriction, the first element of $\mathbf{D}$ equals zero. Upon appropriate substitution, and computing the matrix product, we get

$$\mathcal{W}_u = \left\{ \frac{1}{m\pi_0(1-\pi_0)} + \frac{1}{n\pi_1(1-\pi_1)} \right\}(t - n\pi_1)^2 \tag{5.128}$$

when π_0 and π_1 are evaluated at the restricted mle for α. For a general β, this requires the solution of the quadratic equation (5.76). At $\beta = 0$, the score statistic is given by substituting $\pi_0 = \pi_1 = s/(m+n)$, giving the expression (5.79) given earlier.

****

The LR statistic for β is twice the difference of the log-likelihood evaluated at the unrestricted mle for α and the log-likelihood at its restricted mle.

$$\mathcal{Q}_u = 2\left(\ell(\hat{\alpha}, \hat{\beta}) - \ell(\hat{\alpha}_r, \beta)\right) \tag{5.129}$$

where $\hat{\alpha}_r = \hat{\alpha}_r(\beta)$. Here too in general we need to solve the quadratic equation (5.76). For the null value, $\beta = 0$, upon substitution in the log-likelihood (5.19), this becomes the expression (5.81) given earlier.

The Wald statistic, defined in (5.82), does not need the restricted mle. For it, $\hat{\sigma}^2(\hat{\beta})$ is the $(2,2)$ element of the matrix $\mathbf{\Sigma}$, that is σ_{11}^2, evaluated at the unconditional mles for the two parameters. This leads to expressions (5.83) and (5.84). Note that the Wald statistic is not unique in that it depends on whether the model is parameterized in terms of odds ratio, log-odds ratio or simply probabilities.

- The derivation of such unconditional statistics for the additive and multiplicative risk models for the one margin fixed prospective count data follows similar principles.
- The derivation of these statistics for two group inverse sampling and person time data for additive and multiplicative risk models follows the same outline as well.
- For the total and nothing fixed designs for 2×2 table data, the derivation of such unconditional statistics involves three and four dimensional matrices.
- Unconditional statistics are also used in conditional analysis. For example, we compute an exact conditional *p*-value for a 2×2 table using the chisquare statistic even though the statistic is derived from the unconditional likelihood.
- See Exercise 5.19 for a scheme for computing unconditional score confidence intervals.

5.15 Relevant Literature

5.14.1 Study Design

The design and implementation of clinical, epidemiologic and other studies are covered in many books and review papers. To note a few: Fisher (1935a) is a classic. Hill (1962) is another classic, and in some aspects, a unique work. Jekel, Elmore and Katz (1996) give a succinct perspective on the issues. Pocock (1983) and Piantadosi (1997) thoroughly deal with clinical trials; Hulley and Cummings (1988) cover a variety of biomedical designs; Rothman and Greenland (1998) detail the spectrum of designs in epidemiology while Breslow and Day (1980) and Schesselman (1982) focus on case-control studies. Cohort studies are covered in Breslow and Day (1987). Zolman (1993) is an accessible general work covering this area.

Discrete statistical models for varied sampling models and designs, associated methods for large sample analysis and the relationship between the parameters and estimates from different designs are found in many texts. The principal work, which deals with the subject in a thorough manner, is Bishop, Fienberg and Holland (1975). Other books with somewhat different approaches are Cox and Snell (1989) and Santner and Duffy (1989). See Piegorsch (1994) for an accessible summary of the technical material. A basic exposition of the sampling models for discrete data is in Agresti (1996). Piegorsch and Bailer (1997) also gives a detailed exposition of Poisson count data. For a lucid account of, and the justifications for the Poisson distribution for person time data, see Breslow and Day (1987). They also give the rationale and methods for computing person times values for different data cells. More references on this topic are in Chapter 9. Lui (2004) has a comprehensive coverage of epidemiological designs, including the inverse sampling design.

Faulty design, implementation and interpretation continue to be a major concern. For a description of errors in biomedical studies, see Andersen (1990); a witty account is in Michael, Boyce and Wilcox (1984). Critical evaluation of study quality is dealt with in many books and papers; see, for example, Dawson–Saunders and Trapp (1994), Chapter 15, and Essex–Sorlie (1995), Chapter 20.

5.14.2 Conventional Analysis

Descriptions of the large sample methods for one margin, no margin and nothing fixed designs are found in many texts. Bishop, Fienberg and Holland (1975), Fienberg (1977) and Agresti (1996) cover the models and analytic techniques for discrete data models. The latter also describes conditional exact methods. Sahai and Khurshid (1996) has a detailed listing of exact and large sample methods for analysis of discrete data. These books mostly cover log-linear models and logistic models, with the 2×2 table as the starting point. An elementary introduction to the former models is in Selvin (1995), Chapter 9. The latter is well dealt with in Hosmer and Lemeshow (1989). Fleiss (1981) is in a class by itself. Newman (2001) is to be especially recommended for clarity and coverage.

5.14.3 The 2×2 Table

The literature on the 2×2 table is too extensive for a detailed listing. Almost all introductory books on statistics deal with it. Fleiss (1981) is a fine introduction to the classic methods of analysis with a good bibliography. The following papers, and the discussion they generated, provide a start on conceptual issues relating to the 2×2 table: Fisher (1935b); Berkson (1978); Upton (1982, 1992); Yates (1984); D'Agostino, Chase and Belanger (1988); Little (1989); and Hirji, Tan and Elashoff (1991). The dialogue between Camilli (1990) and Haber (1990) is particularly informative. See also Agresti (1992, 2001), Sahai and Khurshid (1995), and Chapter 15.

The idea of ancillarity was proposed, though in not too clear a manner, by Fisher (1935b). Other papers continue to reflect the confusion (e.g., Gart 1971; Upton 1992). For a meticulous

EXERCISES

exposition, see Basu (1959, 1977). Little (1989) is an excellent elementary exposition. Further references are in Chapter 7, where the Fisher and Irwin exact tests are compared.

There is a sizeable literature on point and interval estimation for the odds ratio in a single 2×2 table. Many approximations have been derived. Fleiss (1981) and Sahai and Khurshid (1996) provide an overall view. Walter and Cook (1991) compared the unconditional and conditional mles for this problem and favored the former. This study is in dispute (Mantel 1992; Walter and Cook 1992). Walter and Cook (1991) reviewed empirical logit type of estimators of the odds ratio. A study of the mue for the odds ratio in a single 2×2 table has yet to be done. Many papers on exact and asymptotic CIs and tests for the odds ratio in a 2×2 table also exist. Cornfield (1956) and Gart (1971) are useful indicators of earlier work. For additional methods, references and alternatives to the odds ratio, see Sahai and Khurshid (1996) and Santner and Duffy (1989).

For an introduction to additive and multiplicative risk models, and large sample and exact unconditional methods for their analysis, see Santner and Snell (1980), Soms (1989a, 1989b), Santner and Duffy (1989) and Newman (2001). See also Hamilton (1979).

5.10.5 Inverse Sampling

The inverse sampling design is rooted in the papers of Haldane (1945a, 1945b) and Finney (1947, 1949). The literature for the one parameter case was noted in Chapter 2. Bishop, Fienberg and Holland (1975), Chapter 13, has a dated but useful review of the negative multinomial distribution (a generalization of the negative binomial). They summarize the works of Steyn (1959) and Rudolph (1967) for two way contingency tables under this model.

Several papers have discussed the application of the negative binomial in complex epidemiological and clinical studies (Bennett 1981, 1986; Kikuchi 1987; Singh and Aggarwal 1991). It is applied in industrial settings as well (Salvia 1984). A series of papers by KJ Lui (Lui 1995a, 1995b, 1996a, 1996b, 1997) systematically developed the 2×2 inverse sampling design for biomedical applications. The references in these papers give a wider view. The conditional distribution for the inverse sampling design was studied by Kudô and Tarumi (1978). Andersen (1982), pages 134–135, gave an exact test for two negative binomials design and noted its applications in the theory of queues and for some genetic models. The reference of choice on inverse sampling in epidemiological studies is Lui (2004).

The negative binomial is also used to model data where "extra Poisson" variation is suspected. Such applications are distinct from the one we cover. For a list of papers in this area, see Ramakrishnan and Meeter (1993).

5.16 Exercises

5.1. T_1 and T_2, the random number of accident victims per day at two emergency units, are independent Poisson variates with means λ_1 and λ_2. On a random day, we observed $t_1 = 3$ patients at Unit 1, and $t_2 = 7$ patients at Unit 2. Test the hypothesis $\lambda_1 = \lambda_2$ and assess the magnitude of the difference between the rates.

5.2. Analyze the clinical trial data in Table 5.2 using asymptotic unconditional and conditional score, LR and Wald methods. Plot the relevant evidence functions and compare them.

5.3. Analyze the HCG data in Table 5.5 using asymptotic score, LR and Wald methods. Compare the results with those give in Table 5.13.

5.4. Investigate the relationship between exposure and gender in the aquatic toxicity data in Table 5.8. Plot the relevant evidence functions and describe your findings.

5.5. Table 5.15 shows data from five hypothetical clinical trials. Analyze them using large sample unconditional, and exact and large sample conditional methods. Plot and compare the relevant evidence functions. Use the probability and CT methods for conditional analysis as well.

Table 5.15 *Five Clinical Trials*

Trial No.	Outcome	Control	Drug	Total
1	Not Improved	4	8	12
	Improved	8	4	12
	Total	12	12	24
2	Not Improved	2	4	6
	Improved	8	6	14
	Total	10	10	20
3	Not Improved	0	3	3
	Improved	7	4	11
	Total	7	7	14
4	Not Improved	3	13	16
	Improved	7	11	18
	Total	10	24	34
5	Not Improved	2	13	15
	Improved	5	5	10
	Total	7	18	25

Note: Hypothetical data.

5.6. Analyze the relationship between gender and exposure from the data in Table 5.9. Use conditional and unconditional approaches and plot the evidence functions.

5.7. Analyze the data in Table 1.1 using conditional and unconditional methods, plot the evidence functions and compare the results. Use the TST, score, probability and CT methods for exact conditional analysis.

5.8. Table 5.16 gives data on gastric ulceration in 82 patients with rheumatic disease on chronic aspirin therapy and 45 controls. Analyze these data using exact and asymptotic conditional approaches; plot the score, LR, Wald, TST, mid-p, probability and CT evidence functions, and write a report on your findings.

5.9. Table 5.17 shows the incidence of carpal tunnel syndrome (CTS) by gender. Does gender affect incidence? Plot relevant evidence functions and write a short report on your findings.

EXERCISES

Table 5.16 *Aspirin and Ulcer*

Gastric Ulcer	Aspirin User	
	No	Yes
No	68	45
Yes	14	0
Total	82	45

Source: Silvoso et al. (1979), Table 1. Used with permission.

Table 5.17 *CTS Data*

Gender	Cases	Person Years
Male	141	44,676
Female	162	44,787

Source: Nordstrom et al. (1998), Table 3. Used with permission.

5.10. Table 5.18 shows hypothetical data on the incidence of a disease by gender. Analyze them using exact and asymptotic methods; plot relevant evidence functions and write a short report on your findings.

Table 5.18 *Disease Incidence Data*

Gender	Cases	Person Years
Male	5	676
Female	1	307

Note: Hypothetical data.

5.11. Prove that the independence relation (5.32) holds for all x, y if and only if $\phi_3 = 1$. Also formally derive the expression (5.39).

5.12. For the total fixed model (5.34), derive the expressions for β's in terms of the π's. State clearly the null hypothesis of no association between the two factors for this model.

5.13. For the Poisson model (5.49), derive the expressions for μ's in terms of the λ's. Extend this to model (5.55). State clearly the null hypothesis of no association between the two factors for the latter model.

5.14. Derive the unconditional distribution (5.65) for the inverse sampling design.

5.15. Give a formal proof of Theorem 5.1.

5.16. Give a formal proof of Theorem 5.2.

5.17. Prove Theorem 5.3 for the overall total fixed and the nothing fixed designs.

5.18. Derive the conditional and unconditional Wald statistics for a one margin fixed design when the model is parameterized in terms of (i) the odds ratio, (ii) the log-odds ratio and (iii) difference of probabilities. Compare the *p*-values and CIs obtained from these variations using relevant datasets given in this chapter, and other actual and hypothetical data. What do you conclude? Compare the effect of using the actual and estimated asymptotic variance for the mle of the relevant parameter.

5.19. The following scheme, extracted from Bender (2001), is used to compute the Wilson unconditional score CI for the risk difference in a two binomials design. For a $100 \times (1-\alpha_*)\%$ confidence interval for $\pi_1 - \pi_0$, suppose that

$$\hat{\pi}_1 = \frac{a}{n} \quad \text{and} \quad \hat{\pi}_0 = \frac{b}{m}$$

Then we implement:

STEP 1:

$$g_0 = \frac{2b + z^2_{1-\alpha_*/2}}{2(m + z^2_{1-\alpha_*/2})} \quad \text{and} \quad g_1 = \frac{2a + z^2_{1-\alpha_*/2}}{2(n + z^2_{1-\alpha_*/2})}$$

$$h_0 = \frac{b^2}{m(m + z^2_{1-\alpha_*/2})} \quad \text{and} \quad h_1 = \frac{a^2}{n(n + z^2_{1-\alpha_*/2})}$$

STEP 2: For i = 0,1

$$l_i = g_i - \sqrt{g_i^2 - h_i} \quad \text{and} \quad u_i = g_i + \sqrt{g_i^2 - h_i}$$

STEP 3: Then compute

$$\epsilon_1 = \sqrt{(\pi_1 - l_1)^2 + (\pi_0 - u_0)^2}$$

$$\epsilon_0 = \sqrt{(\pi_1 - u_1)^2 + (\pi_0 - l_0)^2}$$

STEP 4: And finally, compute

$$L = \hat{\pi}_1 - \hat{\pi}_0 - \epsilon_1 \quad \text{and} \quad U = \hat{\pi}_1 - \hat{\pi}_0 + \epsilon_0$$

The $100 \times (1 - \alpha_*)\%$ CI for the risk difference is then given by (L, U).

(i) Derive this scheme from the unconditional score statistic; (ii) Develop a similar scheme for the odds ratio and the relative risk; (iii) Can you get corresponding schemes for the LR based CIs for these measures; (iv) Extend this scheme to the two Poisson rates case.

5.20. For the one margin fixed design, do the null and nonnull score and LR statistics vary when the models are formulated in terms of (i) the odds ratio, (ii) the log-odds ratio and (iii) the difference between probabilities? Extend this investigation to the other models.

5.21. Show that the null unconditional score and LR statistics for the one margin fixed design and other related designs can be given in terms of observed and expected values.

5.22. How would you construct large sample evidence functions for the unconditional score, Wald and LR statistics? Apply these to selected data sets from this chapter.

5.23. Plot exact TST, mid-*p*, probability, CT, score, LR and Wald statistic evidence functions for a variety of simulated 2×2 tables from a one margin fixed design. What do these plots indicate?

EXERCISES

5.24. Derive in detail conditional and unconditional methods to compare two Poisson rates described in §5.12. What are the implications of using a conditional approach in this setting? What is the implication of making the comparisons in terms of ratio of rates or in terms of difference of rates? Is the evidence function approach relevant here? Apply the methods you derive to analyze a variety of simulated data.

5.25. Derive the unconditional large sample score, LR and Wald statistics for (i) the parameter β_3 in the overall total fixed design and (ii) for the parameter μ_{yx} in the nothing fixed design.

5.26. Derive in detail the conditional distribution (5.97) for the inverse sampling design. Derive the total count, conditional mean and conditional variance for the null conditional distribution with $\phi = 1$, of the inverse sampling design (see Kudô and Tarumi (1978)).

5.27. Derive the null and nonnull conditional and unconditional score, LR and Wald statistics for the inverse sampling design. Compare these with their counterparts for a one margin fixed prospective design.

5.28. Consider a variation of the 2×2 inverse sampling in which subjects are sampled until one specified cell attains a specified value. What is the probability distribution for this study? Is it a PBD? Derive the unconditional and conditional methods to analyze data from this design. Repeat this for a design in which subjects are sampled until the first row total attains a fixed value. Compare the results obtained here with those from the regular 2×2 design and the inverse sampling design with two cells fixed.

5.29. Investigate the computation of exact p-values for the difference of probabilities in a one margin fixed design.

5.30. Consider the two independent binomials design and let $\hat{\beta}$ and $\hat{\beta}_c$ respectively denote the unconditional mle and the cmle for β, the log-odds ratio. Newman (2001) states that $\hat{\beta} \geq \hat{\beta}_c$. Prove or disprove this claim.

5.31. For the regular independent observations nothing fixed, overall total fixed and one margins fixed 2×2 designs, prove that

$$\frac{\mathbf{E}[\mathcal{AC}]}{\mathbf{E}[\mathcal{BD}]} = \frac{\mathbf{E}[\mathcal{A}]\mathbf{E}[\mathcal{C}]}{\mathbf{E}[\mathcal{B}]\mathbf{E}[\mathcal{D}]}$$

Also show that the above does not extend to the both margins fixed case (see Mantel and Hankey 1975). What is the implication for analysis?

5.32. What is the asymptotic distribution of the cmue of the log-odds ratio in a 2×2 table? (see Hirji, Tsiatis and Mehta 1989; Salvan and Hirji 1991).

5.33. Consider the test for equivalence in a two group binary outcome clinical trial. Formulate this in terms of risk difference, relative risk and the odds ratio (see Lui (1996a, 1997) and the references therein). Develop large sample and exact methods for such tests. How would you use the evidence function approach in this context?

5.34. Consider the set of 2×2 tables with overall total equal to 10. Generate all tables from this set in which (i) the unconditional mle and (ii) conditional mle for the log-odds ratio are nonfinite. What is the relationship between these two subsets? Generalize your results to a table with total equal to n.

CHAPTER 6

Two 2 × 2 Tables

6.1 Introduction

Models with three or more discrete variates often occur in scientific research. This chapter extends the two variable models of the previous chapter to the three binary variables case. After illustrating such data in the form of two 2 × 2 tables, we present statistical models for them, and develop unconditional and conditional methods to analyze them. The specific aims are to present:

- The ideas systematic and random variation, and formally define the concept of an exact distribution.
- A rationale for including confounding variables in the study design and/or data analysis.
- Data from a clinical trial with stratified randomization, a cross sectional study, and case-control and incidence density designs.
- Statistical models for these designs, explore the relations between the models, and summarize the usual large sample method for analysis.
- The relevant exact distributions from these models in the form of a PBD, and apply conditional methods to analyze them.
- Justification for the use of the product binomial model in a randomized clinical trial, and the analysis of the odds ratio from case-control data.

6.2 Sources of Variability

In Chapter 1, we noted that a statistical model relates several variables in the context of a random mechanism. To elaborate on this, consider a child who has been treated for an ear infection. A physician examines the affected ear (tympanic) membrane to determine if the problem has resolved or not. Let $\mathcal{Y}$ denote his judgment, with 1 indicating satisfactory resolution and 0, otherwise. Besides the actual outcome, her judgment is influenced by two types of factors, systematic and random. These factors include those affecting the outcome, and those pertaining to the circumstances under which the outcome is judged. The following are among the possible systematic factors in this case:

- Patient related factors such as his or her medical history, nature of the infection (viral or bacterial), infection in one or both ears, and so on.
- Treatment related factors such as the type of treatment given, and the level of compliance with it.
- Examiner related factors like experience and expertise in dealing with childhood ear problems.
- Diagnostic instrument related factors like the type of instrument (otoscope) used.

Systematic factors occur in every discipline, be it education, industry, psychology or public health. Their impact may be a direct or an indirect one, and may itself be subject to random variation. In the ear infection case:

- In a study, the characteristics of the patients will depend on how, where and when the sample was selected. If it was from a specialized clinic, it may have a higher than normal proportion of cases with severe or more persistent form of the condition. The findings may not then apply to ordinary, uncomplicated cases of otitis media.
- Aspects of the treatment, like the doses missed, or times at which it was taken, vary in a random fashion and may affect the outcome. To avoid selection bias and enhance comparability of treatment groups, the treatment assignment in a clinical trial may be random.
- Faced with a borderline case, two physicians may not render the same verdict. While generally more reliable, even an experienced doctor at times makes an erroneous diagnosis.
- Measuring instruments also generate random errors. The lens of the otoscope was cloudy in a particular case, and so on.
- In complex systems comprising of patient biology, social milieu, medical delivery systems and research teams, there is an intrinsic element of random variability which affects both actual and recorded outcomes. For example, a datum of value 0 may be entered as a 1.

Unpredictable factors are always at play, and contribute variability or errors to the process under scrutiny (Coggon and Martyn 2005). A fundamental character of a scientific study is to have a plan to identify the possible sources of systematic and random variability, and use methods to control them and/or adjust for their effects. A good study design, rigorous conduct and appropriate analysis and interpretation are the key ingredients of this task. A valid model relating the pertinent factors to the outcome is required as well.

This chapter deals with models for three binary variables. To exemplify a common model, let $\mathcal{Y} \in \{0, 1\}$ be the binary outcome variable in a clinical study of acute otitis media; x, the therapy indicator (0 = control, 1 = antibiotic); and z, a prognostic factor (0 = absent, 1 = present). Also, let $\pi(x, z)$ denote $P[\mathcal{Y} = 1 \mid x, z]$. We then consider the logistic model

$$\ln \frac{\pi(x, z)}{1 - \pi(x, z)} = \beta_0 + \beta_1 x + \beta_2 z + \beta_{12} xz \tag{6.1}$$

In this model, β_0 is the **constant term**; β_1 is the **main treatment effect** parameter; β_2 is the **main factor effect** parameter; and β_{12} is the **factor and treatment interaction effect** parameter. Further explanation of the parameters is given later. The focus of the study is to estimate the treatment effect parameters, namely, β_1 and β_{12}, in the presence of a possible prognostic factor effect as well as random variability.

When the model in question expresses an underlying cause and effect relationship between the variables, it is called a **causal model**.

6.3 On Stratification

For models with several binary variables, the data are usually given in the form of two or more 2×2 tables. Each table is known as a **stratum**. An analysis that adjusts for the effect of the stratification factors is called a **stratified analysis**. Now take the case of three binary variables. Then, for $i = 0, 1$, Table 6.1 gives the notation for the ith table. Taken as random variables, the ith stratum counts are denoted $\mathcal{A}_i$, $\mathcal{B}_i$, $\mathcal{C}_i$ and $\mathcal{D}_i$, respectively.

ON STRATIFICATION

Table 6.1 *The ith* 2×2 *Table*

Row Variable	Column Variable		
	0	1	Total
0	d_i	c_i	r_i
1	b_i	a_i	s_i
Total	m_i	n_i	N_i

The row and column variables in each stratum commonly are an outcome or response indicator and an explanatory factor like a treatment or exposure indicator (study variable). Though it may affect the outcome, the stratum variable is often not of direct interest. It is then known as a **confounding factor**, a **prognostic factor** or an **effect modifier**. At times, a unique designation of each variable into one of these types is not done. For some analysis, a variable is deemed a response variable, and, later, it becomes a confounding variable.

In modeling the effect of second hand smoke on lung cancer incidence, the age of the subject is a confounding variable. An evaluation of an antibiotic used to treat pneumonia among the elderly needs to adjust for the health status of the patient. Not including such risk, prognostic or confounding factors in the analysis may produce systematic errors. Thus, we may detect an effect where none exists (spurious effect), fail to detect an existent effect (Simpson's paradox) or overlook an interaction effect.

The three scenarios in Table 6.2 illustrate such errors of inference. The data relate to an hypothetical binary response trial comparing treatments A and B. Assume that we can assign the patients into two risk factor groups: "High Risk" and "Low Risk."

Scenario 6.1: In this scenario, the overall data give credence to the hypothesis that treatment A (success rate 74%) is superior to treatment B (success rate 26%). At each level of risk, however, the observed efficacies are equivalent. The discrepancy stems from the fact that among those who received A, 80% were low risk, while among those on B, only 20% were low risk. Ignoring this factor is thereby an invitation to declare a treatment effect that is possibly not there.

Scenario 6.2: Here both the treatments show the same success rates, when viewed overall. But at each level of risk, A seems better than B. This paradox is known as **Simpson's Paradox**. Ignoring the prognostic factor here masks the evidence for a treatment effect.

Scenario 6.3: The final scenario indicates that overall, treatment A is as good as B. Yet, among the low risk patients, it is superior to B, and among the high risk cases, the situation is reversed. The treatment of choice depends on risk level, a phenomenon known as an **interaction between treatment and risk factor**. If the risk factor is ignored, we would miss such an effect.

****

Such considerations hold for all types of studies. Accounting for confounding factors is an essential aspect of drawing inference from data. A simple solution is to analyze the data separately at each level of the factor. Called **subgroup analysis**, this is not an advisable option. First, it

Table 6.2 *Three Clinical Trial Scenarios*

	Scenario 6.1		Scenario 6.2		Scenario 6.3	
Treatment	S	F	S	F	S	F
	Combined					
A	222	78	150	150	180	120
B	78	222	150	150	180	120
	Low Risk					
A	216	24	90	30	162	18
B	54	6	144	96	108	72
	High Risk					
A	6	54	60	120	18	102
B	24	216	6	54	72	48

Note: S = Success; F = Failure.

raises the chance of finding spurious associations. And second, in a randomized trial, it may compare dissimilar subgroups. Clinical trial reports at times claim the efficacy of a treatment for some groups of patients but not others. For example, men may be said to benefit from it but not women. When derived from subgroup analysis, such claims need to be treated with caution. For a series of actual examples of claims based on such analyses that were later disproved, see Rothwell (2005b).

Using an appropriate statistical model to adjust for confounding factors is a better course of action. The potential confounders may also be incorporated into and controlled for through study design. Or, of course, one may do both.

6.4 Data Examples

We give four data examples. The first is from a **clinical trial with stratified randomization**, the second from a **cross sectional sampling** study, the third from a **case-control** study and the fourth from a **follow up or cohort** study. Independent sampling of the study subjects is a feature shared by all of them.

6.4.1 An RCT with Stratified Randomization

Random allocation to treatments helps control selection bias and tends to make treatment groups comparable in terms of known and unknown prognostic factors. But randomization done at the overall level may fail to achieve comparability, especially if the groups are not large. When we need to ensure that the groups are similar in terms of a critical factor, we perform the random allocation at each level of the factor. Such an allocation scheme is known as **stratified randomization**. If the factor and outcome are binary, and there are two treatment arms, we get data in the form of two 2 × 2 tables.

DATA EXAMPLES

Example 6.1: Merigan et al. (1992) report a trial of Ganciclovir for prevention of cytomegalo virus (CMV) disease after a heart transplant. The absence of the disease was called a success ($\mathcal{Y} = 1$), and its occurrence as a failure ($\mathcal{Y} = 0$). Randomization to active treatment or placebo was done separately for the 112 CMV seropositive and 37 CMV seronegative patients. At each of the two levels of serostatus, the numbers allocated to active treatment and placebo were determined by the design. The results are shown in Table 6.3.

Table 6.3 *CMV Prevention Trial*

	Seronegative		Seropositive	
Response	Placebo	Ganciclovir	Placebo	Ganciclovir
Failure	5	7	26	5
Success	12	13	30	51
Total	17	20	56	56

Source: Merigan et al. (1992); Used with permission.

The overall success rate of placebo is 58% and of Ganciclovir, it is 84%. Among seropositive patients, these proportions respectively are 54% and 91%, and among seronegative cases, they are 71% and 65%, a reversal of observed efficacies. Among the questions of interests are: Are these data reflective of an actual difference between the two treatments? And, does the relative effect of the therapies vary according to serostatus? Another way of stating the last question is: Is there an **interaction** or **synergy** between treatment and serostatus?

6.4.2 A Cross Sectional Design

Example 6.2: Table 6.4 shows data on urine bacterial culture done on pregnant women (Orrett and Premanand 1998). The three binary variables are culture result, pregnancy status, and whether the subject was catheterized or not. Does pregnancy status or catheterization promote the presence of bacteria in the urine? Is there an interaction between the two factors? Or, do the findings reflect chance variation?

Table 6.4 *Bacteriuria and Catheterization*

	Primigravida		Multigravida	
Culture	C	NC	C	NC
Negative	14	16	10	70
Positive	2	16	6	34
Total	16	32	114	104

Note: C = Catheterized; NC = Noncatheterized.
Source: Orrett and Premanand (1998), Table 2. Used with permission.

6.4.3 Case-Control Data

Example 6.3: The data from a case-control study of lung cancer by Caporaso et al. (1989) were used by Piegorsch (1994) to model the interaction between a genetic factor (the ability to metabolize the drug debrisoquine) and asbestos exposure in the etiology of lung cancer. We reproduce these data from the latter in Table 6.5. Here the genetic factor is collapsed into two levels: PMM = poor or moderate metabolizer; and EXM = extensive metabolizer.

The observed odds ratio among PMM group is 0.3, and among the EXM group, it is 2.0. Do the data provide evidence that the genetic factor modifies the impact of the environmental exposure on the etiology of lung cancer?

Table 6.5 *Gene and Environment Interaction Data*

	Cases		Controls	
Status	PMM	EXM	PMM	EXM
Unexposed	14	97	55	68
Exposed	1	47	15	17
Total	17	144	70	85

Source: Piegorsch (1994), Table 3; Used with permission from *Environmental Health Perspectives*.

For now, consider two possible designs for such a study. One, the 161 cases were selected from the population of cases, and the 155 controls were selected from a relevant population of controls. This is more in line with how the study was carried out. And two, suppose the list of potential cases and controls had been divided into those with the genetic factor and those without. Sampling was then stratified by the level of the genetic factor. That is, a fixed number of cases and a fixed number of controls are sampled from those with the factor, and the process repeated for those without the factor. Such a design is known as a **group matched case-control** design. (See Chapter 14 for more on matched designs.)

6.4.4 A Follow Up Study

Example 6.4: Table 6.6 shows data on the impact of exposure to arsenic on workers employed in a Montana smelter. The data are from a (retrospective) follow up study of a large cohort workers. Total mortality is given by level of exposure, and two other binary factors, y and z. The person year value for each cell in the table is also shown. The data are extracted from a more complex dataset in Breslow and Day (1987). The factor y is age at death (≤ 59 and ≥ 60) and z is the calendar period ([1938–1949] and [1950–1959]). See the original source for a complete explanation.

One item of interest here is a possible a three way interaction effect of exposure, factor z, and factor y on total mortality. Note, due to our selection, these data may not reflect the relationships in the original data. And, further, see Chapter 15 for critical comments on the type of analysis for these data done in §6.9 below.

STATISTICAL MODELS

Table 6.6 *Mortality and Arsenic Exposure Among Smelter Workers*

	$z=0$		$z=1$	
	Exposure		Exposure	
	Low	High	Low	High
$y=0$	4	4	3	3
person years	5925.03	964.01	3132.34	470.53
$y=1$	5	10	13	9
person years	2919.04	949.34	2649.23	709.37

Source: Breslow and Day (1987), Table 3.14; Used with permission from the International Agency for Research on Cancer.

6.5 Statistical Models

Now, we present statistical models and probability distributions for the designs described above. Choice of a relevant model has to consider both the scientific relationship between the variables and assumptions about what is fixed by design and what is the random element in the data.

6.5.1 The Product Binomial Model

To analyze the stratified clinical trial like that depicted in Table 6.3, we generally take the column sums of each 2×2 table as fixed and apply the product binomial model. This is also what we did for the simple randomized design in Example 5.6. Yet, as noted there, the design (unconditional) distribution is somewhat different. The scheme used for randomization may not fix the number of patients recruited into each treatment, or may fix these totals but allocate patients according to a random sequence. How do we justify using the product binomial distribution?

Consider first the single stratum case with the following design. A patient (a random member of the target population) meeting the eligibility criteria is randomly allocated to treatment $x = 1$ with probability π_*, and to treatment $x = 0$ with probability $1 - \pi_*$. Enrollment continues until a total of N patients are randomized. Clearly, the number of patients in any treatment is not fixed by design.

For the jth patient, let $\mathcal{X}_j$ and $\mathcal{Y}_j$ respectively be the random treatment assignment, and his or her response to treatment. Then

$$P[\mathcal{X}_j = x, \mathcal{Y}_j = y] = P[\mathcal{Y}_j = y \mid \mathcal{X}_j = x]P[\mathcal{X}_j = x] \quad (6.2)$$

which equals

$$\pi_*^x [1 - \pi_*]^{(1-x)} \pi(x)^y [1 - \pi(x)]^{(1-y)} \quad (6.3)$$

where $\pi(x) = P[\mathcal{Y} = 1 \mid \mathcal{X} = x]$. The probability for the entire sample, $P[\mathcal{X}_j = x_j, \mathcal{Y}_j = y_j; j = 1, \ldots, N]$, is then

$$\pi_*^{\Sigma x_j}[1-\pi_*]^{(N-\Sigma x_j)} \prod_{j=1}^{N}\left[\pi(x_j)^{y_j}[1-\pi(x_j)]^{(1-y_j)}\right] \tag{6.4}$$

which simplifies to

$$\pi_*^{(a+c)}[1-\pi_*]^{(b+d)}\pi(1)^a[1-\pi(1)]^c\pi(0)^b[1-\pi(0)]^d \tag{6.5}$$

where $a = \#(x=1, y=1), b = \#(x=0, y=1), c = \#(x=1, y=0), d = \#(x=0, y=0)$. Summing over all realizations, we get the probability $P[\mathcal{A} = a, \mathcal{B} = b, \mathcal{C} = c, \mathcal{D} = d \mid \mathcal{A} + \mathcal{B} + \mathcal{C} + \mathcal{D} = N]$ is equal to

$$\binom{N}{a+c}\pi_*^{(a+c)}[1-\pi_*]^{(b+d)}\binom{a+c}{a}\pi(1)^a[1-\pi(1)]^c\binom{b+d}{b}\pi(0)^b[1-\pi(0)]^d \tag{6.6}$$

For the purpose of treatment comparison, π_* is a nuisance parameter. Further, the total number in any treatment depends only on π_*. Under the definition of §5.2, the total $a+c$ (and $b+d$) is **ancillary** for the parameters of interest. Thus, we condition on the observed treatment totals. With $N = n + m$, the probability $P[\mathcal{A} = a, \mathcal{B} = b \mid \mathcal{A} + \mathcal{C} = n, \mathcal{B} + \mathcal{D} = m]$ comes out as

$$\binom{n}{a}\pi(1)^a[1-\pi(1)]^c\binom{m}{b}\pi(0)^b[1-\pi(0)]^d \tag{6.7}$$

which obviously is a product of two binomials. When the randomization is done separately and independently for each of K strata, the applicability of the K product binomial distribution follows in a similar manner.

More commonly, the number of subjects assigned to treatments in a stratum is fixed and the treatment sequence is randomized separately for each stratum. Such a design and variations of it are called **restricted randomization designs**. Under a population sampling approach, the product binomial model also applies to this and a variety of such designs.

****

Let us return to the trial data shown in Table 6.3 and let $\mathcal{Y} \in \{0, 1\}$ be the outcome variable; x, the therapy indicator (0 = Placebo, 1 = Ganciclovir); and z, the stratum variable (0 = seronegative, 1 = seropositive). With $\pi(x, z)$ equal to $P[\mathcal{Y} = 1 \mid x, z]$, we consider the logistic model (6.1) for this study.

Consider the model parameters. Let $\phi_j = \exp(\beta_j)$. Then the response probabilities for the four combinations of (x, z) are shown in Table 6.7.

Reversing these relationships, we find that

$$\phi_0 = \frac{\pi(0,0)}{1-\pi(0,0)} \tag{6.8}$$

$$\phi_1 = \frac{\pi(1,0)/[1-\pi(1,0)]}{\pi(0,0)/[1-\pi(0,0)]} \tag{6.9}$$

$$\phi_2 = \frac{\pi(0,1)/[1-\pi(0,1)]}{\pi(0,0)/[1-\pi(0,0)]} \tag{6.10}$$

$$\phi_{12} = \frac{\{\pi(1,1)/[1-\pi(1,1)]\}\{\pi(0,0)/[1-\pi(0,0)]\}}{\{\pi(1,0)/[1-\pi(1,0)]\}\{\pi(0,1)/[1-\pi(0,1)]\}} \tag{6.11}$$

STATISTICAL MODELS

Table 6.7 *Success Probabilities*

x	z	$\pi(x,z)$
0	0	$\phi_0/(1+\phi_0)$
1	0	$\phi_0\phi_1/(1+\phi_0\phi_1)$
0	1	$\phi_0\phi_2/(1+\phi_0\phi_2)$
1	1	$\phi_0\phi_1\phi_2\phi_{12}/(1+\phi_0\phi_1\phi_2\phi_{12})$

If $\beta_{12} = \ln(\phi_{12}) = 0$, we get

$$\frac{\pi(1,1)/[1-\pi(1,1)]}{\pi(0,1)/[1-\pi(0,1)]} = \frac{\pi(1,0)/[1-\pi(1,0)]}{\pi(0,0)/[1-\pi(0,0)]} \quad (6.12)$$

meaning that the odds of success for treatment 1 relative to treatment 0 are the same in each stratum. In terms of odds ratios, we say there is no interaction between treatment and stratum factor. The odds ratio for each stratum is $\phi_1 = \exp(\beta_1)$. This is a special case of the **common odds ratio model** we study in detail in Chapter 8.

The probability distribution we apply here is a product of four binomial distributions.

$$P[\mathcal{A}_i = a_i, \mathcal{B}_i = b_i, i \in \{0,1\}] = \prod_{i \in \{0,1\}} \binom{n_i}{a_i}\binom{m_i}{b_i}$$
$$\times \ \pi(1,i)^{a_i}[1-\pi(1,i)]^{(n_i-a_i)}\pi(0,i)^{b_i}[1-\pi(0,i)]^{(m_i-b_i)} \quad (6.13)$$

We write this in a polynomial form by first defining

$$\begin{aligned}
\mathcal{T}_0 &= \mathcal{A}_1 + \mathcal{B}_1 + \mathcal{A}_0 + \mathcal{B}_0 \\
\mathcal{T}_1 &= \mathcal{A}_1 + \mathcal{A}_0; \quad \mathcal{T}_2 = \mathcal{A}_1 + \mathcal{B}_1; \quad \mathcal{T}_3 = \mathcal{A}_1 \\
\mathcal{T} &= (\mathcal{T}_0, \mathcal{T}_1, \mathcal{T}_2, \mathcal{T}_3); \quad t = (t_0, t_1, t_2, t_3) \\
\phi &= (\phi_0, \phi_1, \phi_2, \phi_{12})
\end{aligned}$$

Then we substitute the expressions from Table 6.7 in (6.13), and use the above definitions to find that

$$P[\mathcal{T} = t; \phi] = \frac{c(t)\phi_0^{t_0}\phi_1^{t_1}\phi_2^{t_2}\phi_{12}^{t_3}}{f(\phi)} \quad (6.14)$$

where $f(\phi)$ is the polynomial product

$$(1+\phi_0)^{m_0}(1+\phi_0\phi_1)^{n_0}(1+\phi_0\phi_2)^{m_1}(1+\phi_0\phi_1\phi_2\phi_{12})^{n_1} \quad (6.15)$$

and further where

$$c(t) = \binom{m_0}{t_0+t_3-t_1-t_2}\binom{n_0}{t_1-t_3}\binom{m_1}{t_2-t_3}\binom{n_1}{t_3} \tag{6.16}$$

Note that $c(t)$ is the coefficient of the term

$$\phi_0^{t_0}\phi_1^{t_1}\phi_2^{t_2}\phi_{12}^{t_3}$$

in the expansion of the gp $f(\phi)$. Thus $\mathcal{T}$ has a polynomial based distribution, and is a vector of sufficient statistics for ϕ, or β.

6.5.2 The Multinomial Model

We now turn to Table 6.4. Let the components of (y, x, z) denote bacterial culture status (0 = negative, 1 = positive), pregnancy status (0 = primigravida, 1 = multigravida) and catheterization indicator (0 = no, 1 = yes), respectively. We let $\pi(y, x, z)$ be the probability of the event under (y, x, z), and consider the logit model

$$\ln\left(\frac{\pi(y,x,z)}{\pi(0,0,0)}\right) = \gamma_0 + \gamma_1 y + \gamma_2 x + \gamma_3 z +$$
$$\gamma_{12}yx + \gamma_{13}yz + \gamma_{23}xz + \gamma_{123}yxz \tag{6.17}$$

With $y = 0$, $x = 0$ and $z = 0$, we get that $\gamma_0 = 0$. We have eight possible outcomes here. The overall total is fixed and subjects have been sampled independently. The distribution for this design is then an eight category multinomial. Following the reasoning of §5.5, this is shown to be a PBD. The interpretation of its parameters is similar to that for the model given in §6.5.4 below.

6.5.3 The Retrospective Binomial Model

For the retrospective study of §6.4.3, let $\mathcal{Y} \in \{0, 1\}$ be the case-control indicator (0 = control, 1 = case); x, the exposure indicator (0 = not exposed, 1 = exposed); and z, the genetic factor (0 = PMM, 1 = EXM). With $\pi(x, z)$ denoting $P[\mathcal{Y} = 1 \mid x, z]$, we consider the logistic model

$$\ln\frac{\pi(x,z)}{1-\pi(x,z)} = \beta_0 + \beta_1 x + \beta_2 z + \beta_{12} xz \tag{6.18}$$

The interpretation of the parameters is identical to that for (6.1). As noted in §5.8, retrospective data ordinarily cannot be used to fit such a model. If cases and controls are sampled under the conditions stated there, we can, however, get valid estimates for some of its parameters.

In particular, if the number of cases and number of controls were fixed by design, then we may obtain valid estimates for β_1, β_2 and β_{12}. On the other hand, if the sampling of cases and controls was done similarly but separately at each level of the factor z, then valid estimates for β_1 and β_{12} can be obtained. In either case, a conditional form of analysis is indicated. A formal proof of this claim is given in §6.10 below.

STATISTICAL MODELS   169

6.5.4 The Poisson Person Time Model

Consider a $2 \times 2 \times 2$ table like Table 6.6 in which each count has a person time denominator. We model the data with an independent Poisson variate in each cell. No cell, margin or total count is fixed by design. Let $\lambda(y, x, z)$ be the **mean count per unit person time**, and let $\tau(y, x, z)$ be the person time for cell (y, x, z). Under the usual rationale for such studies, the person time values are taken as fixed quantities even though they are really not so. The mean value in the (y, x, z) cell then is

$$\lambda_*(y, x, z) = \tau(y, x, z)\lambda(y, x, z) \tag{6.19}$$

The probability distribution is a product of eight independent Poisson distributions with mean values $\lambda_*(y, x, z); y, x, z \in \{0, 1\}$. Now consider the model

$$\begin{aligned}\ln(\lambda(y, x, z)) &= \mu_0 + \mu_1 y + \mu_2 x + \mu_3 z + \\ & \quad \mu_{12} yx + \mu_{13} yz + \mu_{23} xz + \mu_{123} yxz\end{aligned} \tag{6.20}$$

By considering all the possibilities, the μ's can be expressed in terms of the λ's. For example, we have that

$$\mu_{123} = \ln\left(\frac{\lambda(1,1,1)\lambda(0,0,1)}{\lambda(0,1,1)\lambda(1,0,1)}\right) - \ln\left(\frac{\lambda(1,1,0)\lambda(0,0,0)}{\lambda(0,1,0)\lambda(1,0,0)}\right) \tag{6.21}$$

If $\mu_{123} = 0$, the cross product ratio of mean rates between y and x is the same at all levels of z. We then say that there is an absence of a **three way interaction** between the variables. An interpretation for each parameter in this model is left to the reader. Further, we can also show that

$$\begin{aligned}\ln(\lambda_*(y, x, z)) &= \mu_{*0} + \mu_{*1} y + \mu_{*2} x + \mu_{*3} z + \\ & \quad \mu_{*12} yx + \mu_{*13} yz + \mu_{*23} xz + \mu_{*123} yxz\end{aligned} \tag{6.22}$$

where the μ_*'s are related to the μ's by

$$\mu_{*1} = \delta_1 + \mu_1; \quad \mu_{*12} = \delta_{12} + \mu_{12} \tag{6.23}$$
$$\mu_{*13} = \delta_{13} + \mu_{13}; \quad \mu_{*123} = \delta_{123} + \mu_{123} \tag{6.24}$$

and so on, and further where

$$\delta_1 = \ln\left(\frac{\tau(1,0,0)}{\tau(0,0,0)}\right) \tag{6.25}$$

$$\delta_{12} = \ln\left(\frac{\tau(1,1,0)\tau(0,0,0)}{\tau(0,1,0)\tau(1,0,0)}\right) \tag{6.26}$$

$$\delta_{13} = \ln\left(\frac{\tau(1,0,1)\tau(0,0,0)}{\tau(0,0,1)\tau(1,0,0)}\right) \tag{6.27}$$

$$\delta_{123} = \ln\left(\frac{\tau(1,1,1)\tau(0,0,1)}{\tau(0,1,1)\tau(1,0,1)}\right) - \ln\left(\frac{\tau(1,1,0)\tau(0,0,0)}{\tau(0,1,0)\tau(1,0,0)}\right) \tag{6.28}$$

As for the simple 2 × 2 person time model in Chapter 5, these relations facilitate analysis of person time data using methods for pure count data. An illustration is given below.

6.5.5 Relations Between Models

Consider an incidence density study with model (6.20). Suppose, for one analysis, we designate y as the response, x as the exposure of interest, and z as a confounder variable. In each stratum, we fix the totals observed at each level of x and then analyze the data as if they were from a product binomial design. Using Properties 1.8, 1.9 and 1.10, we get, for $i = 0, 1$,

$$P[\mathcal{B}_0 = b_0 \mid \mathcal{B}_0 + \mathcal{D}_0 = m_0] = \binom{m_0}{b_0} \frac{\exp(\mu_{*1} b_0)}{\{1 + \exp(\mu_{*1})\}^{m_0}} \tag{6.29}$$

$$P[\mathcal{A}_0 = a_0 \mid \mathcal{A}_0 + \mathcal{C}_0 = n_0] = \binom{n_0}{a_0} \frac{\exp([\mu_{*1} + \mu_{*12}] a_0)}{\{1 + \exp(\mu_{*1} + \mu_{*12})\}^{n_0}} \tag{6.30}$$

$$P[\mathcal{B}_1 = b_1 \mid \mathcal{B}_1 + \mathcal{D}_1 = m_1] = \binom{m_1}{b_1} \frac{\exp([\mu_{*1} + \mu_{*13}] b_1)}{\{1 + \exp(\mu_{*1} + \mu_{*13})\}^{m_1}} \tag{6.31}$$

$$P[\mathcal{A}_1 = a_1 \mid \mathcal{A}_1 + \mathcal{C}_1 = n_1] = \binom{n_1}{a_1} \frac{\exp([\mu_{*1} + \mu_{*12} + \mu_{*13} + \mu_{*123}] a_1)}{\{1 + \exp(\mu_{*1} + \mu_{*12} + \mu_{*13} + \mu_{*123})\}^{n_1}} \tag{6.32}$$

where the μ_*'s were defined previously.

Let us reflect on what this implies. The four parameters expressing the relationship between y and (x, z), namely $\mu_{*1}, \mu_{*12}, \mu_{*13}, \mu_{*123}$, are retained in the restricted model arrived at through partial conditioning. If our focus is on these parameters, we may fit the product binomial type of model

$$\ln \frac{\pi(x, z)}{1 - \pi(x, z)} = \mu_{*1} + \mu_{*12} x + \mu_{*13} z + \mu_{*123} xz \tag{6.33}$$

If the data are pure counts, $\mu_{*j} = \mu_j$. Fitting this model then achieves our purpose. If they are from an incidence density study, fitting this model allows us to draw inference on $\mu_{*1}, \mu_{*12}, \mu_{*13}$ and μ_{*123}. We then derive inference on $\mu_1, \mu_{12}, \mu_{13}$ and μ_{123} using the relationships between μ_{*j} and μ_j given earlier.

The product binomial model is used similarly for the multinomial design and the product multinomial design that is provided the focus is on certain parameters. This is what we saw for a single 2 × 2 table. Further, as seen there, the mles and some null test statistics do not change under partial conditioning. Hence, for all the designs described above, the product binomial model is a valid starting point for certain selected forms of analysis. In that respect, the method of analysis is to an extent independent of the study design.

6.6 Conventional Analysis

Given the centrality of the product binomial design, we consider the large sample conventional analysis for a logistic model for this design. From (6.7) and (6.13), the unconditional likelihood for the data is

CONVENTIONAL ANALYSIS

$$\ell = \sum_{i=0}^{1} b_i \ln[\pi(0,i)] + d_i \ln[1 - \pi(0,i)] + \tag{6.34}$$

$$\sum_{i=0}^{1} a_i \ln[\pi(1,i)] + c_i \ln[1 - \pi(1,i)] \tag{6.35}$$

$$= \beta_0 t_0 + \beta_1 t_1 + \beta_2 t_2 + \beta_3 t_3 - m_0 \ln[1 + \exp(\beta_0)] -$$
$$n_0 \ln[1 + \exp(\beta_0 + \beta_1)] - m_1 \ln[1 + \exp(\beta_0 + \beta_2)] -$$
$$n_1 \ln[1 + \exp(\beta_0 + \beta_1 + \beta_2 + \beta_{12})] \tag{6.36}$$

The first partial derivatives of this, in terms of the β's, are

$$\frac{\partial \ell}{\partial \beta_0} = t_0 - \frac{m_0 \exp(\beta_0)}{1 + \exp(\beta_0)} - \frac{n_0 \exp(\beta_0 + \beta_1)}{1 + \exp(\beta_0 + \beta_1)}$$
$$- \frac{m_1 \exp(\beta_0 + \beta_2)}{1 + \exp(\beta_0 + \beta_2)} - \frac{n_1 \exp(\beta_0 + \beta_1 + \beta_{12})}{1 + \exp(\beta_0 + \beta_1 + \beta_{12})} \tag{6.37}$$

$$\frac{\partial \ell}{\partial \beta_1} = t_1 - \frac{n_0 \exp(\beta_0 + \beta_1)}{1 + \exp(\beta_0 + \beta_1)} - \frac{n_1 \exp(\beta_0 + \beta_1 + \beta_{12})}{1 + \exp(\beta_0 + \beta_1 + \beta_{12})} \tag{6.38}$$

$$\frac{\partial \ell}{\partial \beta_2} = t_2 - \frac{m_1 \exp(\beta_0 + \beta_2)}{1 + \exp(\beta_0 + \beta_2)} - \frac{n_1 \exp(\beta_0 + \beta_1 + \beta_{12})}{1 + \exp(\beta_0 + \beta_1 + \beta_{12})} \tag{6.39}$$

$$\frac{\partial \ell}{\partial \beta_{12}} = t_3 - \frac{n_1 \exp(\beta_0 + \beta_1 + \beta_{12})}{1 + \exp(\beta_0 + \beta_1 + \beta_{12})} \tag{6.40}$$

The mles obtain by setting these first derivatives equal to zero, solving the four equations and checking that a global maximum is achieved. The resulting estimates are

$$\hat{\beta}_0 = \ln\left(\frac{b_0}{d_0}\right) \qquad \hat{\beta}_1 = \ln\left(\frac{a_0 d_0}{b_0 c_0}\right) \tag{6.41}$$

$$\hat{\beta}_2 = \ln\left(\frac{b_1 d_0}{b_0 d_1}\right) \qquad \hat{\beta}_{12} = \ln\left(\frac{a_1 d_1}{b_1 c_1} \times \frac{b_0 c_0}{a_0 d_0}\right) \tag{6.42}$$

The likelihood is used to construct large sample score, LR and Wald test statistics for the parameters of interest along the lines given for the two binomials model in §5.13. When done for a single parameter, these statistics follow an approximately chisquare distribution with one df in large samples. This allows us to compute asymptotic p-values and CIs for the parameter. The details are left to the reader.

The statistic often used to construct confidence intervals for single parameters in this context is an empirical version of Wald statistic. For example, the large sample variance of the mle for β_{12} in this model is derived by inverting the appropriate information matrix. Then we get:

$$\sigma^2(\hat{\beta}_{12}) = \left\{\frac{1}{m_0 \pi(0,0)[1 - \pi(0,0)]} + \frac{1}{n_0 \pi(0,1)[1 - \pi(0,1)]}\right\} +$$
$$\left\{\frac{1}{m_1 \pi(1,0)[1 - \pi(1,0)]} + \frac{1}{n_1 \pi(1,1)[1 - \pi(1,1)]}\right\} \tag{6.43}$$

Large sample methods for the full design models such as the Poisson log-linear model and the multinomial model for two 2×2 tables are developed along the same lines.

Example 6.1 (continued): For illustration, we analyze the CMV prevention study of Table 6.3 using the logistic model. The results, based on the empirical Wald statistic, are in Table 6.8.

Table 6.8 *Conventional Analysis of Ganciclovir Data*

Parameter	p-value	Estimate	95% CI
β_0	0.1000	$+0.8755$	$(-0.1678, +1.9187)$
β_1	0.7177	-0.2564	$(-1.6466, +1.1338)$
β_2	0.2129	-0.7324	$(-1.9004, +0.4356)$
β_{12}	0.0063	$+2.4357$	$(+0.6887, +4.1827)$

This analysis gives evidence for a significant interaction between treatment and serostatus, and indicates that perhaps that is the sole significant effect in the model. It is instructive to analyze these data under a slightly different model. For this analysis, we had set $z = 0$ for a seronegative person, and $z = 1$ for a seropositive one. Now, let $z_+ = 0$ indicate seropositive status and $z_+ = 1$, a seronegative status. Consider then the model

$$\ln \frac{\pi(x, z_+)}{1 - \pi(x, z_+)} = \beta_{+0} + \beta_{+1}x + \beta_{+2}z_+ + \beta_{+12}xz_+ \qquad (6.44)$$

Fitting the new model to the same data, we find the following. First, the *p*-value for the interaction term is unchanged but that for the main treatment effect goes down to 0.0001. Further, while the *p*-value for the stratum effect is unchanged, that for the constant term is higher at 0.5933. The estimates for the interaction parameter and stratum effect parameter have the same absolute magnitude but reversed signs. On the other hand, estimates of the main treatment effect parameter and the constant parameter have different absolute magnitudes. The conclusion now is that both the main treatment and the interaction terms significantly depart from the null.

To explain how such a minor change affects how we interprete the data, we put $z_+ = 1 - z$ in model (6.44) and compare it with (6.7). Equating the corresponding probabilities in these models we find that

$$\beta_0 = \beta_{+0} + \beta_{+2}; \qquad \beta_1 = \beta_{+1} + \beta_{+12}$$

$$\beta_2 = -\beta_{+2}; \qquad \beta_{12} = -\beta_{+12}$$

These explain the discrepant findings. One message is that a simple way of model construction that only retains statistically significant terms is not advisable. A better method uses the **principle of hierarchy.** In a simple form, this is as follows: If a particular term denoting the interaction between some variables is included in a model, then we also include all terms with all lower level interactions and main effects for these variables. Using this principle in the above

CONDITIONAL ANALYSIS

example, the two ways of defining the serostatus variable produce the same model. Each has a constant term, the two main effect terms and the interaction term. We recover one model from the other by substituting the alternate variable definition.

Model construction and variable selection in general involve other considerations. These are, however, beyond the scope of this text. For an elementary but useful overview of such and other issues relating to logistic models, see Hall and Round (1994). Use of an exact approach in this context is discussed in Hirji et al. (1988).

6.7 Conditional Analysis

We first formally define two key terms we have used somewhat loosely thus far.

Exact Distributions

> The exact (unconditional) probability distribution for a study is the distribution emerging from the conjuction of the study design (sampling scheme) and the assumed (causal) probability (statistical) model for the study.
> An exact conditional distribution is a distribution derived from the exact unconditional distribution by treating some aspect of the observed data as fixed.

The exact unconditional distribution is also called the statistical model for the study. The fully conditional analysis of a parameter in a model for a $2 \times 2 \times 2$ table then utilizes the exact distribution of its sufficient statistic conditional on the sufficient statistics for other parameters. This distribution is used to give a conditional evidence function, and then a conditional *p*-value, point estimate and CI.

Let us first consider the product binomial model.

Theorem 6.1: For the product binomial design under model (6.1), and for this theorem let $\phi_{12} = \phi_3 = \exp(\beta_{12}) = \exp(\beta_3)$. Then, for $j = 0, 1, 2, 3$,

$$\mathrm{P}[\mathcal{T}_j = t \mid \mathcal{T}_i = t_i, i \neq j] = \frac{c_j(t)\phi_j^t}{f_j(\phi_j)} \qquad (6.45)$$

where

$$f_j(\phi_j) = \sum_{u=l_{1j}}^{l_{2j}} c_j(u)\phi_j^u \qquad (6.46)$$

and further where

$$\begin{align}
l_{10} &= t_1 + t_2 - t_3 \\
l_{20} &= m_0 + t_1 + t_2 - t_3 \\
l_{11} &= \max(t_3, t_0 + t_3 - t_2 - m_0) \\
l_{21} &= \min(t_3 + n_0, t_0 + t_3 - t_2) \\
l_{12} &= \max(t_3, t_0 + t_3 - t_1 - m_0) \\
l_{22} &= \min(t_3 + l_1, t_0 + t_3 - t_1) \\
l_{13} &= \max(0, t_2 - m_1, t_1 - n_0, t_1 + t_2 - t_0) \\
l_{23} &= \min(n_1, t_2, t_1, m_0 + t_1 + t_2 - t_0)
\end{align} \qquad (6.47)$$

with

$$\begin{align}
c_0(u) &= \binom{m_0}{u + t_3 - t_1 - t_2} \\
c_1(u) &= \binom{m_0}{t_0 + t_3 - u - t_2}\binom{n_0}{u - t_3} \\
c_2(u) &= \binom{m_0}{t_0 + t_3 - t_1 - u}\binom{m_1}{u - t_3} \\
c_3(u) &= \binom{m_0}{t_0 + u - t_1 - t_2}\binom{n_0}{t_1 - u}\binom{m_1}{t_2 - u}\binom{n_1}{u}
\end{align} \qquad (6.48)$$

Proof: For each j, the result follows from appropriate conditioning in (6.14) and canceling common terms in the numerator and the denominator. The range of the associated conditional sample space is derived by considering the values of u for which the combinatorial coefficient $c_j(u)$ is greater than zero. □

The following implications of this result are relevant.

- Provided we focus on selected parameters, conditional inference based on Theorem 6.1 applies to all the models and designs described in this chapter.
- If the data are from a case-control study then depending on the details of the design, valid analysis for only some parameters can be done (see §6.10 below).
- For an incidence density study, the conditional analysis has to be adjusted using the relations between μ_{*j} and μ_j.

6.8 An Example

Example 6.1 (continued): We perform conditional analysis of data from the study of Merigan et al. (1992) using model (6.7). The observed values of the sufficient statistics are $T_0 = 106$, $T_1 = 64$, $T_2 = 81$ and $T_3 = 51$. The conditional distribution relevant for the interaction effect parameter is $P[T_3 = t \mid T_0 = 106, T_1 = 64, T_2 = 81]$. This is a PBD, depicted in Table 6.9, obtained by using Theorem 6.1. The conditional sample space here has 13 points.

The mid-p conditional evidence functions for β_{12} is shown in Figure 6.1. Its rightward shift away

AN EXAMPLE

Table 6.9 *Conditional Generating Polynomial for* T_3

t	$c_3(t)$
44	$0.14673035742888E + 31$
45	$0.28954790532632E + 32$
46	$0.17719715199961E + 33$
47	$0.44984770212667E + 33$
48	$0.52991570345629E + 33$
49	$0.30453931039447E + 33$
50	$0.86821388854278E + 32$
$\Rightarrow$ 51	$0.12178565857388E + 32$
52	$0.81320551932342E + 30$
53	$0.24215592412146E + 29$
54	$0.28576254647284E + 27$
55	$0.10627532720064E + 25$
56	$0.70221227989039E + 21$

Note: $\Rightarrow$ is the observed sample point.

from the null indicates the presence of an interaction effect but its wide horizontal spread shows a low level of precision in estimation of the magnitude of the effect. The evidence functions for β_0, β_1 and β_2 are constructed with similar conditional distributions. All these yield mid-*p*-values, mues and 95% mid-*p* CIs in the usual way, and are shown in Table 6.10.

Table 6.10 *Mid-p Analysis of Ganciclovir Data*

Parameter	*p*-value	Estimate	95% CI
β_0	0.0963	$+0.8571$	$(-0.1475, +2.0248)$
β_1	0.7363	-0.2411	$(-1.7149, +1.1704)$
β_2	0.2298	-0.7085	$(-1.9850, +0.4339)$
β_{12}	0.0087	$+2.3505$	$(+0.5811, +4.2237)$

We used the mues for the parameters to estimate the $\pi(x, z)$'s under the relations of Table 6.7; these are in Table 6.11. The estimated probabilities are close to the observed ones. Overall, there is support for the proposition that among seronegative subjects, the treatment effect is marginal but it is markedly pronounced among seropositive patients.

Before accepting the results of a clinical trial, we also need to critically assess its design, implementation, analysis and reporting method. How was randomization done? Was the study double blind? What was the drop out rate? Did it differ between the treatment arms? Was outcome evaluation conducted in a uniform way? And so on. Major shortfalls at any of these stages can rarely be remedied at the analysis stage, be that an exact or asymptotic analysis.

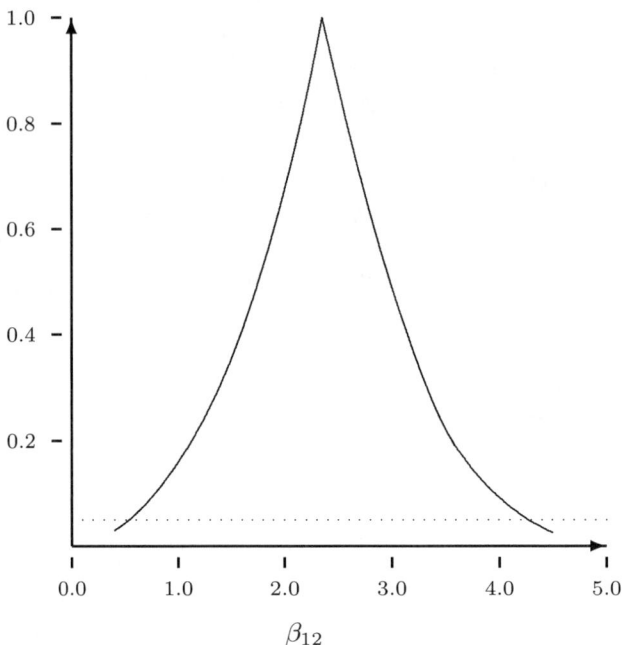

Figure 6.1 *Mid-p Evidence Function for Ganciclovir Data.*

Table 6.11 *Cell Probability Estimates*

		$\pi(x, z)$	
Treatment	Serostatus	Estimate	Observed
Placebo	0	0.70	0.71
Placebo	1	0.54	0.54
Ganciclovir	0	0.65	0.65
Ganciclovir	1	0.91	0.91

6.9 A Second Example

Example 6.4 (continued): We illustrate conditional analysis of incidence density data using the data in Table 6.6. Is there an interactive effect between exposure level, factor z and factor y? To answer this, we first fit model (6.29) using the cell counts only. To analyze μ_{*123}, we employ the conditional distribution of $\mathcal{T}_3$ derived from Theorem 6.1 and construct the TST mid-p evidence function. This is shown in Figure 6.2.

Now, we compute δ_{123} using (6.28). This is

$$\ln\left(\frac{3132.34 \times 709.37}{2649.23 \times 470.53}\right) - \ln\left(\frac{5925.03 \times 949.34}{2919.04 \times 964.01}\right)$$

which is equal to -0.115. To test $\mu_{123} = 0$, we test $\mu_{*123} = \delta_{123}$ from the evidence function. We then obtain a point estimate and confidence limits for μ_{*123} from this function. The corresponding quantities for μ_{123} are obtained by subtracting δ_{123}. The results appear in Table 6.12. As noted previously, because of the selective nature of the data, these conclusions may be at variance with those derived from the original data set.

Table 6.12 *Interaction Analysis of Arsenic Data*

Statistic			Results for μ_{123}	
Form	Name	p-value	Estimate	95% CI
Exact	TST	0.8257	-0.8899	$(-3.9852, +2.1687)$
Mid-p	TST	0.4763	-0.8899	$(-3.6130, +1.8040)$
Asymptotic	Score	0.4123	-0.8920	$(-3.2519, +1.4603)$

This is the case of a **shifted evidence function**. We get it by first constructing the evidence function for the pure count data, and shifting it **leftward** by the corresponding δ_j. These two evidence functions for the arsenic data are shown in Figure 6.2. Note that because $\delta_{123} < 0$, the count data function is shifted to the right.

6.10 On Case-Control Sampling

Suppose $\mathcal{X}$ is a binary exposure (0 = not exposed, 1 = exposed) and $\mathcal{Y}$, the outcome indicator (0 = control, 1 = case). In a cohort study, sampling on the basis of the exposure $\mathcal{X}$ permits a valid analysis of $\pi(x) = P[\mathcal{Y} = 1 \mid \mathcal{X} = x]$ for $x = 0, 1$. Then, for example, we may use the model

$$\ln\frac{\pi(x)}{1 - \pi(x)} = \alpha + \gamma x \qquad (6.49)$$

Subjects in a case-control study are sampled according to their disease status, and the exposure level is assessed subsequently. The validity of fitting (6.49) to data from such a study, thus treating the data if they were from a cohort study, needs careful justification. Towards this end, consider the basic case-control design described in §5.8. The conditions stated therein stated in more precise terms are:

Requirement 6.1: The s cases are randomly selected from an eligible population of cases with sampling fraction $\nu(1)$.

Requirement 6.2: The $N - s$ controls are randomly selected from an eligible population of controls with sampling fraction $\nu(0)$.

Requirement 6.3: The populations from which the cases and the controls are selected are homogeneous in terms of other factors that affect disease occurrence.

Requirement 6.4: For each group, sampling is done independently of exposure status.

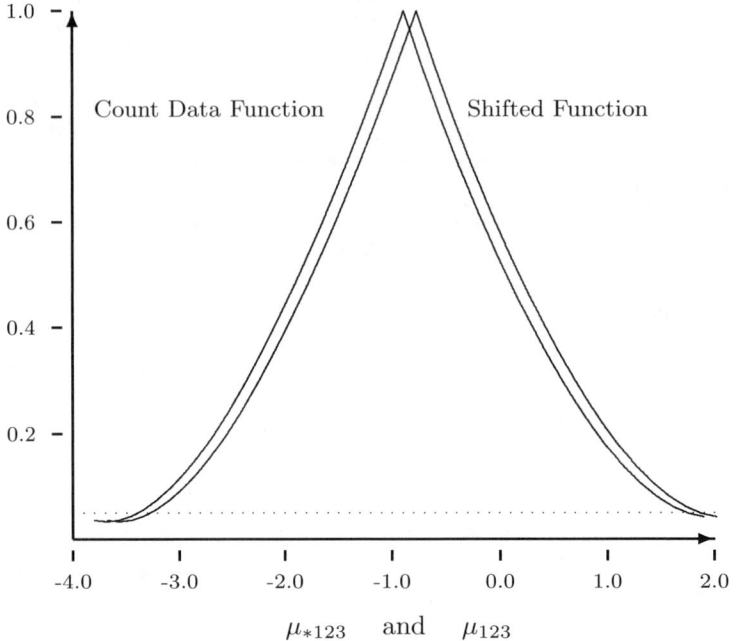

Figure 6.2 *Mid-p Evidence Function for Arsenic Data.*

Let $\mathcal{I}$ be the sampling indicator ($0 =$ not sampled, $1 =$ sampled). The above requirements mean that

$$\nu(y) \;=\; \mathrm{P}[\mathcal{I} = 1 \mid \mathcal{Y} = y] \tag{6.50}$$

and

$$\mathrm{P}[\mathcal{I} = 1 \mid \mathcal{X} = x, \mathcal{Y} = y] \;=\; \mathrm{P}[\mathcal{I} = 1 \mid \mathcal{Y} = y] \tag{6.51}$$

If η is marginal probability of exposure for any subject in the study, then

$$\mathrm{P}[\mathcal{X} = x \mid \mathcal{I} = 1] \;=\; \eta^x (1 - \eta)^{(1 - x)} \tag{6.52}$$

For any subject

$$\mathrm{P}[\mathcal{X} = x \mid \mathcal{Y} = y, \mathcal{I} = 1] \;=\; \frac{\mathrm{P}[\mathcal{X} = x, \mathcal{Y} = y \mid \mathcal{I} = 1]}{\mathrm{P}[\mathcal{Y} = y \mid \mathcal{I} = 1]} \tag{6.53}$$

ON CASE-CONTROL SAMPLING

$$= \frac{P[\mathcal{Y}=y \mid \mathcal{X}=x, \mathcal{I}=1]P[\mathcal{X}=x \mid \mathcal{I}=1]}{P[\mathcal{Y}=y \mid \mathcal{I}=1]} \tag{6.54}$$

$$= \frac{P[\mathcal{Y}=y \mid \mathcal{X}=x, \mathcal{I}=1]\eta^x(1-\eta)^{(1-x)}}{s^y(N-s)^{(1-y)}N^{-1}} \tag{6.55}$$

where the last expression follows from the proportions of cases and controls in the study. Now

$$P[\mathcal{Y}=y \mid \mathcal{X}=x, \mathcal{I}=1] = \frac{P[\mathcal{Y}=y \mid \mathcal{X}=x]P[\mathcal{I}=1 \mid \mathcal{X}=x, \mathcal{Y}=y]}{P[\mathcal{I}=1 \mid \mathcal{X}=x]} \tag{6.56}$$

$$= \frac{P[\mathcal{Y}=y \mid \mathcal{X}=x]P[\mathcal{I}=1 \mid \mathcal{X}=x, \mathcal{Y}=y]}{P[\mathcal{Y}=0, \mathcal{I}=1 \mid \mathcal{X}=x] + P[\mathcal{Y}=1, \mathcal{I}=1 \mid \mathcal{X}=x]} \tag{6.57}$$

$$= \pi(x)^y[1-\pi(x)]^{(1-y)}\nu(y)\delta(x)^{-1} \tag{6.58}$$

where $\delta(x) = \nu(0)[1-\pi(x)] + \nu(1)\pi(x)$. Combining these results, we have

$$P[\mathcal{X}=x \mid \mathcal{Y}=y, \mathcal{I}=1] = \mu(x,y)\pi(x)^y[1-\pi(x)]^{(1-y)} \tag{6.59}$$

where

$$\mu(x,y) = \left(\frac{\nu(y)\eta^x(1-\eta)^{(1-x)}N}{\delta(x)s^y(N-s)^{(1-y)}}\right) \tag{6.60}$$

The joint probability for all the subjects in the study

$$P[\mathcal{X}_j = x_j, j=1,\ldots,N \mid \mathcal{Y}_j = y_j, \mathcal{I}_j = 1, j=1,\ldots,N]$$

$$= \prod_{j=1}^{N} \mu(x_j, y_j)\left(\pi(x_j)^{y_j}[1-\pi(x_j)]^{(1-y_j)}\right) \tag{6.61}$$

$$= \Delta\pi(1)^a[1-\pi(1)]^c\pi(0)^b[1-\pi(0)]^d \tag{6.62}$$

where

$$\Delta = \left(\frac{\nu(1)^s\nu(0)^{(N-s)}\eta^{(a+c)}(1-\eta)^{(N-(a+c))}N^N}{\delta(1)^{(a+c)}\delta(0)^{(N-(a+c))}s^s(N-s)^{(N-s)}}\right) \tag{6.63}$$

Summing the above over all realizations $x_j, j=1,\ldots,N$ for which $\Sigma_j y_j x_j = a$ and $\Sigma_j(1-y_j)x_j = c$, we get

$$P[\mathcal{A}=a, \mathcal{C}=c \mid \mathcal{A}+\mathcal{B}=s, \mathcal{C}+\mathcal{D}=N-s] =$$

$$\Delta\binom{s}{a}\binom{N-s}{c}\pi(1)^a[1-\pi(1)]^c\pi(0)^b[1-\pi(0)]^d \tag{6.64}$$

where $b = s - a$ and $d = N - s - c$. Finally, with $\mathcal{T} = \mathcal{A}$ and conditioning on the number exposed, that is on $\mathcal{A} + \mathcal{C} = n$, $\mathcal{B} + \mathcal{D} = m$, and with $n + m = N$, we get

$$P[\mathcal{T} = t \mid \mathcal{S} = s, \mathcal{A} + \mathcal{C} = n] = \frac{c(t)\phi^t}{f(\phi)} \qquad (6.65)$$

where

$$\phi = \frac{\pi(1)/[1 - \pi(1)]}{\pi(0)/[1 - \pi(0)]} = \exp(\gamma) \qquad (6.66)$$

$$c(u) = \binom{s}{u}\binom{N-s}{n-u}\binom{N}{n}^{-1} \qquad (6.67)$$

$$f(\phi) = \sum c(u)\phi^u \qquad (6.68)$$

$f(\phi)$ is the hypergeometric generating polynomial. We have proved what was expressed earlier through equation (5.63), and restated in the following.

Theorem 6.2: For a case-control study with a binary exposure factor and done under requirements 6.1, 6.2, 6.3 and 6.4 above, a conditional analysis that fixes both margins of the 2×2 table is a valid form of analysis for the odds ratio in model (6.49). □

****

Now consider a group matched case-control study (§6.4.3 and §6.5.3). The populations of cases and controls are divided by the level of a factor $z = 0, 1$. For $z = 0$, s_0 cases and $N_0 - s_0$ controls are sampled and for $z = 1$, s_1 cases and $N_1 - s_1$ controls are sampled. In both cases, this is done under Requirements 6.1, 6.2, 6.3 and 6.4. Then consider, for $z = 0, 1$, the model

$$\ln\frac{\pi(x, z)}{1 - \pi(x, z)} = \alpha_z + \gamma_z x \qquad (6.69)$$

With $\mathcal{T}_0$ and $\mathcal{T}_1$ and other quantities defined for each 2×2 table of the group level data in a corresponding fashion, the following result readily follows:

Theorem 6.3: The conditional gp for the joint distribution of $(\mathcal{T}_0, \mathcal{T}_1)$ upon fixing both margins of each 2×2 table is given by $f_0(\phi_0)f_1(\phi_1)$. □

The group matched design then allows us to perform a valid conditional analysis of the parameters γ_0 and γ_1 in model (6.69). Earlier, in (6.18), the model for a group matched case-control design was stated as

$$\ln\frac{\pi(x, z)}{1 - \pi(x, z)} = \beta_0 + \beta_1 x + \beta_2 z + \beta_{12} xz \qquad (6.70)$$

The two versions are equivalent with $\beta_0 = \alpha_0$, $\beta_1 = \gamma_0$, $\beta_2 = \alpha_1$ and $\beta_{12} = \gamma_1 - \gamma_0$. In other words, fixing all margins of both 2×2 table permits a valid analysis of the main effect of factor x, and the interaction between x and z.

To analyze β_1, we further condition of the sufficient statistic for β_{12} and to analyze β_{12}, we further condition of the sufficient statistic for β_1. This is equivalent to applying relevant portions of Theorem 6.1 to the group matched data.

This result also extends to more complex settings; Theorem 6.2 applies to the situation when the exposure x is a vector of variables.

****

A variant of the above argument is used to justify the application of unconditional logistic regression for case-control data. These further assume that: (i) For each $y = 0, 1$, if $P[\mathcal{X} = x \mid \mathcal{I} = 1, \mathcal{Y} = y] = P[\mathcal{X} = x \mid \mathcal{Y} = y]$, then $P[\mathcal{X} = x \mid \mathcal{I} = 1] = P[\mathcal{X} = x]$. Under this reasoning conditional independence is taken to imply unconditional independence. It is also assumed that $P[\mathcal{X} = x]$ does not contain any information on the odds ratio. Then an unconditional logistic analysis is deemed valid since maximization of the unconditional likelihood provides valid estimates for all the parameters except the constant term.

These conditions are harder to justify. Conditional independence does not necessarily imply unconditional independence. Also, the marginal distribution of the covariates in the sample may not equal the marginal distribution in the population at large but may vary according to the case-control ratio. Unconditional logistic analysis for case-control data is thereby suspect both in large samples and small samples. The use of the conditional likelihood or the exact conditional distribution is called for in either case. However, as the ensuing error may not be that large, an unconditional analysis for large samples can be justified on empirical grounds.

6.11 Anatomy of Interactions

Interactive or synergistic effects are found in medicine, public health, psychology and many other fields. The adverse effects of a medicinal drug may be enhanced by the presence of another drug; an existent HIV infection may make the course of a malarial infection more severe; a genetic factor is suppressed in the presence of certain environmental conditions; the virulence of some microbes in the colon is enhanced by certain types of diets; and so on. In general, two factors are said to exhibit an interactive or synergistic effect when the effect of one varies according to the level of the other. **Heterogeneity** is another term to denote the idea of interaction.

To establish true synergy or actual interaction between two factors, one has to elucidate and verify the causal pathways or mechanism through which it occurs. Often, such type of evidence is either absent or weak. Instead, the presence or absence of synergy is inferred from empirical observations. In this case, we need to distinguish **statistical interaction** from **synergy;** the former does not necessarily imply the latter.

The terms interaction and statistical interaction are, in the context of statistical models, also used in a more restrictive sense. Two factors are then said to interact with one another if the linear effect of one is changed by the level of the other. It is in this sense that the parameter β_{12} in model (6.1) is an interaction effect parameter. When $z = 0$, the treatment log-odds ratio is β_1, and when $z = 1$, the treatment log-odds ratio is $\beta_1 + \beta_{12}$. In this text, the term interaction is synonymous with the term statistical interaction. Biologic interaction or synergy is not equivalent to statistical interaction, and the idea of interaction has substantive and policy dimensions as well.

The existence and magnitude of interaction is also specific to the scale on which it is measured. Consider the data in Table 6.13 relating to a column total fixed design. The first row gives the nonresponse probability, and the second row, the response probability, for the column.

Table 6.13 *Interaction or No Interaction?*

	$z = 0$		$z = 1$	
	$x = 0$	$x = 1$	$x = 0$	$x = 1$
$y = 0$	0.85	0.75	0.70	0.55
$y = 1$	0.15	0.25	0.30	0.45

Note: Cell values are column probabilities.

The odds ratio in the first stratum is $(0.85 \times 0.25)/(0.75 \times 0.15) = 1.9$, and in the second stratum, it is $(0.70 \times 0.45)/(0.55 \times 0.30) = 1.9$. Hence, $\beta_{12} = 0$.

Now let us examine the relative effect of the treatments on two other scales. First consider the difference in the response probabilities within each stratum. In stratum $z = 0$, we have

$$\pi(1, 0) - \pi(0, 0) = 0.25 - 0.15 = 0.10$$

and in stratum $z = 1$, we have

$$\pi(1, 1) - \pi(0, 1) = 0.45 - 0.30 = 0.15$$

Next, consider the ratio of the response probabilities in each stratum. Then we have

$$\pi(1, 0)/\pi(0, 0) = 0.25/0.15 = 1.67$$
$$\pi(1.1)/\pi(0, 1) = 0.45/0.30 = 1.50$$

In terms of comparing the treatments, a summary of these findings is:

- The odds ratio in stratum $z = 1$ is *equal* to the odds ratio in stratum $z = 0$.
- The difference of success probabilities is *greater* in stratum $z = 1$ compared to that in stratum $z = 0$.
- The ratio of success probabilities is *smaller* in stratum $z = 1$ compared to that in stratum $z = 0$.

The odds ratio scale shows an absence of interaction but the other two scales indicate otherwise. The direction of the interaction in the latter two scales is, however, opposite of one another! Hence, we conclude: **The magnitude and direction of interaction effect depends on the scale used to measure it.**

While we focus on logit or odds ratio models, risk difference and relative risk models are also applicable for many forms of discrete data. In some situations, they are preferred over the logit models. This and the fact that interaction is a scale specific entity has to be kept in mind when interpreting a logit analysis. The references noted below and in §6.12 elaborate on this issue.

According to Ottenbacher et al. (2004), checking the presence of and estimating the magnitude of an interaction effect is often not done or reported for logistic regression based multivariate analysis. In part this may be due to the fact that such effects are hard to detect, and also often difficult to ascertain. Three way and higher order interaction effects are also difficult to explain.

For actual examples of the scale specific nature of an interaction effect and comparisons of these scales, see Laupacis, Sackett and Roberts (1988), Hlatky and Whittemore (1991), Sinclair and Bracken (1994) and Cook and Sackett (1995). The illuminating series of articles by Altman and Matthews (1996), Matthews and Altman (1996a, 1996b) and Altman and Bland (2003) is a good starting point on this topic.

When two different treatments are recommended for a disease, clinical trials may be done to see if they exhibit a synergistic effect. Does one treatment enhance the effect of the other, or does it diminish its effect? In the context of model (6.1), let x be an indicator for treatment A and z be an indicator for treatment B. Patients are then randomized to one of four treatment groups: Control, Only A, Only B, and Both A and B. Such a trial design is called a **two way factorial design**. The parameter β_{12} then embodies the logit scale interaction effect between the two treatments. An overall test for $\beta_1 = \beta_2 = \beta_{12} = 0$ is then a test for interaction. The methods developed above also apply to the analysis of individual parameters from such a model. Trials to detect an interactive effect between treatments with a good level of precision generally need large sample sizes.

6.12 Relevant Literature

Analysis of two 2×2 tables is discussed in various books and papers, for example Worcester (1971), Zelterman, Chan and Mielke (1995) and Zelterman (1999). The latter two deal with exact analysis as well. Table 6.2 is based on my recollection from a class at the Harvard School of Public Health in the early 1980s. Simpson's paradox is explained with real examples in many works, among others, Simpson (1951), Wagner (1982), Appleton, French and Vanderpump (1996) and Agresti (1996).

The odds ratio based analysis of interaction in a $2 \times 2 \times 2$ table was formulated by Bartlett (1935). Gart (1972) used an exact probability method based test in this context and derived an approximation to it. Zacks and Solomon (1976) and Haber (1983) studied its properties as well. This test is a special case of the general test of Zelen (1971). Venables and Bhapkar (1978) found it to be highly conservative. A good review of the conceptualizations of interaction is in Rothman and Greenland (1998), Chapter 18. The references they give cover the different views on synergy. In addition to those noted earlier, clear and real data based explanations of the idea are in Agresti (1996), Clayton and Hills (1993) and Piegorsch and Bailer (1997). Darroch and Borkent (1994) focus on models of the type covered in this chapter. Parallels between interaction models for continuous data and discrete data help in understanding the issues. Shuster and van Eys (1983) develop a general method for this purpose. See also Hamilton (1979).

For justifications of the use of logistic models for retrospective data see Prentice (1976), Prentice and Pyke (1979), Hosmer and Lemeshow (1989), Breslow (1996) and Newman (2001). A comprehensive exposition of randomization for clinical trials is in Rosenberger and Lachin (2002). For an introduction to the factorial clinical trial design, see Pocock (1983) and Piantadosi (1997). References on the other topics covered in this chapter are given in several other chapters.

6.13 Exercises

6.1. Plot mid-p evidence functions for β_0 and β_2 for the clinical trial data in Table 6.3. Also plot the conventional exact and asymptotic conditional score evidence functions for all parameters. Also apply the CT method. Do the results from the different functions differ?

6.2. Perform a complete conditional analysis of arsenic exposure data shown in Table 6.6. Plot the exact and asymptotic score shifted evidence functions for all parameters and state your conclusions.

6.3. Analyze the case-control study data given in Table 6.5 using conditional and unconditional methods. How do they compare?

6.4. Kaiser et al. (1996) studied the efficacy of antibiotic treatment for patients with the common cold who had a specific indication for such therapy. This was a randomized, double blind, placebo controlled study. Table 6.14 gives data adapted from their paper.

Table 6.14 *Antibiotics for the Common Cold*

	Culture Positive		Culture Negative	
	Placebo	Co-amoxiclav	Placebo	Co-amoxiclav
Failure	27	22	67	75
Success	1	8	47	41
Total	28	30	114	116

Source: Kaiser et al. (1996), Table 2; © 1996 by The Lancet Ltd. Used with permission.

Analyze the data using conditional methods. Compare your analysis to traditional large sample analysis, and with those obtained by the authors. Critically assess the design, implementation and reporting methods of this study.

6.5. Table 6.15 shows an artificial sparse data version of the type of data often see in studies of the death penalty in the U.S. (For example, Anonymous (1998) gives such data.) They give the race of the victim and the defendant, and whether or not the death penalty was imposed in a total of 257 cases involving murder and with the possibility of capital punishment. Does either the race of the victim, or the race of the defendant influence the chance of the death penalty being imposed? Is there an interaction between these factors? Using these artificial data, perform a comprehensive analysis to address these questions.

Table 6.15 *Race and Outcome of Trial for Murder*

Race of		Total	Death
Defendant	Victim	Cases	Penalty
Black	Black	140	1
Black	White	23	7
White	Black	6	0
White	White	75	5

Note: Artificial data.

EXERCISES

6.6. Nordrehaug et al. (1996) conducted a clinical trial to compare manual and pneumatic artery compression devices for patients undergoing cardiac catheterization. One outcome was occurrence of bleeding. Parts of the data from their study are in Table 6.16 and Table 6.17. The three variables in the former are bleeding, compression device (manual or pneumatic) and blood pressure. In the latter, they are bleeding, aspirin dose and blood pressure. Analyze these data with conditional and unconditional methods and write a short report on your findings.

Table 6.16 *Bleeding and Compression Device*

	Systolic Blood Pressure (mm Hg)			
	< 160		≥ 160	
	Manual	Pneumatic	Manual	Pneumatic
No Bleeding	73	76	27	21
Bleeding	4	1	5	6
Total	77	77	32	27

Source: Nordrehaug et al. (1996), Table 5; Used with permission.

Table 6.17 *Bleeding and Aspirin*

	Systolic Blood Pressure (mm Hg)			
	< 160		≥ 160	
	Low Dose	High Dose	Low Dose	High Dose
No Bleeding	112	37	37	11
Bleeding	1	4	6	5
Total	113	41	43	16

Source: Nordrehaug et al. (1996), Table 4; Used with permission.

6.7. Table 6.18 and Table 6.19 each give five hypothetical data sets with three binary variables. Analyze each of the ten data sets given therein assuming it is from a total fixed design. Use conditional and unconditional methods and compare the *p*-values and confidence intervals for the interaction and main effect terms.

6.8. Table 6.20 gives three artificial data sets for a cohort design. Use appropriate Poisson models to perform evidence function based comparative analyses of these data.

6.9. Where relevant, apply the combined tails method to analyze the data in all the data sets and examples of this chapter; compare the resulting *p*-values and CIs with the TST exact and mid-*p* and asymptotic score methods.

6.10. Extend the arguments in §6.5.1 to demonstrate the relevance of the product binomial model to a clinical trial with randomization done within two strata.

6.11. Extend the arguments in §6.5.1 to demonstrate the relevance of the product binomial model

Table 6.18 *Five 2 × 2 × 2 Data Sets*

			\multicolumn{5}{c}{Study Counts}				
x	y	z	I	II	III	IV	V
0	0	0	3	11	0	2	9
1	0	0	1	0	0	3	11
0	1	0	0	0	1	0	1
1	1	0	2	0	12	0	0
0	0	1	7	4	3	7	2
1	0	1	5	1	3	8	3
0	1	1	0	2	2	6	5
1	1	1	13	3	2	1	0

Note: Hypothetical data.

Table 6.19 *Five More 2 × 2 × 2 Data Sets*

x	y	z	VI	VII	VIII	IX	X
0	0	0	4	1	3	1	19
1	1	0	2	1	0	12	1
0	1	0	1	7	0	5	8
1	1	0	1	3	2	8	0
0	0	1	17	0	13	0	1
1	0	1	15	10	0	18	3
0	1	1	10	2	2	6	5
1	1	1	3	13	0	11	7

Note: Hypothetical data.

to a simple two arm binary outcome clinical trial in which the total number of subjects in each arm is fixed at the outset and the order of entry into the treatment arm is randomized.

6.12. Extend the arguments in §6.5.1 to demonstrate the relevance of the product binomial model to a simple two arm binary outcome clinical trial in which the total number of subjects in each arm is fixed at the outset and the order of entry into the treatment arm is randomized within blocks of a fixed size.

6.13. Extend the arguments in §6.5.1 to demonstrate the relevance of the product binomial model

EXERCISES

Table 6.20 *Three Follow Up Design Data Sets*

			Study I		Study II		Study III	
x	y	z	Count	PY	Count	PY	Count	PY
0	0	0	3	111.0	0	22.5	1	33.3
1	0	0	1	10.0	0	31.2	1	29.4
0	1	0	10	123.0	1	20.0	21	103.8
1	1	0	2	12.5	20	40.6	1	40.7
0	0	1	17	41.4	0	71.2	12	33.0
1	0	1	5	13.3	3	88.0	3	61.0
0	1	1	0	201.3	2	16.6	2	30.2
1	1	1	11	32.1	12	21.3	0	12.0

Note: Hypothetical data; PY = (person years)/100.

to a stratified two arm binary outcome clinical trial in which the total number of subjects in each arm with each stratum is fixed at the outset and the order of entry into the treatment arm for each stratum is randomized.

6.14. Show clearly that expression (6.14) follows from (6.13).

6.15. Consider all 16 submodels of model (6.17). Which sets of subsets contructed from them satisfy the principle of hierarchy and which do not? Can you think of situations where a nonhierarchical model would form a plausible model?

6.16. Derive unconditional mles for (i) model (6.7), (ii) model (6.17) and (iii) models (6.20) and (6.22).

6.17. Derive unconditional score, LR ratio and Wald statistics for each parameter in model (6.7). How would you compute the restricted maximum likelihood estimates?

6.18. Derive unconditional score, LR and Wald statistics for the highest level interaction parameter in (i) model (6.7), (ii) model (6.17) and (iii) models (6.20) and (6.22). What do you conclude?

6.19. Show that appropriate conditioning in the total fixed design with model (6.7) gives a model in which the factor effects are the same as that in the product binomial design.

6.20. Consider a design for two 2×2 tables where subjects are sampled independently with the overall total in each table fixed. Write a probability model for this product multinomial design. Show, in detail, that assuming the data are from a product binomial design does not affect some forms of inference on selected parameters.

6.21. Consider the use of model (6.1) in the context of a factorial clinical trial design. Develop exact and large sample conditional and unconditional methods for the analysis of this model.

6.22. Give a detailed proof of Theorem 6.1.

6.23. Give a detailed proof of Theorem 6.2.

6.24. Extend Theorem 6.2 to the situation with two exposure factors.

6.25. Give a detailed proof of Theorem 6.3 and extend it to the case with multiple exposure factors.

6.26. Formulate a $2 \times 2 \times 2$ design with inverse sampling in which the number of failures in each column of each table is fixed. Develop unconditional and conditional methods to analyze these data.

CHAPTER 7

Assessing Inference

7.1 Introduction

Varied methods of data analysis at times give varied results. A score *p*-value differs from a likelihood ratio based value; an exact unconditional confidence interval is narrower than a conditional one. Which is better? For what type of data, models and designs? If not used with care, the methods of data analysis may themselves unduly distort a scientific investigation.

To guide practice, we need to evaluate and compare the methods. The question then is: How do we gauge the adequacy, accuracy or reliability of the tools of data analysis? Traditionally, such assessments are done along two related but distinct lines. One probes the issue terms of the basic philosophy of drawing inference from data. The other views them at a practical level by specifying desirable properties, and conducting empirical or theoretical checks for them. The latter is, of course, done in the context of a particular philosophy of statistical inference.

For now, we put the foundational issues aside. We restrict our attention to the basic approach adopted thus far, which is a part of what is known as the **frequentist tail area approach** to inference, and assess the tools of analysis based on evidence functions and related methods within this framework. A few desirable and actual properties of the tools of analysis were stated in the previous chapters. For univariate analysis, we also indicated a general preference for the TST mid-*p* and score methods. Now we give a more integrated formulation of the assessment criteria. For a wider comparative perspective, we also describe two other methods of data analysis. The focus remains on polynomial based distributions.

The specific aims here are:

- To describe the exact unconditional and randomized methods of analysis for discrete data.
- To define exact significance level and power of a test, and actual and nominal coverage levels of confidence intervals.
- To detail features of the two versions of the Fisher–Irwin exact test for a 2×2 table.
- To specify desirable features for the tools of analysis, and promote an evidence and power function based approach for assessing these tools.
- To broadly compare conditional and unconditional exact methods, and summarize the case for the mid-*p* based method.

Basic conceptual issues on conditional and exact inference, the relation between study design and methods of analysis, and an introduction to other approaches to statistical inference appear in Chapter 15.

7.2 Exact Unconditional Analysis

The four main avenues for analysis of discrete data specified in Chapter 5 were: exact and asymptotic conditional methods, and exact and asymptotic unconditional methods. So far, we have described all except the exact unconditional method. Now, we fill that lacunae. Since this is done mainly to enrich our comparison of the methods of analysis, our presentation is restricted to key definitions and a simple example. For a thorough explication, see the references in §7.11.1.

Consider a random vector $(\boldsymbol{S}, \mathcal{T})$ with a PBD whose gp is $f(\boldsymbol{\theta}, \phi)$. The vector $\boldsymbol{\theta}$ is a vector of nuisance parameters with sufficient statistic vector $\boldsymbol{S}$. ϕ is the parameter of interest with sufficient statistic $\mathcal{T}$. Suppose Ω is the unconditional sample space, and $(\boldsymbol{s}, t)$ is the observed sample point. And,

$$\mathrm{P}[\,(\boldsymbol{S}, \mathcal{T}) = (\boldsymbol{s}, t); \boldsymbol{\theta}, \phi\,] = \frac{c(\boldsymbol{s}, t)\boldsymbol{\theta}^{\boldsymbol{s}}\phi^{t}}{f(\boldsymbol{\theta}, \phi)} \tag{7.1}$$

Let $\mathcal{W}(\boldsymbol{S}, \mathcal{T}; \phi)$ be a statistic, not dependent on the nuisance parameters, we plan to use to analyze ϕ. Its observed value is $w(\boldsymbol{s}, t; \phi)$, abbreviated w. For any specified ϕ, we compute it at each point $(\boldsymbol{u}, v)$ in Ω, and define an indicator function

$$\begin{aligned}\mathcal{I}(\boldsymbol{u}, v; \phi) &= 1 \quad \text{if} \quad \mathcal{W}(\boldsymbol{u}, v; \phi) \geq w \\ \mathcal{I}(\boldsymbol{u}, v; \phi) &= 0 \quad \text{if} \quad \mathcal{W}(\boldsymbol{u}, v; \phi) < w\end{aligned} \tag{7.2}$$

If $\boldsymbol{\theta}$ is known, then the exact unconditional probability

$$\nu(\boldsymbol{\theta}, \phi; \boldsymbol{s}, t) = \mathrm{P}[\mathcal{I}(\boldsymbol{S}, \mathcal{T}; \phi) = 1; \boldsymbol{\theta}, \phi] \tag{7.3}$$

defines an exact two-sided evidence function for ϕ. Nuisance parameters are rarely, if ever, known. Often the large sample distribution of $\mathcal{W}$ is free from these parameters. Or we substitute an estimate, $\hat{\boldsymbol{\theta}}$, of $\boldsymbol{\theta}$ in (7.3). These options respectively produce an **asymptotic unconditional evidence function,** or an **approximate unconditional evidence function** for ϕ. The latter is written as

$$\mathsf{E}_a(\phi; \boldsymbol{s}, t) = \nu(\hat{\boldsymbol{\theta}}, \phi; \boldsymbol{s}, t) \tag{7.4}$$

An approximate or asymptotic function can be unreliable with sparse data. An **exact unconditional evidence function,** which does not depend on the values of the nuisance parameters, is obtained by maximization.

$$\mathsf{E}(\phi; \boldsymbol{s}, t) = \max_{\boldsymbol{\theta}} \{\nu(\boldsymbol{\theta}, \phi; \boldsymbol{s}, t)\} \tag{7.5}$$

This exact unconditional evidence function is plotted like any evidence function and applied as follows:

- The **exact unconditional p-value** for $H_0 : \phi = \phi_0$ is the value of the function at ϕ_0.
- The horizontal line at α_*, identifies the set of values of ϕ at which the function exceeds α_* and gives an **exact unconditional** $(1 - \alpha_*)100\%$ **confidence region** for ϕ.
- The associated point estimate, if existent and unique, is obtained from the intersection of all possible α_* level confidence regions.

This is a fail safe and conservative approach. If $\mathcal{W}$ has been selected appropriately, then whatever the value of $\boldsymbol{\theta}$, the p-value we get is higher, and the CI wider, than would otherwise be the case.

The following scheme is used in practice to compute such an evidence function. First we select a range of values for $\beta = \ln(\phi)$, say from -2 to 2.

EXACT UNCONDITIONAL ANALYSIS

- Select a particular value for β, say $\beta = 0.5$.
- For this β, and at each point (u, v) in Ω, including the observed point (s, t), get the value of $\mathcal{W}$.
- Obtain the indicator function $\mathcal{I}(u, v; \beta)$ at each point in Ω.
- Use this indicator to get $\nu(\boldsymbol{\theta}, \beta; s, t)$ for an arbitrary, unspecified $\boldsymbol{\theta}$.
- Maximize $\nu(\boldsymbol{\theta}, \beta; s, t)$ over the set of values of $\boldsymbol{\theta}$ to get the evidence function value at the current β.

This process is repeated over a large number of equally spaced values of β in the specified range to get an accurate plot of the function.

****

Example 7.1: We perform exact unconditional analysis for the odds ratio in a two group binary outcome clinical trial. Under the notation of §5.5, this problem has two parameters, (θ, ϕ), with sufficient statistics $(\mathcal{S}, \mathcal{T})$, and with (s, t), the observed point. (5.16) is the design distribution and the unconditional sample space is

$$\Omega = \{(u, v) : 0 \leq u \leq m + n;\ \max(0, u - m) \leq v \leq \min(u, n)\} \qquad (7.6)$$

Suppose the hypothetical data are in Table 7.1. Here $m = n = 3$, $t = 0$ and $s = 3$.

Table 7.1 Hypothetical Data

	$x = 0$	$x = 1$
Not Cured	0	3
Cured	3	0
Total	3	3

Under the product binomial model, the gp is

$$f(\theta, \phi) = (1 + \theta)^3 (1 + \theta\phi)^3$$

with response probabilities

$$\pi_0 = \frac{\theta}{(1 + \theta)} \quad \text{and} \quad \pi_1 = \frac{\theta\phi}{(1 + \theta\phi)}$$

We employ the score statistic (5.78) for $\beta = \ln(\phi)$. This is

$$\mathcal{W}(s, t; \beta) = (t - n\pi_1)^2 \left\{ \frac{1}{m\pi_0(1 - \pi_0)} + \frac{1}{n\pi_1(1 - \pi_1)} \right\} \qquad (7.7)$$

where π_0 and π_1 are evaluated at $\hat{\alpha}_r$, the restricted mle for $\alpha = \ln(\theta)$. This obtains from setting the first partial derivative, with respect to α, of the unconditional likelihood to zero. For the problem at hand,

$$\exp(\hat{\alpha}_r) = \frac{-(1+\phi)(3-s) + \sqrt{(1+\phi)^2(3-s)^2 + 4s\phi(6-s)}}{2\phi(6-s)}$$

With $s = 3$ and $m = n = 3$, we have

$$\hat{\alpha}_r = -0.5\ln\phi = -0.5\beta$$

and

$$\hat{\pi}_{0r} = \frac{\phi^{-0.5}}{1+\phi^{-0.5}} \quad \text{and} \quad \hat{\pi}_{1r} = \frac{\phi^{0.5}}{1+\phi^{0.5}}$$

Using this, the score statistic for $\beta = \ln\phi$ is

$$\mathcal{W}(3, t; \beta) = \frac{2[t + (t-3)\exp(\beta/2)]^2}{3\exp(\beta/2)}$$

We compute the evidence function at $\beta = 0$ for the data in Table 7.1 as follows. The exponents and coefficients of the gp are in the first two columns of Table 7.2. The score statistic at $\beta = 0$ obtains by substituting $\pi_0 = \pi_1 = s/(m+n)$ in (7.7). After simplification, this becomes

$$\mathcal{W}(s, t; 0) = \frac{[tm - (s-t)n]^2(m+n)}{mns(m+n-s)}$$

Table 7.2 *Exact Unconditional Analysis*

(u, v)	$c(u, v)$	$\mathcal{W}(u, v; 0)$	$\mathcal{I}(u, v; 0)$	$P[\mathcal{I}(u, v; 0) = 1]$
$(0, 0)$	1	0	0	
$(1, 0)$	3	6/5	0	
$(1, 1)$	3	6/5	0	
$(2, 0)$	3	3	0	
$(2, 1)$	9	0	0	
$(2, 2)$	3	3	0	
$(3, 0)$	1	6	1	$\theta^3/f(\theta, \phi)$
$(3, 1)$	9	2/3	0	
$(3, 2)$	9	2/3	0	
$(3, 3)$	1	6	1	$\theta^3/f(\theta, \phi)$
$(4, 1)$	3	3	0	
$(4, 2)$	9	0	0	
$(4, 3)$	3	3	0	
$(5, 2)$	3	6/5	0	
$(5, 3)$	3	6/5	0	
$(6, 3)$	1	0	0	

Note: $m = n = 3$ and $\beta = 0$.

This gives the third column of Table 7.2. At observed $s = 3$ and $t = 0$, we have that $w = 6$. Comparing this to the other values in this column gives the fourth column (indicator function) of Table 7.2, the components of $\nu(\alpha, 0; s, t)$, which appear in the final column. Adding these, we get

EXACT UNCONDITIONAL ANALYSIS

$$\nu(\theta, 0; 3, 0) = 2\theta^3 (1+\theta)^{-6}$$

The maximum value of this function, occurring at $\theta = 1$, is $1/32 = 0.03125$. This is the score exact unconditional p-value for $H_0 : \beta = 0$. Repeating this for other values of β values and plotting the result gives the complete evidence function.

The mle for θ is 1.0. Using this in $\nu(\theta, 0; 3, 0)$ gives the score based approximate unconditional p-value for $H_0 : \beta = 0$; this is also equal to $1/32$.

For comparison, the p-values from three other score statistics are also shown in Table 7.3.

****

Table 7.3 *Method and p-values*

Method	p-value
Exact Unconditional Score	0.0313
Exact Conditional Score	0.1000
Approximate Unconditional Score	0.0313
Asymptotic Unconditional Score	0.0143
Asymptotic Conditional Score	0.0253

Note: For the data in Table 7.1.

Other statistics available for exact unconditional analysis are the LR statistic and the probability based statistic. The latter is equivalent to

$$\mathcal{W}(s, t; \beta) = c(s, t)\hat{\theta}^s \exp(\beta t) \qquad (7.8)$$

Another possibility is to use the estimate based statistic

$$\mathcal{W}(s, t; \beta) = (\hat{\beta}(s, t) - \beta)^2 \qquad (7.9)$$

Mehrotra, Chan and Berger (2003) give a list of statistics for the exact unconditional analysis of a 2×2 table. Their focus is on the risk difference. Some of their formulations apply, and some can be extended as well to the odds ratio models. Their comparative results indicate that the choice of a test statistic is a crucial issue for exact unconditional tests.

****

Some important features of the exact unconditional method are:

- The exact unconditional method is not restricted to models with sufficient statistics. It thus has a broader scope than the exact conditional method.
- For a two binary outcome clinical trial, for example, we can get exact unconditional p-values and CIs for the relative risk and the risk difference.

- Computation of exact unconditional evidence functions, p-values and confidence intervals is a two fold effort. First, when not explicitly specified, the exact distribution (7.1) has to be computed. And second, we need to perform the optimization in (7.5).
- In designs with more than one nuisance parameter, often both are complex tasks. The former needs an efficient recursive algorithm, and the latter requires a careful search for the global maximum (Agresti and Min 2002).
- To reduce the computational effort, Berger and Boos (1994) introduced maximization over a confidence set for the nuisance parameter. Their method also ensures control of the type I error rate.
- A mid-p version of exact unconditional analysis has also been devised (Chen 2002).

So far, practically feasible exact unconditional tests exist only for a few discrete data designs. They include the independent and matched 2×2 table, an unordered 2×3 table, two group Poisson data and an ordered $2 \times K$ table. A major hurdle in expanding this list is computational intractability. Also, confidence intervals based on the exact unconditional approach are more difficult to obtain. Methods to compute them exist for the risk difference and relative risk in an independent data and a matched 2×2 table, and for a few other models. The references in §7.11.1 overview the research in this area.

Not all methods fit into the four categories noted above. Thus, a **semiconditional** or **partly conditional** method uses the joint distribution or likelihood for $\mathcal{T}$ and some elements of $\mathcal{S}$ while fixing the remaining elements of $\mathcal{S}$. Another approach takes off from the conditional method and then develops a finer partition of the sample space by adding another criterion (See Kim and Agresti 1994).

Another approach averages the evidentiary index over the space of the nuisance parameters. This requires the specification of an averaging function that is almost like a Bayesian prior. And yet another method is based on marginal distributions.

7.3 Randomized Inference

The p-values in continuous data problems generally have the following properties:

- They are realizations of a random variable that lies between zero to one.
- Their null distribution is $U(0, 1)$, with mean = median = 0.5. That is, the p-value is not inclined to be small or large, *a priori*.
- Each p-value corresponds to the probability of some event in the sample space.
- For a nominal two-sided test, the actual type I error rate is equal to the nominal level.

A singular facet of discrete data analysis is that while the sample space is discrete, the parameter space onto which its results are projected is continuous. For example, in a binomial trial with $n = 20$, the parameter range is $[0, 1]$. Yet, a method based solely on trial outcomes can at most yield twenty one point estimates for π. A method of computing p-values gives at most twenty one distinct values. This disjuncture is what is called the **problem of discreteness**. The p-value does not assume all values between zero and one, may not correspond to the probability of an event in the sample space, and may give an actual type I error rate not equal to the nominal level. Thus the quality of the inference drawn, especially with small samples or sparse data, is compromised.

The mid-p method, which allocates half of the observed probability to each tail, appears to address this problem. In truth, it only deals with some of its consequences. Even asymptotic

RANDOMIZED INFERENCE

methods, including those using the continuity correction, do not rectify it. Unconditional methods reduce the level of discreteness but do not eliminate it.

A method that does resolve the problem of discreteness is the method of randomized inference. Consider again the PBD of the variate $(\mathcal{S}, \mathcal{T})$ with gp $f(\boldsymbol{\theta}, \phi)$. We assess the one-sided hypotheses set up

$$H_0 : \phi \leq \phi_0 \quad \text{versus} \quad H_1 : \phi > \phi_0 \qquad (7.10)$$

Upon observing (s, t), we draw a value from a $U(0, 1)$ random variate Λ that is independent of $(\mathcal{S}, \mathcal{T})$. Let its realized value be λ. Further, we define

$$E_r(s, t, \lambda; \phi) = P[\mathcal{T} > t \mid \mathcal{S} = s; \phi] + \lambda P[\mathcal{T} = t \mid \mathcal{S} = s; \phi] \qquad (7.11)$$

This function, which puts a random portion of the observed point in the right tail, is the one-sided **conditional randomized evidence function** for ϕ.

Next consider the two-sided hypotheses:

$$H_0 : \phi = \phi_0 \quad \text{versus} \quad H_1 : \phi \neq \phi_0 \qquad (7.12)$$

Upon noting that $1 - E_r(s, t, \lambda; \phi)$ is equal to

$$P[\mathcal{T} < t \mid \mathcal{S} = s; \phi] + (1 - \lambda) P[\mathcal{T} = t \mid \mathcal{S} = s; \phi] \qquad (7.13)$$

we define the two-sided **conditional randomized evidence function** for (7.12) as

$$E_{*r}(s, t, \lambda; \phi) = 2 \min\{E_r(s, t, \lambda; \phi), 1 - E_r(s, t, \lambda; \phi)\} \qquad (7.14)$$

This function assigns λth portion of the observed conditional probability to the right tail and $(1 - \lambda)$th portion to the left tail.

Theorem 7.1: For fixed λ and a PBD, the function (7.11) is a continuous monotonically increasing function of ϕ and the function (7.14) is a continuous, bimonotonic (decreasing) function of ϕ.

Proof: Similar to that given for the exact and mid-p evidence functions in Chapters 2 and 3. □

The following remarks apply to one-sided and two-sided randomized evidence functions:

- The randomized evidence function assumes all values between zero and one, thus resolving the conundrum of discreteness.
- At $\lambda = 0.5$, it is the same as the corresponding mid-p evidence function.
- At the realized λ, the **randomized p-value** is the value of the function at ϕ_0.
- At the realized λ, the **randomized $(1-\alpha_*)$100% confidence region** is the parameter region in which the function value is at least α_*.
- Due to the bimonotonicity of the t randomized function, the confidence regions derived from it form single, connected intervals which also satisfy the property of nestedness.

- At the realized λ, the ϕ value at which the t evidence function is maximized is the **randomized point estimate** for ϕ.

****

Other definitions of t randomized evidence functions can also be given. For example, in a univariate setting, we may use

$$E_{*r}(t, \lambda; \phi) = 2\min\{P[\mathcal{T} > t; \phi], P[\mathcal{T} < t; \phi]\} + \lambda P[\mathcal{T} = t; \phi] \quad (7.15)$$

Or we base it on the score, LR or probability statistic. The probability statistic based randomized evidence function is

$$E_{*r}(t, \lambda; \phi) = P[c(\mathcal{T})\phi^{\mathcal{T}} < c(t)\phi^t; \phi] + \lambda P[c(\mathcal{T})\phi^{\mathcal{T}} = c(t)\phi^t; \phi] \quad (7.16)$$

Conditional versions are defined accordingly. Fuchs (2001) provides several varieties of t randomized tests including those based on equal tail probabilities and the CT criteria.

****

Randomized p-values and confidence intervals play a crucial role in the theory of discrete data analysis. Computing them also does not pose any challenge beyond that required for conventional exact analysis. However, they do have two disquieting features:

- In a research study, the randomized p-value and CI depend on a chance mechanism other than, and independent of, the process which generated the data.
- Two investigators using the same method of analysis on the same data may then arrive at different conclusions.

These features limit the use of the randomized method, especially in the biomedical field. The majority of applied statisticians do not consider it a practical option, and some have strongly spoken out against it. A few, however, promote its use. Fuchs (2001), for example, advocated it for genetic studies. This was done with an extended definition of p-values that was devised to partly counter the pitfalls of extraneous randomization.

Randomized p-values and confidence intervals exhibit, even in a conditional setting, some desirable and optimal properties. In that respect, they can serve as a standard for assessing other methods. It is for that reason that we include them in our discourse. We describe these properties later on in this chapter.

7.4 Exact Power

In Chapter 2, we considered the possibility, for a statistical test, of an incorrect decision when the null holds. Suppose, as set out in §2.7, we have a test performed at a nominal type I error rate α_* ($0 < \alpha_* < 1$). The true rejection probability under the null (the actual type I error rate, or actual significance level (ASL)) of the test is α_{*a}. If $\alpha_{*a} \leq \alpha_*$, the test is called a **conservative test**; otherwise, it is a **liberal test**. For more precise definitions of such terms, see Tang and Tony (2004).

Now we consider the case when the null hypothesis is false, that is, when the alternative holds. Then, the following definitions are used:

- The probability of not rejecting the null when it is false, denoted β_*, is called the **type II error rate** of the test.
- The **power** of the test is its ability to detect a factor effect or postulated difference when it exists.
- Power, the probability of rejecting the null when the alternative holds, is thus equal to $1-\beta_*$.

Power is a fundamental criterion for a test. Tests with a higher power are more desirable. Power gain, however, may come at a cost. With other things constant, Type I and type II errors are in general inversely related to one another. Decreasing one tends to increases the other, and *vice versa*. In choosing a test or a test statistic, we therefore need to strike a balance between the two errors.

Under the standard **Neyman–Pearson approach**, this is done by first specifying a desired type I (nominal) error rate. Often this is set at 0.05, and sometimes at 0.01, or 0.10. Then we adopt one of the following options:

- **Option 7.T.I**: Among the tests whose actual type I error rate does not exceed the nominal level, we select those with highest exact power either for the specified alternatives, or over the entire parameter space.
- **Option 7.T.II**: Among the tests whose actual type I error is close to the nominal level, we select those with highest exact power either for the specified alternatives, or over the entire parameter space.

Option 7.T.I limits the choice to conservative tests. Option 7.T.II, on the other hand, allows tests that are neither too conservative nor too liberal. Under this strategy, closeness to, and, where also possible, nonexceedance of the nominal size are favorable properties for a test.

Closeness here is defined in several ways. For the 0.05 level, we may take the range from 0.04 to 0.06. Or we restrict ourselves to being within 10% of the nominal level. We then ask: How often does the actual size fall in the range? Is the mean or median actual size within the range?

****

Power analysis is a central component of study design. Power generally increases with sample size, and also depends on how the subjects are sampled and allocated. After specifying their aims, the investigators normally select a design and analytic methods that enhance the chance of detecting the postulated effect. Within that framework, they strive to attain a sample size that ensures a specified minimal power level at the set nominal significance level. The minimum power is set at either 80% or 90%. Studies with human subjects in which the power to detect a relevant or clinically meaningful effect is small, say less than 50%, are, from an ethical perspective, not recommended.

Many textbooks and review papers contain power formulas for common discrete data tests. They are implemented in many software programs as well. These generally give the approximate power of the asymptotic tests. For example, consider a two arm binary outcome clinical trial under a t setup. An equal number, $n/2$, of subjects are to be assigned to the control and treated groups and the analysis will use a nominal α_* level asymptotic unconditional score test. In this case, the formula linking the approximate power, the nominal level, sample size and the true response rates when the true log-odds ratio is γ is

$$n\gamma^2 = (z_{\alpha_*} + z_{\beta_*})^2 \left\{ \frac{1}{\pi_1(1-\pi_1)} + \frac{1}{\pi_0(1-\pi_0)} \right\}^{-1} \tag{7.17}$$

where π_0 and π_1 are the anticipated response rates in the control and treatment groups respectively, and z_x is the point on the standard normal distribution such that the probability to the right of it is x.

Because they pertain to approximate power for asymptotic tests, such formulae involve two degrees of approximations. These approximations at times have opposing effect on accuracy, but at times, they magnify the effect of one another. The potential for a highly inaccurate result can be substantial. For both exact and asymptotic tests, computing the exact power is usually a better option, even when the sample sizes are moderately large.

****

To formalize the definition of exact power, consider once again $(\mathcal{S}, \mathcal{T})$, a multivariate PBD variate with sample space Ω, gp $f(\boldsymbol{\theta}, \phi)$ and realized observation (s, t). $\mathrm{E}(s, t; \phi)$ is the evidence function we plan to employ to assess $H_0 : \phi = \phi_0$ against $H_1 : \phi \neq \phi_0$. This may be a conditional or an unconditional, an exact or an asymptotic function.

For some $(\boldsymbol{u}, v)$, consider the (nominal) α_* level test indicator function, $\mathcal{I}(\boldsymbol{u}, v; \phi_0, \alpha_*)$:

$$\mathrm{E}(\boldsymbol{u}, v; \phi_0) \leq \alpha_* \Rightarrow \mathcal{I}(\boldsymbol{u}, v; \phi_0, \alpha_*) = 1 \qquad (7.18)$$

$$\mathrm{E}(\boldsymbol{u}, v; \phi_0) > \alpha_* \Rightarrow \mathcal{I}(\boldsymbol{u}, v; \phi_0, \alpha_*) = 0 \qquad (7.19)$$

When $\mathcal{I}(\boldsymbol{u}, v; \phi_0, \alpha_*) = 1$, we reject H_0, otherwise we do not. The subset of Ω for which $\mathcal{I}(\boldsymbol{u}, v; \phi_0, \alpha_*) = 1$ is its **rejection region** or **critical region**. Such a region also obtains by comparing a test statistic to a preset value. In that case, an implicit α_* level is embodied in the test. As the critical region depends on both the null value of the parameter of interest, and the assigned α_* level, we denote it as $\Omega(\phi_0, \alpha_*)$. This extends the univariate formulations given in §2.7.

The **power** or **exact power** of the test, denoted by $\Pi(\boldsymbol{\theta}, \phi, \phi_0, \alpha_*)$, is the probability of rejection under some specified value $(\boldsymbol{\theta}, \phi)$. Note that alternative values of the parameter under study and all the nuisance parameters have to be specified. For the given α_* level and null value,

$$\Pi(\boldsymbol{\theta}, \phi, \phi_0, \alpha_*) = \mathrm{P}[(\mathcal{S}, \mathcal{T}) \in \Omega(\phi_0, \alpha_*); \boldsymbol{\theta}, \phi] \qquad (7.20)$$

Or, for the PBD, we have

$$\Pi(\boldsymbol{\theta}, \phi, \phi_0, \alpha_*) = \{f(\boldsymbol{\theta}, \phi)\}^{-1} \sum_{(\boldsymbol{u}, v) \in \Omega(\phi_0, \alpha_*)} c(\boldsymbol{u}, v) \boldsymbol{\theta}^{\boldsymbol{u}} \phi^v \qquad (7.21)$$

Viewed as a function of the parameters $(\boldsymbol{\theta}, \phi)$, (7.20) and (7.21) are called the **power function**. The **actual significance level** or **actual type I error** of the test then is

$$\Pi(\boldsymbol{\theta}, \phi_0, \phi_0, \alpha_*) = \{f(\boldsymbol{\theta}, \phi_0)\}^{-1} \sum_{(\boldsymbol{u}, v) \in \Omega(\phi_0, \alpha_*)} c(\boldsymbol{u}, v) \boldsymbol{\theta}^{\boldsymbol{u}} \phi_0^v \qquad (7.22)$$

****

Example 7.2: Consider a one dimensional example with two diagnostic procedures of §3.4.

EXACT POWER

Table 7.4 *Two Score Rejection Regions*

t	$\mathsf{E}_a(t;0.5)$	$\mathcal{I}_a(t;0.5,0.05)$	$\mathsf{E}_e(t;0.5)$	$\mathcal{I}_e(t;0.5,0.05)$
0	0.014	1	0.027	1
1	0.066	0	0.094	0
2	0.221	0	0.359	0
3	0.540	0	0.762	0
4	1.000	0	1.000	0
5	0.540	0	0.762	0
6	0.221	0	0.359	0
7	0.066	0	0.094	0
8	0.014	1	0.027	1
9	0.002	1	0.004	1
10	$< .001$	1	$< .001$	1
11	$< .001$	1	$< .001$	1
12	$< .001$	1	$< .001$	1

Note: These regions are for $B(12;1/3)$ and $\alpha_* = 0.05$.

We hypothesize that procedure A is used half as often as B, that is, $\pi_0 = 1/3$ and $\phi_0 = \pi_0/(1-\pi_0) = 0.5$. At $n = 12$, the null distribution is $B(12, 1/3)$.

Consider the asymptotic (subscript = a) and exact (subscript = e) score tests at 0.05 level for this problem. For each test, Table 7.4 gives the *p*-value and test indicator at each sample point. Though the *p*-values for the two tests are not the same, at 0.05 level, they have the same rejection region. This is

$$\Omega(0.5, 0.05) = \{0, 8, 9, 10, 11, 12\}$$

The power of each test at ϕ, $\Pi(\phi, 0.5, 0.05)$, then is

$$(1+\phi)^{-12}\{1 + 495\phi^8 + 220\phi^9 + 66\phi^{10} + 12\phi^{11} + \phi^{12}\}$$

A graph of this function appears in Figure 7.1. The horizontal axis is scaled in terms of $\beta = \ln(\phi)$. If the ratio of the usage of A to B is 2:1 ($\pi_1 = 0.667$), and not 1:2 ($\pi_0 = 0.333$), the power at nominal level $\alpha_* = 0.05$ is 0.6316, obtained when we substitute $\phi = 2$ in the power function.

An asymptotic formula for power of the score test for $H_0 : \pi = \pi_0$ versus $H_0 : \pi = \pi_1$ ($\pi_1 > \pi_0$) in this situation is

$$n(\pi_0 - \pi_1)^2 = \left\{z_{\alpha_*/2}\sqrt{\pi_0(1-\pi_0)} + z_{\beta_*}\sqrt{\pi_1(1-\pi_1)}\right\}^2 \tag{7.23}$$

For the binomial 0.05 level score test with $n = 12$, $\pi_1 = 0.667$ and $\pi_0 = 0.333$, the approximate power is equal to 0.6879. To sum up, in this problem:

- The exact power of the exact score test is 0.6316.
- The exact power of the approximate score test is 0.6316.
- The approximate power of the asymptotic score test is 0.6879.

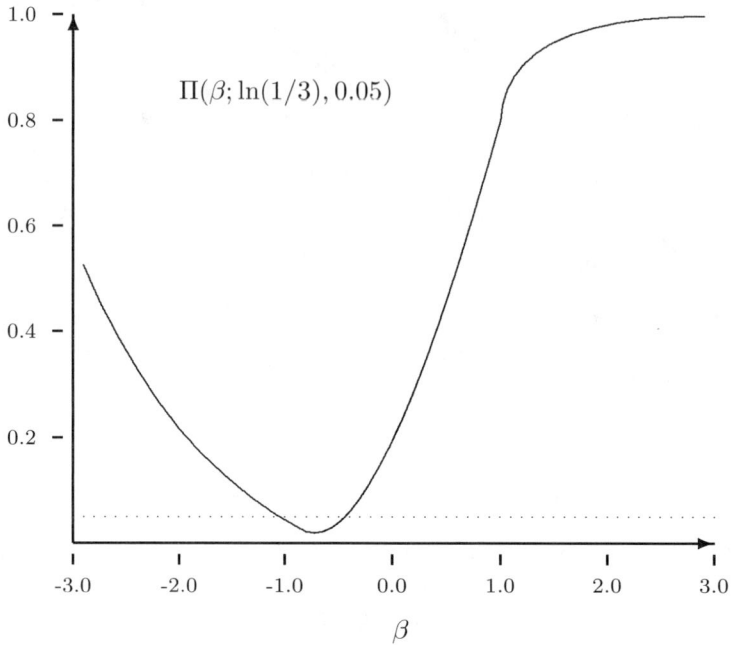

Figure 7.1 *An Exact Score Power Function.*

For the situation in Table 7.4, the probability of rejection under H_0 for the asymptotic and exact score test equals 0.0265. The actual sizes for the mid-*p* TST and exact TST tests in this problem are both also 0.0265, exemplifying the fact that in discrete data problems, the true size normally does not equal the nominal size.

****

Suppose we analyze ϕ with a conditional evidence function. Then power may be computed first at fixed values of the sufficient statistics for the nuisance parameters, and then combined overall as follows.

Given $\boldsymbol{S} = \boldsymbol{u}$, let the conditional rejection region for the nominal α_* level test be $\Omega(\boldsymbol{u}, \phi_0; \alpha_*)$. Then, we write the conditional power as

$$\Pi_c(\boldsymbol{u}; \phi, \phi_0, \alpha_*) = \mathrm{P}[\boldsymbol{T} \in \Omega(\boldsymbol{u}, \phi_0, \alpha_*) \mid \boldsymbol{S} = \boldsymbol{u}; \phi] \qquad (7.24)$$

which is also written as

$$\Pi_c(\boldsymbol{u}; \phi, \phi_0, \alpha_*) = \{f(\boldsymbol{1}, \phi; \boldsymbol{u})\}^{-1} \sum_{v \in \Omega(\boldsymbol{u}, \phi_0; \alpha_*)} c(\boldsymbol{u}, v)\phi^v \qquad (7.25)$$

EXACT POWER

The (unconditional) exact power, $\Pi(\boldsymbol{\theta}, \phi, \phi_0, \alpha_*)$, is then given by

$$\sum_{\boldsymbol{u}} \Pi_c(\boldsymbol{u}; \phi, \phi_0, \alpha_*) \mathrm{P}[\boldsymbol{S} = \boldsymbol{u}; \boldsymbol{\theta}] \qquad (7.26)$$

which is equal to

$$\{f(\boldsymbol{\theta}, 1)\}^{-1} \sum_{\boldsymbol{u}} \Pi_c(\boldsymbol{u}; \phi, \phi_0, \alpha_*) \boldsymbol{\theta}^{\boldsymbol{u}} d(\boldsymbol{u}) \qquad (7.27)$$

where $f(\boldsymbol{\theta}, 1)$ is the gp of $\mathcal{S}$, and $d(\boldsymbol{u})$ is the coefficient of the term of this polynomial in which the exponent of each component of $\boldsymbol{\theta}$ is the respective component of $\boldsymbol{u}$.

Power known to an accuracy of three decimal points suffices for most purposes. For some $\boldsymbol{u}$ in (7.26), the marginal probability $\mathrm{P}[\boldsymbol{S} = \boldsymbol{u}; \boldsymbol{\theta}]$ can be very small, less than, say, 10^{-5}. Ignoring such or smaller margins may still give a power value of desired accuracy. This is implemented as follows.

Let ϵ be a small value, and set $\delta = 0$ at the start. For a margin in (7.26), if $\mathrm{P}[\boldsymbol{S} = \boldsymbol{u}; \boldsymbol{\theta}] < \epsilon$, we do not perform the conditional power computation and set $\delta = \delta + \mathrm{P}[\boldsymbol{S} = \boldsymbol{u}; \boldsymbol{\theta}]$. Suppose the value obtained at the end of this computation is $1 - \beta_{*\epsilon}$. Then the exact power is within the limits given by

$$1 - \beta_{*\epsilon} - \delta \leq 1 - \beta_* \leq 1 - \beta_{*\epsilon} + \delta \qquad (7.28)$$

If this result is not within the desired accuracy, the value of ϵ is reduced. For the margins ignored initially, the computational step in (7.26) is then reconsidered. This strategy often gives an accurate value for exact power with a substantially lower computational effort. A strategy similar to this was adopted by Fu and Arnold (1992).

****

Table 7.5 *Data for a Conditional Test*

	$x = 0$	$x = 1$	Total
$y = 0$	$t + 5$	$10 - t$	15
$y = 1$	$5 - t$	t	5
Total	10	10	20

Example 7.3: For the two independent binomials problem under the notation of Chapter 5, let $m = n = 10$, and select a value of $s = 5$, as done in Table 7.5. The conditional sample space for the analysis of the odds ratio is the set of 2×2 tables with $t \in \{0, 1, 2, 3, 4, 5\}$. Consider the (nominal) level 0.05 conditional score test for $H_0 : \phi = 1$ versus $H_1 : \phi \neq 1$. The conditional rejection region for this is the set $t \in \{0, 5\}$. The conditional power function, $\Pi_c(u = 5; \phi, 1.0, 0.05)$, therefore is

$$\frac{252 + 252\phi^5}{252 + 2100\phi + 5400\phi^2 + 5400\phi^3 + 2100\phi^4 + 252\phi^5}$$

Conditional power is determined in a similar way for all values of $u = 0, 1, \cdots, 20$. Further,

$$f(\theta, 1) = (1 + \theta)^{20} \quad \text{and} \quad d(u) = \binom{20}{u}$$

Hence, the unconditional power function is

$$(1 + \theta)^{-20} \sum_{u=0}^{20} \binom{20}{u} \Pi_c(u; \phi, 1.0, 0.05) \theta^u$$

Suppose we set $\epsilon = 10^{-5}$. Then, when $\theta = 3/7$, in the initial stage, we do not perform the above computation for those values of u for which

$$\binom{20}{u} \left\{\frac{3}{7}\right\}^u \left\{\frac{7}{10}\right\}^{20} < \epsilon$$

****

Example 7.4: Consider a clinical trial with $n = m = 4$. The unconditional sample space is the set of 25 points that satisfy

$$\Omega = \{(a, b) : 0 \leq a \leq 4; 0 \leq b \leq 4\}$$

We test $H_0 : \phi = 1$ against $H_1 : \phi \neq 1$ at $\alpha_* = 0.1$ nominal level with the TST mid-*p* method. Performing this test at all twenty five possible outcomes, we find that the rejection region is

$$\Omega(1, 0.1) = \{(3, 0), (3, 3), (4, 0), (4, 4), (5, 1), (5, 4)\}$$

The power at some postulated values of the parameters is

$$\Pi(\theta, \phi, 1.0, 0.1) = \frac{4\theta^3 + 4\theta^3\phi^3 + 4\theta^4 + \theta^4\phi^4 + 4\theta^5\phi + 4\theta^5\phi^4}{f(\theta, \phi)}$$

with

$$f(\theta, \phi) = (1 + \theta\phi)^4(1 + \theta)^4$$

Assume that $\pi_0 = 0.3$ and $\pi_1 = 0.5$. Plotting power as function of ϕ when θ is set at $3/7$ (not shown), we note that this study has little power to detect even large departures from the null. Only with a very strong effect in favor of the new treatment (i.e., odds ratio greater than 21.4) is there a more than 50% chance of rejecting the null. Minimum power, moreover, is not at the null value but at $\phi = 0.73$. Setting $\phi = 1$, the ASL, as a function of θ, is

$$\frac{2\theta^3 + \theta + 4\theta^2}{(1 + \theta)^8}$$

EXACT POWER

Plotting $\Pi(\theta, 1, 0, 1.0, 0.10)$ as function of θ shows that the ASL of the 10% nominal level test is bell shaped with a maximum of 0.070% at $\theta = 1$, i.e., when the cure rate on the standard treatment is 50%. Earlier, with $\theta = 3/7$, the ASL was noted to be 0.047.

****

Conditioning on values that will be known only at the end of the study is not an option in power analysis for study design. In a clinical trial, for example, the total number of cures is not known in advance. We thus need to look at all possible outcomes, and not a selected subset. Thus, while conditional power is a useful construct for calculations, on its own it has little if any conceptual utility.

Power Analysis

> Study data may be analyzed with a conditional or an unconditional approach. Power analysis is always an intrinsically unconditional level action.

We also note some additional issues about power computation for study design.

- Power analysis needs specification of realistic outcome scenarios, and relevant differences. Preliminary studies and literature review as well as clinical experience may suggest the plausible range of the outcome rates in a proposed clinical trial.
- Computing power at various sample sizes under these scenarios provides an indication of the order of magnitude and plausible range for the sample size needed for the study.
- Power computation is sometimes done at the termination of study, especially when it yields a nonsignificant finding. Low power indicates a need to repeat the study with a larger sample size if the results seem to be clinically significant. The interpretation of such *post hoc* computation is, however, controversial.

Once the relevant parameters are determined, computing approximate power at various sample sizes is usually a straightforward task. In this way, we obtain the minimum sample size needed to attain the specified minimum power level.

On the other hand, sample size determination using exact power is a more intensive exercise. But, given the computing power available now, and provided a user friendly program exists, for a practitioner this need not be an arduous exercise. In such a program, this task is usually done in an iterative fashion:

- Determine an approximate sample size using an asymptotic power formula.
- Determine the exact power for this sample size.
- Decrease or increase the sample size by selected increments until the desired minimum exact power is attained.

****

Finally we provide some comparative cases of exact and approximate power analysis from the literature:

Thomas and Conlon (1992) presented tables of sample sizes for the two independent binomials design analyzed with the conditional probability based test. These are based on exact power computation, and apply to experiments with low event rates in particular. A key finding from them is that in comparison to three different asymptotic power formulae for this problem, **exact power computation often produces a lower sample size.**

Take one example. For $\pi_0 = 0.01$, $\pi_1 = 0.02$, $\alpha_* = 0.05$ and power equal to 90%, the sample size per group determined from t normal approximate computation is about 3800. Exact power computation for the Fisher–Irwin exact test, on the other hand, yields the needed sample size per group at about 3200 (Table 2 of Thomas and Conlon (1992)).

Fu and Arnold (1992) gave tables of sample sizes derived from exact power computation for the TST exact and probability based conditional tests for the 2×2 overall total fixed design. Their results show that when the marginal probabilities are not too far from 0.5, large sample sizes are needed to detect small departures of the odds ratio from unity. Exact computations are then recommended when one of the margins is expected to be small in relative or absolute terms.

For comparing **three binomials,** exact power computation may give sharply different results from approximate power formulae when the true probabilities are unequal (Bennett and Nakamura 1964). Hirji et al. (1994) gave cases of a $K \times 2 \times 2$ design in which asymptotic power formulae give misleading results.

7.5 Exact Coverage

Having dealt with tests, we now consider confidence intervals. For a method of computing such intervals for a parameter, we pose two types of queries: One, how often does the interval contain the true parameter value? Say, does a 95% CI computed in this manner contain it with 95% probability? Second, are the intervals so obtained narrow or wide? The first query relates to the **accuracy**, and the second, to the **precision level**, of the method.

Consider again the parameter ϕ from $(\boldsymbol{\theta}, \phi)$, and the model (7.1). Upon observing $(\boldsymbol{S}, t) = (\boldsymbol{s}, t)$, we compute a $(1 - \alpha_*)100\%$ CI for ϕ using some evidence function. Let $\mathcal{C}(\boldsymbol{s}, t; \alpha_*)$ denote this interval. For some specified value of ϕ, define an indicator function,

$$\phi \in \mathcal{C}(\boldsymbol{s}, t; \alpha_*) \;\Rightarrow\; \Xi(\boldsymbol{s}, t; \phi, \alpha_*) = 1 \tag{7.29}$$

$$\phi \notin \mathcal{C}(\boldsymbol{s}, t; \alpha_*) \;\Rightarrow\; \Xi(\boldsymbol{s}, t; \phi, \alpha_*) = 0 \tag{7.30}$$

The probability that these intervals cover the true parameter value then is

$$\xi(\boldsymbol{\theta}, \phi, \alpha_*) = \sum_{(\boldsymbol{u}, v) \in \Omega} \Xi(\boldsymbol{s}, t; \phi, \alpha_*) \mathrm{P}[\boldsymbol{S} = \boldsymbol{s}, \mathcal{T} = t; \boldsymbol{\theta}, \phi] \tag{7.31}$$

The quantity $(1 - \alpha_*)100\%$ is known as the **nominal, preset or desired coverage level**. The probability of containing the specified parameter value given by (7.31) is the **actual or true coverage level**, abbreviated as **ACL**.

Now consider precision: Let $\| \mathcal{C}(\boldsymbol{s}, t; \alpha_*) \|$ be the length of the CI, $\mathcal{C}(\boldsymbol{s}, t; \alpha_*)$. The mean length of these intervals is

$$\mu(\boldsymbol{\theta}, \phi, \alpha_*) = \sum_{(\boldsymbol{u}, v) \in \Omega} \| \mathcal{C}(\boldsymbol{s}, t; \alpha_*) \| \, \mathrm{P}[\boldsymbol{S} = \boldsymbol{s}, \mathcal{T} = t; \boldsymbol{\theta}, \phi] \tag{7.32}$$

EXACT COVERAGE

When some of the individual intervals are not finite, as often is the case for discrete data under the logit formulation, we compute the median length. Both these give the average precision of that particular method of computing CIs. A better picture is obtained from the distribution of the lengths of the intervals.

The shape and location of t evidence functions derived from a certain method are indicative of the mean length and coverage level of the associated CIs. If the two decreasing segments of the function are close to each other, the method will tend to produce shorter intervals. The function also generally shows a direct relation between length and nominal coverage. As the nominal coverage level increases, so does the length of the individual intervals and thereby also the mean length. A higher coverage level may thus lower precision.

Since the ACL in discrete data models is generally not the same as the preset level, we employ selection standards similar to that used for hypothesis tests. That is, we either admit only methods that guarantee the desired coverage level, the **conservative methods**, or we allow moderately **liberal methods** whose actual coverage level fluctuates near the preset level.

- **Option 7.C.I**: Among the methods whose ACL does not go below the nominal level, select those with shortest mean length either for the specified parameter values, or over the entire parameter space.
- **Option 7.C.II**: Among the methods whose ACL is close to the preset level, select those with shortest mean length either for the specified parameter values, or over the entire parameter space.

Among methods with similar ACL, the one giving shorter (more precise) intervals on average is preferable. We note that, in practice, the focus has more been on coverage levels rather than on lengths of intervals.

****

Conditional coverage and **conditional mean length** of the CIs are formulated in a way similar to that for conditional power. For a PBD, these depend only on the parameter of interest and do not require the specification of the unconditional distribution. The overall ACL and mean length are then the respective weighted averages of the conditional counterparts. The convenience of these conditional entities has, however, to be balanced against their limitation, namely that they relate to their behavior in one portion of the sample space.

****

For entities derived from a well behaved evidence function, the ACL is directly related to the ASL. In a t analysis of a parameter from a continuous, bimonotonic evidence function, a confidence set at any level $(1-\alpha_*)100\%$ is a continuous interval. In this case, a CI contains the true value ϕ if and only if the p-value at the true value as the null value is greater than α_*. In other words, we have

$$\mathcal{I}(s,t;\phi,\alpha_*) = 1 - \Xi(s,t;\phi,\alpha_*) \tag{7.33}$$

Substituting in (7.31) and simplifying, we get a crucial result:

Theorem 7.2: For inference based on a continuous bimonotonic evidence function, the ACL at nominal level $(1-\alpha_*)100\%$ is equal to one minus the ASL of the associated $\alpha_*\%$ t test at the true parameter value, or

$$\xi(\boldsymbol{\theta}, \phi, \alpha_*) = \Pi(\boldsymbol{\theta}, \phi, \phi, \alpha_*) \tag{7.34}$$

□.

In order to obtain the true coverage level, we thereby do not need the CIs at each sample point. This result does not necessarily extend to evidence functions which are not bimonotonic. The exact score or exact LR evidence functions, for example, often violate this property. On the other hand, the asymptotic score evidence function does not.

****

Example 7.5: For the $B(12; \pi)$ problem, Table 7.6 shows the true coverage levels of three methods for computing a 95% CI for β, the log-odds of success. The exact TST method has higher than nominal coverage at all parameter values while the ACL of the other two methods fluctuate around the nominal level. If a larger number of finely distributed parameter values are examined, the asymptotic score and the mid-p methods show a similar overall pattern. But at specific values they give different ACLs.

Table 7.6 *Actual Coverage for Three Methods*

	$\xi(\beta, \alpha_*)$		
β	TST Exact	TST Mid-p	Asymp. Score
−2.0	0.99	0.99	0.95
−1.6	0.99	0.96	0.96
−1.2	0.99	0.92	0.96
−0.8	0.98	0.98	0.94
−0.4	0.97	0.97	0.97
−0.0	0.96	0.96	0.96
+0.4	0.96	0.96	0.96
+0.8	0.98	0.98	0.94
+1.2	0.99	0.92	0.96
+1.6	0.99	0.96	0.96
+2.0	0.99	0.99	0.95

Note: This is for $B(12; \pi)$ and nominal coverage 95%.

A plot of the ACL against the parameter value is called a **coverage function**. Such functions often have a spike shaped pattern with sharp increases and decreases in the coverage levels. Methods that give coverage levels near the nominal level are preferred.

7.6 The Fisher and Irwin Tests

In this section, we illustrate the process of comparing the tools of data analysis with the help of two forms of a commonly used exact test for discrete data. The most popular exact test for categorical data today goes under the name of the **Fisher exact test**. It has several versions which follow the varieties of exact tests noted in Chapter 3. Two of the initial one versions appeared around the same time. The first was given by Fisher (1934) and the other was independently developed by Irwin (1935). Though both utilize the hypergeometric distribution, the contribution made by Irwin is rarely acknowledged. Since at least five versions of the conditional test (one-sided exact and randomized, and t probability, TST and CT) are now and then given the same name, we first clarify the terminology. Then we compare the Fisher and Irwin versions of the test.

Fisher introduced his test as a one-sided test. Here, the type of test statistic is not a critical issue. The probability and score orderings, for example, give the same answer. In the t case, that is not so. Though Fisher usually gave a t p-value when he performed a statistical test, for the conditional hypergeometric test, he never published a t p-value (Cormack 1986). But in a private letter he indicated that a significance level for this test should be obtained by doubling the one-tailed value (Bennett 1990, page 239).

At one point, Fisher illustrated his test with an example in which both margins of the 2×2 table were fixed by design. This concerned an experiment with a lady who said she could distinguish the order of pouring milk and sugar in tea. She was given eight cups of tea. In four, milk had been poured first, and in the rest, sugar had been poured first. The tea cups were presented in a randomized order, though she was aware of the $4:4$ ratio (Fisher 1934; 1935b). The set of five 2×2 tables with all marginal totals equal to four is the unconditional and the conditional sample space here. The lady correctly identified three of the cups in which milk had been poured first. The one-sided exact p-value is then equal to $17/70 = 0.24$.

Irwin, on the other hand, introduced his conditional test as a two-tailed test based on the probability ordering principle, and derived it from the two independent binomials design (Irwin 1935). Thus only one margin was presumed fixed by design. But, for analysis, the other margin was also fixed at its observed value. In the ensuing debates on exact tests, and in other cases, Fisher also employed the product binomial model to illustrate his exact test.

In the statistical and research literature, what is called the Fisher exact test is sometimes the one-sided test, rarely the one-sided randomized test or the t tail probability test, at times the t TST test, but most often the t probability method test. An 1994 survey of 13 textbooks covering this test found that **seven** of them deployed the probability criterion. **Two** books also gave the TST method as an alternative. Of the rest, **five** proposed the TST method as the primary criterion with **one** noting the combined tails method as an alternative. **One** book used a unique criterion not found elsewhere. Further, **one** of each giving the probability or the TST criterion gave it in a mid-p form (Hirji and Vollset 1994c).

In a unimodal setting, the one-sided Fisher and Irwin tests are identical. The one-sided hypergeometric test for a 2×2 table is then best named the **Fisher–Irwin test**. For the t case, the following labels are historically appropriate and come in handy when a distinction is necessary.

- The **Fisher Exact Test** for a 2×2 table is twice the smaller tail (TST) t test based on the conditional hypergeometric distribution.
- The **Irwin Exact Test** for a 2×2 table is the probability method t test based on the conditional hypergeometric distribution.

In general, and especially in applied research, the contribution of both the inventors ought to be acknowledged; the **Fisher–Irwin** exact test is then the appropriate label, whatever the form of the test.

****

Table 7.7 *Comparing Fisher and Irwin Exact Tests*

Dataset	Table Data				p-value	
	a	b	c	d	Irwin Test	Fisher Test
1	0	2	30	28	0.4915	0.4916
2	0	2	31	28	0.2377	0.4754
3	0	2	32	28	0.2300	0.4600
4	0	4	34	31	0.1142	0.1212
5	0	4	35	31	0.1142	0.1142
6	0	4	36	31	0.0539	0.1078
7	10	20	90	80	0.0734	0.0734
8	10	20	91	80	0.0497	0.0690
9	0	5	1000	995	0.0622	0.0622
10	0	5	1000	994	0.0310	0.0620
11	9	20	712	695	0.0400	0.0558
12	9	20	706	695	0.0588	0.0588

Sources: Datasets 1–6 from Cormack (1986).
Datasets 7–10 from Dupont (1986); ©1986 by John Wiley & Sons Limited.
Datasets 11–12 from Vollset and Hirji (1994c) and Hull et al. (1993).
All data used with permission.

Now we compare and contrast the properties of these two exact tests and related entities. One goal of this is to set the stage for a more general discussion of desirable and actual features of the tools of data analysis.

For a PBD, the TST evidence function being a continuous bimonotonic function yields coherent p-values and CIs. The probability ordering principle, on the other hand, may generate a discontinuous evidence function with more than one increasing, and more than one decreasing, segments. The p-values may then decrease and increase repeatedly as one moves away from the null. The CIs may also display similar undesirable characteristics.

Accordingly, several authors have shown that the Irwin test, based on probability ordering, displays several anomalous features. We enumerate them below.

Unstable Performance: The p-value from the Irwin test can be very sensitive to minor changes in the data. This is shown from the five segments in Table 7.7. The counts a, b, c and d are cell values of a 2×2 table as in Table 5.1. In each segment, these counts are minor variations of an underlying data set. For each, the Irwin test p-value shows a big change but the Fisher test value is more stable. Upon a systematic study of 920 pairs of 2×2 tables, Dupont (1986) concluded that for the probability exact test, such *"fluctuations occur frequently over a wide range of sample sizes and significance levels."* In contrast, the TST method *"provides a more consistent measure of inferential strength."*

Counterintuitive Performance: The Irwin test shows three types paradoxical anomalies. (i) Consider an example given by Cormack (1986). Let $a = 0$, $b = 4$, $c = 11$ and $d = 31$. Irwin's p-value for this is 0.5588. Suppose we make a small change in the data: $a = 0$, $b = 4$, $c = 12$

THE FISHER AND IRWIN TESTS

and $d = 31$, which produces a larger observed difference. The Irwin p-value **increases** to 0.5597, in contrast to what would be expected. On the other hand, the TST p-value goes down from 0.6418 to 0.5878. (ii) The power of a test is expected to increases with sample size. The power of the Irwin test can decrease as the sample size increases. (iii) A related paradox occurs in relation to the t p-value. Suppose we test $\phi = \phi_0$ for particular data. As ϕ_0 becomes more extreme compared to 1.0 in a given direction, once the p-value has begun to decrease, we do not then expect it to increase. For the Irwin test, it does at times increase.

For examples of these paradoxes, see Hirji and Vollset (1994c). They do not occur with the TST (Fisher) method.

Confidence Intervals: Exact confidence intervals for the odds-ratio in practice are generally obtained from inverting the TST (Fisher) test while the reported exact p-value is that given by the Irwin test. The two may, however, not be consistent. For the study Hull et al. (1993) with $a = 9$, $b = 20$, $c = 712$ and $d = 695$, for example, the 95% exact TST confidence interval for the odds ratio is $(0.98, 5.72)$. This conflicts with the Irwin p-value, which is 0.0400, but is consistent with the TST (Fisher) p-value, 0.0580 (Vollset and Hirji 1994c).

The probability (Irwin) criterion, which may give a disjointed interval, is never used to compute confidence intervals for real data. The frequent use of this criterion perhaps contributes to the excessive focus on p-values in research reports.

****

The TST Fisher exact test does not exhibit the problems noted above. But, it has some problems of its own. We note three of them below.

Too High p-values: The classical TST method double counts the observed probability. It can then give p-values that are higher than 1.0. This is remedied by defining p-value as the minimum of 1.0 and twice the smaller tail probability, or by using the mid-p correction.

Probabilistic Interpretation: The TST p-value may not correspond to the probability of an event in the sample space. It is thus said to lack a **frequentist** interpretation. This is, however, an expression of a much larger phenomenon. Almost all of the numerous large or small sample methods of computing a p-value for discrete data, including the mid-p method, have the same feature. This is particularly the case in the presence of nuisance parameters. If p-values are seen as indices of inference that assume values between zero and one, then this concern is not a critical one (see Chapter 2).

Conservativeness: Empirical studies show that the Fisher TST test is generally more *conservative* than the Irwin probability based test. It produces larger p-values and is less favorable to the alternative hypothesis than the Irwin test. As such, it is often a very low power test. The mid-p correction adjusts this without an excessive compromise of the type I error. However, the mid-p method produces a better result with the probability criterion (Hirji, Tan and Elashoff 1991).

****

A comparison of the Fisher and Irwin tests also pertains to the manner in which they were derived. In the former, both margins were fixed by design while in the latter, that was not so. Yet, neither of the two test criteria are inextricably related to how the tests were derived. But their disparate derivation raises questions about the validity of conditional inference, and more broadly, about the link between study design and data analysis method.

One approach to analysis of randomized studies assumes responses of treatment for each subject, or at least the total number of responders in each treatment arm to be determinate quantities. The sole mechanism imparting randomness is the randomization scheme. This is used to derive and justify exact conditional tests, especially for a 2×2 table. This approach to inference is called the **sample based permutation approach**, or the **permutation approach**. The two binomials based approach is then called the **model based** or **population based approach**. We reflect on these approaches in Chapter 15.

7.7 Some Features

In this section, we examine some of the actual features of randomized, exact and mid-p values and confidence intervals derived from the respective evidence functions.

First we consider the conditional exact TST evidence function. Below we state results for the t case; the one-sided case is identical in essence.

Theorem 7.3: A univariate TST conditional evidence function for ϕ for a PBD under the usual notation displays the following features: (i) The unconditional ASL of a nominal α_* level exact TST test for $H_0 : \phi = \phi_0$ versus $H_1 : \phi \neq \phi_0$ is less than or equal to α_*; (ii) The unconditional ACL of nominal $(1 - \alpha_*)100\%$ exact TST CIs for ϕ is greater than or equal to $(1 - \alpha_*)100\%$.

Proof: First consider the univariate problem. Suppose there exists $t_+ \in \Omega$ such that

$$\mathsf{F}(t_+; \phi_0) \leq \mathsf{S}(t_+; \phi_0) \tag{7.35}$$

Let t_* denote the largest t_+ that also has the property that $2\mathsf{F}(t_+; \phi_0) \leq \alpha_*$. If no such t_+ exists, then we set $t_* = -\infty$. If any $t \in \Omega$ is such that $t \leq t_*$, then $t \in \Omega(\phi_0; \alpha_*)$.

Similarly, consider the right tail of the rejection region. In this region,

$$\mathsf{S}(t; \phi_0) \leq \mathsf{F}(t; \phi_0) \tag{7.36}$$

with the TST p-value less than α_* at each t. Let t_{**} denote the smallest point in the right tail of rejection region. If this tail region is empty, we let $t_{**} = +\infty$.

It is easy to verify that $t_* < t_{**}$. The rejection region consists of these two points plus all points to the left of t_* and all to the right of t_{**}. By definition, the actual rejection probability under the null is

$$\alpha_{*a} = \mathsf{F}(t_*; \phi_0) + \mathsf{S}(t_{**}; \phi_0) \tag{7.37}$$

Since

$$2\mathsf{F}(t_*; \phi_0) \leq \alpha_* \quad \text{and} \quad 2\mathsf{S}(t_{**}; \phi_0) \leq \alpha_* \tag{7.38}$$

SOME FEATURES

it follows that $\alpha_{*a} \leq \alpha_*$. The coverage property follows from the bimonotonic nature of the TST evidence function and applying Theorem 7.2.

That these properties also hold for exact conditional TST tests and CIs follows by first applying the result to fixed values of the conditioned variables. The unconditional size and coverage level are weighted averages of their respective conditional values. Since each quantity being averaged is bounded, the overall result is bounded accordingly. □

This theorem formally proves what thus far we have stated in passing so far. It applies to all tests conventionally called exact tests, including the univariate exact score, LR, probability and CT based tests and their related CIs. Bimonotonicity is not a required property in that regard. Further, Theorem 7.3 also holds for multiparametric exact tests we develop later.

Conventional Exact Methods

> Conventional exact tests ensure nonexceedance of the nominal type I error. Conventional exact confidence intervals guarantee coverage at least at the nominal level.

For a general definition of the exact method, see Chapter 15. The CT criterion, which combines the probabilities from both tails, always gives smaller p-values than the TST method. Its evidence function does not exhibit all the anomalies noted for the probability criterion above. The following result is easy to prove from the definitions.

Theorem 7.4: The actual size of the exact CT test is not smaller than the actual size of the corresponding exact TST test and the actual coverage levels of the CT based CIs is not higher than the coverage level of the corresponding exact TST intervals. □

A CT based evidence function is potentially a good method for exact analysis. However, its empirical properties, especially in the conditional context, need more investigation (see Blaker 2000).

****

Now we turn to randomized evidence functions. Under the one-sided setup (7.10), an independent $U(0,1)$ random variate Λ, the randomized evidence function (7.11), the one-sided randomized p-value is

$$p_r(t, \Lambda; \phi_0) = \mathsf{E}_r(t, \Lambda; \phi_0) \tag{7.39}$$

Under the t setup (7.12), and randomized evidence function (7.14), the randomized t p-value is

$$p_{*r}(t, \Lambda; \phi_0) = \mathsf{E}_{*r}(t, \Lambda; \phi_0) \tag{7.40}$$

We now show key results for these randomized p-values.

Theorem 7.5: (i) The ASLs of nominal α_* level tests based on p_r and p_{*r} are both equal to

α_*; (ii) The ACLs of one-sided and t nominal level $(1-\alpha_*)100\%$ CIs derived from evidence functions (7.11) and (7.14) respectively are both equal to $(1-\alpha_*)100\%$.

Proof: Consider t randomized tests. Reasoning as in Theorem 7.4 shows that Ω is divided into three regions: Ω_1 where H_0 is always rejected; Ω_2 where it is rejected with some probability; and Ω_3 where it is never rejected.

In the t case we can find $t_* < t_{**}$ (with one or the other possibly not finite) so that Ω_1 consists of all points to the left of t_* and all to the right of t_{**}, $\Omega_2 = \{t_*, t_{**}\}$ and Ω_3 is the set of all points strictly between t_* and t_{**}. At t_*, we reject H_0 with probability ν_* and at t_{**}, we reject it with probability ν_{**}, where ν_* and ν_{**} satisfy

$$P[T < t_*] + \nu_* f(t_*) = \alpha_*/2 \qquad (7.41)$$
$$P[T > t_{**}] + \nu_{**} f(t_{**}) = \alpha_*/2 \qquad (7.42)$$

The sum of the left hand sides of these two expressions, the actual rejection probability, is clearly equal to α_*.

Now consider the coverage levels of t randomized CIs. The evidence function at each realized λ is a continuous bimonotonic function. (ii) follows by combining the previous result with Theorem 7.2. Results for the one-sided case follow readily from these. □

These properties also hold for in the conditional case.

Theorem 7.6: For one and t conditional randomized evidence functions defined respectively by (7.11) and (7.14). (i) The unconditional ASLs of a nominal α_* level conditional test based on the one-sided or t randomized tests are both equal to α_*; (ii) The unconditional ACLs of nominal one-sided and t $(1-\alpha_*)100\%$ CIs derived from conditional randomized evidence functions respectively are both equal to $(1-\alpha_*)100\%$.

Proof: This follows from Theorem 7.5 and by averaging over the conditioned variables. □

****

Now consider the distribution of p-values.

Theorem 7.7: Under the usual definitions for a PBD, (i) The unconditional distribution of one-sided conditional p-values is $U(0,1)$; (ii) The mean of the unconditional distribution of one-sided exact conditional p-values is greater than or equal to 0.5; (iii) The mean of the unconditional distribution of one-sided conditional mid-p-values is equal to 0.5.

Proof: Let $0 < x < 1$. Then, for specified ϕ_0, consider the one-sided univariate p-value

$$P[p_r(T, \Lambda) \leq x] = \sum_{u \in \Omega} P[p_r(T, \Lambda) \leq x, T = u]$$
$$= \sum_{u \in \Omega} P[p_r(u, \Lambda) \leq x \mid T = u] f(u) \qquad (7.43)$$

SOME FEATURES

The last line is equal to

$$= \sum_{u \in \Omega} P[S(u) - \Lambda f(u) \leq x \mid \mathcal{T} = u] f(u)$$
$$= \sum_{u \in \Omega} P\left[\Lambda \geq \{f(u)\}^{-1}(S(u) - x)\right] f(u) \quad (7.44)$$

Let u_x be the largest $u \in \Omega$ such that $S(u) - x \geq 0$. Then for all $u < u_x$,

$$P\left[\Lambda \geq \{f(u)\}^{-1}(S(u) - x)\right] = 0 \quad (7.45)$$

And, for all $u > u_x$,

$$P\left[\Lambda \geq \{f(u)\}^{-1}(S(u) - x)\right] = 1 \quad (7.46)$$

Therefore $P[p_r(\mathcal{T}, \Lambda) \leq x]$ becomes

$$1 - P\left[\Lambda \leq \{f(u_x)\}^{-1}(S(u_x) - x)\right] f(u_x) + \sum_{u > u_x} f(u) \quad (7.47)$$

which in turn is equal to

$$f(u_x) - S(u_x) + x + \{S(u_x) - f(u_x)\} = x \quad (7.48)$$

(ii) Now if $p_c(u)$ is the conventional one-sided p-value, then $p_r(u, \lambda) \leq p_c(u)$ for any λ. That is, for any x,

$$P[p_c(\mathcal{T}) \leq x] \leq P[p_r(\mathcal{T}, \Lambda) \leq x] \quad (7.49)$$

As the distribution of p_c is stochastically greater than that of a $U(0,1)$ variate, its mean is ≥ 0.5. This holds conditionally as well as unconditionally.

(iii) Consider the univariate case first.

$$\mathbf{E}[p_m(\mathcal{T}; \phi)] = \sum_{u \in \Omega} \frac{f(\phi; u)}{f(\phi)} p_m(u; \phi) \quad (7.50)$$

$$\sum_{u \in \Omega} \frac{f(\phi; u)}{f(\phi)} \times \frac{1}{f(\phi)} \left[0.5 f(\phi; u) + \sum_{v > u} f(\phi; v)\right] \quad (7.51)$$

$$= \frac{1}{2f(\phi)^2} \left[\sum_{u \in \Omega} f(\phi; u)^2 + 2 \sum_{u \in \Omega} \sum_{v > u} f(\phi; u) f(\phi; v)\right] \quad (7.52)$$

$$= \frac{1}{2f(\phi)^2} \left[\sum_{u \in \Omega} f(\phi; u)\right]^2 = 0.5 \quad (7.53)$$

$$(7.54)$$

In the multivariate case, this holds conditionally and thereby also unconditionally.

Randomization makes some evidentiary tools for discrete data behave like those for continuous data. Some conditional randomized evidence functions give p-values that uniformly take on all values between zero and one, with mean = median = 0.5; give tests with ASL equal to the nominal level, and confidence intervals with ACL equal to the nominal coverage level. These properties also hold unconditionally.

Randomized tests and confidence intervals are optimal in a way as well. For a PBD, among unbiased tests with ASL not higher than the nominal size, the randomized test is the uniformly most powerful test (see below). This property holds in various one and t setups. Randomized CIs are uniformly more accurate within a class of unbiased intervals. These assertions are proved in Lehmann (1986).

Even though randomized tests and confidence intervals are not used in practice, they provide a standard for judging the other types of p-values, tests and intervals.

When $\lambda = 0.5$ in the randomized setting, we obtain the TST mid-p evidence function. But, the mid-p tests and CIs do not guarantee equality of, or bound, the size or the coverage level. In a variety of settings, empirical studies show that the ASL of a mid-p test tends to fluctuate around the nominal level without deviating too far from it. The actual coverage of mid-p CIs also behaves similarly.

7.8 Desirable Features

How do we identify good methods of generating tools of analysis like p-values, tests, confidence intervals and point estimates? What is an optimal method? Criteria like power and coverage are used for this purpose. We may add some more criteria relating to p-values, estimation and power as well.

From the results of the section, we seek statistics that under the null give a distribution of p-values that is close to $U(0,1)$ and distributions under the alternative that tend to produce values near zero. Or at least, the distribution under the null should have an expected value near 0.5, and expected values under the alternative small, if not smaller than the usual nominal levels.

****

Estimators are assessed in terms of how close the values they produce are to the true parameter value. Suppose $\hat{\beta}(t)$ is an estimator for β. The **mean squared error (mse)** of this estimator then is

$$\sum_{u \in \Omega} [\hat{\beta}(u) - \beta]^2 \mathrm{P}[\mathcal{T} = u] \qquad (7.55)$$

Estimators with a small mse are preferred. When estimates can be nonfinite with a finite probability, as often is the case in discrete data, we need another measure. Let $\delta > 0$ be a small number. The δ level measure of accuracy of the estimator then is

$$\mathrm{P}[\,\|\hat{\beta}(u) - \beta\| \leq \delta\,] \qquad (7.56)$$

The higher this probability, the more accurate the estimator. Methods whose t evidence functions peak near the true estimate value are thus preferred.

****

DESIRABLE FEATURES

Now consider the issue of power. Suppose Figure 7.2 is a power function for a nominal α_* level test for testing $H_0 : \beta = 0$ versus $H_1 : \beta \neq 0$. The dotted horizontal line in this graph is at α_*. This function has the following properties.

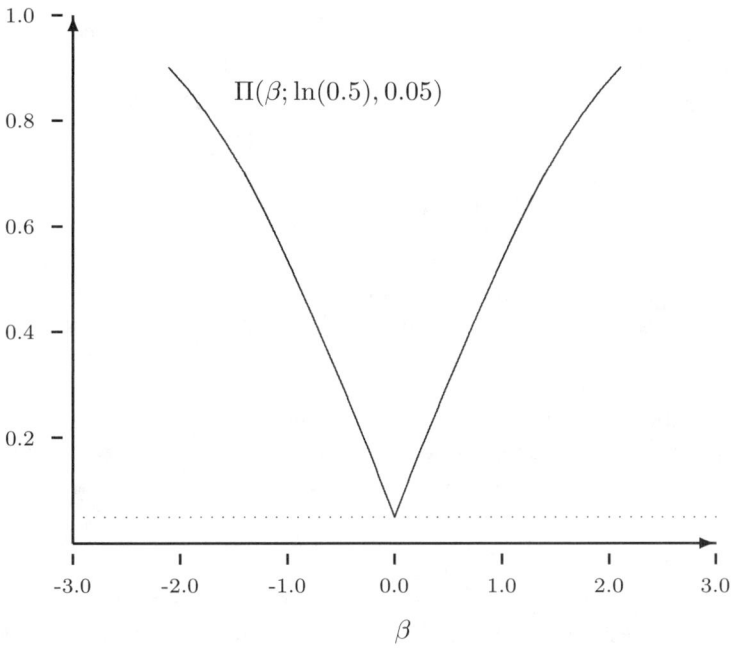

Figure 7.2 *An Ideal Shape for a Power Function.*

- The ASL of the test, that is, the value of the power function at the null parameter value, is equal to the nominal level.

- At no point in the alternative parameter space is the power lower than at the lowest power value in the null space.

- As the distance from the null increases in either direction, power increases in a monotonic manner.

- A high rate of increase of power along the two segments of this function shows that the test has high power to detect even small departures from the null.

A power function with these four properties is an **ideal power function**. Its shape is that of a well behaved (continuous, bimonotonic decreasing) evidence function that has been turned upside down and then pushed up slightly to reach the nominal level.

A Power Function

For a test of $H_0 : \beta = \beta_0$ versus $H_1 : \beta \neq \beta_0$ at nominal level α_*, an ideal power function is a continuous, bimonotonic, increasing function with its minimum value occurring at β_0 and equal to α_*. The closer its two segments, the higher the power in both directions away from the null.

We also use the following related definitions.

- A test for which the lowest power under H_1 is not smaller than the ASL is called an **unbiased test**. (This is another way of stating the second feature in the previous list.)
- In the class of tests with actual size not exceeding α_*, a test with the largest power under all alternative parameter values is called a **uniformly most powerful (UMP)** α_* **level test**.

****

Now we return to the choice of criteria for tools of analysis. We can adopt two basic approaches here. The common one takes each tool of analysis, that is, a test, p-value, CI or estimate, looks at the distinct criteria for evaluating each, applies these criteria to the wide variety of problems and designs, and accordingly gives the recommendations. Literally hundreds, if not more, empirical evaluations of this kind have been done and have generated a lot of relevant information. Yet, all the review papers notwithstanding, that information to date remains in a fragmented state. At times such studies produce controversy rather than clarification since some authors favor one type of evaluative features while others favor another type. A guideline applicable to discrete data analysis in general has yet to be produced.

A related problem is that the tests used in t discrete data problems rarely have some, let alone all, of the desirable features we have noted thus far. For example, the score test for the binomial problem is commonly used. Yet, as we see in Figure 7.1, its power function is not quite ideal. The ASL is about half the nominal size and the test is somewhat biased since the minimum power value is not at $H_0 : \beta = -1.1$ but slightly to its right.

Some theoretical arguments favor UMP tests, if they exist, in the class of unbiased tests. If randomized tests are permitted, then such a test can at times be found (Lehmann 1986). Randomization, as noted above, is not favored in actual studies, and many discrete data tests are biased. Limiting choice to unbiased tests is overly restrictive as the extent of the bias is often not too severe. When power is lowest in a small vicinity of the null, we may use that test since such minor bias at times comes with higher power elsewhere.

At times the power function shows anomalous features. It may begin to decrease after a distance from the null. This counterintuitive behavior occurs with dose-response data and unequal dose-level sample sizes. This shows that for sample size computation, a careful examination of the power function is essential.

Looking at methods confidence interval generation, we also note that a method that has all desirable features does not in general exist. That is, methods that simultaneously produce a connected interval, satisfy the nestedness property, are neither too liberal nor too conservative, and also yield the shortest intervals rarely exist. A method giving short intervals or having good coverage properties may not give connected, or nested intervals. But the method may be used if these violations are rare or minor in nature.

When the tools of analysis are evaluated in a disjointed manner, the recommendations depend on the problem at hand, the level of analysis and the tool employed. A practioner is faced with a bewildering variety of recommendations, one for *p*-values, one for CIs and yet another for estimates, and each varying over different configurations. For one problem, for a 5% nominal level analysis, a score asymptotic test and an adjusted Wald confidence interval by adding 4 to each cell are better; for the 1% nominal level, another set of tools is better. Such an ad hoc approach is often derived from differences between methods that are not practically significant.

****

A second approach to assessing tools of data analysis looks at them in an integrated manner. It examines both the results of the empirical studies as well the overall features of these tools. For the latter task, three basic entities are helpful: (i) the shape of the distribution of *p*-values, (ii) the features of the evidence function and (iii) the features of the power function. We then seek the following type of statistics for a t analysis in particular.

A Good Statistic

> A good evidentiary statistic produces (i) a null
> distribution of p-values close to the $U(0,1)$, and
> alternative distributions with values concentrated
> near zero; (ii) narrow decreasing bimonotonic evidence
> functions and (iii) narrow increasing bimonotonic power
> functions with a low level of bias.

Methods or statistics that generally or more often meet all or most of these requirements are preferred to those that do not. Yet, such an approach to selecting the tools of evidence is in its infancy and needs to be developed and applied. We also note that at this juncture, there is no consensus on how a comprehensive assessment of the tools of analysis of discrete data should be conducted.

Many journals, especially in the biomedical field, now require the authors to report the *p*-value, CI and the point estimate. The evidence function approach is a medium through which this is done in an integrated way. The entire function gives a more complete picture of the evidence. Assessing the tools of analysis in an integrated manner will clarify and simplify the vast maze of recommendations now available for each separate entity.

7.9 On Unconditional Analysis

How does exact conditional analysis for discrete data compare with the exact unconditional analysis? If we have to retain the nominal significance and coverage levels, the restricted sample space in the former tends to make it more conservative. On the other hand, maximization over the parameter space has a similar effect on the latter. Which effect is more pronounced? And how else do these methods compare? The following results, gleaned from the literature, indicate that the answer is not clear cut. (For proofs and empirical details, see the cited papers.)

- The ASL of a test based on exact unconditional *p*-values does not exceed the nominal size (Boschloo 1970).
- The ACL of the CIs obtained from an exact unconditional evidence function is not lower than the nominal level.

- Exact unconditional and exact conditional *p*-values can differ substantially even at moderately large sample sizes (Berkson 1978; Haber 1987; Rice 1988; Routledge 1992; Seneta, Berry and Macaskill 1999).
- For the two binomials problem, exact unconditional *p*-values based on the probability criterion are always smaller than the exact conditional *p*-values based on that criterion. The ASL of tests based on the latter is frequently half that of the tests based on the former (Boschloo 1970; Shuster 1992; Seneta and Phipps 2001).
- For the two binomials problem, several statistics produce exact unconditional tests that are **more conservative** than the exact conditional test based on the probability criterion. Choice of the test statistic is an important factor here (Haber 1987; Mehrotra, Chan and Berger 2003).
- For the two binomials problem, the exact unconditional *p*-values tend to be close to the conditional mid-*p*-values (Routledge 1992).
- The restricted optimization exact unconditional test for the two binomials problem does not compromise power (Berger 1996).
- In the two binomials problem, the ASL of the approximate unconditional test based on the observed difference rarely exceeds the nominal size and its power is comparably higher than that of its exact counterpart (Storer and Kim 1990).
- The approximate unconditional CIs for the risk difference in the two binomials problem are less conservative and more reliable as compared to exact unconditional and mid-*p* unconditional intervals (Newcombe 1998b; Chen 2002).
- The exact unconditional score test based CIs for the odds ratio in the two binomials problem tend to be less conservative and shorter compared to exact TST and exact score conditional intervals (Agresti and Min 2002).
- For the **three binomials** comparison situation, the exact unconditional test based on the chisquare statistic is more conservative and less powerful than the exact conditional test based on the same statistic (Mehta and Hilton 1993).

Comparisons of the exact conditional and exact unconditional methods, and of the variations thereof, have been done only for a limited number of configurations of discrete data. Further, most of the above comparisons did not include the conditional mid-*p* counterparts. More needs to be done. That effort is also predicated upon the development of efficient computational methods for exact unconditional analysis of a wider variety of models and designs. While exact unconditional analysis or some approximation based on it remain a promising option for discrete data analysis, its scope remains limited. On the other hand, exact conditional analysis is currently feasible for a wider class of problems and designs.

7.10 Why the Mid-p?

In this section, we summarize the reasons favoring the mid-*p* approach as a general method of choice for analysis of sparse discrete data. These are as follows:

- Statisticians who hold very divergent views on statistical inference have either recommended or given justification for the mid-*p* method.
- A mid-*p* version has been or can be devised for most of the statistics used in exact conditional and unconditional analysis of discrete data.
- Mid-*p* versions for multiparameter discrete data tests have also been devised.

- The evidence function of the TST mid-*p* method is a continuous bimonotonic function, giving a coherent set of *p*-values and nested, connected confidence intervals.
- The shape of the power function of the TST mid-*p* tests is generally close to the shape of an ideal power function.
- In a wide variety of designs and models, the mid-*p* rectifies the extreme conservativeness of the traditional exact conditional method without compromising the type I error in a serious manner.
- The expected value of the one-sided mid-*p* is equal to 0.5. The t TST mid-*p* has an empirical distribution that is close to the uniform [0,1] distribution with expected value around 0.5.
- Empirical studies show that the performance of the mid-*p* method resembles that of the exact unconditional methods and the conditional randomized methods.
- With the exception of a few studies, most studies indicate that in comparison with a wide variety of exact and asymptotic methods, the mid-*p* methods are among the preferred, if not the preferred ones.
- The median unbiased estimate, derived from the TST mid-*p* evidence function, has good comparative small and large sample properties.
- The progress in computing power and efficient algorithms has made computation of exact distributions, and thus the mid-*p* based indices, a practically feasible option for an extensive array of complex discrete data models and designs.

The references for the above assertions were noted in Chapter 2. The mid-*p* method is thus a widely accepted, conceptually sound, practical and among the better of the tools of data analysis. Especially for sparse or not that large a sample size discrete data, we thereby echo the words of Cohen and Yang (1994) that it is among the *"sensible tools for the applied statistician."* Unless the data are highly sparse, the asymptotic (conditional or unconditional) score based methods also generally perform satisfactorily and are similar to the mid-*p* method. Exceptions to these statements have, of course, to be kept in mind.

7.11 Relevant Literature

7.11.1 Exact Unconditional Analysis

Camilli (1990) and the discussion of Rice (1988) give a readable but critical exposition of the idea of exact unconditional testing. For an overview of the subject, see Mehrotra, Chan and Berger (2003).

Barnard (1945a, 1945b, 1947) introduced exact unconditional tests for discrete data. Pearson (1947) also broached the topic. Barnard later repudiated his own test in favor of the conditional test (Barnard 1949, 1979, 1989). But the idea had taken root and was, after a hiatus, taken up by Boschloo (1970), Garside (1971), Garside and Mack (1971), McDonald, Davis and Milliken (1977). Renewed interest in the subject was sparked by Santner and Snell (1980), Suissa and Shuster (1984, 1985) and Haber (1986a, 1986b). Haber (1987) and Shuster (1992) extended exact unconditional analysis to a 2×2 table generated by an overall total fixed design; Suissa and Shuster (1991) and Royston (1993) similarly tackled the 2×2 matched pairs trial; Chan (1998) devised such a test for equivalence in a comparative trial. Tango (1998) and Hsueh, Liu and Chen (2001) dealt with exact unconditional equivalence analysis in the paired sample and paired binary end points designs (The last paper also notes other related papers). An exact unconditional test for an unordered 2×3 table was evaluated by Mehta and Hilton (1993). Exact unconditional testing for dose-response data was tackled by Tang (2000).

Exact CIs based on the unconditional approach were noted by Garside and Mack (1967, 1968a, 1968b). Starting with Santner and Snell (1980), exact unconditional CIs for the risk difference, risk ratio and odds ratio were constructed by Shuster (1988), Soms (1989a, 1989b), Santner and Duffy (1989), Wallenstein (1997), Newcombe (1998b), Chan and Zhang (1999), Agresti and Min (2001, 2002) and Chen (2002). The last three papers note other work in this field, and give an overall view.

Irwin (1935) considered an approximate exact unconditional test but did not recommend it, deeming it *"impracticable owing to the large number of tables to be enumerated."* An early related work is Wilson (1941). McDonald, Davis and Milliken (1977), Liddell (1978), Upton (1982) and Storer and Kim (1990), among others, also deal with it. Berger and Boos (1994) applied the idea of restricted maximization over a confidence set for the nuisance parameter (see also Berger (1996), Martin and Silva Mato (1994) and Mehrotra, Chan and Berger (2003)). Rice (1988) promoted a test for 2×2 table that averaged over the nuisance parameter space. Kim and Agresti (1995) gave a finer partition conditional test for several 2×2 tables (consult this paper for related references). Basu (1977) is a prime reference on ways of handling nuisance parameters.

On the computational challenge in exact unconditional analysis, see Mehta and Hilton (1993), Berger and Boos (1994), Wallenstein (1997), Chan (1998), Tang (2000), Hsueh, Liu and Chen (2001) and Mehrotra, Chan and Berger (2003).

7.11.2 Randomized Inference

Tocher (1950) devised randomized inference for discrete data; the principal reference on the subject is Lehmann (1986). For comparisons of randomized tests with other tests, see Hirji and Vollset (1994c), Dunne, Pawitan and Doody (1996), Fuchs (2001) and Lydersen and Laake (2003), among others.

Numerous authors have decried randomized tests for practice; many papers cited in this section contain that sentiment. Suissa and Shuster (1984) critiqued UMPU tests while Dunne, Pawitan and Doody (1996) and Fuchs (2001) (and the authors they cite) promote using such tests with modified ways of reporting the results.

7.11.3 Exact Power and Coverage

On general issues relating to sample size determination in clinical and epidemiological studies, and the basic asymptotic formulae for this, see Hulley and Cummings (1988), Newman (2001), Pocock (1983) and Piantadosi (1997).

Computation of exact power and sample size for exact tests for the 2×2 one margin fixed design were done by Bennett and Hsu (1960), Gail and Gart (1973), Casagrande, Pike and Smith (1978a, 1978b), Haseman (1978) and Thomas and Conlon (1992). Extension for exact tests in the 2×2 overall total fixed design was given by Fu and Arnold (1992). See also Gordon (1994).

Exact sample size determination for exact unconditional tests for the 2×2 one margin fixed design was done by Suissa and Shuster (1985). Sample size for exact unconditional analysis for pair matched data were computed by Suissa and Shuster (1991); they were compared with sample sizes based on exact conditional analysis by Royston (1993).

EXERCISES

Exact power and sample size analyses for the three binomials design were done by Bennett and Nakamura (1964), and Mehta and Hilton (1993), and generalized by Tang (1999). Tang, Hirji and Vollset (1995) and Mehta, Patel and Senchaudhuri (1998) dealt with exact power computation for dose-response studies; Tang (2001b) tackled stratified dose-response studies. It was extended by Tang and Hirji (2002) to checking linearity in logit models used in such studies. Hilton and Mehta (1993) showed how to perform exact power computation for two group ordinal outcome studies. Hirji et al. (1994) computed exact power for a series of 2×2 comparative trials. See also Hirji and Tang (1998) and Tang (1998, 2000 and 2001b).

Many papers, including those noted in this and other chapters, have compared the size, power and coverage of exact and other tests and confidence intervals for a vast variety of situations. The notable ones are cited elsewhere in this section and in other chapters.

7.11.4 The 2×2 Table

The literature on the 2×2 table is truly voluminous, covering the philosophy of inference, techniques of analysis and comparative empirical studies. We note some of the main papers (apart from those already noted) on techniques and comparative studies below.

The 2×2 table has also been a medium through which many techniques of data analysis have come into existence. To get a flavor, see Liddle (1978), Upton (1982), D'Agostino, Chase and Belanger (1988), Hirji, Tan and Elashoff (1991), Chen (2002) and Lydersen and Laake (2003). The papers on exact unconditional analysis already cited also provide an additional source. Martin (1991) and Sahai and Khurshid (1995) provide comprehensive reviews.

Periodically, a debate on exact inference has been generated by the publication of a paper on the 2×2 table. These papers also relate to or cite empirical comparisons of different methods of analysis. See the following and the comments they generated: Berkson (1978); Upton (1982); Yates (1984); D'Agostino, Chase and Belanger (1988); Rice (1988); Barnard (1989); Haviland (1990); O'Brien (1994) and Lydersen and Laake (2003). See also the references on the mid-p in Chapter 2, and the related ones in Chapter 5 and Chapter 15.

The Fisher and Irwin exact tests were compared by, among others, Yates (1984), Cormack (1986), Davis (1986), Dupont (1986), Lloyd (1988) and Hirji, Tan and Elashoff (1991). (When reading these and other papers, keep the terminological tangle noted in §7.6 in mind.) Davis (1993) compared the power of exact TST, mid-p and asymptotic score tests for two binomial proportions. The papers noted in the previous paragraph also pertain to this comparison.

7.11.5 Assessing Inference

We have barely scratched the surface of the topic of assessing inference. It also relates to decision theory. Consult Cox and Hinckley (1974), Lehmann (1986, 1987), Santner and Duffy (1989) and other theory oriented works for elaboration. Donahue (1999) and Sackrowitz and Samuel–Cahn (1999) argue for the use of the distribution of p-values in the context; the references they include present a broader picture. Hirji, Tsiatis and Mehta (1989) and its references deal with comparison of estimators when infinite values can occur with a finite probability.

7.12 Exercises

7.1. Each study in Table 7.8 compares treatments generically labeled A and B. Assuming an independent binomials design, compute the exact unconditional score p-value for testing

$\phi = 1$ versus $\phi \neq 1$ for each study. Compare it with the corresponding exact conditional score, asymptotic conditional score, asymptotic unconditional score and approximate unconditional score p-values. Where appropriate, compute the mid-p-value as well.

Compare these with the CSM test of Barnard (1947), and the exact unconditional and conditional probability statistic based analyses.

Table 7.8 *Data from Ten Clinical Trials*

Study No.		A	B	Study No.		A	B
1	Failure	3	1	6	Failure	1	0
	Success	0	2		Success	8	5
2	Failure	4	1	7	Failure	1	3
	Success	0	3		Success	8	2
3	Failure	0	4	8	Failure	1	1
	Success	4	0		Success	8	4
4	Failure	2	3	9	Failure	1	4
	Success	2	1		Success	8	4
5	Failure	5	2	10	Failure	0	2
	Success	1	3		Success	8	4

Note: Hypothetical data.

7.2. For each study in Table 7.8, and under the assumptions of the previous exercise, define and construct the exact unconditional TST evidence function for ϕ. Use these, to obtain, for each study, the 95% exact unconditional CIs and estimates for ϕ. Compare them with the 95% exact conditional TST and TST mid-p CIs and estimates for ϕ.

Repeat this exercise with the exact unconditional and conditional score based evidence functions, and the asymptotic unconditional and conditional score based evidence functions.

7.3. A clinical trial with $n = m = 7$ produces $s = 6$ and $t = 1$. Compute the exact unconditional t p-values for a test of no treatment effect based on the score, LR, CT and probability methods. Compare these with corresponding conditional conventional and mid-p-values. Repeat with $n = m = 10$ and $s = 8$ and $t = 2$.

7.4. In a binomial experiment with $n = 12$, we test $\pi \leq 0.25$ versus $\pi > 0.25$. Determine the conventional, mid-p and asymptotic score p-value for each point in the sample space. Compute their means and variances. What are the true sizes of 0.05 and 0.10 level tests based on these p-values?

Plot the power function of the tests. Are the tests biased? Which tend to have more power?

Repeat the whole exercise for $n = 15, 20, 25$.

7.5. In a binomial experiment with $n = 12$, we test $\pi = 0.5$ versus $\pi \neq 0.5$. Determine the TST exact and mid-p-value for each point in the sample space. Compute their means and variances. What are the true sizes of 0.05 and 0.10 level tests based on these p-values?

EXERCISES

Plot the power function of the tests. Are the tests biased? Which tend to have more power?

Repeat the whole exercise for $n = 15, 20, 25$.

7.6. Repeat the previous exercise for exact score, LR, CT and probability statistics, and the asymptotic score and LR test versions as well.

7.7. In a negative binomial experiment with number of trials before a success, $t = 2$, we test $\pi = 0.5$ versus $\pi \neq 0.5$. Determine the TST exact and mid-p-value for each point in the sample space. Compute their means and variances. What are the true sizes of 0.05 and 0.10 level tests based on these p-values?

Plot the power function of the tests. Are the tests biased? Which tend to have more power?

Repeat the whole exercise for $t = 4, 5, 7$.

7.8. In a two independent binomials experiment with $n = m = 12$, we test $\phi \leq 1.0$ versus $\phi > 1.0$. Determine the (conditional) conventional and mid-p-value for each point in the sample space. Compute their means for various true values of the nuisance parameter. What are the true sizes of 0.05 and 0.10 level tests based on these p-values?

Plot the power function of the tests. Are the tests biased? Which tend to have more power?

Repeat the whole exercise for $n, m = 15, 20, 25$, $n = 10, m = 5$, and $n = 10, m = 15$. Apply the computation reduction method given by expression (7.28).

7.9. In a two independent binomials experiment with $n = m = 12$, we test $\phi = 1.0$ versus $\phi \neq 1.0$. Determine the (conditional) TST conventional and mid-p-value for each point in the sample space. Compute their means for various true values of the nuisance parameter. What are the true sizes of 0.05 and 0.10 level tests based on these p-values?

Plot the power function of the tests. Are the tests biased? Which tend to have more power?

Repeat the whole exercise for $n, m = 15, 20, 25$, $n = 10, m = 5$, and $n = 10, m = 15$. Apply the computation reduction method given by expression (7.28)?

7.10. Repeat the previous exercise for score, LR, CT and probability statistics based on the exact distribution and for the asymptotic score and LR test versions.

7.11. In a clinical trial with $n = m = 25$, determine the power of the exact and mid-p TST tests to detect a difference of a true 50% success rate in treatment A, and a 40% rate in treatment B. Plot the power function, and plot actual size as a function of the null parameter value. Repeat this for the exact probability and CT, and the exact and asymptotic score and LR tests.

Repeat the entire exercise for other sample sizes, both unequal and equal, and other ranges of treatment differences. What conclusions do you draw?

7.12. Describe how exact power for an odds ratio test can be computed for the total fixed and nothing fixed 2×2 designs. Perform the computations for the sample sizes of the studies in Table 7.8.

7.13. Compare the Fisher, Irwin and CT t p-values for the data in Exercise 7.1. Investigate the impact of small changes in these datasets. Does the CT method exhibit the anomalies noted for the probability method? Repeat this for a mid-p version of the CT method.

7.14. Compare the mid-p versions of the Fisher, Irwin and CT t p-values for the data in Table 7.7 and Table 7.8. Investigate the impact of small changes in the datasets. Do these methods exhibit any anomalies? Expand your investigations to other simulated data. What general conclusions do you draw?

7.15. For a binomial trial with $n = 10$, plot the exact coverage of the t 90% and 95% exact TST, mid-p and asymptotic score based CIs for π. Repeat this for $n = 15, 20, 30$. Investigate the distributions of the lengths of the CI.

7.16. For a negative binomial trial with $t = 2$, plot the exact coverage of the t 90% and 95% exact TST, mid-p and asymptotic score, Wald and LR based CIs for π. Repeat this for $t = 4, 5, 7$. Investigate the distributions of the lengths of the CIs.

7.17. For a Poisson count design, plot the exact coverage of the t 90% and 95% exact TST, mid-p and asymptotic score, Wald and LR based CIs for λ. Investigate the distributions of the lengths of the confidence intervals.

7.18. For a two binomials trial with $n = m = 10$, plot, for various true values of the parameters, the exact coverage of the t 90% and 95% exact TST, mid-p and asymptotic score based CIs for ϕ. Repeat this for $n = m = 15, 20$ and $n = 10, m = 15$, $n = 5, m = 15$ and $n = 10, m = 20$. Investigate the distributions of the lengths of the CIs.

7.19. For all the 2×2 tables in this chapter and Chapter 5, and assuming a two binomials design, what is the distribution of the exact, asymptotic and mid-p TST, score, probability, CT, LR and Wald p-values under the null and under various alternatives for the odds ratio? What about the inverse sampling design and risk ratio?

7.20. Repeat the above exercise for a negative binomial and Poisson counts 2×2 design.

7.21. Repeat the above type of exercise for a single binomial, Poisson and negative binomial.

7.22. Prove Theorem 7.4 in a step by step fashion.

7.23. From the tables and methods of Thomas and Conlon (1992) and Fu and Arnold (1992), investigate the difference between sample sizes obtained through exact and asymptotic computations. Which method tends to give higher required sample sizes and under what conditions?

7.24. Construct exact unconditional evidence function based methods for computing CIs and estimates for the risk difference and relative risk for a two binomials design. Apply these to the datasets in Table 7.8. Consult Santner and Duffy (1989) and other papers noted above.

7.25. Investigate the computation of exact unconditional p-values and CIs for the odds ratio in a 2×2 overall total fixed design.

7.26. Investigate the computation of exact unconditional p-values and CIs for relative risk and risk difference in a two group inverse sampling design.

7.27. Investigate the computation of exact unconditional p-values and CIs for rate ratio and rate difference in a two group cohort design.

7.28. Suppose a one-sided p-value for a continuous data problem is given by $p(t) = P[T \geq t; \phi]$. Show that under the null, $p(T)$ is uniform [0,1] with mean = median = 0.5. What is its variance? What is its distribution under the alternative scenarios? (See Donahue (1999) and Sackrowitz and Samuel–Cahn (1999).)

7.29. Suppose T is a discrete PBD variate with gp $f(\phi)$ and probability mass function f(t). Let p_c, p_m and p_r be the exact, mid-p and randomized one-sided p values. What is the variance of p_r? Also show that

$$E[p_c(T)] = \frac{1}{2}\left(1 + \sum_{u \in \Omega} f(u)^2\right)$$

and

$$\text{var}[p_m(T)] = \frac{1}{12}\left(1 - \sum_{u \in \Omega} f(u)^3\right)$$

What is the variance of p_c? (Barnard 1989; Berry and Armitage 1995). Can you simplify

EXERCISES

these expressions for the special cases of the binomial, Poisson, negative binomial and the conditional (2×2) hypergeometric problems?

7.30. T is a discrete PBD variate with gp $f(\phi)$. Let p_c, p_m and p_r be the exact, mid-p and randomized TST t p-values for testing $\phi = \phi_0$ versus $\phi \neq \phi_0$. (i) What is the null and alternative distributions of these p-values? (ii) What are their means and variances? (iii) Apply the results to the binomial, Poisson, negative binomial and conditional (2×2) hypergeometric problems. (iv) Do the results of (i), (ii) and (iii) apply to the score and probability based randomized t p-values? (v) Show by a counterexample that the mean of the null distribution of the TST mid-p is not necessarily equal to 0.5. What if the underlying null distribution of the T is symmetric?

7.31. Suppose T is a discrete PBD variate with gp $f(\phi)$. Let p_r be the randomized TST t p-value for testing $\phi = \phi_0$ versus $\phi \neq \phi_0$. Is the actual size of a nominal α_* test based on these equal to α_*? Does that hold for randomized tests based on score statistic, probability statistic (7.16) and the test based on (7.15)?

7.32. Use large sample theory to derive formulas (7.17) and (7.23).

7.33. Seneta, Berry and Macaskill (1999) define an adjusted mid-p-value with mean 0.5 and variance $1/12$. Investigate its properties.

7.34. Develop a systematic, integrated approach to assess the various tools of data analysis in the context of evidence functions, and formulate empirical investigations to apply it to various designs for the 2×2 table and two 2×2 tables. Conduct these investigations and state your conclusions.

CHAPTER 8

Several 2 × 2 Tables: I

8.1 Introduction

Research data in medicine, public health, sociology, education, law or psychology are often shown as a series of 2×2 tables. Sometimes, as in a multicenter clinical trial, this is their natural form. In other cases, the data are stratified at the analysis stage to adjust for the confounding effect of one or more factors. With the spotlight on logit or odds ratio models for a series of 2×2 tables, the specific aims of this chapter are:

- To present three statistical models for several 2×2 tables arising from a product binomial design. These models are: the common odds ratio model, the heterogenous odds ratios model and the trend in odds ratios model.

- To derive exact unconditional and conditional distributions for these three models and show that they are polynomial based distributions.

- To formulate and illustrate exact conditional and asypmtotic methods for analyzing these models.

First we clarify a term we regularly use. In statistical analysis, the term **sparse data** is used in three ways. One, that the total sample size is small. Two, the total sample size may be moderately large or even large, but some variable configurations have small counts. Cell values in a multiway contingency table then may be near zero or one to make it appear as if there are holes in the data. Three, we have values spread over many strata but the total in any stratum is quite small. This case is relevant for the problems we address in this and the next chapter.

Agresti (1996), pages 190–194, has a lucid account of the sparse data conundrum. Its implications are outlined with specific examples. Traditional methods may fail to converge, or may not be accurate. Such settings thereby call for exact methods.

8.2 Three Models

Consider an hypothetical binary response clinical trial to compare two treatments. Suppose randomization was done within four levels severity of the disease, and the results are in Table 8.1.

The overall response rate among patients with subclinical disease is 80%, while among those with moderate or serious disease, it is 50%. The conditional mues of the odds ratios in the four strata are 1.0, 0.429, 0.490 and 0.0, respectively. Disease severity may affect the relative efficacy of the treatments. So it is advisable to adjust for it when analyzing the data. For this purpose, we utilize three logit models.

Consider a trial with K strata. Let $\mathcal{Y} \in \{0, 1\}$ denote the response (0 = nonresponse or failure, 1 = response or success). Also let $x = 0, 1$ be the therapy indicator, and k, the stratum indicator,

Table 8.1 *Stratified Clinical Trial Data*

Severity	Treatment	Nonresponders	Responders
Subclinical	$x=0$	1	4
	$x=1$	1	4
Mild	$x=0$	1	4
	$x=1$	2	3
Moderate	$x=0$	2	3
	$x=1$	3	2
Serious	$x=0$	0	5
	$x=1$	5	0

Note: Hypothetical data.

$k = 1, \ldots, K$. Let $\pi_k(x) = \mathrm{P}[\mathcal{Y} = 1 \mid x, k]$ be the chance of attaining response level 1 in the kth stratum for treatment x

We first consider the case when the odds ratios do not vary across the K strata. The logit model for this case is:

Model COR: $$\ln\left(\frac{\pi_k(x)}{1 - \pi_k(x)}\right) = \alpha_k + \beta x \qquad (8.1)$$

Model (8.1), very frequently used in data analysis, is the **homogeneous or common odds ratio** (COR) model. It is a special case of the more general logit model

Model HOR: $$\ln\left(\frac{\pi_k(x)}{1 - \pi_k(x)}\right) = \alpha_k + \beta_k x \qquad (8.2)$$

This is the general **heterogenous odds ratios** (HOR) model. Under the notion of interaction described in Chapter 6, it is also called the **qualitative interaction** model.

Now assume that there is some sort of ordering among the strata, and that we can assign a numeric score to each stratum. Let z_k denote the score for stratum k. We have either $z_1 < z_2 < \cdots < z_K$ or $z_1 > z_2 > \cdots > z_K$. For example, $z_k = k - 1$. In Table 8.1, disease severity may be scored so that $z_1 = 0$, $z_2 = 1$, $z_3 = 2$ and $z_4 = 3$, respectively, stand for subclinical, mild, moderate and serious disease. Further, assume that $\beta_k = \beta + \gamma z_k$. Then model (8.2) becomes

Model TOR: $$\ln\left(\frac{\pi_k(x)}{1 - \pi_k(x)}\right) = \alpha_k + \beta x + \gamma x z_k \qquad (8.3)$$

This is the **trend in odds ratios** (TOR), or the **quantitative interaction** model. Some pertinent characteristics of these models are:

- While the HOR model has $2K$, the COR model has $K + 1$, and the TOR model has $K + 2$ parameters.

EXACT DISTRIBUTIONS

- For each model, α_k is the log-odds of success in treatment 0 in stratum k. As the main aim of the study is to compare the two treatments, these K parameters are nuisance parameters in all the three cases.
- The analysis of the HOR model focuses on the parameters β_k, $k = 1, \ldots, K$. We may test the hypothesis $\beta_1 = \beta_2 = \cdots = \beta_K = 0$ against the alternative that at least one among them is not zero. This is known as a test for **conditional independence**.
- If, in the COR model, $\beta = 0$, the therapies are equivalent. If $\beta > 0$, therapy 1 has a higher response rate than therapy 0, and if $\beta < 0$, the reverse holds. The greater the absolute magnitude of β, the greater the disparity between them.
- When $\beta_1 = \beta_2 = \cdots = \beta_K = \beta$, the HOR model turns into the COR model. A test for this condition is called a test for the **homogeneity of the odds ratios** or a test for no interaction.

- The TOR model assumes that the log-odds ratio is a linear function of the stratum score. Then, the parameters of interest are β and γ. If $\gamma = 0$, the TOR model reduces to the COR model.

We also refer to these models in terms of odds ratios instead of log-odds ratios. For this we define

$$\theta_k = \exp(\alpha_k); \quad \phi_k = \exp(\beta_k); \quad \phi = \exp(\beta) \text{ and } \psi = \exp(\gamma) \qquad (8.4)$$

These transformations are used in formulating exact distributions for the three models.

8.3 Exact Distributions

Now we show that the exact distributions for these models have a polynomial form. Then we can factor out the K nuisance parameters by conditioning and apply the conditional methods developed previously to analyze the parameter(s) of interest.

For the clinical trial, assume that responses of subjects within and between centers are statistically independent, and the number of subjects in each treatment at each center, m_k and n_k, is selected by the investigators. Under the rationale given in §6.4, we use the product binomial model as the data generating model for each center. For the kth 2×2 table, we use the notation of Table 6.1. Then

$$P[\mathcal{A}_k = a_k, \mathcal{B}_k = b_k; k = 1, \ldots, K] = \prod_{k=1}^{K} \binom{n_k}{a_k}\binom{m_k}{b_k}$$
$$\times \pi_k(1)^{a_k}[1 - \pi_k(1)]^{(n_k - a_k)} \pi_k(0)^{b_k}[1 - \pi_k(0)]^{(m_k - b_k)} \qquad (8.5)$$

Let $\mathcal{S}_k = \mathcal{A}_k + \mathcal{B}_k$ and $\mathcal{T}_k = \mathcal{A}_k$. In words, $\mathcal{S}_k$ is the total random number of responses, and $\mathcal{T}_k$, the random number of responses in Treatment 1 at the kth center. Further, define

$$\Delta_{1k} = [1 + \theta_k]^{m_k}[1 + \theta_k \phi_k]^{n_k} \qquad (8.6)$$

$$\Delta_{2k} = [1 + \theta_k]^{m_k}[1 + \theta_k \phi]^{n_k} \qquad (8.7)$$

$$\Delta_{3k} = [1 + \theta_k]^{m_k}[1 + \theta_k \phi \psi^{z_k}]^{n_k} \qquad (8.8)$$

We formulate the exact distributions by substituting the logit forms for $\pi_k(0)$ and $\pi_k(1)$ from (8.2) in (8.5). After slight algebra, it becomes

$$\prod_{k=1}^{K} \Delta_{1k}^{-1} c_k(t_k) \theta_k^{s_k} \phi_k^{t_k} \qquad (8.9)$$

where

$$c_k(u) = \binom{n_k}{u}\binom{m_k}{s_k - u} \qquad (8.10)$$

Distribution (8.9) is the exact unconditional distribution under the HOR model. It is evident that it is a multivariate PBD with gp equal to

$$\prod_{k=1}^{K} \Delta_{1k} \qquad (8.11)$$

The sufficient statistics are identified upon examination. Thus $\mathcal{S}_k$ is sufficient for θ_k (and α_k), and $\mathcal{T}_k$ is so for ϕ_k (and β_k). For $k = 1, \ldots, K$, let

$$l_{1k} = \max(0, s_k - m_k) \text{ and } l_{2k} = \min(s_k, n_k) \qquad (8.12)$$

$$\Omega_k = \{l_{1k}, l_{1k} + 1, \ldots, l_{2k}\} \qquad (8.13)$$

$$f_k(\phi_k) = \sum_{v \in \Omega_k} c_k(v) \phi_k^v \qquad (8.14)$$

We need the conditional distribution of $\mathcal{T} = (\mathcal{T}_1, \ldots, \mathcal{T}_K)$ for fixed value of $\mathcal{S} = (\mathcal{S}_1, \ldots, \mathcal{S}_K)$. Suppose the observed vector for $\mathcal{S}$ is $s = (s_1, \ldots, s_K)$ and for $\mathcal{T}$, it is $t = (t_1, \ldots, t_K)$.

Theorem 8.1: The conditional distribution of $\mathcal{T}$ for the HOR model is given by:

$$P[\mathcal{T} = t \mid \mathcal{S} = s] = \prod_{k=1}^{K} \frac{c_k(t_k)\phi_k^{t_k}}{f_k(\phi_k)} \qquad (8.15)$$

Proof: Using between strata independence, we have that

$$P[\mathcal{T} = t \mid \mathcal{S} = t] = \prod_{k=1}^{K} P[\mathcal{T}_k = t_k \mid \mathcal{S}_k = s_k] \qquad (8.16)$$

(8.15) then follows using the results for a single 2×2 table. □

****

This distribution forms the basis for deriving exact distributions for a variety of relevant problems.

EXACT DISTRIBUTIONS

We start with the COR model. Let $\mathcal{T} = \mathcal{T}_1 + \cdots + \mathcal{T}_K$, with t its observed value. Also, let $f(\phi)$ denote the product polynomial

$$f(\phi) = \prod_{k=1}^{K} f_k(\phi) \qquad (8.17)$$

Suppose this polynomial is written in an expanded form as

$$f(\phi) = \sum_{u \in \Omega} c(u) \phi^u \qquad (8.18)$$

In words, Ω is the set of exponents ϕ in $f(\phi)$, and $c(u)$ is the associated coefficient of the term ϕ^u. Obviously,

$$\Omega = \{l_1, l_1 + 1, \cdots, l_2\} \qquad (8.19)$$

where

$$l_1 = \sum_k l_{1k} \quad \text{and} \quad l_2 = \sum_k l_{2k} \qquad (8.20)$$

Equating terms in (8.17) and (8.18), we have that

$$c(t) = \sum_{\boldsymbol{u}} \left(\prod_{k=1}^{K} c_k(u_k) \right) \qquad (8.21)$$

where the summation is performed over all values of $u_1, \cdots, u_K$ that satisfy the conditions

$$u_k \in \Omega_k, \qquad k = 1, \cdots, K \qquad (8.22)$$

and

$$u_1 + \cdots + u_K = t \qquad (8.23)$$

Theorem 8.2: For the COR model, $\mathcal{S}_k$ is sufficient for α_k (and θ_k), and $\mathcal{T}$ is so for β (or ϕ). The conditional distribution of $\mathcal{T}$ is a PBD with gp $f(\phi)$. That is,

$$P[\mathcal{T} = t \mid \boldsymbol{\mathcal{S}} = \boldsymbol{s}] = \frac{c(t)\phi^t}{f(\phi)} \qquad (8.24)$$

Proof: Substituting $\phi_1 = \cdots = \phi_K = \phi$ in (8.9), and simplifying, this joint probability becomes

$$\phi^t \prod_{k=1}^{K} \Delta_{2k}^{-1} c_k(t_k) \theta_k^{s_k} \qquad (8.25)$$

The factorization theorem shows that the sufficient statistics for the parameters are as stated. Now,

$$\begin{aligned} \mathrm{P}[\mathcal{T}=t \mid \mathcal{S}=s] &= \mathrm{P}[\mathcal{T}=t, \mathcal{S}=s]\{Pr[\mathcal{S}=s]\}^{-1} \\ &= \sum_{u} \mathrm{P}[\mathcal{T}=u \mid \mathcal{S}=s] \end{aligned} \qquad (8.26)$$

where the above summation is performed over all values of $u_1, \cdots, u_K$ that satisfy conditions (8.22) and (8.23). The conditional distribution of $\mathcal{T}$ obtains by putting $\phi_1 = \cdots = \phi_K = \phi$ in the rhs of (8.26), and noting the expression for $c(t)$ given by (8.21). $\square$

Since conditionally $\mathcal{T}$ has a univariate PBD with gp $f(\phi)$ and sample space Ω, we may analyze it with the exact methods described previously. For this task, we at times need the conditional mean and variance of $\mathcal{T}$.

Theorem 8.3: For $k = 1, \cdots, K$, let $N_k = m_k + n_k$, and let μ_k and σ_k^2 be the conditional mean and variance of $\mathcal{T}_k$ given $\mathcal{S}_k = s_k$. Then the mean (μ) and variance (σ^2) of $\mathcal{T}$ given $\mathcal{S} = s$ are

$$\mu = \sum_{k=1}^{K} \mu_k \qquad (8.27)$$

and

$$\sigma^2 = \sum_{k=1}^{K} \sigma_k^2 \qquad (8.28)$$

In particular, for $\phi = 1$ (or $\beta = 0$),

$$\mu_0 = \sum_{k=1}^{K} \frac{s_k n_k}{N_k} \qquad (8.29)$$

$$\sigma_0^2 = \sum_{k=1}^{K} \frac{s_k n_k m_k (m_k + n_k - s_k)}{N_k^2 (N_k - 1)} \qquad (8.30)$$

Proof: Follows from stratum-wise independence and the results for a hypergeometric distribution given in Chapter 1. $\square$

****

To construct exact conditional distributions for the TOR model, let

$$\mathcal{W}_k = z_k \mathcal{T}_k \quad \text{and} \quad \mathcal{W} = \sum_{k} \mathcal{W}_k \qquad (8.31)$$

and

$$f(\phi, \psi) = \prod_{k=1}^{K} f_k(\phi \psi^{z_k}) \qquad (8.32)$$

EXACT DISTRIBUTIONS

Note: we use the same symbol f for the TOR model gp as in the COR model. But here $f(.,.)$ is a bivariate function. Now suppose $f(\phi, \psi)$ is expanded as

$$f(\phi, \psi) = \sum_{(u,v) \in \Omega} c(u,v) \phi^u \psi^v \tag{8.33}$$

where Ω is the bivariate set of exponents of ϕ and ψ in $f(\phi, \psi)$ and $c(u,v)$ the coefficient of the term with the exponent of ϕ equal to u, and of ψ equal to v.

Theorem 8.4: In the TOR model, $\mathcal{S}_k$ is sufficient for α_k, $\mathcal{T}$ is sufficient for β, and $\mathcal{W}$ is sufficient for γ. The joint distribution of $(\mathcal{T}, \mathcal{W})$ conditional on $\mathcal{S}$ is given by

$$P[\mathcal{T} = t, \mathcal{W} = w \mid \mathcal{S} = s] = \frac{c(t,w) \phi^t \psi^w}{f(\phi, \psi)} \tag{8.34}$$

Proof: Substitute $\phi_k = \phi \psi^{z_k}$ in (8.9), and simplify to find that this probability is equal to

$$P[\mathcal{T} = t, \mathcal{S} = s] = \phi^t \psi^w \prod_{k=1}^{K} \Delta_{3k}^{-1} \theta_k^{S_k} c_k(t_k) \tag{8.35}$$

The factorization theorem shows that the respective sufficient statistics are as stated. From reasoning given earlier, we write $P[\mathcal{T} = t, \mathcal{W} = w \mid \mathcal{S} = s]$ as

$$\sum_{\boldsymbol{u}} \left\{ \prod_{k=1}^{K} P[\mathcal{T}_k = u_k, \mathcal{W}_k = z_k u_k \mid \mathcal{S}_k = s_k] \right\} \tag{8.36}$$

where the summation is over the set of $\boldsymbol{u} = (u_1, \cdots, u_K)$

$$\{\boldsymbol{u} : u_1 + \cdots + u_K = t;\ z_1 u_1 + \cdots + z_K u_K = w;\ u_k \in \Omega_k\} \tag{8.37}$$

Since, $P[\mathcal{T}_k = a, \mathcal{W}_k = z_k a \mid \mathcal{S}_k = s_k] = P[\mathcal{T}_k = a \mid \mathcal{S}_k = s_k]$, the above is equal to

$$\sum_{\boldsymbol{u}} \left\{ \prod_{k=1}^{K} P[\mathcal{T}_k = u_k \mid \mathcal{S}_k = s_k] \right\} \tag{8.38}$$

which then becomes

$$\sum_{\boldsymbol{u}} \left\{ \prod_{k=1}^{K} \frac{c_k(u_k) \phi^{u_k} \psi^{z_k u_k}}{f_k(\phi \psi^{z_k})} \right\} \tag{8.39}$$

Upon simplifying this, and noting that

$$c(t, w) = \sum_{\boldsymbol{u}} \prod_{k=1}^{K} c_k(u_k) \tag{8.40}$$

we get the desired result. $\square$

$f(\phi, \psi)$ is the bivariate gp of the conditional distribution of $(\mathcal{T}, \mathcal{W})$. The univariate conditional distribution of $\mathcal{T}$ (or of $\mathcal{W}$) is obtained by further conditioning on $\mathcal{W}$ (or on $\mathcal{T}$), that is by selecting appropriate terms of $f(\phi, \psi)$. These latter distributions are univariate PBDs and are amenable to the analytic techniques given previously.

Now we consider details and specific examples of the analysis of these models. We start with the simplest case, the common odds ratio model. This is a popular model not only because of this fact, but also because it is easier to interpret.

8.4 The COR Model

Example 8.1: Consider the data in Table 8.1, and assume a constant odds ratio by disease severity. To then perform inference on ϕ, we need the distribution of $\mathcal{T} = \mathcal{T}_1 + \mathcal{T}_2 + \mathcal{T}_3 + \mathcal{T}_4$ given the fixed marginal totals for each stratum. The total number of responders and the number of responders in treatment $x = 1$ for the four strata respectively are: $s_1 = 8$, $t_1 = 4$; $s_2 = 7$, $t_2 = 3$; $s_3 = 5$, $t_3 = 2$; and $s_4 = 5$, $t_4 = 0$. Further, the hypergeometric gps for these strata are:

$$\begin{aligned}
f_1(\phi) &= 5\phi^3(2 + 5\phi + 2\phi^2) \\
f_2(\phi) &= 10\phi^2(1 + 5\phi + 5\phi^2 + \phi^3) \\
f_3(\phi) &= 1 + 25\phi + 100\phi^2 + 100\phi^3 + 25\phi^4 + \phi^5 \\
f_4(\phi) &= 1 + 25\phi + 100\phi^2 + 100\phi^3 + 25\phi^4 + \phi^5
\end{aligned}$$

The generating polynomial of $P[\mathcal{T} = t \mid \mathcal{S} = s; \phi]$ is

$$f(\phi) = f_1(\phi) f_2(\phi) f_3(\phi) f_4(\phi)$$

For now, we compute this product using simple algebra. (Efficient computation is addressed in Chapter 11.) $f(\phi)$ has 16 terms, and the observed $\mathcal{T}$ is $t = 9$. The exponent and coefficient of each term of $f(\phi)$ are shown in Table 8.2.

Each term of $f(\phi)$ has a common factor, $250\phi^5$. We can ignore it provided we rescale the problem as follow: Since $l_1 = 5; l_2 = 20$, we write $l_{a1} = l_1 - l_1 = 0$; $l_{a2} = l_2 - l_1 = 15$; and $t_a = t - l_1$. Also $f_a(\phi) = f(\phi)/250\phi^2$ and $c_a(u) = c(u + l_1)/250$. Using t_a and $f_a(\phi)$ with observed value $9 - 5 = 4$ instead of the original quantities then has no effect on the inference for ϕ. In Chapter 11, we will see that such rescaling also simplifies computation.

****

From this adjusted gp, we construct the conditional TST exact and mid-p evidence functions in the usual way and extract the corresponding p-values, CIs and the cmue for ϕ. It also provides the basis for conditional score (CS) statistic based inference. This statistic is

$$\mathcal{W}_s = \frac{(\mathcal{T} - \mu)^2}{\sigma^2} \tag{8.41}$$

where μ and σ are the conditional mean and variance of $\mathcal{T}$ computed from $f(\phi)$. At the null,

THE COR MODEL

Table 8.2 *Exponents and Coefficients of* $f(\phi)$

	u	$c(u)/250$
	5	2
	6	115
	7	2437
	8	24662
$\Rightarrow$	9	140490
	10	491931
	11	1110605
	12	1658974
	13	1658974
	14	1110605
	15	491931
	16	140490
	17	24662
	18	2437
	19	115
	20	2

Note: $\Rightarrow$ denotes the observed value of $\mathcal{T}$.

these are given by Theorem 8.3. If w is the observed value of the CS statistic, the exact CS evidence function is

$$P[\mathcal{W}_s \geq w \mid \mathcal{S} = s; \phi] \qquad (8.42)$$

The variance term on both sides of the inequality within (8.43) cancels out. As noted in Chapter 3, the CS statistic is, in such a context, equivalent to the distance from the mean statistic. The associated point estimate is the cmle. If it exists, it maximizes the conditional likelihood, and is contained in all corresponding score CIs for ϕ. To reiterate an earlier formula, it is also the solution to the equation

$$\sum_u uc(u)\phi^u = t \sum_u c(u)\phi^u \qquad (8.43)$$

An asymptotic *p*-value obtains by referring the null CS statistic to a chisquare distribution with one df. The lower and upper $(1-\alpha)100\%$ asymptotic confidence limits for ϕ are obtained by solving the equation

$$[t - \mu(\phi)]^2 = \sigma^2(\phi)\chi^2_{1-\alpha} \qquad (8.44)$$

where $\chi^2_{1-\alpha}$ is the value such that the area to its right under the chisquare curve with one df is α.

****

The Mantel–Haenszel (MH) method is another popular method for the COR model. In the epidemiological literature, the CS statistic evaluated at $\phi = 1$ is generally called the MH statistic. Yet, what is called the MH estimate of the common odds ratio is different from the cmle and is not necessarily contained in all score based CIs for ϕ. We thereby do not call the CS test, the MH test. This label is reserved for quantities defined next.

First consider the MH estimate of the common odds ratio, ϕ. Let

$$Q_{0k} = \frac{a_k + d_k}{N_k}; \quad Q_{1k} = \frac{b_k + c_k}{N_k}; \quad R_{0k} = \frac{a_k d_k}{N_k}; \quad \text{and} \quad R_{1k} = \frac{b_k c_k}{N_k}$$

Further, define

$$R_{0+} = \sum_{k=1}^{K} R_{0k} \quad \text{and} \quad R_{1+} = \sum_{k=1}^{K} R_{1k} \tag{8.45}$$

Then the MH estimates of ϕ and β are given by

$$\hat{\phi}_{MH} = \frac{R_{0+}}{R_{1+}} \quad \text{and} \quad \hat{\beta}_{MH} = \log\left(\frac{R_{0+}}{R_{1+}}\right) \tag{8.46}$$

Various formulae for estimating the variance of the MH estimator of the odds ratio or its logarithm exist. That derived by Robins, Breslow and Greenland (1986) has stood the test of time. It is given by

$$\widehat{var}[\hat{\beta}_{MH}] = \frac{1}{2} \sum_{k=1}^{K} \left[\frac{Q_{0k} R_{0k}}{R_{0+}^2} + \frac{Q_{0k} R_{1k} + Q_{1k} R_{0k}}{R_{0+} R_{1+}} + \frac{Q_{1k} R_{1k}}{R_{1+}^2} \right] \tag{8.47}$$

The MH 95% asymptotic CI for the common log odds ratio then is

$$\hat{\beta}_{MH} \pm 1.96 \sqrt{\widehat{var}[\hat{\beta}_{MH}]} \tag{8.48}$$

The associated 95% CI for ϕ obtains from exponentiating the above.

****

Example 8.1 (continued): We return to the analysis of Table 8.1. With the gp in Table 8.2, the TST exact and mid-p, and asymptotic CS based inference indices are given in Table 8.3.

Table 8.3 *Analysis of the Data in Table 8.1*

Statistic	p-value	Estimate	95% CI
TST Exact	0.0489	0.216	$(0.036, 0.994)$
TST mid-p	0.0284	0.216	$(0.044, 0.856)$
CS Asymptotic	0.0243	0.211	$(0.054, 0.827)$
MH Asymptotic	0.0300	0.239	$(0.066, 0.870)$

Note: These results relate to ϕ.

CONDITIONAL INDEPENDENCE

For these data, the TST exact method has a higher p-value and wider CI compared to the other three methods. The TST mid-p method gives results that are similar to the two conditional asymptotic methods. Under the COR model, all of them point towards the existence of a treatment effect.

8.5 Conditional Independence

Sometimes instead of using the restrictive COR model, we compare the two treatments under the more general HOR model. For this, consider the hypothesis of conditional independence, $H_0 : \phi_1 = \phi_2 = \cdots = \phi_K = 1$ versus the alternative that at least one parameter is not one. This alternative includes the possibility that for some k, $\phi_k > 1$ and for some other k, $\phi_k < 1$, implying that some stratum specific factor reverses the relative efficacies of the treatments. Though such a scenario is rare, it cannot be ruled out a priori. The conditional independence test is a useful but underutilized overall test in stratified data.

We need a statistic for which extreme values are unlikely under the null, and whose conditional distribution is free of the nuisance parameters. Assume for now that we have such a statistic, and that it has a stratum additive form. Let $\mathcal{W}$ denote the statistic with its kth stratum component being $\mathcal{W}_k$. That is,

$$\mathcal{W} = \sum_k \mathcal{W}_k \qquad (8.49)$$

For the purpose of this section, let ψ be a dummy parameter, and define the kth stratum polynomial

$$g_k(\psi, \phi_k) = \sum_{u \in \Omega_k} c_k(u) \phi_k^u \psi^{w_k(u)} \qquad (8.50)$$

With $\boldsymbol{\phi} = (\phi_1, \ldots, \phi_K)$, define the product polynomial

$$g(\psi, \boldsymbol{\phi}) = \prod_{k=1}^{K} g_k(\psi, \phi_k) \qquad (8.51)$$

Now consider specified values of $\phi_k, k = 1, \ldots, K$. Each $g_k(\psi, \phi_k)$ is then a univariate polynomial in ψ. Suppose we perform the multiplication in the rhs of (8.51) and write it as:

$$g(\psi, \boldsymbol{\phi}) = \sum_{v \in \Omega} d(v; \boldsymbol{\phi}) \psi^v \qquad (8.52)$$

with $d(v; \boldsymbol{\phi})$, the coefficient of the term in $g(\psi, \boldsymbol{\phi})$ with exponent of ψ equal to v, and Ω, the set of exponents of ψ.

Theorem 8.5: With fixed margins of each 2×2 table, the exact conditional distribution of $\mathcal{W}$ is

$$P[\mathcal{W} = w \mid \mathcal{S} = s; \boldsymbol{\phi}] = \frac{d(w; \boldsymbol{\phi})}{g(1, \boldsymbol{\phi})} \qquad (8.53)$$

where, under the usual notation,

$$g(1, \boldsymbol{\phi}) = \sum_{v \in \Omega} d(v; \boldsymbol{\phi}) \tag{8.54}$$

In particular, under the null,

$$P[\mathcal{W} = w \mid \mathcal{S} = s; H_0] = d(w; 1) \left(\prod_{k=1}^{K} \binom{N_k}{s_k} \right)^{-1} \tag{8.55}$$

Proof: The conditional distribution of $\mathcal{W}$ obtains from summing expression (8.15) over values of $\boldsymbol{u} = (u_1, \ldots, u_K)$ that range in the set

$$\{ \boldsymbol{u} : w_1(u_1) + \cdots + w_K(u_K) = w, \ u_k \in \Omega_k, \ k = 1, \ldots, K \} \tag{8.56}$$

Under this summation, the required conditional probability becomes

$$\left(\prod_{k=1}^{K} f_k(\phi_k) \right)^{-1} \sum_{\boldsymbol{u}} \left[\prod_{k=1}^{K} c_k(u_k) \phi_k^{u_k} \right] \tag{8.57}$$

(8.53) follows from equating terms in the product (8.51) with those in (8.52) and by noting that $g_k(1, \phi_k) = f_k(\phi_k)$. (8.55) follows from

$$g_k(1, 1) = f_k(1) = \binom{N_k}{s_k} \tag{8.58}$$

□

Computing the distribution of $\mathcal{W}$ is a polynomial multiplication problem similar to that for computing $f(\phi)$ in the COR model. The difference is the set of polynomials being multiplied. As ψ is a dummy parameter, the distribution of $\mathcal{W}$, even though looking like a PBD, is, strictly speaking, not one.

****

Now we derive some test statistics for the hypothesis of conditional independence in the HOR model.

Theorem 8.6: Let $\mathcal{W}_u$ denote the unconditional score (US) statistic, and $\mathcal{W}_c$, the conditional score (CS) statistic. Then, for the HOR model, these statistics under H_0 are given by

$$\mathcal{W}_u = \sum_{k=1}^{K} \mathcal{W}_{uk} \quad \text{and} \quad \mathcal{W}_c = \sum_{k=1}^{K} \mathcal{W}_{ck} \tag{8.59}$$

where

$$\mathcal{W}_{uk} = \frac{(N_k T_k - n_k s_k)^2 N_k}{m_k n_k s_k (N_k - s_k)} \tag{8.60}$$

$$\mathcal{W}_{ck} = \frac{(N_k T_k - n_k s_k)^2 (N_k - 1)}{m_k n_k s_k (N_k - s_k)} \tag{8.61}$$

CONDITIONAL INDEPENDENCE

The unconditional likelihood ratio (ULR) statistic, $\mathcal{Q}_u$, for H_0, is

$$\mathcal{Q}_u = \sum_{k=1}^{K} \mathcal{Q}_{uk} \qquad (8.62)$$

where

$$\begin{aligned}\mathcal{Q}_{uk}/2 &= \mathcal{A}_k \log\left(\frac{\mathcal{A}_k N_k}{s_k n_k}\right) + (n_k - \mathcal{A}_k)\log\left(\frac{(n_k - \mathcal{A}_k)N_k}{(N_k - s_k)n_k}\right) \\ &+ \mathcal{B}_k \log\left(\frac{\mathcal{B}_k N_k}{s_k m_k}\right) + (m_k - \mathcal{B}_k)\log\left(\frac{(m_k - \mathcal{B}_k)N_k}{(N_k - s_k)m_k}\right)\end{aligned} \qquad (8.63)$$

In the above formula, we set $0\log 0 = 0$.

Proof: By definition, the score statistic is

$$\mathcal{W} = \mathbf{D}'\mathbf{I}^{-1}\mathbf{D} \qquad (8.64)$$

where $\mathbf{D}$ is the vector of first partial derivatives, and $\mathbf{I}$, the matrix of the second partial derivatives of the appropriate log-likelihood. The statistic is evaluated at the null values for the parameters of interest and at the restricted mles for the other parameters.

Consider the conditional case first. The relevant likelihood is (8.15). Also, $\mathbf{D}$ is a K dimensional vector while $\mathbf{I}$ is a $K \times K$ matrix. Using results of §5.14, the kth element of $\mathbf{D}$ is seen to be $t_k - \mu_k$, where μ_k is the conditional mean of $\mathcal{T}_k$. $\mathbf{I}$ is a diagonal matrix with the kth element on the main diagonal as σ_k^2, the conditional variance of $\mathcal{T}_k$. Substituting these in the expression for the score statistic, and simplifying we get

$$\mathcal{W}_c = \sum_{k=1}^{K} (t_k - \mu_k)^2 \sigma_k^{-2} \qquad (8.65)$$

(8.61) follows from applying the null conditional mean and variance of the kth hypergeometric distribution. In the unconditional case, $\mathbf{D}$ is a $2K$ dimensional vector while $\mathbf{I}$ is a $2K \times 2K$ block diagonal matrix. The relevant likelihood is (8.9). From this we can show that

$$\mathcal{W}_u = \sum_{k=1}^{K} \mathbf{D}_k \mathbf{I}_k^{-1} \mathbf{D}_k' \qquad (8.66)$$

where $\mathbf{D}_k$ and $\mathbf{I}_k$ are the kth stratum versions of the quantities defined in §5.14 for a single 2×2 table. (8.60) follows by substituting the expression for the stratum specific score statistic derived in §5.14.

The ULR statistic is equal to twice the difference between the log-likelihood evaluated at the unrestricted mles for all the parameters and the log-likelihood evaluated at the restricted mles. (8.63) follows by writing the unconditional log-likelihood from (8.9) as a sum of stratum-wise log-likelihoods and applying the results of §5.14. □

The following remarks are in order:

- The conditional likelihood ratio (CLR) statistic requires the computation of cmle of β_k for each k. When computing exact significance levels, this has to be done for each point in the stratum conditional distribution.
- At large sample sizes, these four statistics have a chisquare distribution with K df.
- All the four statistics have a stratum additive form. This is a critical requirement for efficient computation of exact significance levels and confidence intervals.

****

Example 8.2: Suppose we have two strata with the data shown in Table 8.4.

Table 8.4 *Hypothetical Clinical Data*

	Stratum 1		Stratum 2	
Response	$x=0$	$x=1$	$x=0$	$x=1$
$y=0$	0	4	2	2
$y=1$	4	0	2	2
Total	4	4	4	4

First, we test $\phi_1 = \phi_2 = 1$ using the US statistic $\mathcal{W}_u$. From formula (8.60), we get

$$g_1(\psi, 1) = g_2(\psi, 1) = 36 + 32\psi^2 + 2\psi^8$$

Multiplying the two polynomials, we have

$$g(\psi, 1) = 1296 + 2304\psi^2 + 1024\psi^4 + 144\psi^8 + 128\psi^{10} + 4\psi^{16}$$

From (8.60), we have

$$w_1 = \frac{(8 \times 0 - 4 \times 4)^2 \times 8}{4 \times 4 \times 4 \times 4} = 8$$

$$w_2 = \frac{(8 \times 2 - 4 \times 4)^2 \times 8}{4 \times 4 \times 4 \times 4} = 0$$

The observed statistic is $w = w_1 + w_2 = 8$. At two df, it gives an approximate chisquare *p*-value equal to 0.018. The exact *p*-value obtained from $g(\psi; 1)$, on the other hand, is $(144 + 128 + 4)/4900 = 0.056$.

To test the null with the ULR statistic, we apply (8.63). Then we get

$$g_1(\psi, 1) = g_2(\psi, 1) = 36 + 32\psi^{1.047} + 2\psi^{5.545}$$

Taking their product, we have

$$g(\psi, 1) = 1296 + 2304\psi^{1.047} + 1024\psi^{2.094} + 144\psi^{5.545} + 128\psi^{6.592} + 4\psi^{11.090}$$

TREND IN ODDS RATIOS

The observed ULR statistic is $z = 5.545$. At two df, this gives an approximate chisquare p-value equal to 0.063. The exact p-value obtained from $g(\psi; \mathbf{1})$, on the other hand, is $(144 + 128 + 4)/4900 = 0.056$.

For these data, the score and LR exact p-values are identical. But, the asymptotic LR p is higher than the exact p and the asymptotic score p is the smallest of the four.

8.6 Trend In Odds Ratios

The **main effect** parameter in the TOR model, β, is also the log-odds ratio when $z_k = 0$ or $\gamma = 0$. And, γ, the **interaction** parameter, gives the change in the log-odds ratio for a unit change in z_k. Now we consider exact conditional and large sample analyses of these parameters.

We analyze γ with the distribution of $\mathcal{W} = \Sigma_k z_k t_k$ conditional on the observed value of $\mathcal{T}$. From (8.34), we get

$$P[\mathcal{W} = w \mid \mathcal{S} = s, \mathcal{T} = t; \psi] = \frac{c(t,w)\psi^w}{f(1,\psi;t,.)} \quad (8.67)$$

where $f(1, \psi; t, .)$ is the set of terms in $f(\phi, \psi)$ in which the exponent of ϕ is equal to t, and further where the value of ϕ has been set to 1.

To analyze ϕ, we need the distribution of $\mathcal{T}$ given $\mathcal{S} = s$ and $\mathcal{W} = t$. This is similarly given by

$$P[\mathcal{T} = t \mid \mathcal{S} = s, \mathcal{W} = w; \phi] = \frac{c(t,w)\phi^t}{f(\phi,1;.,w)} \quad (8.68)$$

where $f(\phi, 1; ., w)$ is the set of terms in $f(\phi, \psi)$ in which the exponent of ψ is equal to w, and further where the value of ψ has been set to 1.

The polynomials $f(\phi, \psi; t, .)$ and $f(\phi, \psi; ., w)$ are obtained by extraction from the bivariate polynomial $f(\phi, \psi)$. For now we use an elementary procedure based on Theorem 8.4. An efficient method is given in Chapter 12. These univariate gps are used to perform conditional inference on γ and β in the usual manner.

****

Example 8.3: We apply the TOR model (8.3) to Table 8.1 using the disease severity scores $z_1 = 0$, $z_2 = 1$, $z_3 = 2$ and $z_4 = 3$. The polynomials $f_k(\phi\psi^{z_k})$, for $k = 1, 2, 3, 4$, are in Table 8.5. The product of these four polynomials, $f(\phi, \psi)$, has 142 terms; we do not list them.

The observed values of $\mathcal{T}$ and $\mathcal{W}$ are

$$t = 4 + 3 + 2 + 0 = 9$$
$$w = 0 \times 4 + 1 \times 3 + 2 \times 2 + 3 \times 0 = 7$$

For inference on γ, we select terms with the exponent of ϕ equal to 9. Omitting the factor ϕ^9 from these, we have

Table 8.5 *Stratum-wise gps for Table 8.1*

Stratum	$f_k(\phi \psi^{z_k})$
1	$10\phi^3 + 25\phi^4 + 10\phi^5$
2	$10\phi^2\psi^2 + 50\phi^3\psi^3$ $+ 50\phi^4\psi^4 + 10\phi^5\psi^5$
3	$1 + 25\phi\psi^2 + 100\phi^2\psi^4$ $+ 100\phi^3\psi^6 + 25\phi^4\psi^8 + \phi^5\psi^{10}$
4	$1 + 25\phi\psi^3 + 100\phi^2\psi^6$ $+ 100\phi^3\psi^9 + 25\phi^4\psi^{12} + \phi^5\psi^{15}$

$$f(1, \psi; 9, .) = 250\psi^4(2 + 51\psi + 215\psi^2 + 885\psi^3 + 3475\psi^4 + 4450\psi^5 + 7710\psi^6 + 6100\psi^7 + 4200\psi^8 + 1000\psi^9 + 10\psi^{10})$$

This gp is used to obtain the conditional indices of inference on ψ shown in the upper part of Table 8.6. The estimates for γ obtain by taking logarithms.

For inference on ϕ, we select the terms of $f(\phi, \psi)$ with the exponent of ψ equal to 7. Then we get

$$f(\phi, 1;., 7) = 1250\phi^7(50 + 175\phi + 122\phi^2 + 55\phi^3 + 2\phi^4)$$

This gp is used to obtain the conditional indices of inference on ϕ shown in the lower part of Table 8.6. The results for β are obtained by taking logarithms.

These results provide evidence for an interaction effect but not for the main effect of treatment. A hierarchical approach would include both the terms in the model.

****

Consider a specified stratum j. Suppose we need a CI for its log-odds ratio, namely $\beta_j = \beta + \gamma z_j$. For $k = 1, \ldots, K$, define $z_{jk} = z_k - z_j$. Substituting $\beta = \beta_j - z_j\gamma$ in model (8.3), we get

$$\log\left(\frac{\pi_k(x)}{1 - \pi_k(x)}\right) = \alpha_k + \beta_j x + \gamma x z_{jk} \qquad (8.69)$$

With j fixed, the sufficient statistics for β_j and γ in the new formulation are respectively

$$\mathcal{T} = \sum_k \mathcal{T}_k \quad \text{and} \quad \mathcal{W}_* = \sum_k z_{jk}\mathcal{T}_k \qquad (8.70)$$

TREND IN ODDS RATIOS

Table 8.6 *Analysis of the TOR Model*

Statistic	p-value	Estimate	95% CI
Parameter = ψ			
TST Exact	0.0821	0.278	(0.050, 1.152)
TST mid-p	0.0506	0.278	(0.061, 1.003)
CS Asymptotic	0.0393	0.279	(0.086, 0.948)
Parameter = ϕ			
TST Exact	1.000	1.855	(0.097, 47.187)
TST mid-p	0.6340	1.855	(0.143, 30.775)
CS Asymptotic	0.5825	1.906	(0.245, 18.393)

Note: These results are for Table 8.1.

Let the generating polynomial for the conditional distribution of $(\mathcal{T}, \mathcal{W}_*)$ be $f_*(\phi_j, \psi)$ with $\phi_j = \exp(\beta_j)$. We may obtain the distribution for the analysis of ϕ_j using the procedure outlined above but with the new scores z_{jk}. However, we can also obtain it from $f(\phi, \psi)$, thus obviating a separate computation. Since $\phi = \phi_j \psi^{-z_j}$,

$$f(\phi, \psi) = \sum_{(u,v) \in \Omega} c(u,v) \phi^u \psi^v \qquad (8.71)$$

$$= \sum_{(u,v)} c(u,v) \phi_j^u \psi^{(v-z_j u)} \qquad (8.72)$$

$$= \sum_{(u,v_*)} c(u, v_*) \phi_j^u \psi^{v_*} \qquad (8.73)$$

where $v_* = u - z_j v$. Note also that $w_* = w - z_j t$. Therefore, to get terms of $f_*(\phi_j, \psi)$ at a fixed exponent of ψ equal to w_*, we get terms of the $f(\phi, \psi)$ at fixed values of $w - z_j t$. Then this gp is used to do the analysis of ϕ_j in the usual manner.

******

Example 8.3 (continued): For Table (8.1) data, we want an exact CI for the odds ratio in stratum $j = 4$. With $z_j = 3$, we have $z_{41} = -3$, $z_{42} = -2$, $z_{43} = -1$, $z_{44} = 0$. The observed values of $\mathcal{T}$ and $\mathcal{W}_*$ are

$$t = 4 + 3 + 2 + 0 = 9$$
$$w_* = -3 \times 4 + -2 \times 3 + -1 \times 2 + 0 \times 0 = -20$$

Then from the 142 terms of $f(\phi, \psi)$ computed earlier, we select those that satisfy $v - z_j u = -20$. This gives us the conditional gp for $\mathcal{T}$ at fixed value of $\mathcal{W}_* = -20$. This is shown below.

$$f_*(\phi_4; 1;., -20) = 250\phi_4^8(15 + 885\phi_4^1 + 14475\phi_4^2 + 58127\phi_4^3 +$$
$$73925\phi_4^4 + 35465\phi_4^5 + 6335\phi_4^6 + 275\phi_4^7 + 2\phi_4^8)$$

With observed $t = 9$, the cmue for stratum 4 odds ratio, ϕ_4, derived from this gp is 0.030 with the 95% conditional mid-*p* CI equal to $(0.001, 0.376)$, and the null test mid-*p*-value being 0.005.

****

Before analyzing individual parameters, we may perform a global test of no effect. In that case, the hypothesis of interest is $\beta = \gamma = 0$ in model (8.3). This is akin to a test for conditional independence in the COR model. For this we use the score statistic based on the conditional likelihood ℓ_c from (8.34). To a constant, this is

$$\ell_c = \beta t + \gamma w - \sum_{k=1}^{K} \log\left[\sum_{v \in \Omega_k} c_k(v) \exp(\beta v + \gamma v z_k)\right] \qquad (8.74)$$

The vector of first partial derivatives of ℓ_c is

$$\mathbf{D}_c = \left(\frac{\partial \ell_c}{\partial \beta}, \frac{\partial \ell_c}{\partial \gamma}\right)' = \begin{pmatrix} t - \sum_{k=1}^{K} \mu_k \\ w - \sum_{k=1}^{K} z_k \mu_k \end{pmatrix} \qquad (8.75)$$

and the second partial derivatives are

$$\frac{\partial^2 \ell_c}{\partial \beta^2} = -\sum_k \sigma_k^2 \qquad (8.76)$$

$$\frac{\partial^2 \ell_c}{\partial \gamma^2} = -\sum_k z_k^2 \sigma_k^2 \qquad (8.77)$$

$$\frac{\partial^2 \ell_c}{\partial \beta \partial \gamma} = -\sum_k z_k \sigma_k^2 \qquad (8.78)$$

where

$$\mu_k = \frac{\sum_v v c_k(v) \exp(\beta v + \gamma v z_k)}{\sum_v c_k(v) \exp(\beta v + \gamma v z_k)} \qquad (8.79)$$

$$\nu_k = \frac{\sum_v v^2 c_k(v) \exp(\beta v + \gamma v z_k)}{\sum_v c_k(v) \exp(\beta v + \gamma v z_k)} \qquad (8.80)$$

and further where $\sigma_k^2 = \nu_k - \mu_k^2$. Here, μ_k is the conditional mean and σ_k^2, the conditional variance, of $\mathcal{T}_k$ given $\mathcal{S}_k = s_k$. The conditional information matrix is therefore

$$\mathbf{I}_c = \begin{bmatrix} \sum_k \sigma_k^2 & \sum_k z_k \sigma_k^2 \\ \sum_k z_k \sigma_k^2 & \sum_k z_k^2 \sigma_k^2 \end{bmatrix} \qquad (8.81)$$

TREND IN ODDS RATIOS

The associated variance covariance matrix is therefore

$$\Sigma_c = \|\mathbf{I}_c\|^{-1} \begin{bmatrix} \sum_k z_k^2 \sigma_k^2 & -\sum_k z_k \sigma_k^2 \\ -\sum_k z_k \sigma_k^2 & \sum_k \sigma_k^2 \end{bmatrix} \quad (8.82)$$

where $\|\mathbf{I}_c\|$ is the determinant of the information matrix. By definition, the conditional score statistic is

$$\eta_c(t, w) = \mathbf{D}_c' \Sigma_c \mathbf{D}_c \quad (8.83)$$

Let $\eta_0(t, w)$ denote the value of the score statistic at $\beta = \gamma = 0$. At the null values, μ_k and σ_k^2 are the mean and variance of the central hypergeometric distribution for the kth table. We form an exact test by computing the exact null probability of points in the sample space with a score statistic value greater than or equal to the observed value. Ω, of course, is the bivariate set of exponents of $f(\phi, \psi)$. Consider the subsample space

$$\Omega_0 = \{(u, v) \in \Omega : \eta_0(u, v) \geq \eta_0(t, w)\} \quad (8.84)$$

Then the exact p-value for $\beta = \gamma = 0$ is

$$\left(\prod_{k=1}^{K} \binom{N_k}{s_k} \right)^{-1} \sum_{(u,v) \in \Omega_0} c(u, v) \quad (8.85)$$

****

Methods for analysis of the TOR model using the large sample approach are given in many books and statistical packages. They use the unconditional likelihood

$$\ell = \beta t + \gamma w + \sum_k \{\alpha_k s_k - \log \Delta_{3k}\} \quad (8.86)$$

The unrestricted mles for the parameters are obtained by setting the $K+2$ first partial derivatives of l to zero. The packages provide the inverse of the $(K+2) \times (K+2)$ matrix of second partial derivatives used for computing p-values and CIs for the parameters. The unconditional score and LR tests are constructed in the usual manner.

Another option is to use a conditional likelihood, for example, that from (8.34). The cmles for β and γ obtain by putting the first partial derivatives to zero and solving the equations by an iterative method. Denote these by $\hat{\beta}$ and $\hat{\gamma}$. The asymptotic variance and covariance matrix for the cmles, Σ, obtains by inverting the information matrix above. Estimates for the required variances and covariance obtain by evaluating it at the cmles. Let σ_{ij}^2, $i, j \in \{0, 1\}$ denote the (i, j)th element of the Σ. Then the asymptotic 95% CIs for β and γ are

$$\hat{\beta} \pm 1.96 \hat{\sigma}_{11} \quad \text{and} \quad \hat{\gamma} \pm 1.96 \hat{\sigma}_{22} \quad (8.87)$$

The asymptotic CS test for $\beta = \gamma = 0$ obtains using the bivariate CS statistic, $\eta_0(t, w)$, derived above and comparing it with a chisquare variate at two df. A 95% CI for the log-odds ratio for stratum j, $\beta_j = \beta + \gamma z_j$ is given by

$$\hat{\beta} + \hat{\gamma} z_j \pm 1.96\sqrt{\widehat{\text{var}}(\hat{\beta}_j)} \qquad (8.88)$$

where

$$\widehat{\text{var}}(\hat{\beta}_j) = \hat{\sigma}_{11}^2 + z_j^2 \hat{\sigma}_{22}^2 \qquad (8.89)$$

A third option is to use, as done earlier for the conditional score statistic, the fully conditional likelihoods for the individual parameters, namely, (8.67) and (8.68). For example, for the analysis of γ, we use the statistic

$$\mathcal{W}_* = \frac{[\mathcal{W} - \mu_*(t;\gamma)]^2}{\text{var}_*[t;\gamma]} \qquad (8.90)$$

where $\mu_*(t;\gamma)$ and $\text{var}_*[t;\gamma]$ are the conditional mean and variance of $\mathcal{W}$ given $\mathcal{T} = t$ and $\mathcal{S} = s$. These can be computed from the exact conditional distribution of $\mathcal{W}$. Or, for asymptotic conditional analysis, they can be computed using the backward induction method described in Chapter 13. See also Newman (2001) for some test statistics applicable to the TOR model.

8.7 Recommendations

More comparisons of exact and other methods for the COR model need to be done. In the light of what has been done thus far, the proposals of Emerson (1994) remain reasonable. Accordingly, we also recommend the mid-p based and MH methods for sparse data. In contrast to him, we suggest the cmue instead of the cmle as the associated estimate for mid-p based inference. Conditional methods that further partition the sample space in the COR model may also be used (Agresti 2001).

For testing conditional independence, the exact score test is suggested. In the TOR model, a mid-p evidence function based inference is a prudent option for small samples and sparse data.

8.8 Relevant Literature

Of the numerous papers relevant to the analysis of several 2×2 tables, a few of the seminal early ones are Cochran (1954a), Woolf (1955), Cornfield (1956), Cox (1958), Mantel and Haenszel (1959), Gart (1962, 1970) and Goodman (1964) and Gart (1970). Three excellent reviews of the subject are in Gart (1971), Thomas and Gart (1992) and Emerson (1994); the latter two also cover exact methods, for which Agresti (1992) is a valuable source as well.

Many books deal with the subject. The early material is well presented by Breslow and Day (1980), Fleiss (1981) and Cox and Snell (1989). Sahai and Khurshid (1996) and Rothman and Greenland (1998) give the more recent material.

Mehta and Walsh (1992) compared exact, mid-p and Mantel–Haenszel CIs for stratified 2×2 tables. Related comparisons are in Hirji (1991) and Vollset, Hirji and Afifi (1991). For other comparative views, and other forms of conditional analysis, see Kim and Agresti (1995) and Strawderman and Wells (1998). Emerson (1994) and Agresti (2001) contain good comparative summaries.

The idea of conditional independence in stratified contingency tables is explained by many authors; a succinct and clear presentation is given by Agresti (1996). Yao and Tritchler (1993) developed an exact conditional test for several 2×2 tables; their method and approach was improved by Hirji (1996).

A detailed discussion of the trend in odds ratios or quantitative interaction model for stratified data was given by Zelen (1971). Gart (1977) applied the model to a dose-response study and gave exact and large sample tests for it. It was examined, among others, by Breslow and Day (1980) and Cox and Snell (1989). Brand and Kragt (1992) and Cook and Walter (1997) studied its application to a meta-analysis of clinical trials. See also Newman (2001). But it has not received as much attention as the COR model. Empirical comparisons of size, power and coverage of the exact, mid-p and asymptotic tests and CIs for this model is a useful research area.

Papers on exact analysis of several 2×2 tables also deal with computational issues; we give a comprehensive listing in Chapter 11. Additional papers on heterogeneity in stratified 2×2 tables are noted in Chapter 9.

8.9 Exercises

8.1. Table 8.7 is derived from Table 1.2 by converting infection and ESR into binary variables.

Table 8.7 *ESR and Infection by Age in Kwashiorkor*

Age (months)	Infection	ESR Level ≤ 99	ESR Level $100+$
≤ 12	No	22	0
	Yes	6	1
$12 < \text{Age} \leq 24$	No	12	0
	Yes	4	1
Age > 24	No	5	0
	Yes	2	1

Source: Hameer (1990); Used with permission.

With age as a nominal factor, analyze these data using the COR model. Plot the conditional mid-p and score evidence functions and compare exact and large sample methods, including the MH method.

8.2. Perform the test of conditional independence for the data in Table 8.7. Compare exact and large sample methods.

8.3. Analyze Table 8.7 data using the TOR model and natural scores for age. For the main effect and interaction terms, plot evidence functions, obtain estimates and p-values. Compare exact and large sample methods.

8.4. Vollset (1989) used data from the Norwegian dietary cohort study to examine the relationship between diagnosis of cerebral atrophy and alcohol usage. They were stratified on several factors and only strata where both cases and noncases were present were extracted. We show these data in Table 8.8 with a two level version of (0 and ≥ 1) of his three level alcohol use index. Cases are subjects with cerebral atrophy and noncases, those without. (Because of how they were selected, these data may not reflect the associations in the original.)

Table 8.8 *Cerebral Atrophy Data*

Stratum	Status	AI = 0	AI ≥ 1
1	Noncase	1	4
	Case	1	4
2	Noncase	15	9
	Case	0	1
3	Noncase	3	8
	Case	1	0
4	Noncase	3	7
	Case	0	1
5	Noncase	19	9
	Case	0	1
6	Noncase	6	11
	Case	0	1

Note: AI = Alcohol Index.
Source: Vollset (1989); Used with permission.

Show that the conditional gp of the COR model for these data is

$$f(\phi) = 120(171 + 3054\phi + 19078\phi^2 + 57936\phi^3 + 94496\phi^4 + 83360\phi^5 + 36960\phi^6 + 6400\phi^7)$$

Use this to perform conditional exact, mid-p and score based analysis of ϕ. Plot the evidence functions and test the hypothesis of conditional independence.

8.5. Tables 8.9, 8.10, 8.11 and 8.12 are from a cohort study of the effect of highly active antiretroviral therapy (HAART) for children infected by HIV 1 virus. They are extracted from Table 1 of Johnston et al. (2001). The two outcomes we consider are CD4 level and viral load (VL); both are dichotomized.

Fit the COR model to them and test for conditional independence. Compare conditional exact, mid-p, score and MH methods. Plot the associated evidence functions. For Tables 8.11 and 8.12, use scores for age to assess a trend in the odds ratios and compute exact, mid-p, score and Wald based CIs for the odds ratio for Age > 12 months.

8.6. Perform comprehensive comparative analyses for the COR and TOR models with natural scores, and tests for conditional independence for the four datasets in Table 8.13. Where relevant, plot the evidence functions. Use the CT method as well.

8.7. Give detailed step by step proofs of theorems 8.1, 8.2, 8.3 and 8.4.

8.8. Derive the unconditional score, LR and Wald statistics for β in the COR model. Show how exact and asymptotic CIs are computed using these statistics.

8.9. How would you compute an MH exact p-value for testing $\beta = 0$ in the COR model?

8.10. Compare exact, mid-p, score and MH based p-values, estimates and CIs for the COR ϕ for all the datasets given in Emerson (1994). Compare your results with those obtained by the author.

8.11. Give a detailed proof of Theorem 8.5.

Table 8.9 *HAART Data: A*

Age Group (years)	Treatment	CD4 ≤ 32	CD4 > 32
Age ≤ 6.0	NonHAART	1	5
	HAART	1	1
6 < Age ≤ 7.5	NonHAART	2	0
	HAART	5	3
7.5 < Age ≤ 12.0	NonHAART	1	2
	HAART	2	3
12.0 < Age	NonHAART	4	2
	HAART	2	1

Source: Johnston et al. (2001); Used with permission.

Table 8.10 *HAART Data: B*

Age Group (years)	Treatment	Log-VL ≤ 3.0	Log-VL > 3.0
Age ≤ 6.0	NonHAART	1	5
	HAART	1	1
6 < Age ≤ 7.5	NonHAART	0	2
	HAART	5	3
7.5 < Age ≤ 12.0	NonHAART	1	2
	HAART	4	1
12.0 < Age	NonHAART	2	4
	HAART	3	0

Source: Johnston et al. (2001); Used with permission.

8.12. Give a detailed proof of Theorem 8.6 and also derive the conditional LR and Wald statistics for testing conditional independence in several 2×2 tables.

8.13. Perform exact and asymptotic score, LR and Wald tests for conditional independence for all the datasets in Emerson (1994). Use the CT method as well.

8.14. Design and perform a simulation study to compare exact and asymptotic score, LR and Wald tests for conditional independence for data in $K \times 2 \times 2$ tables. Compare the results in terms of size and power. Which method(s) do you recommend and under what circumstance?

8.15. Conduct a simulation study to compare exact and asymptotic methods for computing p-values, estimates and CIs for the COR in $K \times 2 \times 2$ tables. Compare, as appropriate, the results in terms of size, power, bias, precision, length and coverage. Which method(s) do you recommend and under what circumstance?

8.16. Construct conditional and unconditional LR and score tests for the individual parameters in

Table 8.11 *HAART Data: C*

Group	Treatment	CD4 $\leq$ 32	CD4 $>$ 32
Female, AA	NonHAART	5	5
	HAART	6	1
Male, AA	NonHAART	1	3
	HAART	3	4
Female, C	NonHAART	2	0
	HAART	0	1
Male, C	NonHAART	1	0
	HAART	2	1

Note: AA = African American; C = Caucasian.
Source: Johnston et al. (2001); Used with permission.

Table 8.12 *HAART Data: D*

Group	Treatment	Log-VL $\leq$ 3.0	Log-VL $>$ 3.0
Female, AA	NonHAART	1	9
	HAART	6	1
Male, AA	NonHAART	2	2
	HAART	5	2
Female, C	NonHAART	1	1
	HAART	0	1
Male, C	NonHAART	0	1
	HAART	2	1

Note: AA = African American; C = Caucasian.
Source: Johnston et al. (2001); Used with permission.

the TOR model (see Newman 2001). How would you construct exact tests based on these statistics?

8.17. Apply, where relevant, the conditional exact and asymptotic score test for $\beta = \gamma = 0$ in the TOR model to the datasets given in this chapter and in Emerson (1994).

8.18. Derive in detail the probability, score and LR test statistics for the hypothesis $\beta = \gamma = 0$ in the TOR model. Apply these to the data sets used in the previous exercise.

8.19. Execute a simulation study to compare unconditional asymptotic, conditional asymptotic and exact analyses of the TOR model. Compare, as appropriate, the results in terms of size, power, bias, precision, length and coverage. Which method(s) do you recommend and under what circumstance?

8.20. In a binary response two arm trial with stratified randomization, are the $s_k, k = 1, \cdots, K$

EXERCISES

Table 8.13 *Data for Comparative Analyses*

		Data A		Data B		Data C		Data D	
K	y	$x=0$	$x=1$	$x=0$	$x=1$	$x=0$	$x=1$	$x=0$	$x=1$
1	0	1	4	1	12	1	3	0	7
	1	1	2	0	12	1	0	3	5
2	0	5	9	0	12	0	1	0	3
	1	0	1	0	12	1	0	3	1
3	0	3	8	0	12	1	2	0	3
	1	1	0	2	12	1	1	3	1
4	0	2	7	0	12	0	2	1	3
	1	0	1	0	12	1	7	4	1
5	0	8	3	0	12	1	3	2	3
	1	0	1	0	12	1	0	0	3
6	0	6	2	0	12	1	3	0	1
	1	0	1	2	12	0	9	5	1

Note: Hypothetical data.

ancillary statistic for β? If they are not, how do you justify the use of conditional analysis in this context?

CHAPTER 9

Several 2 × 2 Tables: II

9.1 Introduction

In this chapter, we continue our study of models for several 2×2 tables. After a brief comment on two other types of models, we return to the logit models with a focus on checking their assumptions and extending their scope to designs other than the prospective product binomials design. The specific aims in these pages are for a series of 2×2 contingency tables:

- To introduce the stratified risk difference and risk ratio models.
- To derive the exact distributions of the stratum additive statistics used for testing the assumption of a common odds ratio, and of a linear trend on the log-odds scale.
- To present a number of such test statistics, demonstrate how to perform exact conditional tests, and formulate the exact power of the tests for stratified binary data.
- To show the applicability of exact methods based on the product binomial design to the multinomial, Poisson count, incidence density, case-control, randomized trials and inverse sampling designs.
- To describe random effect models and contrast them with fixed effect models.

9.2 Models for Combining Risk

The analysis of several 2×2 tables, as noted in the last chapter, occurs in a variety of research fields. Recently, it has gained a foothold in the conduct of systematic reviews of clinical trials. The statistical portion of such an endeavor is called a **meta-analysis**. The variation in risk over the set of trials is modeled in terms of additive risk, multiplicative risk or multiplicative odds. Further, the models used are either **fixed effect** or **random effect** models. The former assumes that each trial assesses an underlying true effect while the latter postulates that each trial is a random event, and its observed effect is a realization from an underlying distribution of effects. For now, we consider fixed effect models; the latter will be described in §9.10.

Let $\mathcal{Y}$ denote a binary response variable, and x, the treatment indicator, with $\mathcal{Y}, x \in \{0, 1\}$. Let $\pi_k(x) = P[\mathcal{Y} = 1 \mid x, \text{trial} = k]$. Three regularly used fixed effect models for this setting are:

I. The Additive Risk Model

Assume that the probabilities of the event of interest relate to each other on the additive scale. Then

$$\pi_k(x) = \alpha_{1k} + \delta_k x \qquad (9.1)$$

A simpler model is the **pooled, common or homogeneous risk difference model**. Then $\delta_k = \delta$ for all k, and

$$\pi_k(x) = \alpha_{1k} + \delta x \qquad (9.2)$$

II. The Multiplicative Risk Model

Alternatively, assume that the response probabilities relate to one another on the multiplicative scale.

$$\pi_k(x) = \alpha_{2k}\rho_k^x \qquad (9.3)$$

Written in another form, the logarithms of these probabilities change on the additive scale. A simpler version of (9.3) is the **common or homogeneous risk ratio model**. Then $\rho_k = \rho$ for all k, and

$$\pi_k(x) = \alpha_{2k}\rho^x \qquad (9.4)$$

III. The Odds Ratio or Logit Model

Finally, we have the logit model introduced in the last chapter. In terms of odds, it is

$$\frac{\pi_k(x)}{1 - \pi_k(x)} = \alpha_k \phi_k^x \qquad (9.5)$$

In the logarithmic form, this is model (8.1), and with $\phi_k = \phi$ for all k, it is the COR model (8.2).

****

One method of estimation of an assumed common effect, used regularly in meta-analysis and which was not covered in Chapter 8, is based on **weighted averages**. According to this method, the estimated common risk difference, for example, is

$$\hat{\delta} = \left\{\sum w_k \hat{\delta}_k\right\} \left\{\sum w_k\right\}^{-1} \qquad (9.6)$$

where $\hat{\delta}_k$ is an estimate of δ_k, and the w_k's are the weights. The weights applied are the **variance weights** and the **Mantel–Haenszel (MH) weights**. In the former, w_k is the inverse of an estimate of variance of the estimator of the treatment effect in the kth stratum. Let $\hat{\delta}_k$ be the unconditional mle of δ_k. Then the kth variance weight is

$$w_k = \left\{\widehat{\text{var}}[\hat{\delta}_k]\right\}^{-1} \qquad (9.7)$$

and the kth MH weight is

$$w_k = \frac{m_k n_k}{N_k} \qquad (9.8)$$

An asymptotic CI for δ, the common risk difference, using the variance based weights is computed with an estimate of the variance $\hat{\delta}$ given by

$$\widehat{\text{var}}[\hat{\delta}] = \left\{\sum_k w_k\right\}^{-1} \tag{9.9}$$

This formulation applies to the risk ratio and odds ratio models as well. In these two cases, however, the weighted average is taken for the logarithm of the estimate of the risk ratio, or the odds ratio. See the material on random effect models in §9.10, and the references in §9.12.

****

Some pertinent points regarding the above three models in the context of meta-analysis of clinical trials are:

- Empirical investigations show that risk differences across trials are more heterogenous than the risk ratios or odds ratios. The latter two measures are preferred for estimating a common effect. More studies of this issue are, however, needed.
- The odds ratio has more shortfalls in terms of clinical relevance and ease of communication. In case-control studies, it is a stand in for the risk ratio, the entity of interest. When event rates are small, the two are nearly equal.
- The risk ratio tends to exaggerate the comparative benefit or risk of therapy. The message it conveys also depends on whether one looks at the ratio for benefit or that for harm.
- Practicing clinicians favor the risk difference. When it is put in an integerized form, it conveys the effect of therapy in a more meaningful manner.
- The logit or odds ratio models are more amenable to multivariate analyses.
- The odds ratio is the sole effect measure that can be estimated in the case of certain study designs.
- Exact methods exist in a more mature form for the odds ratio model.

In an empirical study of 125 papers dealing with combining data from binary response clinical trials, Engels et al. (2000) found that whatever the scale one used, the finding of heterogeneity was common. Noting a few exceptions, they go on to state:

Odds Ratio or Risk Difference?

> [F]or most meta-analyses, summary odds ratios or risk differences agreed in statistical significance, leading to similar conclusions regarding whether treatments affected outcome. ... However, risk differences displayed more heterogeneity than odds ratios. Engels et al. (2000).

The choice of a risk outcome measure and model for clinical trials is a complex matter. This needs to be kept in mind as we continue to study the odds ratio models. These models are not necessarily the best in any absolute sense. In particular, they may not convey health information in an optimal way. At the same time, there are sound grounds for using in data analysis. Hence, when we use such models, we need to translate the final results into a more user friendly format. The references on meta-analysis given in §9.12 indicate how that is done.

9.3 Testing for Homogeneity

Among the three odds ratio models for stratified 2×2 tables introduced in Chapter 8, the COR model is the one most often applied in practice. But the assumption of a uniform odds ratio does not always hold. It is thus advisable to test it prior to using the COR model. A **test for homogeneity** of the odds ratio is also called a **test for (qualitative) interaction** or a **test for heterogeneity**.

Under a general principle of model selection, we embed a simpler model within a more general one, and test if the data support the reduction of model complexity. The HOR model provides such a general context here. The TOR model may also be invoked. In the latter context, testing whether the stratum and treatment interaction parameter is zero is a test for the constancy of the odds ratio. However, since the TOR model is a special case of the HOR model, is appropriate only if the strata are ordered, and is itself based on assumptions that need testing, the former context is preferable and used in practice.

Tests for homogeneity are often done in health and other studies. They are crucial in a meta-analysis of clinical trials. Prior to generating an estimate of a common treatment effect from a series of trials we need to check that an underlying effect is present.

Example 9.1: Consider data from the four hypothetical clinical trials shown in Table 9.1. The mles of the odds ratios for the first two trials are very different from those for the last two. The COR model seems suspect here.

Table 9.1 *Data from Four Clinical Trials*

Trial No.	Outcome	$x=0$	$x=1$	$\hat{\phi}_k$
1	Nonresponse	0	2	
	Response	4	2	0.0
2	Nonresponse	0	2	
	Response	4	2	0.0
3	Nonresponse	4	2	
	Response	0	2	$+\infty$
4	Nonresponse	4	2	
	Response	0	2	$+\infty$

Note: $\hat{\phi}_k$ is the mle of kth trial odds ratio.

With reference to the HOR model in (8.2), testing homogeneity is to test the hypothesis H_0:

$$\beta_1 = \cdots = \beta_K = \beta, \quad \text{or equivalently,} \quad \phi_1 = \cdots = \phi_K = \phi.$$

A variety of statistics applicable for this purpose exist. A statistic based on stratum weights, frequently applied in meta-analysis, is

$$\mathcal{Q} = \sum_{k=1}^{K} w_k (\hat{\beta} - \hat{\beta}_k)^2 \qquad (9.10)$$

where w_k is the appropriate kth stratum weight defined in §9.2. This is referred to a chisquare distribution with $K-1$ df to obtain an asymptotic p-value. As the mean of a χ^2_{K-1} variate is

TESTING FOR HOMOGENEITY

$K-1$, the quantity $Q/(K-1)$ is taken as a measure of the degree of heterogeneity among the trials. A value much larger than 1 is deemed indicative of substantial heterogeneity.

****

We turn to exact tests for H_0. Let $\mathcal{W}$ denote a generic test statistic for this purpose. Also we reparametrize (8.2) by writing $\beta_k = \beta + \kappa_k$ with $\kappa_1 = 0$, or $\phi_k = \phi\rho_k$, with $\rho_1 = 1$ and $\rho_k = \exp(\kappa_k)$. Then H_0 becomes:

$$\kappa_2 = \cdots = \kappa_K = 0, \quad \text{or equivalently,} \quad \rho_2 = \cdots = \rho_K = 1.$$

Assume that responses within and between strata are independent and the data are from a product binomial design. With w_k now denoting its kth stratum component, let the test statistic $\mathcal{W}$ have the stratum additive form

$$\mathcal{W} = \sum_{k=1}^{K} w_k(\mathcal{T}_k) \tag{9.11}$$

where $\mathcal{T}_k = \mathcal{A}_k$. Several specific test statistics are given in the next section. A key requirement is that extreme values of the statistic are unlikely under H_0.

As before, $\mathcal{S}_k = \mathcal{A}_k + \mathcal{B}_k$ and $\mathcal{T} = \Sigma_k \mathcal{T}_k$. The realized value of $\mathcal{S}_k$ is s_k, of $\mathcal{T}$ is t and of $\mathcal{W}$ is w. The earlier reparametrization shows that for the task at hand, α_k, $k = 1, \cdots, K$, and β are nuisance parameters with sufficient statistics $\mathcal{S} = (\mathcal{S}_1, \cdots, \mathcal{S}_K)$ and $\mathcal{T}$, respectively. These parameters are removed by conditioning. The exact significance level for H_0 is thereby

$$\mathrm{P}[\mathcal{W} \geq w \mid \mathcal{S} = s, \mathcal{T} = t; H_0] \tag{9.12}$$

We need the exact distribution of $\mathcal{W}$ when both margins of each stratum and the sum, over all strata, of a particular cell in the 2×2 set up, are fixed. With $c_k(u)$ and Ω_k as in (8.10) and (8.14), respectively, let ψ stand for a dummy parameter. Consider the polynomial

$$g_k(\phi, \psi, \phi_k) = \sum_{u \in \Omega_k} c_k(u) \rho_k^u \phi^u \psi^{w_k(u)} \tag{9.13}$$

At fixed values of $\boldsymbol{\rho} = (\rho_1, \ldots, \rho_K)$, this is a bivariate polynomial. Also, define the product

$$g(\phi, \psi, \boldsymbol{\rho}) = \prod_{k=1}^{K} g_k(\phi, \psi, \rho_k) \tag{9.14}$$

Suppose we perform the multiplication in the rhs of (9.14) and write it in an expanded form as:

$$g(\phi, \psi, \boldsymbol{\rho}) = \sum_{(u,v)} d(u, v; \boldsymbol{\rho}) \phi^u \psi^v \tag{9.15}$$

This is also a bivariate polynomial when the values of ρ_k, $k = 1, \cdots, K$, are specified. Let $g(\phi, \psi, \boldsymbol{\rho}; t, .)$ comprise all the terms of $g(\phi, \psi, \boldsymbol{\rho})$ in which the exponent of ϕ is equal to t.

Theorem 9.1: The exact conditional distribution of $\mathcal{W}$ is

$$P[\mathcal{W} = w \mid \mathcal{S} = s, \mathcal{T} = t] = \frac{d(t, w; \boldsymbol{\rho})}{g(1, 1, \boldsymbol{\rho}; t, .)} = \frac{d(t, w; \boldsymbol{\rho})}{\sum_v d(t, v; \boldsymbol{\rho})} \qquad (9.16)$$

Proof: From Theorem 8.1, we know that

$$P[\mathcal{T} = \boldsymbol{u} \mid \mathcal{S} = s] = \prod_{k=1}^{K} \left\{ \frac{c_k(u_k)\phi_k^{u_k}}{f_k(\phi_k)} \right\} \qquad (9.17)$$

where $\boldsymbol{u} = (u_1, \cdots, u_K)$. Substituting $\phi_k = \phi\rho_k$, the conditional distribution of $(\mathcal{T}, \mathcal{W})$ obtains by summing (9.17) over the set

$$\left\{ \boldsymbol{u} : \begin{array}{c} u_1 + \cdots + u_K = t;\ w_1(u_1) + \cdots + w_K(u_K) = w \\ u_k \in \Omega_k,\ k = 1, \ldots, K \end{array} \right\}$$

Doing this and rearranging terms, we get that the joint probability $P[\mathcal{T} = t, \mathcal{W} = w \mid \mathcal{S} = s]$ is equal to

$$\phi^t \left(\prod_{k=1}^{K} f_k(\phi_k) \right)^{-1} \sum_{\boldsymbol{u}} \left\{ \prod_k c_k(u_k)\rho_k^{u_k} \right\} \qquad (9.18)$$

Equating coefficients from (9.14) and (9.15), we see that

$$d(t, w; \boldsymbol{\rho}) = \sum_{\boldsymbol{u}} \left\{ \prod_k c_k(u_k)\rho_k^{t_k} \right\} \qquad (9.19)$$

Hence

$$P[\mathcal{T} = t, \mathcal{W} = w \mid \mathcal{S} = s] = \phi^t d(t, w; \boldsymbol{\rho}) \left(\prod_{k=1}^{K} f_k(\phi_k) \right)^{-1} \qquad (9.20)$$

The required result follows upon conditioning further on $\mathcal{T}$. □

Note, since ψ is a dummy parameter, the conditional distribution of $\mathcal{W}$ is not truly a PBD. But, as will be seen later, this form is useful for efficient computation.

9.4 Test Statistics

The weight based statistic for testing the homogeneity of the odds ratio across several of 2×2 tables was noted in (9.10). Now, we give three other commonly used statistics. The question of derivation is addressed in §9.11.

One other type of statistic used for this purpose has the quantities $w_k(\mathcal{T}_k)$ in (9.11) given by

$$w_k(\mathcal{T}_k) = (\mathcal{T}_k - \mu_k)^2 \sigma_k^{-2} \qquad (9.21)$$

where μ_k and σ_k^2 are respectively the mean and variance of the kth hypergeometric distribution,

TEST STATISTICS

or their estimates, or related entities. What further differentiates the statistics is the type of estimate of ϕ used for computing the conditional mean and variance.

First, consider the conditional score (CS) statistic. For this, let $\hat{\phi}$ denote the conditional mle of ϕ, as computed from (8.43). Then, for $k = 1, \cdots, K$, we compute μ_k and σ_k^2 with

$$\mu_k = E[\mathcal{T}_k \mid \mathcal{S}_k = s_k; \hat{\phi}] \qquad (9.22)$$
$$\sigma_k^2 = \text{var}[\mathcal{T}_k \mid \mathcal{S}_k = s_k; \hat{\phi}] \qquad (9.23)$$

Then, for $(t_1, \cdots, t_K)$, we substitute these values in (9.21) and apply them to (9.11) to obtain the needed CS statistic for H_0.

Another relevant statistic is the **Tarone–Breslow–Day (TBD) statistic**. Let $\hat{\phi}_{MH}$ be the MH estimate of ϕ given by (8.46). Then, for $k = 1, \cdots, K$, we first compute μ_k as the solution of the equation

$$\mu_k(m_k - s_k + \mu_k) = \hat{\phi}_{MH}(n_k - \mu_k)(s_k - \mu_k) \qquad (9.24)$$

and use this to get σ_k^2 from

$$\frac{1}{\sigma_k^2} = \frac{1}{\mu_k} + \frac{1}{(n_k - \mu_k)} + \frac{1}{(s_k - \mu_k)} + \frac{1}{(m_k - s_k + \mu_k)} \qquad (9.25)$$

These are substituted in (9.21) and subsequently in (9.11), and further reduced by a correction factor. The correction for the TBD statistic is as follows. Let $\mu_+ = \Sigma_k \mu_k$ and $\sigma_+^2 = \Sigma_k \sigma_k^2$. Then with the $\mathcal{W}$ as above, the TBD statistics for heterogeneity is

$$\mathcal{W}_{TBD} = \mathcal{W} - \frac{(t - \mu_+)^2}{\sigma_+^2} \qquad (9.26)$$

This correction is usually very small and can be ignored (Agresti 1996; Breslow 1996; Newman 2001).

The Zelen statistic is often used to perform an exact test of homogeneity. However, it is not of the form (9.21). Based on the ordering principle of the Irwin test for a 2×2 table, its p-value is the null probability, for fixed $\mathcal{S}$ and $\mathcal{T}$, of all sets of K 2×2 tables whose null conditional probability does not exceed that of the observed set of tables. This is equivalent to setting

$$w_k(t_k) = -\log\{c_k(t_k)\} = -\log\left\{\binom{n_k}{t_k}\binom{m_k}{s_k - t_k}\right\} \qquad (9.27)$$

Substituting these in (9.11) gives the Zelen statistic.

A crucial difference between the CS statistic and TBD statistics is their applicability to exact conditional analysis. $\hat{\phi}$, used for the former, depends on the data only through the sufficient statistics for the nuisance parameters. $\hat{\phi}_{MH}$, used in the latter, does not have this property. The TBD statistic, even without the correction factor, is not appropriate for exact conditional analysis. Although both appear to be stratum additive, the TBD statistic is not. Stratum additiveness is critical as well for efficient computation. For this reason, the latter is not used for computing an exact p-value. The Zelen statistic, on the other hand, is a stratum additive statistic, and until recently, was the only one used for this purpose.

Other statistics for homogeneity are similar to the CS statistic but use an estimate other than the cmle. For example, the unconditional mle, or the mue for ϕ may be used. So long as the estimate depends on the data only through the sufficient statistics of the nuisance parameters, computational efficiency for exact analysis is not compromised. Other statistics for homogeneity include the LR and the chisquare statistics.

With the exception of the Zelen statistic, they all approximately follow a chisquare distribution with $(K-1)$ df in large samples.

9.5 A Worked Example

A rudimentary scheme for computing the exact conditional distribution of $\mathcal{W}$ is derived from Theorem 9.1. It gives the exact distribution at any values for $(\rho_2, \cdots, \rho_K)$, and not just at the null values.

Step 1. For each k, specify the set Ω_k and values of ρ_k, $k = 1, \cdots, K$.

Step 2. For each k, and each $u \in \Omega_k$, compute $c_k(u)$, ρ_k^u, and the value of w_k. These determine the polynomials $g_k(\phi, \psi, \boldsymbol{\rho})$, $k = 1, \cdots, K$.

Step 3. Multiply these polynomials to get $g(\phi, \psi, \boldsymbol{\rho})$.

Step 4. Select terms with the exponent of ϕ equal to t to get $g(\phi, \psi, \boldsymbol{\rho}; t, .)$. Apply Theorem 9.1 to get the exact p-value.

****

Example 9.1 (continued): We illustrate this scheme with the data in Table 9.1. First let $\mathcal{W}$ be the Zelen statistic. Under Step 1 and Step 2, the hypergeometric gps for the 4 strata are

$$f_1(\phi) = f_2(\phi) = 2\phi^2(3 + 8\phi + 3\phi^2)$$
$$f_3(\phi) = f_4(\phi) = 2(3 + 8\phi + 3\phi^2)$$

Assuming H_0, and under Step 2 and Step 3, we obtain

$$g(\phi, \psi, 1) = 2^4 \phi^4 \left(3\psi^{\log 6} + 8\phi\psi^{\log 16} + 3\phi^2\psi^{\log 6}\right)^4$$

As the observed $\mathcal{T}$ is $t = 3+3+1+1 = 8$, we need terms of $g(\phi, \psi, 1)$ with exponent of ϕ equal to 8. In this example, we apply the expression

$$(a+b+c)^4 = \sum_{u+v+z=4} \binom{4}{u\ v\ z} a^u b^v c^z$$

This gives $g(\phi, \psi, 1)$ to be

$$2^4 \sum_{0 \le u+v \le 4} \binom{4}{u\ v\ (4-u-v)} 3^{4-v} 8^v \phi^{(12-2u-v)} \psi^{((4-v)\log 6 + v \log 16)}$$

Terms of the above with exponent of $\phi = 8$ are terms where $2u + v = 4$. Since $u, v \in \{0, 1, 2, 3, 4\}$, there are three such terms. These are shown in Table 9.2.

CHECKING THE TOR MODEL

Table 9.2 *Exact Computation Results*

u	v	t	w	$d(t,w)$
0	4	8	$4\log 16$	2^{16}
1	2	8	$2\log 6 + 2\log 16$	$2^{12}3^3$
2	0	8	$4\log 6$	$2^5 3^5$

The observed $\mathcal{W}$ is $w = 4\log 6$. With (9.21), (9.11) and (9.12), the exact Zelen test significance level is

$$\frac{243}{2048 + 3456 + 243} = 0.042$$

Now consider the TBD statistic. The MH estimate of ϕ for Table 9.1 is 1.0. Then, for $k = 1, 2, 3, 4$, we solve (9.24) and use (9.25) to get

$$\mu_1 = \mu_2 = 3;\ \mu_3 = \mu_4 = 1$$

and

$$\sigma_1^{-2} = \sigma_2^{-2} = \sigma_3^{-2} = \sigma_4^{-2} = \frac{8}{3}$$

The correction factor (9.26) here is equal to zero. With (9.21) and (9.11), the observed TBD statistic then is $w = 32/3$. Referring this to the chisquare distribution with three df, the asymptotic *p*-value is 0.0137. The exact distribution of the TBD statistic requires enumeration of all sets of 2×2 tables with the given margins. In this case, there are $3 \times 3 \times 3 \times 3 = 81$ sets of tables. And we cannot remove the nuisance parameter ϕ by conditioning. For the TBD statistic, a **semiconditional** form of analysis that will optimize over the nuisance parameters is a possible option. But such a test has not been developed.

Application of the CS statistic, either for exact or asymptotic analysis, faces the quandary that for some strata the cmle of ϕ is not finite.

9.6 Checking the TOR Model

Now we develop an exact test for the assumption of a linear trend in the log-odds ratio made in the TOR model. To restate, this model is

$$\ln\left(\frac{\pi_k(x)}{1-\pi_k(x)}\right) = \alpha_k + \beta x + \gamma x z_k \tag{9.28}$$

Assume, without loss of generality, that $z_1 = 0$. Otherwise, rewrite the problem with new scores $z_{*k} = z_k - z_1$. Consider the HOR model with $\beta_1 = \beta$; $\beta_2 = \beta + \gamma z_2$; $\beta_k = \beta + \gamma z_k + \kappa_k$, $k = 3, \ldots, K$, with $\kappa_1 = \kappa_2 = 0$. Testing the validity of the TOR model is to test, in the context of HOR model, the hypothesis H_0:

$$\kappa_3 = \cdots = \kappa_K = 0, \quad \text{or equivalently,} \quad \rho_3 = \cdots = \rho_K = 1$$

where $\rho_1 = \rho_2 = 1$ and $\rho_k = \exp(\kappa_k)$. Assume the independent product binomial design. Let $\mathcal{Q}$ denote a test statistic for H_0, and assume it has the form

$$\mathcal{Q} = \sum_{k=1}^{K} q_k(\mathcal{T}_k) \tag{9.29}$$

As before, $\mathcal{T} = \Sigma_k \mathcal{T}_k$ and $\mathcal{W} = \Sigma_k z_k \mathcal{T}_k$. The realized value of $\mathcal{S}_k$ is s_k, $\mathcal{T}$ is t, $\mathcal{W}$ is w and $\mathcal{Q}$ is q. For testing H_0, α_k, $k = 1, \ldots, K$, β and γ are nuisance parameters whose sufficient statistics are $\mathcal{S} = (\mathcal{S}_1, \ldots, \mathcal{S}_K)$, $\mathcal{T}$ and $\mathcal{W}$, respectively. The exact p-value for H_0 is thereby

$$P[\mathcal{Q} \geq q \mid \mathcal{S} = s, \mathcal{T} = t, \mathcal{W} = w; H_0] \tag{9.30}$$

Recall the definitions of $c_k(u)$ and Ω_k, let $\psi = \exp(\gamma)$, and let δ stand for a dummy parameter. Then define the polynomial

$$g_{*k}(\phi, \psi, \delta, \phi_k) = \sum_{u \in \Omega_k} c_k(u) \rho_k^u \phi^u \psi^{z_k u} \delta^{q_k(u)} \tag{9.31}$$

With $\boldsymbol{\rho} = (\rho_1, \ldots, \rho_K)$, define the product polynomial

$$g_*(\phi, \psi, \delta, \boldsymbol{\rho}) = \prod_{k=1}^{K} g_{*k}(\phi, \psi, \delta, \rho_k) \tag{9.32}$$

Suppose, with specified values of $\rho_k, k = 1, \ldots, K$, we perform the multiplication in the rhs of (9.32) and write it in an expanded form as:

$$g_*(\phi, \psi, \delta, \boldsymbol{\rho}) = \sum_{(u,v,q)} d_*(u, v, q; \boldsymbol{\rho}) \phi^u \psi^v \delta^q \tag{9.33}$$

Let the polynomial $g_*(\phi, \psi, \delta, \boldsymbol{\rho}; t, w, .)$ comprise all the terms of $g_*(\phi, \psi, \delta, \boldsymbol{\rho})$ in which the exponent of ϕ is equal to t, and that of ψ is equal to w.

Theorem 9.2: The conditional distribution, $P[\mathcal{Q} = q \mid \mathcal{S} = s, \mathcal{T} = t, \mathcal{W} = w; H_0]$, is

$$\frac{d_*(t, w, q; \boldsymbol{\rho})}{g_*(1, 1, 1, \boldsymbol{\rho}; t, w, .)} = \frac{d_*(t, w, q; \boldsymbol{\rho})}{\sum_v d_*(t, w, v; \boldsymbol{\rho})} \tag{9.34}$$

Proof: This follows the proof for Theorem 9.1. □

One form of statistic used to test H_0 has the quantities $q_k(\mathcal{T}_k)$ in (9.29) given by

$$q_k(\mathcal{T}_k) = (\mathcal{T}_k - \mu_k)^2 \sigma_k^{-2} \tag{9.35}$$

where μ_k and σ_k^2 are the mean and variance of the kth conditional hypergeometric distribution under an estimate of β_k based on the TOR model. For example, let $\hat{\beta}$ and $\hat{\gamma}$ denote the cmles

AN INCIDENCE DENSITY STUDY

of β and γ computed from (8.35), or from (8.67) and (8.68). Then, for $k = 1, \cdots, K$, we set $\hat{\beta}_k = \hat{\beta} + z_k \hat{\gamma}$ and compute μ_k and σ_k^2 with

$$\mu_k = E[\mathcal{T}_k \mid \mathcal{S}_k = s_k; \hat{\beta}_k] \tag{9.36}$$

$$\sigma_k^2 = \text{var}[\mathcal{T}_k \mid \mathcal{S}_k = s_k; \hat{\beta}_k] \tag{9.37}$$

Then, for $(t_1, \cdots, t_K)$, we substitute these values in (9.35) and then in (9.30) to obtain the $\mathcal{Q}$ statistic for testing H_0.

Unlike tests for the homogeneity of odds ratios, the literature on this matter is sparse. A version of the Zelen statistic for homogeneity is applicable, provided attention is restricted to the conditional sample space of the generating polynomial $g_*(\phi, \psi, \delta, \boldsymbol{\rho}; t, w, .)$. We may also derive various forms of score, LR and Wald statistics for this problem.

9.7 An Incidence Density Study

We now consider $K \times 2 \times 2$ data from pure count or incidence density designs. In both, a Poisson distribution is assumed to generate the data. With each cell, we have two values: a nonnegative integer count and a person time value. In the pure count case, the person time value is identical in all the cells.

Example 9.2: Consider a study in which subjects in a given age group were followed up until the end of the study or death. Suppose the loss to follow up was negligible. Table 9.3 shows hypothetical data with counts of death from a specific cause broken down by region, gender and occupation. Each cell has the total follow up time for all persons in that category as well.

For these data we inquire: Adjusting for the regional effect, if any, do gender and occupation affect mortality from this cause in an associated fashion? Is the level of their association constant across the regions? For these variables, the observed cross product ratios of the death rates in the Eastern, Central and Western regions respectively are 1.50, 1.33 and 1.25.

****

To address the queries posed, assume each count is the realization of an independent Poisson variable and that its person time value is a fixed entity. Let $y = 0, 1$ denote female or male; $x = 0, 1$, clerical or production worker and $k = 1, 2, 3$, Eastern, Central and Western region, respectively. Further, let $\lambda(y, x, k)$ denote the mean event rate per unit person time, and $\tau(y, x, k)$, the person time for cell (y, x, k). The expected count in this cell is

$$\lambda_*(y, x, k) = \lambda(y, x, k)\tau(y, x, k) \tag{9.38}$$

The overall unconditional probability is the product of $4K$ Poisson probabilities with means $\lambda_*(y, x, k)$. The pure count Poisson model is a special case of this with $\tau(y, x, k) = \tau$ for all y, x, k. Now consider the model

$$\ln(\lambda(y, x, k)) = \mu_0 + \mu_1 y + \mu_2 x + \mu_{3k} + \mu_{13k} y + \mu_{23k} x + \mu_{123k} yx \tag{9.39}$$

with $\mu_{31} = \mu_{131} = \mu_{231} = 0$. This is a log-linear model in which the μ's can be directly expressed in terms of the λ's. For example,

Table 9.3 *Death by Region, Gender and Occupation*

		Occupation	
Region	Gender	Clerical	Production
Eastern	Female	1	1
	PY	50.0	25.0
	Male	2	3
	PY	100.0	50.0
Central	Female	1	4
	PY	25.0	50.0
	Male	3	8
	PY	100.0	100.0
Western	Female	1	2
	PY	25.0	25.0
	Male	1	5
	PY	25.0	50.0

Note: Gender cell values are counts; PY = (person years)/1000.

$$\mu_{123k} = \ln(\phi_k) = \ln\left[\frac{\lambda(1,1,k)\lambda(0,0,k)}{\lambda(0,1,k)\lambda(1,0,k)}\right] \tag{9.40}$$

If $\mu_{123k} = \mu$, $k = 1, \cdots, K$, the cross product ratio is equal for all k. If $\mu_{123k} = 0$, $k = 1, \cdots, K$, there is an absence of a three way interaction between these variables.

Theorem 9.3: Let $\mathcal{T}_k$ denote the count in cell $(1, 1, k)$. Conditioning on the marginal counts in stratum k, ϕ_k as in (9.40), and otherwise with the usual notation, we have

$$P[\mathcal{T}_k = u \mid s_k, m_k, n_k] = \frac{c_{*k}(u)\phi_k^u}{f_{*k}(\phi_k)} \tag{9.41}$$

where

$$c_{*k}(u) = c_k(u)\tau_k^u \tag{9.42}$$

$$f_{*k}(\phi_k) = \sum_{u \in \Omega_k} c_{*k}(u)\phi_k^u \tag{9.43}$$

and

$$\tau_k = \left(\frac{\tau(1,1,k)\tau(0,0,k)}{\tau(0,1,k)\tau(1,0,k)}\right) \tag{9.44}$$

AN INCIDENCE DENSITY STUDY

Proof: This is left as an exercise. □

The implications of this result are:

- Distribution (9.41) can be a starting point for some form of conditional analyses of stratified incidence density data. In particular, it can be used for testing conditional independence, analyzing a constant cross products ratio, a trend in the cross product ratio or for testing the homogeneity of the cross product ratio.
- These analyses can be done following the methods developed for $K \times 2 \times 2$ tables in Chapter 8 and Theorems 9.1 and 9.2. The difference is that now we use the polynomials $f_{*k}(\phi_k)$ instead of $f_k(\phi_k)$.
- Ω_k is the set of hypergeometric exponents for kth stratum.
- The gp $f_{*k}(\phi_k)$, though, is NOT a hypergeometric polynomial.

For example, the conditional analysis of the common cross product ratio is based on the product

$$f_*(\phi) = \prod_{k=1}^{K} f_{*k}(\phi) \qquad (9.45)$$

****

Example 9.2 (continued): We illustrate Theorem 9.3 with the data in Table 9.3. Assume the cross product ratios associating gender and occupation are constant across the strata, that is, $\phi_k = \phi$ for all k. In that case, the hypergeometric gps for the three regional strata are

$$f_1(\phi) = 3\phi^2(2 + 4\phi + \phi^2)$$
$$f_2(\phi) = 12\phi^7(66 + 165\phi + 110\phi^2 + 22\phi^3 + \phi^4)$$
$$f_3(\phi) = 7\phi^4(5 + 6\phi + \phi^2)$$

Using (9.43), we get $\tau_1 = 1.0$, $\tau_2 = 0.5$ and $\tau_3 = 2.0$. Applying these in (9.42), we get the adjusted gps

$$f_{*1}(\phi) = 3\phi^2(2 + 4\phi + \phi^2)$$
$$f_{*2}(\phi) = 12\phi^7 2^{-11}(1056 + 1320\phi + 220\phi^2 + 44\phi^3 + \phi^4)$$
$$f_{*3}(\phi) = 7\phi^4 2^4(5 + 12\phi + 4\phi^2)$$

We multiply the three polynomials to get

$$f_*(\phi) = 852\phi^{13} 2^{-7}(10560 + 59664\phi + 124696\phi^2 + 120208\phi^3 +$$
$$56550\phi^4 + 14168\phi^5 + 2173\phi^6 + 204\phi^7 + 4\phi^8)$$

The observed exponent of ϕ for these data is $t = 3 + 8 + 5 = 16$, and $\Omega = \{13, 14, \cdots, 21\}$. Using this gp and observed value, the data are analyzed in the usual way for a univariate PBD

Table 9.4 *Analysis of the Data in Table 9.3*

Statistic	p-value	Estimate	95% CI
TST Exact	0.6795	1.416	(0.294, 6.652)
TST mid-p	0.9958	1.443	(0.168, 9.263)
CS Asymptotic	0.6862	1.443	(0.220, 7.584)

Note: These results relate to ϕ.

to obtain the results in Table 9.4. These results show weak evidence for the proposition that the assumed constant cross product ratio between gender and occupation is different from unity.

The details of other analyses for stratified incidence density data that parallel that developed for the usual $K \times 2 \times 2$ table is left as an exercise. Of course, plotting the complete evidence function is also a good idea.

9.8 Other Study Designs

The basic multinomial, product multinomial, cohort, case-control, binary response randomized trial and inverse sampling designs were described in Chapter 5 and Chapter 6. Using the formulations of Chapter 6, they are readily extended to the general stratified data case. The question then is: Can we apply methods developed for the stratified product binomial design in the previous and present chapter to these designs?

Consider a controlled clinical trial where the subjects are randomized to treatments in a stratified fashion. Consider two schemes: (i) Within each stratum, patient allocation to treatment is done using an equivalent of a coin toss, and (ii) Within each stratum, a random sequence for allocation has been generated. In the first scheme, the number of patients in any treatment within any stratum is not fixed while in the second scheme, it is. The justification for using the product binomial model here was given in §6.5.1.

Now consider the case when the $K \times 2 \times 2$ tables are from a design in which subjects are sampled in a cross sectional manner with either the overall total fixed or the total within each stratum fixed. The justification for using a product binomial based conditional analysis here follows the arguments given in §5.6, §5.7 and §6.5.2.

Now consider case-control retrospective designs with a binary exposure factor. Suppose either a basic group or frequency matched study of §5.8 was conducted and the data stratified with an unmatched factor, or the frequency matching was done in stratified fashion. In either case, we have data in a $K \times 2 \times 2$ table. The justification for using the conditional product binomial design based method to analyze the odds ratio model here follows from an extension of the arguments given in §6.10. The estimable parameters for the two cases are, however, not the same. The details are left as an exercise.

Finally, consider a stratified variant of the inverse sampling design of §5.9. The adverse effects of two drugs are studied in K localities. At each locality, for the drugs $x = 0$ and $x = 1$, subjects are sampled from populations of patients getting the two drugs until design fixed respective numbers d_k and c_k with the reaction in question are identified. Let b_k and a_k be the (random) corresponding numbers of patients who did not get the reaction and $\pi_k(x)$ is the underlying rate of not contracting the adverse for drug x at locality k.

EXACT POWER

Each 2 × 2 table here has two internal cells fixed by design. Let $\mathcal{A}_k$ and $\mathcal{B}_k$ denote the random number of subjects needed to obtain c_k and d_k adverse reactions from $x = 1$ and $x = 0$, respectively. As before, let $\mathcal{T}_k = \mathcal{A}_k$, and $\mathcal{S}_k = \mathcal{A}_k + \mathcal{B}_k$. Consider the model

$$\pi_k(x) = \theta_k \phi_k^x \qquad (9.46)$$

This is a stratified version of the relative risk model of §5.9. ϕ_k in this case is the relative risk of not having the effect in stratum k.

Theorem 9.4: Under the stratified inverse sampling design,

$$P[\mathcal{T}_k = t_k \mid \mathcal{S}_k = s_k] = \frac{c_k(t)\phi_k^{t_k}}{f_k(\phi_k)} \qquad (9.47)$$

where

$$\phi_k = \frac{\pi_k(1)}{\pi_k(0)} \qquad (9.48)$$

$$f_k(\phi_k) = \sum_{v \in \Omega_k} c_k(v) \phi_k^v \qquad (9.49)$$

$$c_k(t_k) = \binom{t_k + c_k - 1}{t_k} \binom{s_k - t_k + d_k - 1}{s_k - t_k} \qquad (9.50)$$

and further where

$$\Omega_k = \{v : 0 \le v \le s_k\} \qquad (9.51)$$

Proof: This follows from using expressions (5.64) and (5.97) for each stratum and then applying the arguments like those of Theorem 8.1. □.

This distribution is a PBD but, as noted for (5.97), $f_k(\phi_k)$ is NOT a hypergeometric gp. The parameter ϕ_k is the relative "risk" of not having the adverse effect. Given the PBD form, we may analyze the parameters ϕ_k along the same lines as for the stratified binary data developed earlier with the proviso that the parameter of interest, the conditional distribution, and the relevant statistics are all different. The details are left as an exercise.

9.9 Exact Power

Now we give a specific formulation for the exact power of conditional tests on stratified binary data. This is based on the general formulation for multivariate PBD given in §7.4. We consider the COR model (8.1) under the product binomial design and test $H_0 : \phi = \phi_0$ against the $H_1 : \phi \ne \phi_0$ at level α_* ($0 < \alpha_* < 1$). Suppose we use a conditional test based on fixing the margins of all the 2 × 2 tables. We continue with the notation of Chapter 8, except that now we explicitly indicate the dependence of some quantities on the margins.

For some kth stratum, define

$$f_k(\theta_k, \phi) = (1 + \theta_k)^{m_k}(1 + \theta_k\phi)^{n_k} \qquad (9.52)$$

Then, we have that

$$f_k(1, \phi; s_k, .) = \sum_{\max(0, s_k - m_k) \leq x \leq \min(n_k, s_k)} \binom{n_k}{x}\binom{m_k}{s_k - x}\phi^x \qquad (9.53)$$

Note, $f_k(1, \phi; s_k, .)$ is the same as the polynomial $f_k(\phi)$ defined in (8.14). Under the COR model,

$$P[S_k = s_k; \theta_k, \phi] = \frac{\theta_k^{s_k} f_k(1, \phi; s_k, .)}{f_k(\theta_k, \phi)} \qquad (9.54)$$

Now consider all the K strata, and let $\Omega(s)$ be the sample space of $\mathcal{T}$ given $\mathcal{S} = s$. Then for the test of H_0 versus H_1, we denote the conditional rejection region for the nominal α_* level as $\Omega(s; \phi_0, \alpha_*)$. Also, let

$$f(\boldsymbol{\theta}, \phi) = \prod_{k=1}^{K} f_k(\theta_k, \phi) \qquad (9.55)$$

Then

$$f(\mathbf{1}, \phi; s, .) = \prod_{k=1}^{K} f_k(1, \phi; s_k, .) = \sum_{v \in \Omega(s)} c(s, v)\phi^v \qquad (9.56)$$

Then, with the definitions in §7.4, the conditional power, $P[\mathcal{T} \in \Omega(s; \phi_0, \alpha_*) \mid \mathcal{S} = s; \phi]$, is written as

$$\Pi_c(s; \phi, \phi_0, \alpha_*) = \frac{\eta(s; \phi, \phi_0, \alpha_*)}{f(\mathbf{1}, \phi; s, .)} \qquad (9.57)$$

where

$$\eta(s; \phi, \phi_0, \alpha_*) = \sum_{v \in \Omega(s; \phi_0, \alpha_*)} c(s, v)\phi^v \qquad (9.58)$$

Theorem 9.5: The exact (unconditional) power, $\Pi(\boldsymbol{\theta}, \phi, \phi_0, \alpha_*)$, for the conditional test of H_0 versus H_1 is given by

$$\sum_{s} \Pi_c(s; \phi, \phi_0, \alpha_*) \left[\prod_{k=1}^{K} P[S_k = s_k; \theta_k, \phi]\right] = \frac{1}{f(\boldsymbol{\theta}, \phi)} \sum_{s} \eta(s; \phi, \phi_0, \alpha_*) \prod_{k=1}^{K} \theta_k^{s_k} \qquad (9.59)$$

where the summation is over $s = (s_1, \cdots, s_K)$ that satisfy $0 \leq s_k \leq m_k + n_k$, $k = 1, \cdots, K$.

Proof: Left as an exercise. □

ADDITIONAL ISSUES

Computing exact power then is a two step exercise. First, for each s, we identify the conditional rejection region. Then, for each s, we perform the summation in (9.58). Since the number of distinct vectors s is

$$\prod_{k=1}^{K} (1 + m_k + n_k) \qquad (9.60)$$

this is clearly an onerous task. In general, computing exact power requires much more effort than computing an exact *p*-value. We provide more details on this in Chapter 11.

Exact power for the tests for conditional independence, homogeneity of odds ratios and trend in the odds ratios for $K \times 2 \times 2$ tables is formulated in a similar manner. Exact power can, of course, also be computed for asymptotic tests.

9.10 Additional Issues

This section deals with two topics: **random effect models** and choice of statistics for testing the heterogeneity of the odds ratios in several 2×2 tables. We start with the former.

The models for several 2×2 tables given so far in this and the previous chapter are called **fixed effect** models. In the context of clinical trials, they assume the existence of a true unknown treatment effect. The trial is designed to assess its magnitude. The trials are assumed to have followed a common protocol. The patient sample and its allocation to treatment are thus the random entities here. In the random effect formulation, any study is a random realization from infinitely many studies. The observed effect is a random entity not just because of the sample and randomization scheme but also because of an intrinsic between trials variability. **Random effect models** incorporate the heterogeneity between the trials, and are used for meta-analysis of such trials. These models are given in terms of risk difference, risk ratio and odds ratio measures.

The fixed effect COR and HOR models were given in (8.1) and (8.2). Their random effect counterpart, on a logarithmic scale, is

$$\ln\left(\frac{\pi_k(x)}{1 - \pi_k(x)}\right) = \alpha_{rk} + (\beta + \zeta_k)x \qquad (9.61)$$

where ζ_k is normally distributed with mean zero and variance τ. A large sample theory based method for the analysis of the mean effect β in this model uses, in a modified form, the idea of weights given in §9.2. This is done as follows:

The fixed effect estimate (mle) of the log-odds ratio in the kth trial is

$$\hat{\beta}_k = \ln(a_k d_k) - \ln(b_k c_k) \qquad (9.62)$$

and the corresponding fixed effect MH weights is

$$w_k = \frac{b_k c_k}{N_k} \qquad (9.63)$$

Now define

$$R = \left\{\sum_k w_k\right\}^{-1} \left\{\left[\sum_k w_k\right]^2 - \sum_k w_k^2\right\} \quad (9.64)$$

Then we first compute the MH estimate of the fixed effects common log-odds ratio as:

$$\hat{\beta} = \left\{\sum_k w_k\right\}^{-1} \sum_k w_k \hat{\beta}_k \quad (9.65)$$

This gives the fixed effects MH heterogeneity statistic, $\mathcal{Q}$,

$$\mathcal{Q} = \sum_k w_k \left[\hat{\beta}_k - \hat{\beta}\right]^2 \quad (9.66)$$

If $\mathcal{Q} < K - 1$, then we set

$$D = 0 \quad (9.67)$$

or else we set

$$D = R^{-1}[\mathcal{Q} - (K-1)] \quad (9.68)$$

Using these, we obtain the new (random effects) weights as

$$w_{rk} = \frac{1}{D + w_k^{-1}} \quad (9.69)$$

Finally compute the DerSimonian–Laird (random effect) estimate of the mean log-odds ratio as

$$\hat{\beta}_r = \left\{\sum_k w_{rk} \hat{\beta}_k\right\} \left\{\sum_k w_{rk}\right\}^{-1} \quad (9.70)$$

To compute a 95% CI for β under the random effect model, let

$$s_r = \left\{\sum_k w_{rk}\right\}^{-1} \quad (9.71)$$

Then we get the required interval as

$$\hat{\beta}_r \pm 1.96 s_r \quad (9.72)$$

The approximate 95% CI for $\phi = \exp(\beta)$, is

$$[\exp(\hat{\beta}_r - 1.96 s_r), \exp(\hat{\beta}_r + 1.96 s)] \quad (9.73)$$

****

There are three distinct views on the use of random effect models in meta-analysis of clinical trials.

- Some authors argue that heterogeneity between the trials is quite common. Yet, the power of fixed effect model tests of heterogeneity, even with several moderately sized trials, is low. Random effect models incorporate such heterogeneity in an intrinsic way and should thereby be in regular use.

- Other authors say that random effect models should be avoided, especially for estimating a common treatment effect. First, they are based on an unrealistic assumption, namely the random sampling of studies. Using them in an automatic fashion will moreover mask the existence of actual heterogeneity, deter an analysis of its sources, give wider CIs, accord an undue weight to small studies and amplify the distortions of publication bias, when present.

- Other authors, while acknowledging their limitations, nevertheless advocate their use in situations where they may be appropriate. In particular, comparing the results of random and fixed effects analyses is deemed a useful endeavor.

Consult the references in §9.12 for a flavor of this debate. We do not explore such models any further because exact methods have so far not been developed for them.

****

We return to the fixed effect model for the odds ratio in several 2×2 tables and the tests for heterogeneity. One statistic commonly used in the meta-analytic context which we have not mentioned is the Peto statistic. The Peto method is also used for estimation and CIs. Many other statistics that can be used here have also not been noted (see the references in §9.12). Which ones should be used and which ones should be avoided? The research to date provides the following guide:

- Empirical studies indicate that the Peto method performs poorly in both the small strata and sparse data cases. It should be avoided.
- Some asymptotic tests, especially the LR tests for heterogeneity, exhibit a very unstable performance.
- For this problem, the exact tests hold *"their size close to the prespecified nominal levels while the asymptotic tests [are] usually more conservative."* (Reis, Hirji and Afifi 1999). This runs counter to the conventional wisdom on exact tests.
- For small sized strata, the power of most tests is generally low even for detecting moderate to high degree of heterogeneity among the odds ratios.
- For small strata and sparse data situations, we recommend the conditional score and the Zelen exact tests for homogeneity.
- Among the asymptotic tests, we favor the Tarone–Breslow–Day test and the Mantel–Haenszel test for the heterogeneity.

After performing a detailed empirical review, Petitti (2001) called for a more consistent, universal testing for heterogeneity in meta-analysis, and the provision of a clear rationale for the random effect model when it is used. Given their low power a nominal level of 0.1 was suggested as the standard level for such tests.

9.11 Derivation

Consider the derivation of the CS statistic for homogeneity given in §9.4. The relevant conditional likelihood is

$$\ell_c = \beta t + \sum_{k \geq 2} \kappa_k t_k - \sum_{k=1}^{K} \ln[h_k(\beta, \kappa_k)] \qquad (9.74)$$

where

$$h_k(\beta, \kappa_k) = \sum_{u \in \Omega_k} c_k(u) \exp[u(\beta + \kappa_k)] \qquad (9.75)$$

The first and second partial derivatives in this likelihood are:

$$\frac{\partial \ell_c}{\partial \beta} = t - \mu_+ \qquad (9.76)$$

$$\frac{\partial \ell_c}{\partial \kappa_k} = t_k - \mu_k \qquad (k \geq 2) \qquad (9.77)$$

$$\frac{\partial^2 \ell_c}{\partial \beta^2} = -\sigma_+^2 \qquad (9.78)$$

$$\frac{\partial^2 \ell_c}{\partial \kappa_k \partial \beta} = -\sigma_k^2 \qquad (k \geq 2) \qquad (9.79)$$

$$\frac{\partial^2 \ell_c}{\partial \kappa_k^2} = -\sigma_k^2 \qquad (k \geq 2) \qquad (9.80)$$

$$\frac{\partial^2 \ell_c}{\partial \kappa_k \partial \kappa_j} = 0 \qquad (k \neq j) \qquad (9.81)$$

where

$$\mu_+ = \sum_{k=1}^{K} \mu_k \quad \text{and} \quad \sigma_+^2 = \sum_{k=1}^{K} \sigma_k^2 \qquad (9.82)$$

and μ_k and σ_k^2 as usual are the conditional mean and variance of $\mathcal{T}_k$. The information matrix, **I**, is therefore

$$\mathbf{I} = \begin{bmatrix} \sigma_+^2 & \sigma_2^2 & \sigma_3^2 & \cdots & \sigma_K^2 \\ \sigma_2^2 & \sigma_2^2 & 0 & \cdots & 0 \\ \sigma_3^2 & 0 & \sigma_3^2 & \cdots & 0 \\ \cdot & \cdot & \cdot & \cdots & \cdot \\ \sigma_K^2 & 0 & 0 & \cdots & \sigma_K^2 \end{bmatrix} \qquad (9.83)$$

The inverse of the information matrix is

DERIVATION

$$\mathbf{I}^{-1} = \begin{bmatrix} \sigma_1^{-2} & -\sigma_1^{-2} & -\sigma_1^{-2} & \cdots & -\sigma_1^{-2} \\ -\sigma_1^{-2} & \sigma_1^{-2} + \sigma_2^{-2} & \sigma_1^{-2} & \cdots & \sigma_1^{-2} \\ -\sigma_1^{-2} & \sigma_1^{-2} & \sigma_1^{-2} + \sigma_3^{-2} & \cdots & \sigma_1^{-2} \\ \cdot & \cdot & \cdot & \cdots & \cdot \\ -\sigma_1^{-2} & \sigma_1^{-2} & \sigma_1^{-2} & \cdots & \sigma_1^{-2} + \sigma_K^{-2} \end{bmatrix} \quad (9.84)$$

This can be verified by direct multiplication. A method of computing the inverse is given in Exercise 9.17. The score vector here is

$$\Delta = \begin{pmatrix} t - \mu_+ \\ t_2 - \mu_2 \\ \cdots \\ \cdots \\ t_K - \mu_K \end{pmatrix} \quad (9.85)$$

The conditional score statistic then is

$$\mathcal{W} = \Delta' \mathbf{I}^{-1} \Delta \quad (9.86)$$

For inference on H_0, $\mathcal{W}$ is evaluated at the restricted mles. Under the conditional likelihood, this means evaluation at $\kappa_k = 0\,\forall k$ and $\beta = \hat{\beta}$. In this case, $\mu_+ = t$. Thus the first element of the score vector is zero. Bearing this in mind, we get

$$\mathcal{W} = \sum_{k \geq 2}^{K} (t_k - \mu_k)^2 (\sigma_k^{-2} + \sigma_1^{-2}) + \sigma_1^{-2} \sum_{\substack{i \neq j \\ i,j \geq 2}}^{K} (t_i - \mu_i)(t_j - \mu_j) \quad (9.87)$$

which, upon simplification and noting that $\Sigma_k t_k = \Sigma_k \mu_k$, yields, as previously stated, that

$$\mathcal{W} = \sum_{k=1}^{K} (t_k - \mu_k)^2 \sigma_k^{-2} \quad (9.88)$$

The conditional score statistic for testing linearity on a log-odds scale in the TOR model is based on the conditional likelihood

$$\ell_{*c} = \beta t + \gamma w + \sum_{k \geq 3} \kappa_k t_k - \sum_{k=1}^{K} \ln[h_{*k}(\beta, \gamma, \kappa_k)] \quad (9.89)$$

where

$$h_{*k}(\beta, \gamma, \kappa_k) = \sum_{u \in \Omega_k} c_k(u) \exp[u(\beta + \gamma z_k + \kappa_k)] \quad (9.90)$$

The derivation of the CS statistic proceeds along the same lines as above but is somewhat more involved. The details of this and other derivations are left as an exercise.

9.12 Relevant Literature

There has been an explosive growth of meta-analytic papers in health and biomedical literature. None is typical. Interesting instances of an early and a recent one are Yusuf et al. (1985) and Geddes et al. (2000), respectively. A journey into the principles, models, statistical methods and recent controversies of meta-analysis of clinical trials and other health studies may begin with Altman (2000), Davey Smith, Egger and Phillips (1997), Egger, Davey Smith and Phillips (1997), Glasziou et al. (2001), Mosteller and Chalmers (1992), Petitti (2000), Rothman and Greenland (1998), Sutton, Abrahams and Jones (2001) and Wolf (1986). These works also cover the spectrum of statistical methods used in this context.

The foundational paper on random effect models in meta-analysis is DerSimonian and Laird (1986). Comparative perspectives on them appear in Berlin et al. (1989), Glasziou et al. (2001), Greenland (1994), Mosteller and Chalmers (1992), Petitti (2000, 2001), Rosenfeld (2003) and Sutton, Abrams and Jones (2001). The method of computing estimates for these models in §9.10 is taken from these works.

For early approaches to the analysis of heterogeneity of the odds ratios in stratified binary data, see Norton (1945), Cochran (1954a), Woolf (1955), Mantel and Haenszel (1959), Gart (1971), Mantel, Brown and Byar (1977) and Radhakrishna (1965). A plethora of tests for homogeneity have been devised; several are tailored for the sparse data (many small strata) case. See Breslow and Day (1980), Liang and Self (1985), Tarone (1985), Berlin et al. (1989), Jones et al. (1989), Paul and Donner (1989, 1992), Sahai and Khurshid (1996), Reis (1996) and Reis, Hirji and Afifi (1999), and also the references therein, and the noted works on meta-analysis for a comprehensive picture. Engels et al. (2000) give an empirical study of odds ratio and risk difference models for testing heterogeneity in this context; Petitti (2001) provides another perspective in this regard.

An exact test for the homogeneity of the odds ratios was devised by Zelen (1971). Parts of this paper were critiqued by Halperin et al. (1977). More exact tests were given by Reis (1996) and Hirji et al. (1996). An evaluation of exact and asymptotic tests was done by Reis (1999) and Reis, Hirji and Afifi (1999). Several evaluations of sparse data and large strata asymptotic tests have also been done; see Jones et al. (1989) and Paul and Donner (1989, 1992), Sahai and Khurshid (1996) and the references therein. Papers on algorithms for doing exact tests for the homogeneity of the odds ratios are noted in Chapter 12.

Work on exact analysis of risk difference and relative risk models $K \times 2 \times 2$ tables appears nonexistent. Overviews of the large sample methods for them are in Newman (2001), Petitti (2000) and Sahai and Khurshid (1996). These works also deal with the weight based (weighted least squares) methods.

References on analysis of inverse sampling data, including the stratified data case, are in Chapter 5. Analysis of Poisson count and incidence density data was referenced in Chapter 2. More references, especially those relating to exact methods for stratified data, are in Chapter 10.

9.13 Exercises

9.1. Table 9.5 is from an hypothetical meta-analysis of a series of clinical trials. Perform exact and asymptotic tests of the homogeneity of the odds ratios for these data.

9.2. Perform exact and asymptotic tests of the homogeneity of the odds ratios for the relevant $K \times 2 \times 2$ tables given in Chapter 8.

9.3. Table 9.6 gives several data sets from hypothetical binary response clinical trials. Perform exact and asymptotic tests of the homogeneity of the odds ratios for these data.

EXERCISES

Table 9.5 *Hypothetical Meta-Analysis Data*

Study	Status	Placebo	Drug
I	Failure	20	19
	Success	0	3
II	Failure	22	26
	Success	1	1
III	Failure	38	39
	Success	2	0
IV	Failure	18	20
	Success	3	1

Table 9.6 *Data for Comparative Analyses*

		Data A		Data B		Data C		Data D	
K	y	$x=0$	$x=1$	$x=0$	$x=1$	$x=0$	$x=1$	$x=0$	$x=1$
1	0	3	4	0	12	0	3	2	3
	1	0	4	1	12	10	1	4	0
2	0	5	2	0	12	0	1	1	1
	1	1	0	2	12	3	1	1	1
3	0	3	8	0	12	2	2	0	1
	1	2	0	1	12	0	1	1	0
4	0	3	7	1	12	0	1	0	3
	1	1	1	0	12	1	3	2	0
5	0	1	9	1	12	0	3	0	3
	1	3	0	1	12	0	4	0	3
6	0	3	1	0	12	2	3	1	3
	1	0	1	0	12	0	3	0	9

Note: Hypothetical data.

9.4. Perform exact and asymptotic tests of the homogeneity of the odds ratios for the relevant $K \times 2 \times 2$ tables in Emerson (1994) and Reis, Hirji and Afifi (1999).

9.5. Perform asymptotic analyses for the risk difference and risk ratio models including testing for homogeneity and fitting common effect models on the relevant datasets given in this chapter and Chapter 8.

9.6. Table 9.7 gives data from two hypothetical cohort studies with no loss to follow up. Subjects, categorized by three factors ($\mathcal{X}$, $\mathcal{Y}$ and $\mathcal{Z}$), were monitored for the occurrence of an event (say, a particular disease). Each cell has the number of events and the total duration of follow up for all persons in that category. Use a log-linear model to analyze the relationship between $\mathcal{X}$ and $\mathcal{Y}$ adjusting for $\mathcal{Z}$. (i) Fit a common cross product ratio model. (ii) Formulate a model with $\mathcal{Z}$ an ordinal factor and a linear trend in the logarithm of the cross product ratio. (iii) Fit this model to these data.

9.7. Consider stratified data in which one margin in some of the 2×2 tables is zero. How do such tables affect exact and large sample analysis? (See Agresti (1996), pages 191–194.)

Table 9.7 *Events by Study Factors*

		Study A		Study B	
z	y	$x=0$	$x=1$	$x=0$	$x=1$
0	0	(0;150.0)	(1;25.0)	(1;50.0)	(0;75.0)
	1	(2;100.0)	(3;250.0)	(1;50.0)	(1;25.0)
1	0	(1;125.0)	(0;150.0)	(2;75.0)	(1;25.0)
	1	(0;100.0)	(2;100.0)	(1;50.0)	(0;25.0)
2	0	(1;25.0)	(2;150.0)	(2;100.0)	(2;125.0)
	1	(0;100.0)	(3;100.0)	(1;50.0)	(0;125.0)
3	0	(1;125.0)	(2;150.0)	(1;50.0)	(2;75.0)
	1	(0;100.0)	(3;100.0)	(1;50.0)	(1;25.0)
4	0	(0;0.0)	(0;0.0)	(3;50.0)	(1;50.0)
	1	(0;0.0)	(0;0.0)	(2;50.0)	(1;25.0)

Note: Cell values are count and (person years)/1000.

9.8. Formulate in detail the weight based methods for the odds ratio. Show how to compute a 95% CI for the COR for each method. Prove the formula (9.9).

9.9. Can the weight based statistic of the form (9.6) be applied in an exact conditional analysis of the odds ratio model? Why or why not? If applicable, analyze the data sets in this and the previous chapters and compare the results with those for the other statistics. Give the results of corresponding large sample analysis.

9.10. Derive the conditional and unconditional score, LR and Wald statistics for testing homogeneity of the odds ratios in a $K \times 2 \times 2$ table (see Newman 2001). Which of them can be used for exact conditional analysis? Apply them, for exact and approximate analysis, to the data sets in this chapter and Chapter 8.

9.11. Consider the use of the conditional mue of ϕ in (9.22) and (9.23) and then for substitution in the heterogeneity statistic (9.21). Apply this to the data sets in this chapter and Chapter 8, and compare your results with those obtained from using the cmle.

9.12. Show that the Zelen test for homogeneity of the odds ratios is equivalent to using (9.27).

9.13. Can an exact conditional test for homogeneity be done with the TBD statistic? Why or why not? Investigate an exact unconditional test with this statistic.

9.14. Investigate the use of the CS score test for homogeneity in cases when some of the odds ratios are zero or infinity.

9.15. For the stratified incidence density design and log-linear model, derive the score, LR, Wald and Zelen type of statistics for testing conditional independence and homogeneity of cross product ratios. Apply them, using exact and approximate distributions, to the relevant data sets in this and previous chapter.

9.16. Use the following from Rao (1973) to invert the information matrix $\mathbf{I}$ of §9.11. (i) Suppose $\mathbf{A}_{11}$ and $\mathbf{A}_{22}$ are symmetric matrices. Then

$$\begin{pmatrix} \mathbf{A}_{11} & \mathbf{A}_{12} \\ \mathbf{A}'_{12} & \mathbf{A}_{22} \end{pmatrix}^{-1} = \begin{pmatrix} \mathbf{B}_{11} & \mathbf{B}_{12} \\ \mathbf{B}'_{12} & \mathbf{B}_{22} \end{pmatrix}$$

EXERCISES

where $\mathbf{B}_{11} = \mathbf{A}_{11}^{-1} + \mathbf{F}\mathbf{E}^{-1}\mathbf{F}'$, $\mathbf{B}_{12} = -\mathbf{F}\mathbf{E}^{-1}$, $\mathbf{B}_{22} = \mathbf{E}^{-1}$, $\mathbf{E} = \mathbf{A}_{22} - \mathbf{A}_{12}'\mathbf{A}_{11}^{-1}\mathbf{A}_{12}$, $\mathbf{F} = \mathbf{A}_{11}^{-1}\mathbf{A}_{12}$, when all the inverses in these expressions exist.

(ii) Suppose $\mathbf{A}$ has an inverse. If $\mathbf{U}$ and $\mathbf{V}$ are column vectors, then

$$(\mathbf{A} + \mathbf{U}\mathbf{V}')^{-1} = \mathbf{A}^{-1} - \frac{(\mathbf{A}^{-1}\mathbf{U})(\mathbf{V}'\mathbf{A}^{-1})}{1 + \mathbf{V}'\mathbf{A}^{-1}\mathbf{U}}$$

9.17. Design and execute a simulation study to compare the results asymptotic and exact tests for homogeneity of odds ratios in a product binomial design. Compare the results in terms of actual size and power. Which method(s) do you recommend and under what circumstance?

9.18. Derive conditional and unconditional score, LR and Wald statistics for testing the linearity assumption in the TOR model. Apply them, using exact and approximate distributions, to the relevant datasets in this and the last chapter.

9.19. Consider model (9.39) for the stratified incidence density design. (i) In a model without any interaction terms, derive an exact method for estimation of the main effects of $\mathcal{X}$ and $\mathcal{Y}$; (ii) In a model with all interaction terms present, derive an exact method for estimation of the main effects of $\mathcal{X}$ and $\mathcal{Y}$.

9.20. Consider the frequency matched case-control design. For the variants where matching is done in stratified fashion and where the stratification is done at the analysis stage, write the appropriate models, indicate which parameters are estimable and which are not, and justify the use of a conditional analysis for the estimable parameters.

9.21. Derive the unconditional and conditional Wald, LR and score statistics for the common risk ratio model for the stratified inverse sampling design.

9.22. Compute the null conditional mean and variance of $\mathcal{T}$ for the stratified inverse sampling design. How do they compare with the null conditional mean and variance of $\mathcal{T}$ for the stratified product binomial design?

9.23. Derive the unconditional and conditional score, LR and Wald statistics for testing conditional independence and the homogeneity of risk ratio in the stratified inverse sampling design. How would you get their exact distributions? What are their approximate distributions?

9.24. Construct a trend in risk ratios model for the stratified inverse sampling design and develop methods for exact conditional analysis of this model.

CHAPTER 10

The 2 × K Table

10.1 Introduction

The primary explanatory factor in the models studied so far was binary. Now we examine models with one binary and one multivalued ordered or unordered factor. For example, in a trial comparing surgery, a chemotherapy regimen and placebo, treatment is a three level nominal factor. But if the treatments are three doses of a drug, treatment is a three level ordinal factor. The specific aims of this chapter are:

- To present logit models for an unordered or ordered $2 \times K$ table of counts from a product binomial design, and to derive the related exact conditional distributions.

- To develop exact conditional and asymptotic methods of analysis for these models.

- To extend the exact method to models for $2 \times K$ table data from multinomial, Poisson count, incidence density, case-control, randomized trial and inverse sampling designs.

- To examine some general issues arising in dose-response analysis, give a formulation for exact power in this context and deal with performing exact pair-wise comparisons.

- To introduce exact analysis of several $2 \times K$ tables, and the general logistic model.

Table 10.1 gives the notation for a $2 \times K$ table. The binary row variable, $\mathcal{Y} \in \{0, 1\}$, is a response variable. The column variable, x_k, is a factor, treatment or exposure variable. Assume that the column totals are fixed by design, and the responses within and between columns are independent of one another. At factor level k, n_k is the number of units, $\mathcal{T}_k$, the random number of units expressing $\mathcal{Y} = 1$ and t_k, the observed number of such units.

The sampling distribution here is a product of K independent binomial distributions. Let $\mathcal{T} = (\mathcal{T}_1, \cdots, \mathcal{T}_K)$ with observed vector $t = (t_1, \cdots, t_K)$. Then, with $\pi_k = \mathrm{P}[\mathcal{Y} = 1 \mid x_k]$,

Table 10.1 *Notation for a 2 × K Table*

	Treatment		
Response	x_1	$\cdots$	x_K
$\mathcal{Y} = 0$	$n_1 - t_1$	$\cdots$	$n_K - t_K$
$\mathcal{Y} = 1$	t_1	$\cdots$	t_K
Total	n_1	$\cdots$	n_K

$$P[\mathcal{T} = t] = \prod_{k=1}^{K} \binom{n_k}{t_k} \pi_k^{t_k}(1-\pi_k)^{(n_k - t_k)} \qquad (10.1)$$

$$= \prod_{k=1}^{K} \frac{c_k(t_k)\theta_k^{t_k}}{f_k(\theta_k)} \qquad (10.2)$$

where

$$\theta_k = \pi_k/(1-\pi_k) \text{ and } c_k(u) = \binom{n_k}{u} \qquad (10.3)$$

and where, with $\Omega_k = \{0, 1, \ldots, n_k\}$,

$$f_k(\theta_k) = (1+\theta_k)^{n_k} = \sum_{u \in \Omega_k} c_k(u)\theta_k^u \qquad (10.4)$$

Distribution (10.2) is a multivariate PBD. We also define three quantities for later use.

$$N = \sum_{k=1}^{K} n_k, \quad \mathcal{T} = \sum_{k=1}^{K} \mathcal{T}_k, \quad \mathcal{W} = \sum_{k=1}^{K} \mathcal{T}_k x_k \qquad (10.5)$$

N is the grand total, $\mathcal{T}$ is the random number of responders for all groups, with observed value t and $\mathcal{W}$ is a statistic with observed value w. At times, we will write π_k as $\pi(x_k)$.

10.2 An Ordered Table

Ordered $2 \times K$ tables occur in epidemiological, ecological and toxicity studies. In the latter, cells, microorganisms, plants or animals are exposed to varied doses of a chemical agent and observed for the manifestation of some toxic effect.

Example 10.1: A chemical is added to the diet of rats to evaluate how it affects the intestinal mucosa. The data from a hypothetical study are in Table 10.2; the zero dose group is a control group. We pose two questions: (i) Does the agent affect the risk of acquiring the pathology? (ii) How then does the risk probability vary by dose? A statistical analysis conducted along these lines is a **dose-response**, or a **trend** analysis.

Many models for dose-response analysis have been constructed; some derive from positing an underlying mechanism of disease causation, and others are more empirical in nature. The linear version of the logit model is the most commonly used model in practice. Exact conditional methods, furthermore, have been developed only for the logit model.

Let $\pi(x)$ be the probability of response at dose x. Then the **linear logit model**, which assumes that the logit of the response varies linearly by dose, is

$$\log\left(\frac{\pi(x)}{1-\pi(x)}\right) = \beta + \gamma x \qquad (10.6)$$

The two parameters in this model are: β, the log-odds of response at $x = 0$, or at zero dose. And,

AN ORDERED TABLE

Table 10.2 *Hypothetical Dose Response Data*

Response	Dose (mg/kg)			
	0	10	20	40
No	2	2	2	1
Yes	0	0	1	2

$$\gamma = \log\left(\frac{\pi(x+1)}{1-\pi(x+1)}\right) - \log\left(\frac{\pi(x)}{1-\pi(x)}\right) \tag{10.7}$$

In words, γ is the change in log-odds of response due to a unit change in the dose, or the incremental log-odds ratio. When $\gamma = 0$, the probability of response does not vary by dose. The sign of γ shows the nature of the dose-response relationship. If the outcome is a toxic effect, then $\gamma > 0$ means that higher doses are more toxic. If $\gamma < 0$, then higher doses are less toxic. The magnitude of γ indicates the strength of that effect.

For a study with K dose levels, let x_k denote the dose in the kth group, with $x_1 < \cdots < x_K$. With $\theta = \exp(\beta)$ and $\phi = \exp(\gamma)$, we have $\theta_k = \theta\phi^{x_k}$. Also define

$$f(\theta, \phi) = \prod_{k=1}^{K} f_k(\theta\phi^{x_k}) \tag{10.8}$$

Suppose this bivariate polynomial is expanded as

$$f(\theta, \phi) = \sum_{(u,v) \in \Omega} c(u,v)\theta^u \phi^v \tag{10.9}$$

where Ω is the bivariate set of exponents of θ and ϕ in $f(\theta, \phi)$ and $c(u, v)$, the coefficient of the term with the exponent of θ equal to u, and of ϕ equal to v.

Theorem 10.1: Under the product binomial design and model (10.6), $\mathcal{T}$ is sufficient for β, and $\mathcal{W}$ is sufficient for γ. The joint distribution of $(\mathcal{T}, \mathcal{W})$ is

$$P[\mathcal{T} = t, \mathcal{W} = w] = \frac{c(u,v)\theta^t \phi^w}{f(\theta, \phi)} \tag{10.10}$$

Proof: We substitute $\theta_k = \theta\phi^{x_k}$ in (10.2), and simplify to get that

$$P[\boldsymbol{\mathcal{T}} = \boldsymbol{t}] = \exp\{\beta t + \gamma w\} \prod_{k=1}^{K} \frac{c_k(t_k)}{f_k(\theta\phi^{x_k})}$$

The factorization theorem demonstrates sufficiency. Summing this probability over the set of $\boldsymbol{t} = (t_1, \ldots, t_K)$ for which $\{\boldsymbol{t} : t_1 + \cdots + t_K = t;\ t_1 x_1 + \cdots + t_K x_K = w;\ t_k \in \Omega_k\}$ we get that

$$P[\mathcal{T}=t, \mathcal{W}=w] = \frac{\theta^t \phi^w \sum_t \left\{ \prod_{k=1}^K c_k(t_k) \right\}}{\prod_{k=1}^K (1+\theta\phi^{x_k})^{n_k}}$$

which yields the desired result. □

$f(\theta, \phi)$ is the gp of the distribution of $(\mathcal{T}, \mathcal{W})$. The conditional distribution of $\mathcal{W}$ obtains by selecting appropriate terms of $f(\theta, \phi)$. Hence,

$$P[\mathcal{W}=w \mid \mathcal{T}=t] = \frac{c(t,w)\phi^w}{f(1,\phi;t,.)} \qquad (10.11)$$

where

$$f(\theta,\phi;t,.) = \theta^t \sum_u c(t,u)\phi^u \qquad (10.12)$$

(10.11) is a univariate PBD amenable to the analytic techniques given previously.

Theorem 10.2: Let μ_0 and σ_0^2 be the mean and variance, respectively, of the conditional distribution of $\mathcal{W}$ given $\mathcal{T}$ when $\phi = 1$. Then

$$\mu_0 = \frac{t}{N} \sum_{k=1}^K n_k x_k \qquad (10.13)$$

$$\sigma_0^2 = \frac{t(N-t)}{N^2(N-1)} \left[N \sum_{k=1}^K n_k x_k^2 - \left\{ \sum_{k=1}^K n_k x_k \right\}^2 \right] \qquad (10.14)$$

Further

$$f(1,1;t,.) = \sum_{u \in \Omega(t,.)} c(t,u) = \binom{N}{t} \qquad (10.15)$$

Proof: This is left to the reader. □

The above formulation resembles that given for the TOR model in Chapter 8. Instead of hypergeometric coefficients and sample space, now we have binomial coefficients and sample space. The analytic approach and formulas are otherwise the same.

****

These results facilitate conditional analyses of an ordered $2 \times K$ table. γ is the parameter of interest. A test of $\gamma = 0$ versus the alternative $\gamma \neq 0$ is known as a **test for trend**. For exact analysis, we obtain the conditional distribution of $\mathcal{W}$ by implementing the steps of Theorem 10.1.

Exact or mid-p TST evidence functions derived from the conditional distribution of $\mathcal{W}$ provide

Table 10.3 *Dose Level Generating Polynomials*

k	Dose	$f_k(\theta, \phi)$
1	0	$1 + 2\theta + \theta^2$
2	10	$1 + 2\theta\phi^{10} + \theta^2\phi^{20}$
3	20	$1 + 3\theta\phi^{20} + 3\theta^2\phi^{40} + \theta^3\phi^{60}$
4	40	$1 + 3\theta\phi^{40} + 3\theta^2\phi^{80} + \theta^3\phi^{120}$

respective *p*-values, estimates and CIs for γ. The score statistic based on this distribution is given by

$$\mathcal{W}_{cs} = \frac{(\mathcal{W} - \mu)^2}{\sigma^2} \qquad (10.16)$$

where μ and σ^2 are the conditional mean and variance, respectively, of (10.11). This statistic also provides exact, mid-*p* and large sample evidence functions for γ. A test based on $\mathcal{W}_{cs}$, known as the **Cochran–Armitage (CA) test** or the **Mantel extension test** for trend, is the most popular test for trend. We will call it the **CAM test for trend**. Note that for the purpose of exact or mid-*p* tests, the variance in (10.16) is not needed, making it equivalent to the distance from the mean method test.

The unconditional score test for model (10.6), denoted as $\mathcal{W}_{us}$, is at times called the CA test. The two statistics, however, differ by the factor $N/(N-1)$ with

$$\mathcal{W}_{cs} = \frac{N}{N-1}\mathcal{W}_{us} \qquad (10.17)$$

The two statistics, however, yield identical exact conditional *p*-values and CIs, and are thus, for the purpose of exact analysis, equivalent. Except in small samples, they also produce virtually identical asymptotic *p*-values.

The unconditional Wald and LR statistics are derived in the usual way; their conditional counterparts are also readily available. At sufficiently large cell counts, and under $\gamma = 0$, $\mathcal{W}_{cs}$, $\mathcal{W}_{cl}$, $\mathcal{W}_{cw}$, $\mathcal{W}_{us}$, $\mathcal{W}_{ul}$ and $\mathcal{W}_{uw}$ (with the self explanatory subscript labels) follow an approximately chisquare distribution with one df.

Asymptotic score CIs obtain from the score based asymptotic evidence function. If $\hat{\beta}$ and $\hat{\gamma}$ be the unconditional mles for β and γ, respectively, then estimates for the variances and covariance of these estimates obtain from the appropriate variance covariance matrix (see §10.12). An asymptotic 95% Wald CI for γ is

$$\hat{\gamma} \pm 1.96\widehat{\mathrm{var}}(\hat{\gamma}) \qquad (10.18)$$

****

Example 10.1 (continued): For the data in Table 10.2, the four doses are: $x_1 = 0$, $x_2 = 10$, $x_3 = 20$ and $x_4 = 40$. For exact analysis, we first construct f_k for each dose level. These are shown in Table 10.3. Then we multiply the four polynomials. The result, $f(\theta, \phi)$, which has 87 terms, is shown below.

$$\begin{aligned}
& 1 + \theta(2 + 2\phi^{10} + 3\phi^{20} + 3\phi^{40}) \\
& + \theta^2(1 + 4\phi^{10} + 7\phi^{20} + 6\phi^{30} + 9\phi^{40} + 6\phi^{50} + 9\phi^{60} + 3\phi^{80}) \\
& + \theta^3(2\phi^{10} + 5\phi^{20} + 12\phi^{30} + 12\phi^{40} + 18\phi^{50} + 22\phi^{60} + \\
& \qquad 18\phi^{70} + 15\phi^{80} + 6\phi^{90} + 9\phi^{100} + \phi^{120}) \\
& + \theta^4(\phi^{20} + 6\phi^{30} + 9\phi^{40} + 18\phi^{50} + 20\phi^{60} + 38\phi^{70} + 30\phi^{80} + \\
& \qquad 30\phi^{90} + 24\phi^{100} + 18\phi^{110} + 11\phi^{120} + 2\phi^{130} + 3\phi^{140}) \\
& + \theta^5(3\phi^{40} + 6\phi^{50} + 10\phi^{60} + 22\phi^{70} + 28\phi^{80} + 42\phi^{90} + 30\phi^{100} + \\
& \qquad 42\phi^{110} + 28\phi^{120} + 22\phi^{130} + 16\phi^{140} + 3\phi^{160}) \\
& + \theta^6(3\phi^{60} + 2\phi^{70} + 11\phi^{80} + 18\phi^{90} + 24\phi^{100} + 30\phi^{110} + 30\phi^{120} + \\
& \qquad 38\phi^{130} + 20\phi^{140} + 18\phi^{150} + 7\phi^{160} + 6\phi^{170} + \phi^{180}) \\
& + \theta^7(\phi^{80} + 9\phi^{100} + 6\phi^{110} + 15\phi^{120} + 18\phi^{130} + 22\phi^{140} + \\
& \qquad 18\phi^{150} + 12\phi^{160} + 12\phi^{170} + 5\phi^{180} + 2\phi^{190}) \\
& + \theta^8(3\phi^{120} + 9\phi^{140} + 6\phi^{150} + 9\phi^{160} + 6\phi^{170} + 7\phi^{180} + 4\phi^{190} + \phi^{200}) \\
& + \theta^9(3\phi^{160} + 3\phi^{180} + 2\phi^{190} + 2\phi^{200}) + \theta^{10}\phi^{200}
\end{aligned}$$

The observed values of $\mathcal{T}$ and $\mathcal{W}$ are

$$\begin{aligned}
t &= 0 + 0 + 1 + 2 = 3 \\
w &= 0 \times 0 + 0 \times 10 + 1 \times 20 + 2 \times 40 = 100
\end{aligned}$$

For inference on γ, select terms of $f(\theta, \phi)$ with the exponent of θ equal to 3. Then we have

$$\begin{aligned}
f(1, \phi; 3, .) &= 2\phi^{10} + 5\phi^{20} + 12\phi^{30} + 12\phi^{40} + 18\phi^{50} + \\
& \quad 22\phi^{60} + 18\phi^{70} + 15\phi^{80} + 6\phi^{90} + 9\phi^{100} + \phi^{120}
\end{aligned}$$

This is the gp for the conditional distribution of $\mathcal{W}$ given $\mathcal{T} = 3$. For example,

$$\mathrm{P}[\mathcal{W} = 80 \mid \mathcal{T} = 3] = 15\phi^{80}/f(1, \phi; 3, .)$$

As expected, the sum of the coefficients of $f(1, \phi; 3, .)$ is

$$f(1, 1; 3, .) = 120 = \binom{10}{3}$$

The null mean and standard deviation of this distribution are 60 and 22.66. The two-sided exact TST and score *p*-values are 0.1667 and 0.1447, and their respective mid-*p* versions are 0.0917 and 0.0833. The asymptotic CAM *p*-value is 0.0776 while the asymptotic unconditional score *p*-value is 0.0627.

AN UNORDERED TABLE

Table 10.4 *Outcome at Five Years*

	Treatment			
Outcome	A	B	C	D
Dead	0	0	0	3
Alive	2	2	3	0

Note: Hypothetical data.

The mle for γ is 0.1108 and its mue is 0.0945. The 95% asymptotic Wald CI for γ is (-0.0236,0.2453) and the exact TST 95% interval is (-0.0251,0.2949).

****

Suppose we need to estimate the probability of response at a specified dose level x_j. We do this in a way analogous to that for estimation of a stratum specific odds ratio in §8.6, that is, by using a reparametrized model of the type (8.69).

We also ask: What is the dose at which half of the units are expected to show the effect? Is there a dose level below which the effect is too small to be of concern? These questions involve specifying a probability π_*, say $\pi_* = 0.5$ or $\pi_* = 0.001$, and solving for x_* in the equation

$$\frac{\exp(\beta + \gamma x_*)}{1 + \exp(\beta + \gamma x_*)} = \pi_* \qquad (10.19)$$

which gives

$$x_* = \frac{1}{\gamma}\left(\ln\frac{\pi_*}{(1-\pi_*)} - \beta\right) \qquad (10.20)$$

Large sample methods of estimating x_* are avaliable. Direct exact methods have yet to be devised. Indirectly, we obtain say, the cmues, or the cmles for β and γ from the conditional distributions of their sufficient statistics and substitute them in (10.20) to get an estimate for x_*. The properties of such an estimate have not been studied, and the computation of an exact CI for x_* remains an open question.

10.3 An Unordered Table

Now we turn to a $2 \times K$ table with unordered columns, or an ordered table where we ignore the ordering.

Example 10.2: Artificial data from a four treatment binary response clinical trial are shown in Table 10.4. The ordering of the columns is arbitrary, or a matter of convention. The outcome is survival at or beyond five years after treatment. Assume that no patients were lost to follow up.

Let π_k be the five year survival rate in treatment k. We test $H_0 : \pi_1 = \pi_2 = \cdots = \pi_K = \pi$. This is a test for **homogeneity** or **association** in the $2 \times K$ table. In terms of (10.3), we test $\theta_1 = \cdots = \theta_K = \theta$. Rewrite the parameters θ_k: $\theta_1 = \theta$ with $\beta = \ln(\theta)$, and for $k \geq 2$, $\ln(\theta_k) = \beta + \gamma_k$, or $\theta_k = \theta\phi_k$, with $\gamma_1 \equiv 0$, $\phi_1 \equiv 1$ and $\phi_k = \exp(\gamma_k)$. We then have

$$\ln\left(\frac{\pi_k}{1-\pi_k}\right) = \beta + \gamma_k \tag{10.21}$$

The numeric value of x_k does not figure in the model. The hypothesis of treatment equivalence is $\gamma_2 = \ldots = \gamma_K = 0$ or $\phi_2 = \cdots = \phi_K = 1$. This formulation facilitates construction of exact tests for H_0. Consider a test statistic of the form

$$\mathcal{W}_* = \sum_{k=1}^{K} w_k(\mathcal{T}_k) \tag{10.22}$$

Several specific statistics are given in the next section. For now, we require that it has an additive form and that extreme values are unlikely under H_0. As before, the realized value $\mathcal{T} = \Sigma \mathcal{T}_k$ is t, and of $\mathcal{W}_*$ is w_*.

Assume that responses within and between each treatment are independent and that the data derive from the product binomial design (10.1). Under this, β or θ is a nuisance parameter with sufficient statistics $\mathcal{T}$. An exact conditional p-value for H_0 then is

$$P[\mathcal{W}_* \geq w_* \mid \mathcal{T} = t; H_0] \tag{10.23}$$

To compute the exact distribution of $\mathcal{W}_*$ when the sum of responses is fixed, recall the definitions of $c_k(u)$ and Ω_k, and let ψ be a dummy parameter. Then define the polynomial

$$g_k(\theta, \psi, \phi_k) = \sum_{u \in \Omega_k} c_k(u) \phi_k^u \theta^u \psi^{w_k(u)} \tag{10.24}$$

Also, with specified values of $\phi = (\phi_1, \ldots, \phi_K)$, define the product

$$g(\theta, \psi, \phi) = \prod_{k=1}^{K} g_k(\theta, \psi, \phi_k) \tag{10.25}$$

Suppose we perform the multiplication in the rhs of (10.25) and write it in an expanded form as:

$$g(\theta, \psi, \phi) = \sum_{(u,v)} d(u, v; \phi) \theta^u \psi^v \tag{10.26}$$

At fixed values of ϕ_k, $k = 1, \cdots, K$, this is a bivariate polynomial. Let $g(\theta, \psi, \phi; t, .)$ denote all the terms of $g(\theta, \psi, \phi)$ in which the exponent of θ is equal to t.

Theorem 10.3: The exact conditional distribution of $\mathcal{W}_*$ is

$$P[\mathcal{W}_* = w_* \mid \mathcal{T} = t] = \frac{d(t, w; \phi)}{g(1, 1, \phi; t, .)} \tag{10.27}$$

Proof: Substitute $\theta_k = \theta \phi_k$ in (10.24). The distribution of $(\mathcal{W}_*, \mathcal{T})$ obtains from summing the product binomial probability over values of $t = (t_1, \ldots, t_K)$ over the set

TEST STATISTICS

$$\left\{ t : \begin{array}{c} t_1 + \cdots + t_K = t; \ w_1(t_1) + \cdots + w_K(t_K) = w_* \\ t_k \in \Omega_k, \ k = 1, \ldots, K \end{array} \right\}$$

Note that the polynomials involved here have the same form as those for testing the homogeneity of odds ratios in Chapter 9. The proof then follows the arguments in Theorem 9.1. □

Since ψ is a dummy parameter, the conditional distribution of $\mathcal{W}_*$ is like a PBD but actually is not so. Under the null hypothesis,

$$g(1, 1, \mathbf{1}; t, .) = \binom{N}{t} \tag{10.28}$$

and the exact p-value is the right tail probability

$$\binom{N}{t}^{-1} g(1, 1, \mathbf{1}; t, \geq w_*) \tag{10.29}$$

10.4 Test Statistics

The exact test commonly used for testing the equality of the proportions (as well as for testing association) in an $R \times K$ table in general, and a $2 \times K$ table in particular, is the **Freeman–Halton exact test**, also known as the Fisher exact test, the minimum likelihood test or the exact probability based test. It is a generalization of the Fisher–Irwin exact test for a 2×2 table. We will refer to it as the **FH exact test**.

Here, the null conditional probability of the table is the test statistic. Let us denote it as $\mathcal{W}_F$. The exact p-value is the null probability of the set of all tables with given row and column margins whose null conditional probability does not exceed that of the observed table. Formally,

$$\mathcal{W}_F = P[\mathcal{T} = t \mid \mathcal{T} = t; H_0] \tag{10.30}$$

$$= \binom{N}{t}^{-1} \left[\prod_{k=1}^{K} \binom{n_k}{t_k} \right] \tag{10.31}$$

The FH exact p-value is defined by

$$P[\mathcal{W}_F \leq w_F \mid \mathcal{T} = t; H_0] \tag{10.32}$$

where w_F is the observed value of $\mathcal{W}_F$.

The **power divergence class** of statistics is also relevant for contingency table data. The chisquare and LR statistics, often applied to the table at hand, belong to this class. We define them with the following notation. Let

$$\mathcal{O}_{0k} = n_k - \mathcal{T}_k \quad \& \quad \mathcal{O}_{1k} = \mathcal{T}_k \tag{10.33}$$

$$e_{0k} = \frac{(N-t)n_k}{N} \quad \& \quad e_{1k} = \frac{tn_k}{N} \tag{10.34}$$

For cell (i, k), $\mathcal{O}_{ik}$ and e_{ik} are respectively the observed and expected counts under H_0. The chisquare statistic, $\mathcal{W}_X$, and the LR statistic, $\mathcal{W}_L$, are:

$$\mathcal{W}_X = \sum_{k=1}^{K} \sum_{i=0}^{1} \frac{(\mathcal{O}_{ik} - e_{ik})^2}{e_{ik}} \tag{10.35}$$

$$\mathcal{W}_L = 2 \sum_{k=1}^{K} \sum_{i=0}^{1} \mathcal{O}_{ik} \ln\left(\frac{\mathcal{O}_{ik}}{e_{ik}}\right) \tag{10.36}$$

The chisquare statistic is the unconditional score statistic for model (10.21). Another member of the power divergence class is the **Cressie–Read statistic**, denoted by $\mathcal{W}_R$. This is

$$\mathcal{W}_R = \frac{9}{5} \sum_{k=1}^{K} \sum_{i=0}^{1} \mathcal{O}_{ik} \left\{ \left(\frac{\mathcal{O}_{ik}}{e_{ik}}\right)^{2/3} - 1 \right\} \tag{10.37}$$

The **Zelterman statistic**, which is not in the power divergence class, has an improved asymptotic approximation for sparse tables. Its version for a $2 \times K$ table needs the following definitions.

$$\mu_D = \frac{2K - N(1+K)}{N-1} \tag{10.38}$$

$$\tau_R = \frac{1}{N-2} \left\{ \frac{N^2}{t(N-t)} - 4 \right\} \tag{10.39}$$

$$\tau_C = \frac{1}{N-2} \left\{ N \sum_{k=1}^{K} n_k^{-1} - K^2 \right\} \tag{10.40}$$

$$\sigma_D^2 = \frac{2N}{N-3} \left\{ \frac{N-2}{N-1} - \tau_R \right\} \left\{ \frac{(K-1)(N-K)}{N-1} - \tau_C \right\} + \frac{4\tau_R \tau_C}{N-1} \tag{10.41}$$

The Zelterman statistic, $\mathcal{W}_Z$, is then defined as

$$\mathcal{W}_Z = \sigma_D^{-2} \left\{ \sum_{k=1}^{K} \sum_{i=0}^{1} \left\{ \frac{(\mathcal{O}_{ik} - e_{ik})^2}{e_{ik}} - \frac{\mathcal{O}_{ik}}{e_{ik}} \right\} - \mu_D \right\}^2 \tag{10.42}$$

In large samples, and under the null, $\mathcal{W}_X$, $\mathcal{W}_L$ and $\mathcal{W}_R$ roughly follow a chisquare distribution with $K-1$ df, and $\mathcal{W}_Z$ roughly has a chisquare distribution with 1 df. This gives asymptotic p-values. These statistics also generate exact p-values. Deriving their equivalent versions facilitates exact computation.

Theorem 10.4: For an exact conditional p-value, the statistics $\mathcal{W}_F$, $\mathcal{W}_X$, $\mathcal{W}_L$, $\mathcal{W}_R$ and $\mathcal{W}_Z$ are respectively equivalent to $\mathcal{W}_{eF}$, $\mathcal{W}_{eX}$, $\mathcal{W}_{eL}$, $\mathcal{W}_{eR}$ and $\mathcal{W}_{eZ}$, defined by

AN ILLUSTRATION

$$\mathcal{W}_{eF} = \sum_{k=1}^{K} \ln\left(\frac{n_k}{\mathcal{T}_k}\right) \tag{10.43}$$

$$\mathcal{W}_{eX} = \sum_{k=1}^{K} \frac{\mathcal{T}_k^2}{n_k} \tag{10.44}$$

$$\mathcal{W}_{eL} = \sum_{k=1}^{K} \left[\mathcal{T}_k \ln(\mathcal{T}_k) + (n_k - \mathcal{T}_k)\ln(n_k - \mathcal{T}_k)\right] \tag{10.45}$$

$$\mathcal{W}_{eR} = \sum_{k=1}^{K} \left[\mathcal{T}_k \left(\frac{\mathcal{T}_k}{tn_k}\right)^{2/3} + (n_k - \mathcal{T}_k)\left(\frac{(n_k - \mathcal{T}_k)}{(N-t)n_k}\right)^{2/3}\right] \tag{10.46}$$

$$\mathcal{W}_{eZ} = \sum_{k=1}^{K} \frac{(\mathcal{T}_k - N + 2\mathcal{T})\mathcal{T}_k}{n_k} \tag{10.47}$$

Under these equivalent forms, the exact p-value is computed as follows: (i) For the FH statistic, we compute the left tail conditional probability; (ii) for the chisquare, LR and Cressie–Read, we compute the respective right tail conditional probability; and (iii) for the Zelterman statistic, we compute, under the null,

$$P[\mathcal{W}_{eZ} \geq \mu_e + \tau_e \mid \mathcal{T} = t] + P[\mathcal{W}_{eZ} \leq \mu_e - \tau_e \mid \mathcal{T} = t] \tag{10.48}$$

where w_{eZ} is the observed value of $\mathcal{W}_{eZ}$ and

$$\mu_e = \frac{t(N-t)}{N}\left[\frac{N(K+t)}{N-t} + \mu_D\right] \text{ and } \tau_e = \|w_{eZ} - \mu_e\| \tag{10.49}$$

Proof: Consider the chisquare statistic. In the case of a $2 \times K$ table, after some algebra, this is written as

$$\mathcal{W}_X = \frac{N^2}{\mathcal{T}(N-\mathcal{T})}\left\{\sum_{k=1}^{K}\frac{\mathcal{T}_k^2}{n_k}\right\} - \frac{\mathcal{T}N}{N-\mathcal{T}}$$

$$= \frac{N^2}{\mathcal{T}(N-\mathcal{T})}\mathcal{W}_{eX} - \frac{\mathcal{T}N}{N-\mathcal{T}}$$

As the exact p-value is computed at a fixed value of $\mathcal{T}$, $\mathcal{W}_{eX}$ is equivalent to $\mathcal{W}_X$. The proof for the other statistics is given in a similar fashion. □

Note that exact computation for the Zelterman statistic is a more complex task than for the other statistics. Mid-p-values are computed with any of these statistics by deducting half the null conditional probability of attaining its observed value.

Table 10.5 *Tables with Same Margins as Table 10.4*

No.	A	B	C	D	Null Prob.	No.	A	B	C	D	Null Prob.
1	0 2	0 2	0 3	3 0	1/120	10	1 1	1 1	3 0	1 2	12/120
2	0 2	0 2	1 2	2 1	9/120	11	0 2	0 2	3 0	0 3	1/120
3	0 2	1 1	0 3	2 1	6/120	12	1 1	0 2	2 1	0 3	6/120
4	1 1	0 2	0 3	2 1	6/120	13	0 2	1 1	2 1	0 3	6/120
5	0 2	0 2	2 1	1 2	9/120	14	0 2	2 0	1 2	0 3	3/120
6	0 2	2 0	0 3	1 2	3/120	15	1 1	1 1	1 2	0 3	12/120
7	2 0	0 2	0 3	1 2	3/120	16	2 0	0 2	1 2	0 3	3/120
8	0 2	1 1	1 2	1 2	18/120	17	1 1	2 0	0 3	0 3	2/120
9	1 1	0 2	1 2	1 2	18/120	18	2 0	1 1	0 3	0 3	2/120

Note: Null Prob. is the null conditional probability of the table.

10.5 An Illustration

Example 10.2 (continued): For Table 10.4, the observed value of T is $t = 2 + 2 + 3 + 0 = 7$. The 18 2×4 tables with the same margins as Table 10.4 are shown in Table 10.5.

FH Test: The observed values of the original and equivalent versions of the FH statistic are

$$w_F = \binom{2}{2}\binom{2}{2}\binom{3}{3}\binom{3}{0}\binom{10}{72}^{-1} = \frac{1}{120}$$

$$w_{eF} = \ln 1 + \ln 1 + \ln 1 + \ln 1 = 0$$

The null category level gps are

$$g_1(\phi, \psi, 1) = g_2(\phi, \psi, 1) = 1 + 2\phi\psi^{\ln 2} + \phi^2$$
$$g_3(\phi, \psi, 1) = g_4(\phi, \psi, 1) = 1 + 3\phi\psi^{\ln 3} + 3\phi^2\psi^{\ln 3} + \phi^3$$

We multiply the four polynomials and select the terms where the exponent of ϕ is 7. Then

AN ILLUSTRATION

Table 10.6 *Expected Values for Table 10.4*

Outcome	Treatment			
	A	B	C	D
Dead	0.6	0.6	0.9	0.9
Alive	1.4	1.4	2.1	2.1

$$g(1,\psi,\mathbf{1};7,.) = 2 + 4\psi^{\ln 2} + 12\psi^{\ln 3} + 24\psi^{\ln 6}$$
$$+ 18\psi^{\ln 9} + 24\psi^{\ln 12} + 36\psi^{\ln 18}$$

The exact *p*-value is

$$P[\mathcal{W}_{eF} \leq 0 \mid \mathcal{T} = 7; H_0] = 2/120 = 0.0167$$

Chisquare Test: The null expected values for Table 10.4 are in Table 10.6.

The observed values of the equivalent and original versions of the chisquare statistic are

$$w_{eX} = \frac{2^2}{2} + \frac{2^2}{2} + \frac{3^2}{3} + \frac{0^2}{3} = 7$$

$$w_X = \frac{10 \times 10}{7 \times 3} w_{*X} - \frac{7 \times 10}{3} = 10$$

The null category level gps are

$$g_1(\phi,\psi,1) = g_2(\phi,\psi,1) = 1 + 2\phi\psi^{1/2} + \phi^2\psi^2$$
$$g_3(\phi,\psi,1) = g_4(\phi,\psi,1) = 1 + 3\phi\psi^{1/3} + 3\phi^2\psi^{4/3} + \phi^3\psi^3$$

We multiply the four polynomials and select the terms where the exponent of ϕ is 7. Then

$$g(1,\psi,\mathbf{1};7,.) = 36\psi^{31/6} + 24\psi^{32/6} + 18\psi^{34/6} + 24\psi^{35/6}$$
$$+ 12\psi^{38/6} + 4\psi^{39/6} + 2\psi^{42/6}$$

The exact *p*-value is

$$P[\mathcal{W}_{eX} \geq 7 \mid \mathcal{T} = 7; H_0] = 2/120 = 0.0167$$

and, at 3 df, the asymptotic *p*-value is 0.0186.

LR Test: The observed values of the original and equivalent versions of the LR statistic are

$$0.5w_L = 2\ln\frac{2}{1.4} + 2\ln\frac{2}{1.4} + 3\ln\frac{3}{2.1} + 0\ln\frac{0}{2.1}$$
$$+ 0\ln\frac{0}{0.6} + 0\ln\frac{0}{0.6} + 0\ln\frac{0}{0.9} + 3\ln\frac{3}{0.9}$$
$$= 7\ln\frac{10}{7} + 3\ln\frac{10}{3} = 6.1086$$

Hence, $w_L = 12.2172$.

$$w_{eL} = 2\ln2 + 2\ln2 + 3\ln3 + 0\ln0 + 0\ln0 + 0\ln0 + 0\ln0 + 3\ln3$$
$$= 4\ln2 + 6\ln3 = \ln 11664$$

The null category level gps are

$$g_1(\phi,\psi,1) = g_2(\phi,\psi,1) = \psi^{2\ln2} + 2\phi + \phi^2\psi^{2\ln2}$$
$$g_3(\phi,\psi,1) = g_4(\phi,\psi,1) = \psi^{3\ln3} + 3\phi\psi^{2\ln2} + 3\phi^2\psi^{2\ln2} + \phi^3\psi^{3\ln3}$$

We multiply the four polynomials and select the terms where the exponent of ϕ is 7. Then

$$g(1,\psi,\mathbf{1};7,.) = 36\psi^{\ln64} + 24\psi^{\ln108} + 18\psi^{\ln256} + 24\psi^{\ln432}$$
$$+ 12\psi^{\ln1728} + 4\psi^{\ln2916} + 2\psi^{\ln11664}$$

The exact p-value thus is

$$P[\mathcal{W}_{eL} \geq \ln 11664 \mid \mathcal{T} = 7; H_0] = 2/120 = 0.0167$$

and, at 3 df, the asymptotic p-value is 0.0067.

Cressie–Read Test: The observed values of the equivalent and original versions of the Cressie–Read statistic are

$$w_{eR} = 2\left(\frac{2}{7\times2}\right)^{2/3} + 0\left(\frac{0}{3\times2}\right)^{2/3} + 2\left(\frac{2}{7\times2}\right)^{2/3} + 0\left(\frac{2}{3\times2}\right)^{2/3}$$
$$+ 3\left(\frac{3}{7\times3}\right)^{2/3} + 0\left(\frac{0}{3\times3}\right)^{2/3} + 0\left(\frac{0}{7\times3}\right)^{2/3} + 3\left(\frac{3}{3\times3}\right)^{2/3}$$
$$= 7^{1/3} + 3^{1/3} = 3.3552$$

$$w_R = \frac{9\times10}{5}\left(10^{-1/3}w_{eR} - 1\right) = 10.0320$$

The null category level gps are

$$g_1(\phi,\psi,1) = g_2(\phi,\psi,1) = \psi^{0.9615} + 2\phi\psi^{0.4751} + \phi^2\psi^{0.5466}$$
$$g_3(\phi,\psi,1) = g_4(\phi,\psi,1) = \psi^{1.4422} + 3\phi\psi^{0.8651} + 3\phi^2\psi^{0.6483} + \phi^3\psi^{0.8198}$$

AN ILLUSTRATION

We multiply the four polynomials and select the terms where the exponent of ϕ is 7. Then

$$g(1, \psi, 1; 7, .) = 36\psi^{2.3183} + 24\psi^{2.4183} + 18\psi^{2.6066} + 24\psi^{2.7066}$$
$$+ 12\psi^{2.9762} + 4\psi^{3.0762} + 2\psi^{3.3552}$$

The exact p-value thus is

$$P[\mathcal{W}_{eR} \geq 3.3552 \mid \mathcal{T} = 7; H_0] = 2/120 = 0.0167$$

and, at 3 df, the asymptotic p-value is 0.0183.

Zelterman Test: The observed value of the equivalent and original Zelterman statistic is

$$w_{eZ} = 24 + 24 + 34 + 0 = 82$$

$$w_Z = \frac{1}{2.0856 \times 2.0856} \left(\frac{10 w_{eZ}}{7 \times 3} - \frac{10 \times 11}{3} + \frac{14}{3} \right)^2 = 3.3792^2 = 11.4190$$

giving an asymptotic p-value equal to 0.0007. For exact computation, the null category level gps for the equivalent version are

$$g_1(\phi, \psi, 1) = g_2(\phi, \psi, 1) = 1 + 2\phi\psi^7 + \phi^2\psi^{24}$$
$$g_3(\phi, \psi, 1) = g_4(\phi, \psi, 1) = 1 + 3\phi\psi^{14/3} + 3\phi^2\psi^{16} + \phi^3\psi^{34}$$

We multiply the four polynomials and select the terms where the exponent of ϕ is 7. Then

$$g(1, \psi, 1; 7, .) = 36\psi^{63} + 24\psi^{64} + 18\psi^{206/3} + 24\psi^{209/3} + 12\psi^{74} + 4\psi^{75} + 2\psi^{82}$$

Further, $\mu_e = 67.2$. Then using formula (10.48), the exact p-value is

$$P[\mathcal{W}_{eZ} \geq 82 \mid \mathcal{T} = 7; H_0] + P[\mathcal{W}_{eZ} \leq 52.4 \mid \mathcal{T} = 7; H_0] = 2/120 = 0.0167$$

The highly discretized conditional sample space in this artificial case made all the five exact p-values the same. The asymptotic ones, though, are not, the Zelterman statistic asymptotic value being the smallest and furthest apart from the exact p-value(s).

****

A point of note is that for statistics that are not integer valued, the number of significant digits of accuracy for the statistic components may impact exact computation. In most problems, four decimal digits accuracy provide acceptable accuracy. See Chapter 12 for elaboration.

10.6 Checking Linearity

The assumption in model (10.6) of a linear effect on the logit scale may not always hold. A test for checking this is called a **goodness-of-fit test** for the linear model. Under the notation of §10.2, assume, without loss of generality, that $x_1 = 0$. Otherwise, rewrite the problem with new scores $x_{*k} = x_k - x_1$. (10.6) is a special case of the general model (10.21).

Note that in (10.21), $\gamma_1 \equiv 0$. Further, in this model and for $k \geq 3$, we let $\gamma_2 = \gamma x_2$; $\gamma_k = \gamma x_k + \delta_k$, with $\delta_1 = \delta_2 = 0$. Testing if the linear model (10.6) holds is the same as to test, in the context of (10.21), the hypothesis H_0:

$$\delta_3 = \cdots = \delta_K = 0, \quad \text{or equivalently,} \quad \rho_3 = \cdots = \rho_K = 1$$

where $\rho_k = \exp(\delta_k)$. A test of H_0 is then a goodness-of-fit test for the linear logit model. Let $\mathcal{Q}$ denote a test statistic for this hypothesis. Score and LR statistics for H_0 are constructed by taking the difference between the respective statistic for homogeneity and that for linear trend. In that case, with reference to (10.17) and (10.35),

$$\mathcal{Q}_X = \mathcal{W}_X - \mathcal{W}_{us} \quad \text{and} \quad \mathcal{Q}_L = \mathcal{W}_L - \mathcal{W}_{ul} \tag{10.50}$$

These are referred to a chisquared distribution with $K - 2$ df to get an asymptotic p-value. Now consider exact tests for H_0. We assume that $\mathcal{Q}$ has the form

$$\mathcal{Q} = \sum_{k=1}^{K} q_k(\mathcal{T}_k) \tag{10.51}$$

Under the product binomial setting, the sufficient statistics for β and γ respectively are $\mathcal{T} = \Sigma_k \mathcal{T}_k$ and $\mathcal{W} = \Sigma_k z_k \mathcal{T}_k$. Thus, an exact significance level for the H_0 is

$$P[\,\mathcal{Q} \geq q \mid \mathcal{T} = t,\, \mathcal{W} = w;\, H_0\,] \tag{10.52}$$

where t, w and q are the respective realized values of $\mathcal{T}$, $\mathcal{W}$ and $\mathcal{Q}$. Let $\phi = \exp(\gamma)$, and let ψ be a dummy parameter. Then define the polynomial

$$g_{*k}(\theta, \phi, \psi, \phi_k) = \sum_{u \in \Omega_k} c_k(u) \rho_k^u \theta^u \phi^{z_k u} \psi^{q_k(u)} \tag{10.53}$$

With $\boldsymbol{\rho} = (\rho_1, \ldots, \rho_K)$, define the product polynomial

$$g_*(\theta, \phi, \psi, \boldsymbol{\rho}) = \prod_{k=1}^{K} g_{*k}(\theta, \phi, \psi, \rho_k) \tag{10.54}$$

At fixed values of ρ_k, $k = 1, ..., K$, we perform the multiplication in the rhs of (10.54) and write it in an expanded form as:

$$g_*(\theta, \phi, \psi, \boldsymbol{\rho}) = \sum_{(u,v,q)} d_*(u, v, q; \boldsymbol{\rho}) \theta^u \phi^v \psi^q \tag{10.55}$$

Let the polynomial $g_*(\theta, \phi, \psi, \boldsymbol{\rho}; t, w, .)$ comprise the terms of $g_*(\theta, \phi, \psi, \boldsymbol{\rho})$ in which the exponent of θ is equal to t, and that of ϕ is equal to w.

OTHER SAMPLING DESIGNS

Theorem 10.5: The conditional distribution of $\mathcal{Q}$ is

$$P[\mathcal{Q} = q \mid \mathcal{T} = t, \mathcal{W} = w; H_0] = \frac{d_*(t, w, q; \boldsymbol{\rho})}{g_*(1, 1, 1, \boldsymbol{\rho}; t, w, .)} \quad (10.56)$$

Proof: This follows the proofs of Theorem 9.1 and Theorem 9.2. □

The unconditional score statistic is a candidate for $\mathcal{Q}$. This is obtained as the difference between the chisquare statistic for homogeneity and the null unconditional score statistic for trend. A second possibility is the unconditional LR statistic, obtained in a likewise manner. Their conditionally equivalent versions, for use in exact analysis, are as follows:

Theorem 10.6: For the purpose of exact conditional tests, the unconditional score, $\mathcal{Q}_X$, and LR, $\mathcal{Q}_L$, statistics given by (10.50) for H_0 are respectively equivalent to $\mathcal{Q}_{eX}$ and $\mathcal{Q}_{eL}$ given by

$$\mathcal{Q}_{eX} = \sum_{k=1}^{K} n_k^{-1} \mathcal{T}_k^2 \quad (10.57)$$

$$\mathcal{Q}_{eL} = \sum_{k=1}^{K} \mathcal{T}_k \ln \mathcal{T}_k + (n_k - \mathcal{T}_k) \ln(n_k - \mathcal{T}_k) \quad (10.58)$$

Proof: Left as an exercise. □

Note that these equivalent versions of the goodness-of-fit tests are identical to the equivalent versions for the respective homogeneity tests. The conditional sample spaces in the two test procedures are, however, distinct. Chapman and Nam (1968) gave another statistic for testing linearity. It is, though, not of the form (10.51), and thus not readily amenable to efficient exact computation.

10.7 Other Sampling Designs

Besides the prospective column margins fixed (CMF) design, unordered and ordered $2 \times K$ tables also occur in other types of studies. They include the row margins fixed product multinomial design, the total fixed multinomial design, Poisson designs with count or incidence density data, case-control studies, binary outcome K and treatments randomized trial, two treatment randomized trials with a K level outcome and the inverse sampling product negative binomial study. Their statistical models depend on whether the column factor is ordered or not, and on other aspects of sampling and allocation.

In this section, we consider some more designs, and give relevant logit models for nominal and ordinal scale data. The presentation is less detailed than before. Data examples and other elaborations are given in the exercises.

10.7.1 Unordered Tables

The first case we consider is the counts with nothing fixed (CNF) design. Counts of an event of interest are recorded within a $2 \times K$ layout. Assuming them as $2K$ independent Poisson variates, the mean count in cell (i,k), λ_{ik}, $i = 0, 1; k = 1, \ldots, K$, is modeled as

$$\ln(\lambda_{ik}) = \mu + \alpha_k + (\beta + \gamma_k)i \qquad (10.59)$$

where $\alpha_1 \equiv \gamma_1 \equiv 0$. To assess the relationship, if any, between the row and column factors (denoted respectively by the labels i and k), we test the hypothesis $H_{01} : \gamma_k = 0 \,\forall\, k$.

Second, consider the multinomial total fixed (MTF) design with nominal scale data. A random sample of subjects are classified by gender and one of K diseases. The overall total, N, is fixed by design. Let i be the gender indicator (0 = male, 1 = female), and π_{ik}, $i = 0, 1$, $k = 1, \cdots, K$, the probability of a person of gender i having disease k, with $\Sigma_{ik}\pi_{ik} = 1$. Assume that the observations follow a $2K$ category multinomial distribution. An equivalent logistic formulation then is

$$\ln\left(\frac{\pi_{ik}}{\pi_{01}}\right) = \alpha_k + (\beta + \gamma_k)i \qquad (10.60)$$

where $\alpha_1 \equiv \gamma_1 \equiv 0$. These restrictions ensure that the number of free parameters is $2K - 1$. We test the hypothesis of absence of association between row and column factors, that is, $H_{02} : \pi_{ik} = \pi_{i+}\pi_{+k} \,\forall\, i, k$ where π_{i+} and π_{+k} are the row and column marginal probabilities, respectively. The equivalent hypothesis for model (10.60) is $\gamma_k = 0 \,\forall\, k$.

Next, consider the row margins fixed (RMF) design. Subjects from two regions (row factor) are sampled and classified by a K level nominal (column) outcome. With the row margins fixed, and independence, we model the cell counts in each row as a K category multinomial.

Let $i \in \{0, 1\}$ indicate the region, and π_{*ik}, the probability of the kth outcome for region i. By definition, $\Sigma_k \pi_{*ik} = 1$. To compare the outcomes by region, we test $H_{03} : \pi_{*0k} = \pi_{*1k}$, $k = 1, \cdots, K$. In a logistic form, the relevant model is

$$\ln\left(\frac{\pi_{*ik}}{\pi_{*1k}}\right) = \alpha_k + \gamma_k i \qquad (10.61)$$

where $\alpha_1 \equiv \gamma_1 \equiv 0$. H_{03} is then equivalent to $\gamma_k = 0$, $k = 1, \cdots, K$.

In line with what we earlier observed about 2×2 and $2 \times 2 \times 2$ tables, the above models are related under conditioning. We state these relations in a schematic fashion below and leave the more detailed formulation to the reader:

- If we fix the overall count in the Poisson CNF design (10.59), we obtain the MTF model (10.60).
- If we fix the row margin totals in MTF design (10.60), we obtain the RMF model (10.61).
- If we fix the column margin totals in MTF design (10.60), we get the CMF model (10.21) which was anlayzed in §10.4 and §10.5.
- If we fix the row margins in the CMF design, or fix the column margins in the RMF design, we get the same $K - 1$ dimensional hypergeometric distribution.

OTHER SAMPLING DESIGNS

10.7.2 Ordered Tables

Now we turn to designs for an ordered $2 \times K$ table. First, consider the CNF design with an ordinal column factor, $x_k, k = 1, \cdots, K$. The counts are $2K$ independent Poisson variates with the mean count in cell (i,k), λ_{ik}, $i = 0, 1; k = 1, \cdots, K$. These are modeled as

$$\ln(\lambda_{ik}) = \mu + \alpha x_k + (\beta + \gamma x_k)i \tag{10.62}$$

Note that when $\gamma \neq 0$, the column factor effect differs with the level of the row factor. The parameter of interest here is γ.

Now examine the MTF design with two factors: one binary and one ordinal. Under the notation of model (10.62) and mutual independence, a logistic model for the $2K$ category multinomial is

$$\ln\left(\frac{\pi_{ik}}{\pi_{01}}\right) = \alpha(x_k - x_1) + (\beta + \gamma x_k)i \tag{10.63}$$

The parameter of interest again is γ.

Now consider $2 \times K$ data from an RMF design with an ordinal outcome. Using earlier notation, we employ the model:

$$\ln\left(\frac{\pi_{*ik}}{\pi_{*1k}}\right) = (\alpha + \gamma i)(x_k - x_1) \tag{10.64}$$

which is also written as

$$\ln\left(\frac{\pi_{*ik}}{\pi_{*i,k-1}}\right) = (\alpha + \gamma i)(x_k - x_{k-1}) \tag{10.65}$$

If we set $x_k = k$, the model converts to

$$\ln\left(\frac{\pi_{*ik}}{\pi_{*i,k-1}}\right) = \alpha + \gamma i \tag{10.66}$$

which is the **adjacent categories model** for ordinal response. To state in schematic fashion, and in a manner similar to that for unordered tables, the above ordered data models are related as follows:

- If we fix the overall count in the Poisson CNF design (10.62), we obtain the MTF model (10.63).
- If we fix the row margin totals in the MTF design (10.63), we obtain the RMF model (10.64).
- If we fix the column margin totals in the MTF design (10.63), we get the CMF model (10.6) which was anlayzed in §10.2.
- If we fix the row margins in the CMF design or fix the column margins in the RMF design, we obtain the same one dimensional conditional distribution of the sufficient statistic for the parameter γ.

10.7.3 Conditional Analysis

The observations made above, for both unordered and ordered tables, apply in part to group or frequency matched case-control studies. Also, the unconditional and conditional distributions for the models and designs stated above are PBDs. For example, for the RMF design for an unordered table, the unconditional gp is

$$\prod_{i=0}^{1}\left(1+\sum_{k>1}\theta_k\phi_k^i\right)^{n_i} \qquad (10.67)$$

and for the RMF design for ordered tables, it is

$$\prod_{i=0}^{1}\left(1+\sum_{k>1}(\theta\phi^i)^{(x_k-x_1)}\right)^{n_i} \qquad (10.68)$$

and for the MTF unordered tables, it is

$$\left(1+\sum_{i,k>1}\theta_i\phi_k\psi_{ik}\right)^{N} \qquad (10.69)$$

Analysis of the models given in this section, especially for key parameters, is facilitated by the observations made above, and by the results stated below.

Theorem 10.7: For statistical analysis of a $2 \times K$ table:

- The respective unconditional (at the design level in question) score and LR statistics for the hypotheses, H_{01}, H_{02}, and H_{03} for the RMF, MTF and CNF unordered data designs, are identical to those derived from testing the equality of proportions in the CMF design in §10.2, namely $\mathcal{W}_X$ and $\mathcal{W}_L$, and follow, as before, a chisquare distribution with $K-1$ df at large sample sizes.
- The respective unconditional (at the design level in question) likelihood based large sample analyses of the parameter γ in the ordered data models (10.6), (10.63), (10.64) and (10.65) give identical mles, and also identical *p*-values for some tests.
- For the unordered and ordered data models noted in (i) and (ii), and where not fixed by design, the row and/or column sum totals are sufficient for the relevant nuisance parameters. The exact conditional distributions of the test statistic (in the unordered data case), or the sufficient statistic (in the ordered data case), are identical and are respectively (10.27) and (10.11).

Proof: Left as an exercise. □

Theorem 10.7 also extends to higher dimensional contingency tables. Such results simplify statistical analysis by showing that under a broad range of situations, for the purpose of computing exact conditional and asymptotic *p*-values, estimates and CIs, we can, to an extent, set aside how the data arose. Data analysis, in that sense, is design invariant.

INCIDENCE DENSITY DATA

But we should not overgeneralize from such results as many exceptions to the rule also exist. In the first place, these results are restricted to the logistic model for the odds ratio, and do not apply as broadly to relative risk or risk difference models. Even in the case of the odds ratio, the mle of cell probabilities, and the conditional score and LR statistics for related hypothesis differ from the above when we fix both margins. The conditional and unconditional score statistics for the CMF design for ordered data, for instance, differ by a small factor. Exact conditional analysis is then not affected but exact unconditional analysis is. In other settings large sample analyses are less affected by design than the exact methods.

The invariance of design to data analysis for ordered $2 \times K$ tables holds for models consistent with the general adjacent categories model. For other models, the situation is more varied. Consider, for example, the **proportional odds model** for ordinal response in a $2 \times K$ layout with $x_k = k$. Under the RMF design, let $\pi_{+ik} = \pi_{*i1} + \cdots + \pi_{*ik}$. Instead of using (10.67), let

$$\log\left(\frac{\pi_{+ik}}{1-\pi_{+ik}}\right) = \alpha + \gamma i \tag{10.70}$$

If the categories are reversed in this model, the mle of γ is unchanged, a property the adjacent categories model does not have. On the other hand, in (10.70), sufficient statistics do not exist. Conditioning is then harder to justify, and exact analysis, not a straightforward task. Asymptotic analysis of equivalent models under the MTF or RMF designs moreover does not necessarily produce the same inference on the parameters. Both asymptotic and exact power are also computed in distinct ways for the distinct designs even when the models and data analysis are related.

10.8 Incidence Density Data

Cohorts of healthy persons from K environments are tested for a genetic marker ($i = 0$ for marker absent, and $i = 1$ for marker present) and then monitored for the occurrence of a particular disease. Let τ_{ik} be the total follow up time, and λ_{ik}, the rate of disease occurrence per unit person time for marker status i and environment k.

For environment k, let $\mathcal{N}_k$ and $\mathcal{T}_k$ respectively be the random total number of cases, and the random number of cases with marker present. Their observed values are n_k and t_k. Further,

$$\mathcal{N} = (\mathcal{N}_1, \cdots, \mathcal{N}_K) \quad \text{and} \quad \boldsymbol{n} = (n_1, \cdots, n_K) \tag{10.71}$$

Also, define, for $k = 1, \cdots, K$,

$$\phi_k = \frac{\lambda_{1k}}{\lambda_{0k}} \quad \text{and} \quad \tau_k = \frac{\tau_{1k}}{\tau_{0k}} \tag{10.72}$$

Theorem 10.8: With the usual Poisson model assumptions for person time data under the above set up, and conditioning on the total number of cases within each environment,

$$P[\boldsymbol{\mathcal{T}} = \boldsymbol{t} \mid \boldsymbol{\mathcal{N}} = \boldsymbol{n}] = \prod_{k=1}^{K} \frac{c_{*k}(u)\phi_k^u}{f_{*k}(\phi_k)} \tag{10.73}$$

where

$$c_{*k}(u) = \binom{n_k}{t_k} \tau_k^u \quad \text{and} \quad f_{*k}(\phi_k) = (1 + \tau_k \phi_k)^{n_k} \qquad (10.74)$$

Proof: Follows from the single environment case and independence. □

- The above formulation is similar with the stratified incidence density models of Chapter 9.
- Thereby, (10.73) forms the basis for testing conditional independence, analyzing a constant rate ratio, trend in the rate ratios or testing homogeneity of rate ratios for a $2 \times K$ setting.
- Further, these analyses follow the scheme we have laid out for a $2 \times K$ table of counts, the difference being that now we use $f_{*k}(\phi_k)$ instead of the binomial gp $f_k(\phi_k)$.

We give the relevant analysis details under an ordered and an unordered factor.

10.8.1 An Ordered Factor

Let x_k denote the level of a chemical X in environment k. In this case, we apply model (10.62) for λ_{ik} here.

$$\ln(\phi_k) = \beta + \gamma x_k \qquad (10.75)$$

Let $\mathcal{T} = \sum_k \mathcal{T}_k$ and $\mathcal{W} = \sum_k \mathcal{T}_k x_k$ with respective realized values t and w. The joint distribution of $(\mathcal{T}, \mathcal{W})$ given the column totals is

$$P[\mathcal{T} = t, \mathcal{W} = w \mid \mathcal{N} = n] = \frac{c_*(t, w) \theta^t \phi^w}{f_*(\theta, \phi)} \qquad (10.76)$$

where $\theta = \exp(\beta)$, $\phi = \exp(\gamma)$ and

$$f_*(\theta, \phi) = \prod_{k=1}^K (1 + \tau_k \theta \phi^{x_k})^{n_k} \qquad (10.77)$$

and $c_*(t, w)$ is the coefficient of the term of $f_*(\theta, \phi)$ with the exponent of θ equal to t and that of ϕ equal to w, as in (10.10). To analyze γ, we use the distribution of $\mathcal{W}$ obtained from (10.76) by further conditioning on $\mathcal{T}$. This yields exact, mid-p and score evidence functions, and associated analytic entities in the usual manner.

Another option is to perform semiconditional analysis based on distribution (10.76). Following the derivation in §10.12, the score statistic for γ based on this is

$$\mathcal{W}_{*s} = \sigma^{-2} \left\{ \sum_k (t_k - n_k \pi_k) x_k \right\}^2 \qquad (10.78)$$

where

$$\pi_k = \frac{\tau_k \exp(\beta + \gamma x_k)}{1 + \tau_k \exp(\beta + \gamma x_k)} \qquad (10.79)$$

INCIDENCE DENSITY DATA

and

$$\sigma^2 = \sigma_{22}^2 - \sigma_{12}^2 \sigma_{11}^{-2} \tag{10.80}$$

with the σ's are as given in §10.12 with the further proviso that π_k is given by (10.79) and evaluated at the unrestricted mle for β or θ. Denote this by $\hat{\theta}_r = \exp(\hat{\beta}_s)$. Then it is the solution to the equation

$$\sum_{k=1}^{K} \frac{n_k \tau_k \theta}{[1 + \tau_k \theta]} = t \tag{10.81}$$

The iterative method of Clayton (1982) described in (4.73) solves (10.81) efficiently.

In relation to the CMF model in §10.2, we see that the unconditional score statistic (10.78) reduces to (10.17) when $\tau_k = \tau$ for all k. However, the simple relation between the conditional and unconditional score statistics stated there does not hold here.

10.8.2 Unordered Factor

Now suppose there is no ordering between the k environments. Then we apply the general model (10.21). Suppose we need to test for the homogeneity of the rate ratio across the K environments, that is to test $\gamma_k = 0$, $k = 1, \cdots, K$. Fixing the totals in each environment, let a relevant test statistic be of the form:

$$\mathcal{W}_* = \sum_{k=1}^{K} w_{*k}(\mathcal{T}_k) \tag{10.82}$$

With $\phi_k = \exp(\gamma_k)$, and for fixed values of ϕ_k, $k = 1, \cdots, K$, and the usual notation otherwise, the gp for the distribution of $(\mathcal{T}, \mathcal{W}_*)$ is

$$f_*(\theta, \boldsymbol{\phi}, \psi) = \prod_{k=1}^{K} \left\{ \sum_{u=0}^{n_k} c_{*k}(u) \theta^u \phi_k^u \psi^{w_k(u)} \right\} \tag{10.83}$$

where ψ is a dummy parameter. This polynomial is obtained in a way analogous to that done earlier for a $2 \times K$ table of counts, and exact and asymptotic analyses are then performed in a similar fashion.

A relevant statistic for homogeneity is the score statistic obtained from conditioning on the number of cases in each environment. We can show that this is

$$\mathcal{W}_{*X} = \sum_{k=1}^{K} \frac{[t_k + (t_k - n_k)\tau_k \hat{\theta}_r]^2}{n_k \tau_k \hat{\theta}_r} \tag{10.84}$$

where $\hat{\theta}_r$ is the solution to equation (10.81). $\mathcal{W}_{*X}$ reduces to the chisquare statistic for testing homogeneity, (10.35), when $\tau_k = \tau$ for all k. The LR and other evidentiary statistics are constructed in the usual way.

10.8.3 An Example

Example 10.3: Consider a cohort study in four work environments, labeled A, B, C and D, in which persons classified by the presence or absence of a genetic marker were followed. The outcome of interest was onset symptoms of a chronic disease. Hypothetical data from the study are in Table 10.7. Assume that the exposure levels of the chemical suspected to induce the disease in environments A, B, C and D are given by $0, 1, 2, 3$, respectively.

Table 10.7 *Cases by Genetic Marker and Environment*

Marker		A	B	C	D
Absent	Cases	0	1	0	1
	PY	1.6	1.6	1.0	0.8
Present	Cases	2	3	4	2
	PY	1.2	0.8	0.6	0.4

Note: Hypothetical data; PY = (person years)/1000.

For these data, we have $\tau_1 = 0.75$; $\tau_2 = 0.5$; $\tau_3 = 0.6$; and $\tau_4 = 0.5$. Also the four environment level gps of the form (10.74) are

$$f_{*1}(\phi_1) = (1 + 0.75\phi_1)^2 \qquad f_{*2}(\phi_2) = (1 + 0.50\phi_2)^4$$
$$f_{*3}(\phi_3) = (1 + 0.60\phi_3)^4 \qquad f_{*4}(\phi_4) = (1 + 0.50\phi_4)^3$$

With $x_1 = 0$; $x_2 = 1$; $x_3 = 2$; and $x_4 = 3$, we use model (10.75) to analyze these data. The observed sufficient statistics are $t = 11$ and $w = 17$. To get the conditional distribution of $\mathcal{W}$ given $\mathcal{T} = 11$, we substitute

$$\phi_k = \theta \phi^{x_k}$$

in the kth polynomial, multiply the polynomials and select the terms with the exponent of θ equal to 11. This polynomial is

$$f_*(1, \phi; 11, .) = 0.006834\phi^{15} + 0.022781\phi^{16} + 0.036830\phi^{17} + $$
$$0.039487\phi^{18} + 0.023794\phi^{19} + 0.012150\phi^{20} + 0.001013\phi^{21}$$

The conditional mid-*p* evidence functions for γ is shown in Figure 10.1. It gives a mid-*p*-value for testing $\gamma = 0$ to be 0.6723. The cmue for γ is -0.332 and the 95% mid-*p* CI is $(-2.275, 1.211)$. Under the linear model, evidence for an exposure effect is thus not persuasive. The wide horizontal spread of this function, however, implies a low level of precision associated with this conclusion.

The gp $f_*(1, \phi; 11, .)$ is used to compute the null mean and variance of the conditional distribution of $\mathcal{W}$. These, respectively, are 17.638 and 1.765. The conditional score statistic then is

AN INVERSE SAMPLING DESIGN

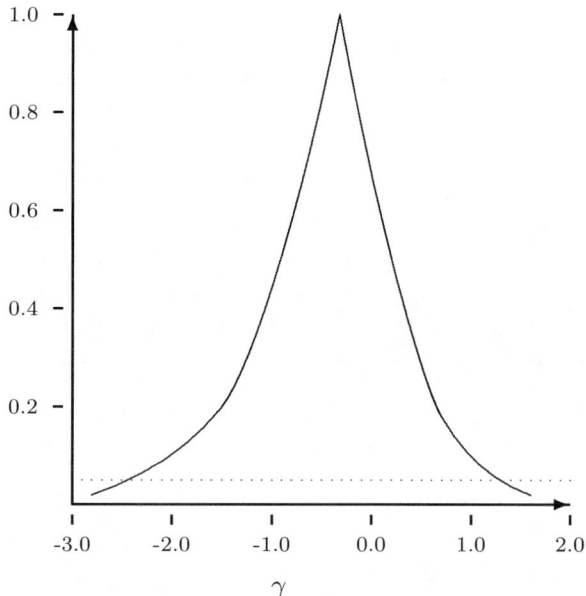

Figure 10.1 *Mid-p Evidence Function for Table 10.7 Data.*

$$\frac{(17 - 17.638)^2}{1.765} = 0.231$$

giving an asymptotic *p*-value equal to 0.6312.

Suppose we ignore the ordering between the environments. The homogeneity of the rate ratios is tested by first solving (10.81). This gives $\hat{\theta}_r = 9.843$. Substituting this in (10.84), we get $\mathcal{W}_{*X} = 7.43$. At 3 df, this gives a chisquare *p*-value equal to 0.059. We leave the exact computations to the reader.

10.9 An Inverse Sampling Design

A study compared the manifestation of an adverse effect in a standard therapy and two new therapies, respectively indicated by $k = 0, 1, 2$. Let π_k be the probability that a patient on treatment k will not have the effect. Recruitment for treatment k is continued until r_k, a fixed number of patients with the adverse effect, has been observed.

Let $\mathcal{T}_k$ be the random number of subjects **not** having the effect under treatment k. The total on treatment k is $t_k + r_k$. With the usual notation, $\mathcal{T}$ follows a product of three independent negative binomial distributions

$$P[\mathcal{T} = t] = \prod_{k=0}^{2} c_k(t_k)(1-\pi_k)^{r_k}\pi_k^{t_k} \qquad (10.85)$$

where

$$c_k(u) = \binom{u+r_k-1}{r_k-1} \qquad (10.86)$$

Consider now, for $k = 0, 1, 2$, the model

$$\ln(\pi_k) = \beta + \gamma_k \qquad (10.87)$$

with $\gamma_0 =$. Also let $\theta = \exp(\beta)$ and $\phi_k = \exp(\gamma_k)$. Then, $\theta = \pi_0$, and ϕ_k is the relative risk, compared to treatment 0, of **not** experiencing the effect on treatment k. To analyze the risks, we condition on the total number not experiencing the effect, or equivalently, the total sample size. Let $T = \sum_k \mathcal{T}_k$ with observed value equal to t. Also, with $\phi_0 = 1$, we define

$$f_k(\theta, \phi_k) = \sum_{u=0}^{\infty} c_k(u)\theta^u \phi_k^u \qquad (10.88)$$

Let $f(\theta, \phi_1, \phi_2)$ be the product of these polynomials. Then

$$P[\mathcal{T} = t \mid T = t] = \frac{c(t, t_1, t_2)\phi_1^{t_1}\phi_2^{t_2}}{f(1, \phi_1, \phi_2; t, ., .)} \qquad (10.89)$$

When multiplying the polynomials from (10.88), we consider only those terms in which the exponent of θ is less than or equal to t. The sample space of the conditional distribution of $(\mathcal{T}_1, \mathcal{T}_2)$ then is

$$\Omega = \{(u_1, u_2) : 0 \leq u_1, u_2, \leq t; 0 \leq u_1 + u_2 \leq t\} \qquad (10.90)$$

This distribution, the starting point for conditional analysis of the relative risks ϕ_1 and ϕ_2, is **not** a hypergeometric distribution.

Example 10.4: For illustration, consider a study with $r_0 = r_1 = r_2 = 3$ and $t_0 = 4, t_1 = 0, t_2 = 0$. Hence $t = 4 + 0 + 0 = 4$. The three gps (written up to the maximal needed exponent of θ) are

$$\begin{aligned}
f_0(\theta, 1) &= 1 + 3\theta + 6\theta^2 + 10\theta^3 + 15\theta^4 + \ldots \\
f_1(\theta, \phi_1) &= 1 + 3\theta\phi_1 + 6\theta^2\phi_1^2 + 10\theta^3\phi_1^3 + 15\theta^4\phi_1^4 + \ldots \\
f_2(\theta, \phi_2) &= 1 + 3\theta\phi_2 + 6\theta^2\phi_2^2 + 10\theta^3\phi_2^3 + 15\theta^4\phi_2^4 + \ldots
\end{aligned}$$

Multiplying these polynomials and selecting the terms with the exponent of θ equal to 4 gives

$$\begin{aligned}
f(1, \phi_1, \phi_2; 4, ., .) &= 15 + 30\phi_1 + 36\phi_1^2 + 30\phi_1^3 + 15\phi_1^4 + 30\phi_2 + 54\phi_1\phi_2 + 54\phi_1^2\phi_2 \\
&\quad + 30\phi_1^3\phi_2 + 36\phi_2^2 + 54\phi_1\phi_2^2 + 36\phi_1^2\phi_2^2 + 30\phi_2^3 + 30\phi_1\phi_2^3 + 15\phi_2^4
\end{aligned}$$

ADDITIONAL TOPICS

At observed $u = v = 0$, $c(4, u, v) = 15$, we have $f(1, 1, 1) = 525$. The probability based exact conditional *p*-value is $(15+15+15)/525 = 0.0857$. Under the usual both margins fixed analysis, the hypergeometric based probability method *p*-value for these data is $64/715 = 0.0895$.

The conditional and unconditional score and LR statistics for an unordered $2 \times K$ inverse sampling design are derived in the usual way.

Now consider an ordered $2 \times K$ inverse sampling design using the model

$$\ln(\pi_k) = \beta + \gamma x_k \tag{10.91}$$

where $0 < x_1 < \cdots < x_K$. The unconditional score statistic for trend in the relative risk, that is, for testing $\gamma = 0$ is

$$\mathcal{W}_{**} = \frac{R s_{**}^{-2}}{t(t+R)} \left\{ \left(Rw - t \sum_k r_k x_k \right)^2 \right\} \tag{10.92}$$

where

$$s_{**}^2 = R \sum_k r_k x_k^2 - \left(\sum_k r_k x_k \right)^2 \tag{10.93}$$

This resembles the score statistic for trend in the CMF design in form but is distinct in substance. Further details are left as an exercise.

10.10 Additional Topics

In this section, we consider four topics relating to a $2 \times K$ table: (i) other models for trend analysis; (ii) exact power analysis; (iii) pair-wise comparisons of treatments; and (iv) general recommendations.

10.10.1 Aspects of Trend Analysis

In addition to those noted in §10.2, other key issues in relating a binary outcome to an ordinal or a continuous factor include: (A) What model should one use? (B) How is the factor variable to be scored? (C) Should it be categorized? If so, how? Below we briefly reflect on these matters.

Many models are applied to analyze ordered $2 \times K$ table data. In addition to the logit model, they include the probit, linear, one hit and extreme value models. With $\pi(x)$ the probability of response at dose x, several of these models are of the general form

$$\pi(x) = h(\beta + \gamma x) \tag{10.94}$$

In the case of the logit model

$$h(z) = \frac{\exp(z)}{1 + \exp(z)} \tag{10.95}$$

In the probit model, $h(z)$ is the normal cumulative distribution function, and in the one hit model it is

$$h(z) = 1 - \exp(-z) \qquad (10.96)$$

Many books deal with models for dose-response, and give asymptotic methods for analysis. Methods that are not explicitly model based, such as isotonic regression, are also used in this context.

Among all, the logit model is more frequently used today. In the applied literature. it has replaced the previously popular probit model. Several reasons account for this change. The logit model is considered 'less parametric' in nature, is easier to analyze and interprete, and is also applicable to case-control studies where other models do not give a valid estimate. For a number of alternatives, it yields robust results. The CAM test we drived from it is also derivable from other models. It also has good general power properties. Importantly, unlike other models, the logit model allows exact conditional analysis.

Yet, the logit model may not reflect the underlying biologic relation between the factor and outcome, and is often used by convention and convenience. Empirical studies indicate that apart from certain extreme conditions, broad interpretation of the data is generally not drastically affected by the choice of a particular model. In that sense, the logit model is as good as any other.

Analysis based on the linear logit model, like the CAM test for trend, has at times been interpreted in a misleading fashion. Evidence for a factor effect is then taken as evidence for the linearity of the effect, that is, of the validity of the model. Yet, the CAM test assumes the linear model. Performing an exact analysis does not alter this basic fact. Therefore, separate tests for linearity, like those in §10.6, need to be done for this purpose. Like tests for the homogeneity of the odds ratios in several 2×2 tables, the power of such tests tends to be low, especially with sparse data. A quadratic effect logit model, described in §10.11, also gives a less restrictive context for modeling dose-response data.

Another related issue is the choice of factor score. In addition to the actual values, or mid-points of categorized data groups, scores such as

$$\text{Linear Scores:} \qquad x_k = k \qquad (10.97)$$
$$\text{Quadratic Scores:} \qquad x_k = k^2 \qquad (10.98)$$
$$\text{Exponential Scores:} \qquad x_k = 2^k \qquad (10.99)$$

are also used. These scores are independent of the observed data. Rank or midrank scores, which are data based, are also used for such data. However, their use in trend or dose-response data requires appropriate theoretical justification. They also turn estimation and CI construction into a conceptually complex issue. In our view, their application in discrete data problems context is quite limited; and we do not consider them in this text.

Scores are best chosen on the basis of a biologic model, or indications from earlier studies. When such indicators are unavailable, midpoint scores or linear scores, the basis for the CAM test, are often used. A seminal paper by Graubard and Korn (1987) recommends using the linear or natural scores, especially when there is uncertainty. This is also one of the two proposals of Moses, Emerson and Hosseini (1984).

A related matter is whether a continuous factor should be categorized, a practice thought

ADDITIONAL TOPICS

to enhance interpretability and increase power. This view is not accepted by several authors. It assumes homogeneity of effect within a category, and it may distort the relationship between the outcome and the factor. Use of splines and polynomial regression are advocated as alternatives. The relevant references in §10.13 provide an introduction to the issues involved.

10.10.2 Exact Power

Formulae for approximate power of approximate tests for a $2 \times K$ table are widely available. We consider the exact power of conditional tests such data. Under the CMF product binomial design and model (10.6) for ordered data, we test $H_0 : \phi = \phi_0$ against the $H_1 : \phi \neq \phi_0$ at level α_* $(0 < \alpha_* < 1)$ with a test that fixes the margins of the table.

We continue the usual notation for the $2 \times K$ table; in particular, $f(\theta, \phi)$ is the bivariate polynomial (10.9). Further, let the conditional rejection region for the test be $\Omega_c(t; \phi_0, \alpha_*)$. With the definitions in §7.4, the conditional power $P[\mathcal{W} \in \Omega_c(t; \phi_0, \alpha_*) \mid \mathcal{T} = t; \phi]$ of this test is

$$\Pi_c(t; \phi, \phi_0, \alpha_*) = \frac{\eta(t; \phi, \phi_0, \alpha_*)}{f(1, \phi; t, .)} \tag{10.100}$$

where

$$\eta(t; \phi, \phi_0, \alpha_*) = \sum_{v \in \Omega_c(t; \phi_0, \alpha_*)} c(t, v) \phi^v \tag{10.101}$$

Next we need the marginal distribution of $\mathcal{T}$ at specified values of θ and ϕ. $\mathcal{T}$ is the sum K binomial variates with odds equal to $\theta_k = \theta \phi^{x_k}$. To determine its distribution, let ψ be a dummy parameter and let the product in the lhs side of the equation below be expanded as

$$\prod_{k=1}^{K} (1 + \theta_k \psi)^{n_k} = \sum_{v=1}^{N} d(v; \theta, \phi) \psi^v \tag{10.102}$$

With specified values of θ and ϕ, we can then show that

$$P[\mathcal{T} = t; \theta, \phi] = \frac{d(t; \theta, \phi)}{f(\theta, \phi)} \tag{10.103}$$

Theorem 10.9: The exact unconditional power of the conditional test of $\phi = \phi_0$ versus $\phi \neq \phi_0$ is

$$\Pi(\theta, \phi, \phi_0, \alpha_*) = \frac{1}{f(\theta, \phi)} \sum_{t=0}^{N} \Pi_c(t; \phi, \phi_0, \alpha_*) d(t; \theta, \phi) \tag{10.104}$$

Proof: Follows from combining the preceding results. □

Computing exact power is thereby a two step exercise. First, we compute, for each t, the marginal probability (10.103). Then, for each $t = 0, \cdots, N$, we identify the conditional rejection region, compute conditional power and add its contribution to the sum in (10.104). As this has to be done $(1 + N)$ times, this is also an intensive exercise.

Formulations of exact power for homogeneity tests and tests for linearity are done in a similar way. There are more nuisance parameters in these cases, making the task more complex. The principles of computation, however, are the same. Exact power is computed in a similar way for the corresponding asymptotic tests. These formulations also extend to case-control, incidence density, multinomial, RMF and inverse sampling designs, both for ordered and unordered factors.

10.10.3 Pair-wise Comparisons

Suppose in a clinical trial comparing K treatments an overall test for homogeneity shows a significant difference. Which treatment(s) differs from the others? The need for such pair-wise comparisons arises in many circumstances. And a large body of works on the subject exists, most of which deals with asymptotic tests. See Piegorsch and Bailer (1997), Chapter 5, for a good introduction to the subject, especially in relation to discrete data. In this section, we deal with exact pair-wise comparisons for a few binary response data situations.

A direct way of dealing with the question set above is to compare the treatments two at a time. As the number of comparisons are $K(K+1)/2$, this strategy raises the actual type I error rate in a dramatic fashion. The **Bonferroni method** is a crude way of controlling this error. Each pair-wise comparison in this case is done at an adjusted nominal level equal to the overall nominal level divided by the total number of comparisons (see Problem 1.40 for a justification). This approach, though, is highly conservative and has low power.

Consider a specific pair-wise comparison situation. Experiments to screen chemicals for cancer or mutation causing potential at times test groups of animals that are given one of a number of chemicals with the same control group. Suppose the groups are indexed $k = 0, \cdots, K$ with group 0, the control group. Instead of comparing each treated group with the control group, we prefer a global test indicative of an effect.

To develop this test, for $k \geq 0$, let π_k be the probability of response, n_k, the group size, and t_k, the number of responders in the kth group. Otherwise, we have the usual notation for the $2 \times (K + 1)$ table, with the modifications indicative of the control group. Also define:

$$\theta = \frac{\pi_0}{(1 - \pi_0)} \quad \text{and} \quad \phi_k = \frac{\pi_k/(1 - \pi_k)}{\pi_0/(1 - \pi_0)} \qquad (10.105)$$

One global hypothesis is: Does at least one chemical induce the effect in question? In this case, we test $H_0 : \phi_k = 1, k = 1, \cdots, K$ against the alternative $H_1 : \phi_k \neq 1$ for at least one k. Or we ask: Do all the chemicals induce the effect in question? Then we test $H_0 : \phi_k = 1, k = 1, \cdots, K$ against the alternative $H_2 : \phi_k \neq 1$ for all k.

Suppose $w_k(t_k, t_0) \geq 0$ is a statistic we use for a pair-wise comparison of the kth group with control, that is, for testing $\phi_k = 1$ versus $\phi_k \neq 1$. This may, for example, be the 2×2 conditional score statistic. For testing H_0 versus H_1, a relevant global statistic then is

$$\mathcal{W}_x(\boldsymbol{t}; t_0) = \max\{w_k(t_k, t_0), k = 1, \cdots, K\} \qquad (10.106)$$

And for testing H_0 versus H_2, a relevant global statistic is

ADDITIONAL TOPICS

$$\mathcal{W}_n(\boldsymbol{t}; t_0) = \min\{ w_k(t_k, t_0),\ k = 1, \cdots, K \} \tag{10.107}$$

Under the product binomial design, and with $\mathcal{T} = \mathcal{T}_1 + \cdots + \mathcal{T}_K$, we know that $\mathcal{T}_+ = \mathcal{T} + \mathcal{T}_0$ is sufficient for θ. Dropping the subscript in $\mathcal{W}$, the exact conditional p-value for either of the above alternatives is given by

$$P[\mathcal{W} \geq w \mid \mathcal{T}_+ = t_+; H_0] \tag{10.108}$$

The task then is to determine the exact conditional distribution of $\mathcal{W}$.

****

Each component of $\mathcal{W}$ depends on the value of $\mathcal{T}_0$. Let us at first fix this value. Using the product binomial probabilities for the K treated groups, we can then show that

$$P[\mathcal{T} = t, \mathcal{W} = w \mid \mathcal{T}_0 = t_0] = \frac{c(t, w, t_0; \boldsymbol{\phi})\theta^t}{g(\theta, \phi)} \tag{10.109}$$

where

$$g(\theta, \phi) = \prod_{k \geq 1}(1 + \theta\phi_k)^{n_k} \tag{10.110}$$

and

$$c(t, w, t_0; \boldsymbol{\phi}) = \sum_{\boldsymbol{u}} \left[\prod_{k=1}^{K} c_k(u_k)\phi_k^{u_k} \right] \tag{10.111}$$

where the summation is over the values of $\boldsymbol{u} = (u_1, \cdots, u_K)$ that satisfy $0 \leq u_k \leq n_k$, $k = 1, \cdots, K$, $u_1 + \cdots + u_K = t$ and $w(\boldsymbol{u}, t_0) = w$.

Next, consider the relation

$$P[\mathcal{W} = w \mid \mathcal{T}_+ = t_+]P[\mathcal{T}_+ = t_+] = P[\mathcal{W} = w, \mathcal{T}_+ = t_+] \tag{10.112}$$

The rhs of the above is equal to

$$\sum_{u=0}^{n_0} P[\mathcal{T} = t_+ - u, \mathcal{W} = w \mid \mathcal{T}_0 = u]P[\mathcal{T}_0 = u] \tag{10.113}$$

Applying (10.109) to this, it becomes

$$\frac{g_+(\theta, \boldsymbol{\phi})}{\theta^{t_+}} \sum_{u=0}^{n_0} \binom{n_0}{u} c(t_+ - u, w, u; \boldsymbol{\phi}) \tag{10.114}$$

where

$$g_+(\theta, \boldsymbol{\phi}) = g(\theta, \phi)(1 + \theta)^{n_0} \tag{10.115}$$

Next we consider the marginal probability $P[\mathcal{T}_+ = t_+]$. Let $N_+ = N + n_0$. Suppose at specified values of ϕ_k, the product (10.115) is written in an expanded form as

$$g_+(\theta, \boldsymbol{\phi}) = \sum_{u=0}^{N_+} d_+(u; \boldsymbol{\phi})\theta^u \qquad (10.116)$$

Then we can readily show that

$$P[\mathcal{T}_+ = t_+] = \frac{g_+(\theta, \boldsymbol{\phi})}{d_+(t_+; \boldsymbol{\phi})\theta^{t_+}} \qquad (10.117)$$

Theorem 10.10: The exact conditional distribution $P[\mathcal{W} = w \mid \mathcal{T}_+ = t_+]$ required for computing the exact p-value (10.108) is given by

$$\frac{1}{d(t_+; \boldsymbol{\phi})} \sum_{u=0}^{n_0} \binom{n_0}{u} c(t_+ - u, w, u; \boldsymbol{\phi}) \qquad (10.118)$$

Proof: Follows from combining the above arguments. □

At the null value $\phi = 1$, we have

$$d(t_+; \mathbf{1}) = \binom{N_+}{t_+}$$

****

The task now is to compute the coefficients of the form $c(v, w, u; \boldsymbol{\phi})$. For fixed values of ϕ_k, $k \geq 1$, and with ψ a dummy parameter, we consider the bivariate polynomials

$$f_k(\theta, \psi, \phi_k; t_0) = \sum_{u=0}^{n_k} \binom{n_k}{u} \theta^u \phi_k^u \psi^{w_k(u, t_0)} \qquad (10.119)$$

Now consider the polynomial "product" defined in a specific manner with respect to the coefficient of ψ. We define the product

$$f_k(\theta, \psi, \phi_k; t_0) \otimes f_{k+1}(\theta, \psi, \phi_k; t_0) \qquad (10.120)$$

In this product everything is done in the usual way except when we "multiply" the term ψ^a with ψ^b. In that case, with the maximizing statistic $\mathcal{W}_x$, the product is given as

$$\psi^a \otimes \psi^b = \psi^{\max(a, b)}$$

and with the minimizing statistic $\mathcal{W}_n$, the product is

$$\psi^a \otimes \psi^b = \psi^{\min(a, b)}$$

ADDITIONAL TOPICS

Then we perform the product below done with this modification.

$$f(\theta, \psi, \boldsymbol{\phi}; t_0) = \prod_{k=1}^{K} f_k(\theta, \psi, \phi_k; t_0) \qquad (10.121)$$

Then we can show that if the above product is written in an expanded form as a bivariate polynomial in θ and ψ, then it is equal to

$$\sum_{(u,v)} c(u, v, \boldsymbol{\phi}; t_0) \theta^u \psi^v \qquad (10.122)$$

This provides the coefficients needed for (10.118).

The computation of exact p-values for the global tests arising in pair-wise comparisons of K response probabilities with a control is then, with the modifications indicated, a polynomial multiplication problem. Some additional points of relevance here are:

- These formulations readily extend to a comparison of one exposed group with several control groups. Epidemiological studies in which several distinct control groups are needed for improved control of confounding are relevant here.
- Incidence density designs where several Poisson rates are compared to a control rate, or one Poisson rate is compared with several control rates, may also use these formulations. Both one- and two-sided comparisons can be done in this setting.
- A comparison of multinomial probabilities can also be done under this approach.
- Controlling the type I error rates when comparing several prognostic factors in an analysis of clinical trial data is another area where it can be applied.
- The references given in §10.13 elaborate on the specific aspects of these matters.
- For the type of problems stated above and, in general, performing exact pair-wise comparisons remains a promising area for research.

10.10.4 Recommendations

The accuracy of asymptotic tests, especially the chisquare test, for homogeneity in tables with small cell counts or total has been a long standing concern. Several rules stating when to avoid asymptotic tests exist. The best known is Cochran's rule, which warns against the use of the asymptotic chisquare test if any expected value is less than 1.0, or if more than 5% of cells have expected values less than 5.0 (Cochran 1954b). A simpler form suspects the asymptotic p-value when any expected cell value is less than 5.0. Another version, specific to a $2 \times K$ table, suspects the approximation when either $N/2K$ or $N/(K-1)$ is less than 5, or at times 10. Other versions of such rules also exist.

These rules were devised in an era when exact analysis was generally infeasible. Today efficient algorithms and available software make exact analysis feasible for even relatively large tables. In our view, these rules are largely redundant now. The exact method, with the FH or chisquare statistic, should be used whenever possible. Among the asymptotic tests, the score or chisquare test is a robust alternative.

Moses, Lincoln and Hosseini (1984) noted the tendency, in the medical literature, to use the tests for unordered data even when one factor is ordered. This practice reduces power. Appropriate ordered data tests are generally the better alternatives.

For trend analysis of the logit model, we recommend reporting the mid-p-value, the cmue and mid-p CI. The asymptotic conditional score statistic based analysis is a reasonable alternative. This means using the asymptotic CAM test with the cmue and score based CI. In either case, plot the evidence function when possible. This applies to incidence density studies as well.

Exact or asymptotic goodness-of-fit tests for the linear model should be performed on a regular basis. Nonlinear models like the quadratic logit model (see below) need consideration. Also, in general, categorization is to be done with care, and collapsing small sample size categories into a single category should be avoided.

10.11 Extensions

This section briefly overviews generalizations of the models for a $2 \times K$ table we have dealt with thus far.

10.11.1 A Quadratic Model

Under the notation of §10.2, we model the probability of response at dose x in a quadratic fashion as

$$\ln\left(\frac{\pi(x)}{1-\pi(x)}\right) = \beta + \gamma x + \rho x^2 \tag{10.123}$$

In the product binomial design, the respective sufficient statistic for β, γ and ρ are T, W and $Q = \sum_k T_k x_k^2$. The joint distribution of (T, W, Q) is a PBD with the gp

$$f(\theta, \phi, \psi) = \prod_{k=1}^{K} \left(1 + \theta \phi^{x_k} \psi^{x_k^2}\right)^{n_k} \tag{10.124}$$

where $\theta = \exp(\beta)$, $\phi = \exp(\gamma)$ and $\psi = \exp(\rho)$. Exact conditional and large sample analytic methods for γ and ρ are then derived from (10.124). For example, to check for a dose effect, we test $\gamma = \rho = 0$ by extracting terms of this polynomial with the exponent θ equal to the observed value of its sufficient statistic. Analysis of γ and ρ is done individually by further conditioning. Exact analysis for a quadratic model for incidence density and other designs is done in an analogous fashion.

A quadratic curve allows a broader range of shapes of the relationship between the factor and outcome, and overcomes in part the limitations associated with the linear logit model. A test of $\rho = 0$ is a test of the validity of the linear model in this context.

10.11.2 Several $2 \times K$ Tables

Consider a binary response design with J ordered $2 \times K$ strata. Within each stratum, the column totals are fixed. π_{jk} is the probability of response at dose x_{jk} in stratum j. Note that $0 < x_{j1} < \cdots < x_{jK}$. Consider the model

$$\ln\left(\frac{\pi_{jk}}{1-\pi_{jk}}\right) = \beta_j + \gamma x_{jk} \tag{10.125}$$

EXTENSIONS

$\mathcal{T}_j = \sum_k \mathcal{T}_{jk}$ is sufficient for β_j and $\mathcal{W} = \sum_j \sum_k \mathcal{T}_{jk} x_{jk}$ is so for γ. The gp for the joint distribution of $(\mathcal{T}_1, \ldots, \mathcal{T}_J, \mathcal{W})$ is

$$f(\theta_1, \cdots, \theta_J, \phi) = \prod_{j=1}^{J} \prod_{k=1}^{K} \left(1 + \theta_j \phi^{x_{jk}}\right)^{n_{jk}} \quad (10.126)$$

where $\theta_j = \exp(\beta_j)$ and $\phi = \exp(\gamma)$. This is the common linear trend effect model for several $2 \times K$ tables, a counterpart of the common odds ratio model for several 2×2 tables. Exact analysis of the parameter γ proceeds as follows. First we remove (as in §10.2) the parameter β_j by conditioning on the row totals for stratum j. This gives the conditional distribution of the jth stratum component, $\mathcal{W}_j$, of $\mathcal{W}$. Then (as for several 2×2 tables), we multiply the J stratum specific conditional polynomials to get the conditional distribution of $\mathcal{W}$. Large sample analysis proceeds by using the unconditional and conditional likelihoods, with the derivation of the test statistics being slightly more involved.

We may also consider the more general model

$$\ln\left(\frac{\pi_{jk}}{1 - \pi_{jk}}\right) = \beta_j + \gamma_j x_{jk} \quad (10.127)$$

and perform a test for homogeneity of the trend parameter, namely, a test for $\gamma_j = \gamma$ for all j. In principle, upon conditioning on the row margins of each table, the test is done in the same way as for several 2×2 tables; the details are more involved.

For unordered J $2 \times K$ tables, we can use the model

$$\ln\left(\frac{\pi_{jk}}{1 - \pi_{jk}}\right) = \beta_j + \gamma_{jk} \quad (10.128)$$

where $\gamma_{j1} = 0$ for all j. To test the independence of the row and column factors in all the strata, we test $\gamma_{jk} = 0$ for all j, k. Large sample score and other tests are avaliable. Formulation of exact distributions of the test statistics is done using a polynomial formulation; exact computation, though, is a formidable task.

10.11.3 Logistic Regression

Now consider a general binary response model for n_j units with a K dimensional covariate vector $\boldsymbol{x}_j = (x_{j1}, \ldots, x_{jK})$, $j = 1, \ldots, J$. Let $\pi(\boldsymbol{x}_j)$ be the probability of response at covariate $\boldsymbol{x}_j$. Consider the model

$$\ln\left(\frac{\pi(\boldsymbol{x}_j)}{1 - \pi(\boldsymbol{x}_j)}\right) = \beta + \gamma_1 x_{j1} + \cdots + \gamma_K x_{jK} \quad (10.129)$$

$\mathcal{T} = \sum_j \mathcal{T}_j$ is sufficient for β and $\mathcal{W}_k = \sum_j \mathcal{T}_j x_{jk}$ is so for γ_k. The gp for the joint distribution of $(\mathcal{T}, \mathcal{W}_1, \cdots, \mathcal{W}_K)$ is

$$f(\theta, \phi_1, \ldots, \phi_K) = \prod_{j=1}^{J} \left(1 + \theta \phi_1^{x_{j1}} \cdots \phi_K^{x_{jK}}\right)^{n_j} \quad (10.130)$$

where $\theta = \exp(\beta)$ and $\phi_k = \exp(\gamma_k)$. Inference on γ_k for any given k is based on the terms of this polynomial in which the exponents of θ and ϕ_i, $i \neq k$, are fixed. This gives a univariate conditional PBD on which exact conditional and large sample conditional analysis is done in the usual manner. Unconditional large sample analysis of (10.129) is described in many standard text books. Exact conditional analysis was formulated by Cox (1958) and efficiently implemented by Hirji, Mehta and Patel (1987).

10.12 Derivation

In this section, we derive the unconditional score statistic for γ in model (10.6) under the product binomial design. This lays a framework from which the other statistics given in this chapter are derived. The vector of first partial derivatives of the unconditional likelihood based on (10.10), ℓ_u, is

$$\mathbf{\Delta}_u = \left(\frac{\partial \ell_u}{\partial \beta}, \frac{\partial \ell_u}{\partial \gamma} \right)' = \begin{pmatrix} t - \mu_1 \\ w - \mu_2 \end{pmatrix} \tag{10.131}$$

where

$$\mu_1 = \sum_k n_k \pi_k \quad \text{and} \quad \mu_2 = \sum_k n_k x_k \pi_k \tag{10.132}$$

The unconditional mles for β and γ obtain by putting the components of (10.131) to zero and solving the equations in an iterative manner. The second partial derivatives of ℓ_u, needed for this purpose, are

$$\frac{\partial^2 \ell_u}{\partial \beta^2} = -\sigma_{11}^2; \quad \frac{\partial^2 \ell_u}{\partial \gamma^2} = -\sigma_{22}^2; \quad \frac{\partial^2 \ell_u}{\partial \beta \partial \gamma} = -\sigma_{12} \tag{10.133}$$

where

$$\sigma_{11}^2 = \sum_k n_k \pi_k (1 - \pi_k) \tag{10.134}$$

$$\sigma_{12} = \sum_k n_k x_k \pi_k (1 - \pi_k) \tag{10.135}$$

$$\sigma_{22}^2 = \sum_k n_k x_k^2 \pi_k (1 - \pi_k) \tag{10.136}$$

The information matrix is therefore

$$\mathbf{I}_u = \sum_k n_k \pi_k (1 - \pi_k) \mathbf{I}_{uk} \tag{10.137}$$

where

$$\mathbf{I}_{uk} = \begin{bmatrix} 1 & x_k \\ x_k & x_k^2 \end{bmatrix} \tag{10.138}$$

RELEVANT LITERATURE

The inverse of $\mathbf{I}_u$ is the variance covariance matrix, $\boldsymbol{\Sigma}_u$. The unconditional score statistic, $\mathcal{W}_{us}$, is

$$\mathcal{W}_{us} = \boldsymbol{\Delta}_u' \boldsymbol{\Sigma}_u \boldsymbol{\Delta}_u \qquad (10.139)$$

Simplifying this we get

$$\mathcal{W}_{us} = \sigma^{-2} \left\{ \sum_k (t_k - n_k \pi_k) x_k \right\}^2 \qquad (10.140)$$

where

$$\sigma^2 = \sigma_{22}^2 - \sigma_{12}^2 \sigma_{11}^{-2} \qquad (10.141)$$

At $\gamma = 0$, and using the restricted mle for β, this gives the unconditional score statistic $\mathcal{W}_{us}$ given in (10.17).

Estimates for the variances and covariance of the mles obtain from the respective components of the variance covariance matrix. These are used for computing asymptotic 95% CIs.

This derivation is readily extended to the incidence density and inverse sampling designs with an ordered factor. Derivation for score statistics when there is an unordered factor follows similar principles. The LR statistics in these cases are straightforward to derive.

10.13 Relevant Literature

Conventional analysis of ordered and unordered $2 \times K$ tables is found in general categorical data analysis books. For an overview, see Bishop, Fienberg and Holland (1975), Breslow and Day (1980), Fleiss (1981), Andersen (1982), Gart et al. (1986), Collet (1991), Agresti (1996), Williams (2005) and Newman (2001). Lancaster (1969) details the properties of the chisquare distribution. Agresti (1992, 2001) and Mehta (1994) overview the exact analysis of ordered and unordered contingency tables.

Among the foundational papers for the ordered $2 \times K$ models we consider are Yates (1948), Cochran (1954b), Armitage (1955), Cox (1958) and Mantel (1963). Exact analysis of logit models was outlined by Cox (1970); for actual examples, see Mehta and Patel (1995). The papers on exact computation are noted in Chapter 12.

For a discussion of categorization and interpretation of trend analysis, see Maclure and Greenland (1992), Greenland (1995a, 1995b) and Weinberg (1995). Graubard and Korn (1987) is a seminal paper on the choice of scores in the analysis of an ordered $2 \times K$ table. See also Moses, Emerson and Hosseini (1984). On $2 \times K$ tables with both nominal and ordered categories, see Gautam (2002).

Williams (2005) describes the large sample tests for checking linearity in dose-response data. Another test is in Chapman and Nam (1968). Exact analysis was developed by Tang and Hirji (2002) who also gave two new exact tests. See also Tang (2000, 2001a).

Freeman and Halton (1951) extended the Fisher–Irwin exact test to an $R \times C$ table. Zelterman (1999) and Cressie and Read (1984) gave the tests bearing their names noted in §10.3. Papers on exact tests for such tables are noted in Chapter 12 and Chapter 13.

For comparisons of some tests for trend, see Tarone and Gart (1980), Portier and Hoel (1984) and Cohen and Sackrowitz (1992). Other papers also exist. Exact and asymptotic tests were compared by Hirji and Tang (1998). While the robustness of the asymptotic chisquare test in an unordered $2 \times K$ table was stressed by Lewontin and Felsenstein (1965), Haberman (1988) and Zelterman, Chan and Mielke (1995), among others, decry its use in a table with small expected values. Many views on this topic exist. Exact and large sample tests for sparse $2 \times K$ tables with large K, which occur in some types of gene studies, were done by Adams and Skopek (1987) and Piergorsch and Bailer (1994). Piergorsch (1994) and Tang (1998, 1999) are also relevant.

Breslow and Day (1987) is a comprehensive text on the design and analysis of cohort data. Miettinen and Nurminen (1985) has valuable material related to the topic; Selvin (1995), Sahai and Khurshid (1996), Rothman and Greenland (1998) and Newman (2001) give good introductions. The latter three books cover, in varying degrees, the exact analysis of such data. A polynomial formulation for exact analysis of a series of incidence rates was given by Martin and Austin (1996); Liddle (1984), Guess and Thomas (1990), Guess et al. (1987) and the relevant papers noted in Chapter 2 and Chapter 5 are earlier works on exact analysis.

Dunnet (1955) addressed the problem of comparing several exposed groups with a control. Piergorsch and Bailer (1994) is an accessible introduction to the pair-wise comparisons problem. An exact test for multinomial probabilities was given by Shaffer (1971). Brown and Fierce (1981) dealt with exact multiple binomial tests and applied them to screening carcinogens; see also Farrar and Crump (1988). Exact multiple comparisons for Poisson rates was dealt with by Suissa and Salami (1989). Liu, Li and Boyett (1997) gave exact tests for pair-wise comparisons in the context of adjusting for the prognostic factors in a binary response trial. The polynomial formulation given in this chapter has not been published previously.

10.14 Exercises

10.1. Consider the artificial data from six toxicology experiments shown in Table 10.8. Use the linear logit model for asymptotic and exact analyses of the trend parameter for each study. Repeat the analyses with quadratic and exponential scores for the factor x. Compare your results. Also, plot the evidence functions.

10.2. Perform FH exact, and exact and asymptotic chisquare, LR, Cressie–Read and Zelterman tests of homogeneity for each study in Table 10.8 and compare your findings.

10.3. Table 10.9 gives data from six hypothetical clinical trials. Perform FH exact, and exact and asymptotic chisquare, LR, Cressie–Read and Zelterman tests of homogeneity for each study and compare your findings.

10.4. Check the goodness-of-fit for the linear logit model with natural scores for each dataset in Table 10.8. Use asymptotic and exact chisquare and LR tests. Repeat the same exercise for Table 10.9 by assigning natural scores to treatments A to E, in that order.

10.5. Rats were given four doses of a chemical in their diet to assess its effect on liver function. Table 10.10 shows the data from an hypothetical study, which also incudes a control group. Use the logit model to analyze if the chemical affects the risk for the pathology in question. Plot the mid-p and score evidence functions.

10.6. As an example of the MTF design with nominal scale data, Table 10.11 shows the gender of 40 patients with three forms of spinal disease. The overall total, N, is presumed fixed by design. Give a logit model for the data and perform exact and asymptotic analyses of the association between gender and disease type.

10.7. Table 10.12 has data from a study of mutation of a specific gene (XX) in lung cancer. The row margins are assumed fixed. Analyze the data using exact and asymptotic methods.

EXERCISES

Table 10.8 *Toxicology Datasets*

Study	Response y	Exposure Level $x = 0$	$x = 1$	$x = 2$	$x = 3$	$x = 4$	$x = 5$
1	0	3	4	2	0	0	0
	1	0	1	1	0	0	0
2	0	4	4	3	1	0	0
	1	0	0	1	3	0	0
3	0	0	1	1	2	0	0
	1	2	0	0	1	0	0
4	0	3	1	2	1	0	0
	1	0	2	0	1	1	1
5	0	0	4	0	2	0	0
	1	4	0	3	2	0	0
6	0	3	2	1	2	0	0
	1	0	1	2	1	3	3

Note: Hypothetical data.

Table 10.9 *Clinical Trial Datasets*

Study	Response y	Treatment Control	Drug A	Drug B	Drug C	Drug D	Drug E
1	0	4	4	1	0	0	0
	1	0	0	3	0	0	0
2	0	4	2	1	2	0	0
	1	0	2	3	2	0	0
3	0	5	5	5	1	5	0
	1	0	0	0	4	0	0
4	0	9	4	3	4	1	0
	1	0	0	1	0	3	1
5	0	7	0	0	1	0	0
	1	0	4	4	3	0	0
6	0	3	2	1	3	3	0
	1	0	1	2	0	0	3

Note: Hypothetical data.

10.8. Table 10.13 shows hypothetical data on 55 boys from a neighborhood who were followed for a period of one year. All the boys were four years old at the start. The table shows whether or not the child had an ear infection during this period in terms of the average

Table 10.10 *Liver Function Data*

Liver Function	Chemical Dose (mg/kg)				
	0	10	20	40	80
Normal	5	2	1	0	0
Abnormal	3	6	7	8	8

Note: Hypothetical data.

Table 10.11 *Gender and Spinal Disease*

Gender	Type of Disease		
	I	II	III
Male	10	0	0
Female	23	1	6

Note: Hypothetical data.

Table 10.12 *XX Gene Mutations in Lung Cancer*

	Mutation Type					
	GC	AT	GC	GC	AT	
	⇓	⇓	⇓	⇓	⇓	
Cancer	TA	GC	CG	TA	CG	Total
None	0	1	1	0	0	2
Lung	12	2	2	4	2	22

Note: Hypothetical data.

number of hours he spent at day care. Does longer attendance at day care affect the probability of acquiring an ear infection? Plot the relevant evidence functions.

Table 10.13 *Ear Infection and Hours at Day Care*

Ear Infection	Hours Per Week			
	0	1–10	11–20	≥ 20
No	5	14	16	12
Yes	0	1	2	5
Total	5	15	18	17

Note: Hypothetical data.

Table 10.14 *Deaths by Gender and Exposure*

			Exposure Level		
Region	Gender	Outcome	0	1	3
North	Male	Deaths	1	3	2
		PY	3.2	1.6	0.8
	Female	Deaths	1	1	1
		PY	2.4	1.2	0.6
South	Male	Deaths	2	2	2
		PY	4.0	1.6	0.8
	Female	Deaths	1	0	1
		PY	2.4	1.2	0.6

Source: Hypothetical data; PY = (person years)/1000.

10.9. Construct three 2×4 tables from the data in Table 1.2 by separating each age group and making infection a binary variable with levels "infection" or "no infection." For each table obtained, analyze the linear logit model trend parameter using exact and asymptotic methods. Plot the evidence functions and also perform exact and asymptotic tests of homogeneity.

10.10. Table 10.14 gives hypothetic data from an incidence density study. For each region, use a log-linear model and exact and asymptotic methods to analyze how gender and exposure affect mortality. How would you analyze the combined data? Develop a model and exact method for this, and apply them to these data.

10.11. The data in Table 10.15 are from six hypothetical clinical studies with an inverse sampling design. In each study, the number of responders ($y = 1$) for each treatment was fixed by design at the stated level. Perform exact and asymptotic analyses for each study. Repeat the exercise if the treatments A to D were four doses of a drug, scored from 1 to 4, respectively. In both cases, are the results different if we assume a CMF design for the study?

10.12. Give a detailed proof of Theorem 10.2.

10.13. Derive the statistics $\mathcal{W}_{cl}$, $\mathcal{W}_{cw}$, $\mathcal{W}_{ul}$ and $\mathcal{W}_{uw}$ for the trend parameter γ in model (10.6).

10.14. For the linear logit model (10.6), construct a method for computing an exact CI for the probability of response at a specified dose x_j. Apply this to dose level $x = 2$ for the datasets in Table 10.8.

10.15. Develop large sample and exact methods for estimation of x_*, the dose level at which the probability of response is π_*, given by (10.20). Locate other formulae in the literature. Study the properties of these estimators. How would you construct exact and large sample CIs for x_*?

10.16. Fill in all the details in the proof of Theorem 10.3 and derive the relation (10.28).

10.17. For the logit formulation and the product binomial design, derive the unconditional and conditional score, LR and Wald statistic for tests of homogeneity in an unordered $2 \times K$ table.

10.18. Complete the proof of Theorem 10.4.

Table 10.15 *Inverse Sampling Trial Datasets*

Study	Response y	Control	Drug A	Drug B	Drug C	Drug D
1	0	3	1	0	0	0
	1	1	1	1	1	0
2	0	4	3	1	2	3
	1	1	1	1	1	1
3	0	7	1	1	3	5
	1	1	1	1	1	1
4	0	9	2	2	4	1
	1	1	1	1	1	1
5	0	5	0	0	1	0
	1	1	1	1	1	1
6	0	3	2	0	3	1
	1	1	1	1	1	0

Note: Hypothetical data.

10.19. Give a detailed proof of Theorem 10.5.

10.20. Derive the expressions for the goodness-of-fit score and LR statistics given in (10.50), and derive their equivalent versions given in Theorem 10.6. Locate other statistics that may be used for this purpose in the literature.

10.21. Prove the identity of the score tests for an unordered $2 \times K$ tables under four sampling designs of §10.7: RMF, CMF, total fixed and nothing fixed. Repeat this for the LR statistic. Does this hold for the Wald statistic?

10.22. Develop exact and large sample methods for a $2 \times K$ table under a quadratic effect logit model parameter. Apply this to the data in Table 10.8 and other relevant datasets in this chapter. Plot evidence functions for the dose effect parameters. How would you check for linearity in this context?

10.23. Develop exact and large sample methods for analysis of several $2 \times K$ tables with a common linear logit model trend effect. From the $2 \times K$ subtables in Table 10.8, select two subtables at a time and apply your method. Repeat this with three subtables at a time.

Develop exact and asymptotic tests of homogeneity of the trend parameter in the multi-strata case. Apply them to the datasets constructed previously.

10.24. Prove Theorem 10.7.

10.25. Derive, for an incidence density design with one binary and one ordered factor, and log-linear model, the unconditional score and LR statistics, the interaction effect between the factors (refer to §10.8.1). Repeat the exercise after (i) fixing the number of responses at each level of the ordered factor and (ii) further fixing the total number at each level of the binary factor. How would you perform exact and asymptotic analyses of the main effect parameters? Derive exact tests to check the linearity assumption for the ordered factor. Extend the methods to a quadratic effect model.

EXERCISES

10.26. Repeat the previous exercise to an incidence density design with a binary factor and a K level unordered factor.

10.27. Extend the results of the previous two exercises to a J level stratified incidence density design.

10.28. For a binary outcome inverse sampling design with one ordered factor, derive the unconditional and conditional score and LR statistics for the factor effect (refer to §10.9). Derive exact tests of the linearity assumption. Extend your results to a design with a K level unordered factor. Further extend these results to a J level stratified inverse sampling design.

10.29. Repeat all aspects of the previous exercise for (i) stratified multitreatment randomized clinical trials and (ii) case-control studies.

10.30. Prove Theorem 10.9. Formulate exact power for the goodness-of-fit tests for the linear logit model under the product binomial design.

10.31. Formulate exact power of conditional tests of homogeneity for a $2 \times K$ table under (i) a product binomial and (ii) a multinomial design.

10.32. Formulate exact power of the exact and asymptotic conditional tests on the interaction parameter in an incidence density study with one binary and one K level (ordered or unordered) factor.

10.33. Fill in all the details of Theorem 10.10.

10.34. Perform global pair-wise comparisons, under alternatives H_1 and H_2 in §10.10.3, for the datasets in Table 10.9 and other relevant data in this chapter.

10.35. Extend the pair-wise comparison method of §10.10.3 to comparing one treatment with several control groups.

10.36. Develop methods for performing exact pair-wise comparisons in an incidence density with a single $K+1$ level factor. The problem is to compare the control level with the other levels. Extend this to the $2 \times (K+1)$ situation. In both cases, examine the situations with ordered and unordered factors.

CHAPTER 11

Polynomial Algorithms: I

11.1 Introduction

An important lesson we have learned from the previous chapters is that computing the product of a series of polynomials is central to the exact analysis of common discrete data models. Often, only some terms from the product are needed. For the purpose of illustrating the main ideas, the sample sizes in the worked examples of these chapters were very small. The multiplications were thereby done with ease. But such an effort grows fast with sample size. Often the data structure may remain sufficiently sparse to preclude an asymptotic alternative. Barring an efficient method of computation, exact analysis can then be infeasible even on modern computers. Data that are too sparse for the asymptotic method, but in which the numbers are sufficiently large to make computational feasibility a barrier to exact analysis are not uncommon. An efficient computational strategy is therefore imperative.

In this chapter, we continue the development of computational methods for exact analysis. The specific aims for the moment are:

- To describe the exhaustive enumeration, randomization and recursive polynomial approaches for exact analysis of discrete data.
- To demonstrate the efficiency of the recursive approach.
- To introduce the notion of an exponent check, and show its utility for computing exact conditional distributions and their tail areas.
- To apply these techniques to the common odds ratio model and the test for conditional independence in a series of 2×2 tables.
- To introduce the Fast Fourier Transform and assess its utility for exact analysis of discrete data.
- To discuss aspects of computer implementation of the recursive polynomial algorithm.

11.2 Exhaustive Enumeration

As defined in Chapter 6, the distribution deriving from the design and scientific model of a study is called its exact (unconditional) distribution. Imposing additional restrictions gives us an exact conditional distribution. For various models for a 2×2 table, or a $2 \times 2 \times 2$ table, we can write down such distributions in explicit terms. That is, with readily implementable formulas for the exponents and coefficients of its gp, we directly produced the whole distribution. This method is equivalent to generating all the tables under the given design. Determining the exact distribution by a direct formula or other systematic means is called **complete or exhaustive enumeration (EE)** of the distribution.

With the notation of Table 6.1, we consider a $2 \times 2 \times 2$ table under the product binomial design and model (6.1). The formulas in Theorem 6.1 allow us to directly obtain exact conditional distributions for the data. Or, we can write a scheme to generate the unconditional distribution (6.14). That is, we enumerate the gp given by (6.15), and, from it, extract the conditional distribution we need.

To visualize this process, suppose we analyze the interaction parameter in (6.1). We first select an n at least as large as any value of $\mathcal{T}_3$. Thus, let $n = n_1$. Then the desired distribution is obtained by an exhaustive enumeration scheme, presented below as a generic computer pseudo code.

Algorithm 11.EE.1

Step 01.   For $u = 0, 1, \cdots, n$, $c(u) \Leftarrow 0$
Step 02.   For $b_0 = 0, 1, \cdots, m_0$ do
Step 03.   For $a_0 = 0, 1, \cdots, n_0$ do
Step 04.   For $b_1 = 0, 1, \cdots, m_1$ do
Step 05.   For $a_1 = 0, 1, \cdots, n_1$ do
Comment: Perform conditionality checks:
Step 06.   If $b_0 + a_0 + b_1 + a_1 = t_0$ then
Step 07.   If $a_0 + a_1 = t_1$ then
Step 08.   If $b_1 + a_1 = t_2$ then
Comment: Compute coefficient:
Step 09.
$$c_* = \binom{m_0}{b_0}\binom{n_0}{a_0}\binom{m_1}{b_1}\binom{n_1}{a_1}$$
Step 10. $c(a_1) \Leftarrow c(a_1) + c_*$
Step 11.   End if; End if; End If
Step 12.   End do; End do; End do; End do

□□□□□□

At the completion of this process, we let $\Omega = \{u : c(u) > 0\}$. The sets Ω and $\{c(u) : u \in \Omega\}$ allow us to construct the gp, and thus the conditional distribution of $\mathcal{T}_3$ for any ϕ_{12}.

The EE technique is a term by term multiplication of the polynomials that comprise the gp, $f(\phi)$, in (6.15). The effort required under it grows rapidly as the column totals increase.

****

Now consider a somewhat different problem. For $k = 1, \ldots, K$, we consider the univariate polynomials

$$f_k(\phi) = \sum_{u \in \Omega_k} c_k(u) \phi^u \qquad (11.1)$$

where $c_k(u) \geq 0$ and $\Omega_k = \{0, 1, \ldots, e_k\}$. The product of these K polynomials is

$$f(\phi) = \prod_{k=1}^{K} f_k(\phi) = \sum_{u \in \Omega} c(u) \phi^u \qquad (11.2)$$

where $\Omega = \{0, 1, \ldots, e_+\}$ and $e_+ = e_1 + \cdots + e_K$. If we need the value of $f(\phi)$ at a specific parameter point $\phi = \phi_0$, we evaluate each polynomial in the product at ϕ_0 using an efficient method (see Chapter 4), and multiply the K values. The task at hand is to obtain the generic polynomial, $f(\phi)$. This, we note, is completely specified by the sets Ω and $\{c(u) : u \in \Omega\}$.

EXHAUSTIVE ENUMERATION

For illustration, consider the case with $K = 4$. As in the $2 \times 2 \times 2$ data above, we obtain $f(\phi) = f_1(\phi) f_2(\phi) f_3(\phi) f_4(\phi)$ through term by term multiplication. An EE algorithm tailored for this problem appears below as Algorithm 11.EE.2.

Algorithm 11.EE.2

Step 01. For $k = 1, \ldots, 4$, compute $c_k(u)$ for all $u = 0, 1, \ldots, e_k$
Step 02. $e_+ = e_1 + e_2 + e_3 + e_4$
Step 03. For $u = 0, 1, \ldots, e_+$, $c(u) \Leftarrow 0$
Step 04. For $i_1 = 0, 1, \ldots, e_1$ do
Step 05. For $i_2 = 0, 1, \ldots, e_2$ do
Step 06. For $i_3 = 0, 1, \ldots, e_3$ do
Step 07. For $i_4 = 0, 1, \ldots, e_4$ do
Step 08. $u = i_1 + i_2 + i_3 + i_4$
Step 09. $x = c_1(i_1) * c_2(i_2) * c_3(i_3) * c_4(i_4)$
Step 10. $c(u) \Leftarrow c(u) + x$
Step 11. End do; End do; End do; End do

□□□□□□□

Consider the computational effort under the EE method. First, we need to compute the K coefficient arrays. This is common to all methods. The number of multiplications for the coefficients under Algorithm 11.EE.2 is

$$\nu_{EE.2} = (K-1)(1+e_1)(1+e_2) \cdots (1+e_K) \qquad (11.3)$$

The dominant term in (11.3) is $e_1 e_2 \cdots e_K$ which grows rapidly as K and $\min\{e_k, k = 1, \ldots, K\}$ increase. One way of reducing the effort is to not do all the multiplications in the innermost loop. The modified algorithm is shown in Algorithm 11.EE.3.

Algorithm 11.EE.3

Step 01. For $k = 1, \ldots, 4$, compute $c_k(u)$ for all $u = 0, 1, \ldots, e_k$
Step 02. $e_+ = e_1 + e_2 + e_3 + e_4$
Step 03. For $u = 0, 1, \ldots, e_+$, $c(u) \Leftarrow 0$
Step 04. For $i_1 = 0, 1, \ldots, e_1$ do
Step 05. For $i_2 = 0, 1, \ldots, e_2$ do
Step 06. $u_2 \Leftarrow i_1 + i_2$; $x_2 \Leftarrow c_1(i_1) * c_2(i_2)$
Step 07. For $i_3 = 0, 1, \ldots, e_3$ do
Step 08. $u_3 \Leftarrow u_2 + i_3$; $x_3 \Leftarrow x_2 * c_3(i_3)$
Step 09. For $i_4 = 0, 1, \ldots, e_4$ do
Step 10. $u_4 \Leftarrow u_3 + i_4$; $x_4 \Leftarrow x_3 * c_4(i_4)$
Step 11. $c(u_4) \Leftarrow c(u_4) + x_4$
Step 12. End do; End do; End do; End do

□□□□□□□

The number of multiplications in Algorithm 11.EE.3 is

$$\begin{aligned} \nu_{EE.3} = & (1+e_1)(1+e_2)[1 + (1+e_3) + (1+e_3)(1+e_4) + \\ & \cdots + (1+e_3)(1+e_4) \ldots (1+e_K)] \end{aligned} \qquad (11.4)$$

As shown later, Algorithm 11.EE.3 can reduce the multiplicative effort by 50% or more. The term $e_1 e_2 \cdots e_K$, however, remains as the dominant term. Also, the EE method, as implemented above, needs large processor memory stacks.

11.3 Monte-Carlo Simulation

An alternative when complete enumeration is too demanding is to adopt a **Monte–Carlo (MC) simulation** or a **randomization** approach. Instead of considering all outcomes, this approach draws random samples from the unconditional or conditional distribution, thereby adapting a strategy used in data collection to data analysis. Manly (1997) delineates the complete or systematic enumeration approach from the randomization approach, and notes that in many practical problems, *"the large number of permutations makes complete enumeration extremely difficult (if not impossible) with present-day computers"* (page 14).

Monte–Carlo methods form a major branch of statistics. Many efficient, sophisticated methods from this area have been applied to expand the horizons of exact analysis of discrete data in the recent years. This approach is founded upon having techniques of generating random numbers, or random observations, from common distributions like the uniform, binomial and Poisson distributions.

Suppose we have a method for generating random observations from a binomial $B(n, \pi)$ distribution. Consider again the $2 \times 2 \times 2$ problem of the previous section. As this involves four independent distributions, we draw a random value from each, combine them to see if they yield a realization in the conditional sample space and repeat it many times to get an approximation of the desired exact conditional distribution. Let R be the specified maximal number of random generations. The following basic MC algorithm illustrates the approach.

Algorithm 11.MC.1

Step 01. Set $r = 0$, $u = 0, 1, \cdots, n$, $c_*(u) \Leftarrow 0$
Step 02. $r \Leftarrow r + 1$
Step 03. Generate an observation b_0 from $B(m_0, \pi_{00})$
Step 04. Generate an observation a_0 from $B(n_0, \pi_{10})$
Step 05. Generate an observation b_1 from $B(m_1, \pi_{01})$
Step 06. Generate an observation a_1 from $B(n_1, \pi_{11})$
Comment: Perform conditionality checks:
Step 07. If $b_0 + a_0 + b_1 + a_1 \neq t_0$, go to Step 11
Step 08. If $a_0 + a_1 \neq t_1$, go to Step 11
Step 09. If $b_1 + a_1 \neq t_2$, go to Step 11
Comment: Update the count:
Step 10. $c_*(a_1) \Leftarrow c_*(a_1) + 1$
Step 11. If $r \leq R$ go to Step 02
Step 12. Stop

◻◻◻◻◻◻◻

We then let $\Omega_* = \{u : c_*(u) > 0\}$, and $c_{*+} = \Sigma c_*(u)$. The former is the random construct of the conditional sample space. The conditional probabilities are approximated by $c_*(u)/c_{*+}$. It can be shown that as R increases, these converge to the true conditional probabilities.

When we need to compute tail areas, we add the appropriate conditions after Step 09 in Algorithm 11.MC.1.

With a random generator of the binomial handy, this method is easy to implement. As given, it generates the conditional distribution for specified cell probability values. If the distribution is needed in a generic form, that is, as a set of exponents and coefficients of the gp, we set $\pi_{ij} = 0.5$, $i, j = 0, 1$.

We can also employ a somewhat different strategy that uses a random generator from a uniform

U$[0, 1]$ distribution. When not directly available, it is obtainable with a simple formula (see §11.6.5). The uniform discrete variate is obtained from its continuous counterpart as follows.

Lemma 11.1: Assume that $0 < \mathcal{U} < 1$ has a U$[0, 1]$ distribution. Let n be a fixed integer. Define $\mathcal{U}_d$ to be a nonnegative integer variate such that $\mathcal{U}_d < \mathcal{U}(n+1) \leq \mathcal{U}_d + 1$. $\mathcal{U}_d$ is uniformly distributed over the discrete sample space $\{0, 1, \ldots, n\}$. We denote this distribution by U$_d[0, n]$.

Proof: Left to the reader. □

We return to the problem of multiplying K polynomials in (11.2). This gp is completely determined if the values of $c(u)$ up to a proportionality constant are known. Consider a randomization algorithm that multiplies randomly selected terms from the polynomials, depicted as follows.

<pre>
                   Algorithm 11.MC.2
   Step 01.   For $k = 1, \cdots, K$, compute $c_k(u)$ for all $u = 0, 1, \ldots, e_k$
   Step 02. $e_+ = e_1 + e_2 + e_3 + e_4$
   Step 03.   For $u = 0, 1, \ldots, e_+$, $c_*(u) \Leftarrow 0$
   Step 04.   Specify R, and $r \Leftarrow 0$
   Step 05. $r \Leftarrow r + 1$
   Step 06.   For $k = 1, \cdots, K$ do
   Step 07.   Generate an u_k from U$_d[0, e_k]$
   Step 08.   End do
   Step 09. $u = \sum_k u_k$
   Step 10. $z = \prod_k c_k(u_k)$
   Step 11. $c_*(u) \Leftarrow c_*(u) + z$
   Step 12.   If $r \leq R$ go to Step 05
   Step 13.   Stop
</pre>

□□□□□□□

To demonstrate the properties of the MC algorithm, we need a preliminary result.

Lemma 11.2: Suppose $\mathcal{Y} \in \{0, 1\}$ and $\mathcal{X}$ are discrete random variables and P$[\mathcal{Y} = 1]$ does not depend on the value of $\mathcal{X}$. Then

$$\mathbf{E}[\mathcal{Y}\mathcal{X}] = \mathrm{P}[\mathcal{Y} = 1]\mathbf{E}[\mathcal{X} \mid \mathcal{Y} = 1] \qquad (11.5)$$

Proof: Left to the reader. □

This results allows us to show the following.

Theorem 11.1: Consider Algorithm 11.MC.2 for the polynomial multiplication (11.2). For any $u \in \Omega$, let $\mathcal{C}_{*R}(u)$ denote the random coefficient after R generations. Then (i) $\mathcal{C}_{*R}(u)/\mathcal{C}_{*+}$ converges to $c(u)/c_+$ as R increases, and (ii)

$$\mathbf{E}[\mathcal{C}_{*R}(u)] = \frac{Rc(u)}{\nu} \qquad (11.6)$$

where $\nu = \nu_{EE.2}$ as defined in (11.3).

Proof: The terminology and proof of the first part involve ideas beyond the scope of this book. See Ross (2002), Manly (1997) and Edgington (1995). We consider part (ii) here.

Let $\mathcal{U}_{ik}$ denote the kth discrete variate generated in simulation r, $k = 1, \cdots, K$ and $r = 1, \cdots, R$. Then, for $u = 0, 1, \cdots, e_k$, we have that

$$P[\mathcal{U}_{rk} = u] = \frac{1}{1 + e_k} \qquad (11.7)$$

Now let $\mathcal{U}_r = (\mathcal{U}_{r1}, \cdots, \mathcal{U}_{rK})$, and define

$$\begin{aligned} I_r(\mathcal{U}_r, u) &= 1 \quad \text{if} \quad \sum_k \mathcal{U}_{rk} = u \\ &= 0 \quad \text{otherwise} \end{aligned} \qquad (11.8)$$

It then follows that

$$\mathcal{C}_{*R}(u) = \sum_{r=1}^{R} I_r(\mathcal{U}_r, u) \left\{ \prod_k c_k(\mathcal{U}_{rk}) \right\} \qquad (11.9)$$

Taking expectations and using Lemma 11.2, we have that $\mathbf{E}[\mathcal{C}_{*R}(u)]$ is equal to

$$\sum_{r=1}^{R} P[I_r(\mathcal{U}_r, u) = 1] \mathbf{E}[\prod_k c_k(\mathcal{U}_{rk}) \mid \sum_k \mathcal{U}_{rk} = u] \qquad (11.10)$$

Let $d(u)$ be the number of integer solutions to the equation $u_1 + \cdots + u_K = u$ and $0 \le u_k \le e_k$, $k = 1, \cdots, K$. For any $u \in \Omega$, this also denotes the number of distinct random generations producing the exponent value equal to u. Then looking at the two components under the summation in (11.10), we see that

$$\mathbf{E}[\mathcal{C}_{*R}(u)] = R \frac{d(u)}{\nu} \times \frac{c(u)}{d(u)} = \frac{Rc(u)}{\nu} \qquad (11.11)$$

□

The randomization algorithm gives **unbiased estimates** of the coefficients of the gp up to a proportionality factor. The mean of the values from the R generations is equal to $c(u)/\nu$. As for any sampling method, the variance of the sample mean goes down as we increase R, the number of points sampled (Exercise 11.30). The effort required also depends on R.

As before, let $\Omega_* = \{u : c_*(u) > 0\}$. Then we use Ω_* and $\{c_*(u) : u \in \Omega_*\}$ to approximate the conditional PBD and so to compute evidence functions, p-values, estimates and CIs in the usual way. The p-value estimate tends to be precise. A large number of generations are usually needed to get accurate confidence limits. A test based on a **randomization distribution** is known as a **randomization test**. Note, this is distinct from the randomized test defined in Chapter 7.

There are several ways of improving randomization algorithms. A simple way is to do the multiplication in a sequential way. Also EE and MC algorithms improve with application of **exponent checks** described in the next section. Also, especially for conditional distributions, many efficient versions of the MC approach exist.

This book focuses on computing exact distributions exactly. We show that for a major class of

RECURSIVE MULTIPLICATION

discrete data problems with an expansive sample space, a simple recursive approach together with certain checks can generate exact conditional distributions in a rapid manner even on personal computers. While problems in which partial randomization is the only feasible method still remain, a more extensive coverage of Monte–Carlo methods is beyond the scope of this book.

Therefore we deal with Monte–Carlo methods only in very basic terms. Our main application is exact power computation, which is covered in Chapter 12.

11.4 Recursive Multiplication

A third approach to computing the product (11.2) is to do it in a step-wise or recursive manner. Define $F_1(\phi) = f_1(\phi)$. Then, for $k = 2, \ldots, K$, we implement

$$F_k(\phi) = f_k(\phi) * F_{k-1}(\phi) \qquad (11.12)$$

This is called the **Recursive Polynomial Multiplication (RPM)** algorithm. The number of multiplications it performs is

$$\nu_{RPM.1} = \sum_{k=1}^{K-1} (1 + e_{k+1})(1 + e_1 + \cdots + e_k) \qquad (11.13)$$

A generic implementation of this method appears in Algorithm 11.RPM.1.

```
                   Algorithm 11.RPM.1
```

Step 01.   For $k = 1, \ldots, K$, compute $c_k(u)$ for all $u = 0, 1, \ldots, e_k$
Step 02. $e_+ = e_1 + \cdots + e_K$
Step 03.   For $u = 0, 1, \ldots, e_+$ and $j = 0, 1$, $c(j, u) \Leftarrow 0$
Step 04.   Set $j = 0$, $k = 1$, $n = e_1$
Step 05.   For $u = 0, 1, \ldots, n$, $c(j, u) \Leftarrow c_k(u)$
Step 06. $k \Leftarrow k + 1$
Step 07.   For $u = 0, 1, \ldots, n$ do
Step 08.   For $v = 0, 1, \ldots, e_k$ do
Step 09. $w = u + v$
Comment: Perform exponent checks here (when needed)
Step 10. $z = c(j, u) * c_k(v)$
Step 11. $c(1 - j, w) \Leftarrow c(1 - j, w) + z$
Step 12.   End do; End do
Step 13.   For $u = 0, 1, \ldots, n$, $c(j, u) \Leftarrow 0$
Step 14. $n \Leftarrow n + e_k$
Step 15. $j \Leftarrow 1 - j$
Step 16.   If $(k < K)$ go to Step 06

□□□□□□

At the end of above, the array with values $c(j, u) > 0$ gives the exponents and coefficients we need.

For an illustration of the efficiency of the RPM approach in relation to the EE approach, consider the case with $e_k = n$ for all k. Then,

Table 11.1 *A Comparison of EE and RPM Algorithms (K = 4)*

	Multiplications			Efficiency Ratio	
n	11.EE.2	11.EE.3	11.RPM.1	EE.3/EE.2	RPM.1/EE.3
1	48	28	18	.583	0.643
6	7203	2793	273	.388	0.098
12	85683	30927	975	.361	0.031
18	390963	137541	2109	.352	0.015
24	1171875	406875	3675	.347	0.009

$$\begin{aligned}
\nu_{EE.2} &= (K-1)(1+n)^K \\
\nu_{EE.3} &= (1+n)^2[1 + (1+n) + \cdots + (1+n)^{K-2}] \\
\nu_{RPM.1} &= (K-1)(1+n)[1 + nK/2]
\end{aligned} \qquad (11.14)$$

In computer science terms, the EE method is called a **polynomial time**, written as $O(n^K)$, algorithm, and the RPM method is a **quadratic time**, or $O(n^2)$, algorithm. These determinations reflect the dominant term in each method. Consider a comparison of these methods when $K = 4$, and $e_k = n$ for $k = 1, 2, 3, 4$. Table 11.1 shows that the efficiency gap between RPM and EE grows rapidly with n. The ratio of EE.3 to EE.2 approaches 0.333 but the ratio of RPM.1 to EE.3 fast tends to zero. As n and K increase, the EE approach becomes computationally infeasible.

11.5 Exponent Checks

Suppose we are multiplying K polynomials with nonnegative exponents. If we need only the final terms with exponent value equal to at least 7, and if a current product term has exponent equal to 9, then we can delete it from consideration without affecting the final result. In the process, we also reduce the number of multiplications we have to perform. In this section, we give more general forms of such checks on the exponents that help reduce the computational effort in the RPM, EE and MC approaches.

Let e_k^* and e_k respectively be the smallest and largest exponents of ϕ in $f_k(\phi)$ in (11.1). The latter is assumed to be finite. Obviously, $e_k^* \geq 0$. For $k = 1, \ldots, K$, define

$$\begin{aligned}
e_{k,+}^* &= e_k^* + \cdots + e_K^* & (11.15) \\
e_{k,+} &= e_k + \cdots + e_K & (11.16)
\end{aligned}$$

$e_{k,+}^*$ and $e_{k,+}$ are respectively the smallest and largest exponents of ϕ in $f_k(\phi) \cdots f_K(\phi)$. For the problem in the previous section, $e_{k,+}^* = 0$.

Suppose we are multiplying the polynomials in (11.12), and are at the kth stage. In particular, we are about to multiply the terms $d_{k-1}(u)\phi^u$ in $F_{k-1}(\phi)$ and $c_k(v)\phi^v$ in $f_k(\phi)$, that is, to perform the operation

EXPONENT CHECKS

$$d_{k-1}(u) \times c_k(v) \tag{11.17}$$

The polynomials not yet multiplied are $f_{k+1}(\phi), \ldots, f_K(\phi)$. We now state two self evident but critical rules relating to the exponents of these polynomials. They allow us to determine, to a degree, the values of exponents in the final product on the basis of the current information.

Rule 11.01: If the exponents of any two terms being multiplied are added to the maximum value of the exponent of ϕ in the product of all polynomials not yet multiplied and the result is smaller than (smaller than or equal to) t, then all the final stage terms resulting from the current multiplication will have exponents smaller than (smaller than or equal to) t.

Rule 11.02: If the exponents of any two terms being multiplied are added to the minimum value of the exponent of ϕ in the product of all polynomials not yet multiplied and the result is larger than (larger than or equal to) t, then all the final stage terms resulting from the current multiplication will have exponents larger than (larger than or equal to) t.

These rules enable us to formulate **exponent checks (EC)** that allow direct computation of the final stage coefficient of a single term, coefficients of all terms in a left or right tail segment, or the value of a tail segment at a particular value of ϕ. They often enhance computational efficiency to a major degree, and can turn an infeasible task into an easy effort. Below we detail the checks for specific situations.

11.5.1 A Single Term

Suppose we need the coefficient of the term of $f(\phi)$ given by (11.2) in which the exponent of ϕ equals t. Though we can extract it from the final result of Algorithm 11.RPM.1, a more efficient way is to apply appropriate exponent checks to this algorithm so that the Kth stage operation yields just this particular term.

Assume that at the kth stage of Algorithm 11.RPM.1, we are about to multiply the term $d_{k-1}(u)\phi^u$ in $F_{k-1}(\phi)$ with the term $c_k(v)\phi^v$ in $f_k(\phi)$. Then we apply the following checks.

Check EC 11.01

$$\langle \text{ If } u + v + e^*_{k+1,+} > t \rangle$$

Do not perform the multiplication. Continue on to the next term.

$$\langle \text{ If } u + v + e_{k+1,+} < t \rangle$$

Do not perform the multiplication. Continue on to the next term.

$$\langle \text{ Otherwise } \rangle$$

Perform the multiplication $d_{k-1}(u) \times c_k(v)$, accumulate the result in the $(u+v)$th position in $F_k(\phi)$ and continue.

□□□□□□

The first check deletes current terms which only produce terms at the final stage with exponents strictly greater than t. The second check deletes the terms which always yield a final set of terms with exponents strictly smaller than t. At the Kth stage, the RPM algorithm with these modifications yields the desired term $c(t)\phi^t$. The deletion of terms at intermediate stages reduces the number of multiplications. Only the terms surviving the checks are stored and used at the next stage. The efficiency gain from these reductions generally outweighs the additional effort entailed by the use of the checks.

As indicated in Algorithm 11.RPM.1, these and similar checks are implemented after Step 09. In the literature, such checks are also known as **infeasibility criteria**, **trimming and augmentation criteria** or **smallest and longest path criteria**.

11.5.2 A Tail Segment

Suppose we need the coefficients of all terms of $f(\phi)$ in which the exponent of ϕ is larger than or equal to t. That is we need the terms in the right segment

$$\sum_{u \geq t} c(u)\phi^u \qquad (11.18)$$

These terms are obtained with the following checks.

Check EC 11.02

⟨ If $u + v + e_{k+1,+} < t$ ⟩
Do not perform the multiplication. Continue on to the next term.

⟨ Otherwise ⟩
Perform the multiplication $d_{k-1}(u) \times c_k(v)$, accumulate the result in the $(u+v)$th position in $F_k(\phi)$ and continue.

◻◻◻◻◻◻◻

The explanation of and the efficiency gain resulting from these checks is similar to that given earlier. If we need the terms in the left tail segment,

$$\sum_{u \leq t} c(u)\phi^u \qquad (11.19)$$

we use the following checks.

Check EC 11.03

⟨ If $u + v + e^*_{k+1,+} > t$ ⟩
Do not perform the multiplication. Continue on to the next term.

⟨ Otherwise ⟩
Perform the multiplication $d_{k-1}(u) \times c_k(v)$, accumulate the result in the $(u+v)$th position in $F_k(\phi)$ and continue.

◻◻◻◻◻◻◻

EXPONENT CHECKS

11.5.3 A Tail Sum

Now consider a different problem. Suppose we need, for a particular value of $\phi = \phi_0$, the value of the sum of all terms of (11.2) in which the exponent of ϕ is larger than or equal to t. That is, we need the value of

$$\Delta = \sum_{u \geq t} c(u)\phi_0^u \qquad (11.20)$$

For this purpose, define, for $k = 1, \ldots, K$,

$$\delta_k = f_k(\phi_0) \cdots f_K(\phi_0) \qquad (11.21)$$

Then we proceed with the RPM algorithm as follows. At the start, set $\Delta = 0$, and implement Algorithm 11.RPM.1 with the following checks and changes.

Check EC 11.04

$$\langle \text{ If } u + v + e^*_{k+1,+} \geq t \rangle$$

Perform the multiplication $\rho = d_{k-1}(u) \times c_k(v) \times \phi_0^{(u+v)}$ and set $\Delta = \Delta + \rho \delta_{k+1}$. But do not store anything in $F_k(\phi)$ and continue on to the next term.

$$\langle \text{ If } u + v + e_{k+1,+} < t \rangle$$

Do not perform the multiplication. Continue on to the next term.

$$\langle \text{ Otherwise } \rangle$$

Perform the multiplication $d_{k-1}(u) \times c_k(v)$, accumulate the result in the $(u+v)$th position in $F_k(\phi)$ and continue.

□□□□□□□

The first check identifies terms that contribute values to the desired tail segment, computes their total contribution, and adds the result to Δ. To avoid double counting, these terms are dropped from further consideration. The second check identifies the terms that can never generate a term in the desired tail segment. They are deleted without adding to the value of Δ. The value of Δ at the end of the last stage is what we need. Note, when $\phi_0 = 1$, Δ is the sum of the coefficients in the right tail.

The left tail sum at $\phi = \phi_0$,

$$\Delta = \sum_{u \leq t} c(u)\phi_0^u \qquad (11.22)$$

is obtained with the following checks.

Check EC 11.05

⟨ If $u + v + e^*_{k+1,+} > t$ ⟩

Do not perform the multiplication. Continue on to the next term.

⟨ If $u + v + e_{k+1,+} \leq t$ ⟩

Perform the multiplication $\rho = d_{k-1}(u) \times c_k(v) \times \phi_0^{(u+v)}$ and set $\Delta = \Delta + \rho \delta_{k+1}$. But do not store anything in $F_k(\phi)$ and continue on to the next term.

⟨ Otherwise ⟩

Perform the multiplication $d_{k-1}(u) \times c_k(v)$, accumulate the result in $(u+v)$th position in $F_k(\phi)$ and continue.

☐☐☐☐☐☐☐

11.6 Applications

We apply the RPM algorithms with exponent checks to two models for the analysis of several 2×2 tables. Let the kth stratum data be as in Table 6.1, with the table index being k instead of i.

With $e^*_k = \max(0, s_k - m_k)$ and $e_k = \min(s_k, n_k)$, consider the kth polynomial

$$f_k(\phi_k) = \sum_{v \in \Omega_k} c_k(v) \phi_k^v \qquad (11.23)$$

where

$$\Omega_k = \{e^*_k, e^*_k + 1, \ldots, e_k\} \quad \text{and} \quad c_k(v) = \binom{n_k}{v}\binom{m_k}{s_k - v} \qquad (11.24)$$

$f_k(\phi_k)$ is the gp for the conditional distribution of $\mathcal{A}_k$.

11.6.1 The COR Model

First consider the common odds ratio model with $\phi_k = \phi$ for all k. The gp for the conditional distribution of $\mathcal{T} = \mathcal{A}_1 + \cdots + \mathcal{A}_K$, the sufficient statistic for ϕ, is

$$f(\phi) = \prod_{k=1}^{K} f_k(\phi) \qquad (11.25)$$

Since $f(\phi)$ completely specifies the conditional distribution of $\mathcal{T}$, the following observations hold for computing this distribution.

- Algorithm 11.RPM.1 applies to computing the complete distribution.
- Algorithm 11.RPM.1 with Check EC 11.01 apply to computing a point probability.
- Algorithm 11.RPM.1 with Check EC 11.02 apply to computing a right tail probability segment.

APPLICATIONS

- Algorithm 11.RPM.1 with Check EC 11.03 apply to computing a left tail probability segment.
- Algorithm 11.RPM.1 with Check EC 11.04 apply to computing the total probability mass in a right tail at a specified parameter value.
- Algorithm 11.RPM.1 with Check EC 11.05 apply to computing the total probability mass in a left tail at a specified parameter value.

The implementation is simplified by a reformulation.

$$e_{*k} = e_k - e_k^* \tag{11.26}$$
$$c_{*k}(v) = c_k(v - e_k^*) \tag{11.27}$$
$$\Omega_{*k} = \{0, 1, \cdots, e_{*k}\} \tag{11.28}$$
$$f_{*k}(\phi) = \sum_{v \in \Omega_{*k}} c_{*k}(v)\phi^v \tag{11.29}$$

Then we implement the multiplication

$$f_*(\phi) = \prod_{k=1}^{K} f_{*k}(\phi) \tag{11.30}$$

and use the observed value $t_* = t - e_+^*$ where $e_+^* = e_1^* + \cdots + e_K^*$. Also note that

$$f_*(\phi) = \phi^{-e_+^*} f(\phi) \tag{11.31}$$

For probability computation, we also need the normalizing factor $f(\phi)$. Depending on the objectives in (i) to (vi) above, we need this either for unspecified ϕ, or for a specified value, ϕ_0. For the latter, we use either the product value $f_{*1}(\phi_0) \cdots f_{*K}(\phi_0)$, or the value $f_*(\phi_0)$ obtained by evaluating the outcome of Algorithm 11.RPM.1 at $\phi = \phi_0$. Note that when $\phi_0 = 1$,

$$f_k(1) = \binom{n_k + m_k}{s_k} \tag{11.32}$$

which directly give the required normalizing factor, and also gives the values of δ_k for use with Check EC 11.04 and Check EC 11.05.

The iterative techniques developed by Liao and Hall (1995) to improve efficiency in computing exact confidence limits for the COR model are also relevant. They tend to reduce the number multiplications by about 25%.

11.6.2 Conditional Independence

Now consider the hypothesis of conditional independence, $H_0 : \phi_1 = \phi_2 = \cdots = \phi_K = 1$, in model (8.2). Let $\mathcal{W}$ be a test statistic for this purpose, with

$$\mathcal{W} = \sum_k \mathcal{W}_k \tag{11.33}$$

For specified values of ϕ_k, consider the polynomial

$$g_k(\psi, \phi_k) = \sum_{v \in \Omega_k} c_{*k}(v) \psi^{w_k(v)} \tag{11.34}$$

where $c_{*k}(u) = c_k(u) \phi_k^u$ and where ψ is a dummy parameter. Two score and one likelihood versions of the statistic applicable here were given in Theorem 8.6. The product polynomial

$$g(\psi, \boldsymbol{\phi}) = \prod_{k=1}^{K} g_k(\psi, \phi_k) = \sum_{w} c_*(w; \boldsymbol{\phi}) \psi^w \tag{11.35}$$

with $\boldsymbol{\phi} = (\phi_1, \ldots, \phi_K)$ is the nominal gp for the conditional distribution of $\mathcal{W}$. As the values of ϕ_k are specified, (11.35) is a multiplication of K univariate polynomials with parameter ψ. This is obtainable with Algorithm 11.RPM.1.

Now suppose we need the right tail area, $\text{P}[\mathcal{W} \geq w \mid \boldsymbol{S} = \boldsymbol{s}; \boldsymbol{\phi}]$. Let ϵ_k^* be the smallest and ϵ_k, the largest exponent of ψ in $g_k(\psi, \phi_k)$. These two exponents can be obtained either by inspection, or in some cases, by an explicit formulae. For example, for the score statistics, we need the minimum and maximum values of $(N_k u - n_k s_k)^2$ subject to $u \in \Omega_k$.

Noting that $g_k(1, \phi_k) = f_k(\phi_k)$, we define

$$\delta_{*k} = f_k(\phi_k) \cdots f_K(\phi_K) \tag{11.36}$$

To compute the required right tail area of the conditional distribution of the test statistic, we proceed as follows. At the start, set $\Delta = 0$, and implement the algorithm with the following checks. Assume that at the kth stage, we are about to multiply the term $d_{*k-1}(u) \psi^u$ in (the appropriately defined) $G_{k-1}(\psi)$ with the term $c_k(v) \phi_k^v \psi^{w_k(v)}$ in $g_k(\psi)$. Then

Check EC 11.06

$$\langle \text{ If } u + w_k(v) + \epsilon_{k+1,+} \geq w \rangle$$

Perform the multiplication $\rho = d_{*k-1}(u) \times c_k(v) \times \phi_k^v$ and set $\Delta = \Delta + \rho \delta_{*k+1}$. Do not store anything in $G_k(\psi)$ and continue on to the next term.

$$\langle \text{ If } u + w_k(v) + \epsilon_{k+1,+} < w \rangle$$

Do not perform the multiplication. Continue on to the next term.

$$\langle \text{ Otherwise } \rangle$$

Perform the multiplication $d_{*k-1}(u) \times c_k(v) \times \phi_k^v$, accumulate the result in the $(u + w_k(v))$th position in $G_k(\psi)$ and continue.

□□□□□□

Note that when $\phi_k = 1$ for all k,

$$g_k(1) = \binom{n_k + m_k}{s_k} \tag{11.37}$$

(11.37) directly gives the null normalizing factor, and the values of δ_{*k} for use with Check EC 11.06. In particular, the former is

APPLICATIONS

$$\delta_{*1} = g(1,\mathbf{1})$$

We note two important points. The exponent values, $w_k(u)$, $u \in \Omega_k$, may not be distinct. So, we may collate the terms in $g_k(\psi, \phi_k)$ prior to the kth stage multiplication. For some statistics and especially with symmetric data, this reduces the number of multiplications by a noticeable amount. That reduction has, however, to be balanced by the effort required to compare and combine the values of $w_k(u)$. Further, unlike in the COR model, these tend to be sparse polynomials, an issue to keep in mind when implementing the method on a computer.

11.6.3 Conditional Power

Let $t_1 \in \Omega$ and $t_2 \in \Omega$, with $t_1 < t_2$, and suppose, for a given ϕ_0, we need to compute

$$\Delta = \sum_{u \leq t_1} c(u)\phi_0^u + \sum_{u \geq t_2} c(u)\phi_0^u \tag{11.38}$$

Then we start with $\Delta = 0$, and apply modified checks shown below in Algorithm 11.RPM.1 along the same lines as for Check EC 11.04.

Check EC 11.07

$$\langle \text{ If } u + v + e^*_{k+1,+} > t_1 \rangle$$

And

$$\langle \text{ If } u + v + e_{k+1,+} < t_2 \rangle$$

Do not perform the multiplication. Continue on to the next term.

$$\langle \text{ If } u + v + e_{k+1,+} \leq t_1 \rangle$$

Or

$$\langle \text{ If } u + v + e_{*k+1,+} \geq t_2 \rangle$$

Perform the multiplication $\rho = d_{k-1}(u) \times c_k(v) \times \phi_0^{(u+v)}$ and set $\Delta = \Delta + \rho\delta_{k+1}$. But do not store anything in $F_k(\phi)$ and continue on to the next term.

$$\langle \text{ Otherwise } \rangle$$

Perform the multiplication $d_{k-1}(u) \times c_k(v)$, accumulate the result in $(u+v)$th position in $F_k(\phi)$ and continue.

◻◻◻◻◻◻◻

These checks are used to compute the conditional two-sided p-value for some test statistics for the COR model. For example, for the conditional score statistic, we let $\phi_0 = 0$ and get the values of t_1 and t_2 from the boundary values of $u \in \Omega$ that satisfy

$$(u - \mu_0)^2 \geq (t - \mu_0)^2 \tag{11.39}$$

In this case, if $t \geq \mu_0$, then we set $t_1 = 2\mu_0 - t$ and $t_2 = t$, and if $t \leq \mu_0$, we set $t_1 = t$ and $t_2 = 2\mu_0 - t$.

We can also use these checks to compute the conditional power for a given $\phi = \phi_0$. In the latter case, t_1 and t_2 are the boundary values of the critical region for the test in question. These are determined by appropriately applying the test in question to a series of points in the conditional sample space.

Further, these criteria can be extended to compute conditional power for the tests of conditional independence for several 2×2 tables.

11.6.4 Other Remarks

The EE Approach: The exponent checks, when used in the EE method, are applied each time a term is generated in a stratum specific loop. Even though they improve efficiency, their impact is generally not as pronounced as in the RPM context where they are collectively applied to a group of terms. A single deletion in the RPM context thus corresponds to a large number of individual deletions under the EE approach. Also, such checks can also be applied with the MC approach.

Point & Tail probabilities: The computation of a point and a tail probability may be combined in a single recursive pass. For this, we need to harmonize the exponent checks for these two objectives.

Recursive Randomization: The MC approach can also be implemented in a recursive manner. For example, we may generate say five random points from each polynomial, multiply them in a recursive fashion and repeat that many times.

Multiple Tasks: A conditional distribution in the COR model is often needed for several tasks: to compute an estimate, a *p*-value, a CI or the entire evidence function. In that case, it is generally not efficient to implement the recursive process separately for each task. It is better to generate the complete distribution with the Algorithm 11.RPM.1 without using any checks. Efficient methods to evaluate a polynomial at one or many parameter values are relevant when computing point estimates, CIs or the evidence function.

Exact Power: Exact unconditional power computation is dealt with in the next chapter.

11.6.5 Computer Implementation

The algorithms in this and the following chapters are presented in a traditional manner that facilitates comprehension of the underlying idea. For efficient computer implementation, we need to use an object oriented modularized approach. The nested 'do loops' in the EE and MC algorithms can also be replaced with a simulated version (see Chapter 13).

The other key issues relevant for computer implementation are: (i) control of numerical overflow or underflow, (ii) rounding, (iii) storing polynomial arrays, and (iv) generating random numbers. The first three issues bring together several points we made in Chapter 4.

First, to reduce the danger of numeric overflow or underflow, double precision arithmetic and the logarithmic scale with a scale sensing test as described in §4.5 are suggested.

Two, when the components of the test statistic (or equivalently, the exponents of the polynomials) are not integer valued, they are maintained at a finite level of precision. Too low a rounding factor may adversely impact accuracy, and very high accuracy affects efficiency. Rounding the exponents in each polynomial to three or four decimal digits is generally adequate.

Three, for noninteger valued statistics, a good storage technique is essential for maintaining memory efficiency. A hash function is recommended. See (4.13) and the associated references.

The extent of aggregation in the cumulative polynomial of the RPM method is less pronounced if the statistic is not integer valued. This reduces the efficiency of the RPM method relative to EE. When there are only a few strata, the difference between the two approaches, especially in conjuction with the exponent checks, may not be noticeable.

Four, Ross (2002), pages 37–38, provides a good quality but simple device for generating random numbers. Note, these are more accurately called pseudo random numbers. In a machine with word size of 32 bits, we set

$$a = 7^5 = 16807 \quad \text{and} \quad m = 2^{31} - 1 \tag{11.40}$$

Using an initial seed z_0, and for $i = 1, 2, \ldots$, we then successively compute

$$z_i = az_{i-1} \mod m \tag{11.41}$$

Putting a decimal point in front of the generated random digits is then (almost) equivalent to generating independent values from a $U[0, 1]$ distribution. Such values are needed for the randomized procedures of Chapter 7, for the MC algorithms of this chapter and for exact power computation (see Chapter 12).

11.7 The Fast Fourier Transform

Over the years, algorithms for exact analysis of several 2×2 tables have been constructed in a variety of ways. The initial efforts enumerated all tables with fixed margins. Since each term in $f_k(\phi)$ corresponds to such a kth 2×2 table, this is identical to the EE method given above.

Early strategies to devise more efficient algorithms took two different directions. One formulated the problem as that of processing a directed network, and applied checks based on shortest and longest paths in the network. Efficiency was vastly improved when a stage-wise instead of a depth-wise processing of the network was formulated. We describe this approach in the next chapter. For now, note that network algorithms are pictorial representations of equivalent polynomial algorithms, and the shortest and longest path checks in the former correspond to the exponent checks of the latter.

Prior to the network method, one construction of an efficient algorithm for exact analysis of several 2×2 was based on the **fast Fourier transform (FFT)**. We describe its basic ideas below.

Let $\omega = \exp(2\pi i/n)$ where $i = \sqrt{-1}$. It is readily shown that

$$\omega^n = 1 \tag{11.42}$$

ω is called the principal nth root of unity. The set $\{\omega^0, \omega^1, \cdots, \omega^{n-1}\}$ consists of the n complex nth roots of unity. These roots satisfy further properties which are applied to construct the FFT. These properties are:

Property 11.1: These roots are periodic in n. That is, for integer a and b, and $0 \leq b < n$, $\omega^{an+b} = \omega^b$.

Property 11.2: For even n, (i) $\omega^{(n/2)} = -1$, (ii) ω^2 is the principal $(n/2)$th root of unity, (iii) ω^{-1} is an nth root of unity and (iv) for $v = 0, 1, \ldots, n/2 - 1$, we have $\omega^v = -\omega^{(n/2)+v}$.

Property 11.3: For $v = 0, 1, \ldots, n-1$,

$$\sum_{u=0}^{n-1} \omega^{uv} = 0 \qquad (11.43)$$

****

With the above definitions in mind, the **discrete Fourier transform (DFT)** of the set of (possibly complex) coefficients, $\{\, c(u) : u = 0, 1, \ldots, n-1 \,\}$, is defined by

$$f_{n,v} = \sum_{u=0}^{n-1} c(u) \omega^{uv} \qquad (11.44)$$

for $v = 0, 1, \ldots, n-1$. We note that there are only n distinct values of ω^{uv} in (11.44). Further, taking the DFT is the same as evaluating the polynomial

$$f(\phi) = \sum_{u=0}^{n-1} c(u) \phi^u \qquad (11.45)$$

at these n roots. This transform has an important inversion property. If the n transform values $\{\, f_{n,0}, f_{n,1}, \ldots, f_{n,n-1} \,\}$ are known, the coefficients of the associated polynomial can be recovered by the **inverse DFT**

$$c(u) = \frac{1}{n} \sum_{v=0}^{n-1} f_{n,v} \omega^{-uv} \qquad (11.46)$$

Apart from the multiplicative factor $1/n$, and replacement of ω by ω^{-1}, the formulas for the inverse DFT and the DFT are the same. A method of computation for the DFT is then readily adapted for the inverse DFT, especially if n is even.

****

Results (11.44) and (11.46) provide us another method for evaluating the product of polynomials. Suppose we want to multiply the polynomial $f(\phi)$ given by (11.45) with the polynomial

$$g(\phi) = \sum_{u=0}^{m-1} c_*(u) \phi^u \qquad (11.47)$$

Let $N = m + n - 1$ and $\omega = \exp(2\pi i/N)$. Also for $u \geq n$, let $c(u) = 0$, and for $u \geq m$, let $c_*(u) = 0$. Then we proceed as follows:

THE FAST FOURIER TRANSFORM

Algorithm 11.FFT.1

Step 01. For $v = 0, 1, \ldots, N-1$, determine the DFT of the set $\{\, c(u) : u = 0, 1, \ldots, N-1 \,\}$. Denote this by $\{\, f_{N,0}, f_{N,1}, \ldots, f_{N,N-1} \,\}$.

Step 02. For $v = 0, 1, \ldots, N-1$, determine the DFT of the set $\{\, c_*(u) : u = 0, 1, \ldots, N-1 \,\}$. Denote this by $\{\, g_{N,0}, g_{N,1}, \ldots, g_{N,N-1} \,\}$.

Step 03. For $v = 0, 1, \ldots, N-1$, let $h_{N,v} = f_{N,v} g_{N,v}$.

Step 04. For $u = 0, 1, \ldots, N-1$, apply the inverse DFT on $\{\, h_{N,0}, h_{N,1}, \ldots, h_{N,N-1} \,\}$, to get the set $\{\, d(u) : u = 0, 1, \ldots, N-1 \,\}$.

□□□□□□□

The required product then is

$$f(\phi)g(\phi) = \sum_{u=0}^{N-1} d(u)\phi^u \tag{11.48}$$

This method readily extends to the product of K polynomials. In this case, K DFTs are done and after multiplication of the results, the inverse DFT is applied in the same way as above.

****

The efficiency of this approach to multiplying polynomials is dependent on having an efficient method of performing the DFT (which would also be an efficient method for the inverse DFT). This method is the **Fast Fourier Transform (FFT)**.

We explain it by a specific example. Let $n = 8$, and $\omega = \exp(-2\pi i/8)$. Then the DFT of $\{\, c_0, c_1, \ldots, c_7 \,\}$ is

$$f_{8,v} = c_0 + c_1 \omega^v + c_2 \omega^{2v} + c_3 \omega^{3v} + c_4 \omega^{4v} + c_5 \omega^{5v} + c_6 \omega^{6v} + c_7 \omega^{7v}$$

which is done for $v = 0, 1, \ldots, 7$. Ordinarily, using Horner's scheme, we would need n operations for each v. Thus n^2 operations would be needed for the complete transform. To improve this, we subdivide the problem into smaller but similar subproblems, and work upwards in a recursive manner when we reach a trivial stage.

Assume that n is a power of 2. Divide the associated polynomial into two parts: one containing terms with even exponents, and the other, terms with odd exponents. Denote these parts as f^e and f^o, respectively. Then

$$f_{n,v}(\omega) = f^e_{n/2,v}(\omega) + \omega^v f^o_{n/2,v}(\omega) \tag{11.49}$$

For $n = 8$, we have

$$f_{8,v}(\omega) = (c_0 + c_2 \omega^{2v} + c_4 \omega^{4v} + c_6 \omega^{6v}) + \omega^v (c_1 + c_3 \omega^{2v} + c_5 \omega^{4v} + c_7 \omega^{6v})$$

Let $\rho = \omega^2$ and $m = n/2$. Then for $v = 0, 1, \ldots, m-1$,

$$f_{n,v}(\omega^v) = f^e_{n/2,v}(\rho^v) + \omega^v f^o_{n/2,v}(\rho^v) \qquad (11.50)$$

$$f_{n,v}(\omega^{v+m}) = f^e_{n/2,v}(\rho^v) - \omega^v f^o_{n/2,v}(\rho^v) \qquad (11.51)$$

A key observation here is that $\{\rho^v, v = 0, 1, \ldots, m-1\}$ are the mth roots of unity. Therefore, the original problem has now been converted into two half the size problems, namely, evaluating the two m degree polynomials, $f^e_v(\rho^v)$ and $f^o_v(\rho^v)$, at the m roots of unity. Because n is a power of two, we can continue this process until the problem becomes that of evaluating a constant.

To program this, we need a subroutine that divides a polynomial of even degree into the odd and even components and forms the relations (11.50) and (11.51). The subroutine is called from within itself to process the two half the degree polynomials. We continue this until the degree of the polynomial to be processed is 1, where we have a direct solution. Such a technique of performing the FFT is called the recursive approach. Note that the term recursion has a slightly different meaning here. It needs a programming language in which a subroutine can be called within itself and which can handle large memory stacks in an efficient manner.

Another approach to implement the FFT is developed by looking at the details of the subdivisions of the coefficients at each stage of the recursive process. Thus, as we divide f^e and f^o into their even and odd coefficients segments, and continue similarly, after four steps we get (under a loose use of the above notation)

$$\begin{aligned} f_v &= \left[f^{ee}_v + \omega^{2v} f^{eo}_v \right] + \omega^v \left[f^{oe}_v + \omega^{2v} f^{oo}_v \right] \\ &= \left\{ \left[f^{eee}_v + \omega^{4v} f^{eeo}_v \right] + \omega^{2v} \left[f^{eoe}_v + \omega^{2v} f^{eoo}_v \right] \right\} + \\ &\quad \omega^v \left\{ \left[f^{oee}_v + \omega^{4v} f^{oeo}_v \right] + \omega^{2v} \left[f^{ooe}_v + \omega^{4v} f^{ooo}_v \right] \right\} \end{aligned}$$

Applying this to the polynomial with $n = 8$, we have

$$f_v = \left\{ \left[c_0 + \omega^{4v} c_4 \right] + \omega^{2v} \left[c_2 + \omega^{4v} c_6 \right] \right\} + \omega^v \left\{ \left[c_1 + \omega^{4v} c_5 \right] + \omega^{2v} \left[c_3 + \omega^{4v} c_7 \right] \right\}$$

Equating the two, we have that, for all $v = 0, 1, \ldots, 7$,

$$f^{eee}_v = c_0; \; f^{eeo}_v = c_4; \; f^{eoe}_v = c_2; \; f^{eoo}_v = c_6$$
$$f^{oee}_v = c_1; \; f^{oeo}_v = c_5; \; f^{ooe}_v = c_3; \; f^{ooo}_v = c_7$$

For $n = 8$, we thus have directly identified the final stage polynomials of degree 1. For a general n, determining the correspondence between a coefficient c_v and a final step polynomial $f^{eooe\cdots}_v$ is done by the bit reversal process. For a given $f^{eooe\cdots}_v$, we reverse the pattern of e's and o's, and let $o = 1$ and $e = o$. This gives the binary value of v and helps to simplify the implementation of the FFT.

Under an alternative approach, we rearrange the coefficients according to the bit reversed scheme and, starting from the so identified innermost loops, implement a sequential evaluation of the loops and proceed outwards. Such an approach for the FFT is called the iterative approach.

THE FAST FOURIER TRANSFORM

(Again, this term has a different meaning here from that in Chapter 4.) Implementation is further simplified by using, in the innermost loop, the fact that for an even n, $\omega^{(n/2)} = -1$. When $n = 8$, we get $\omega^{4v} = (-1)^v$.

For both the recursive and iterative FFT approaches, there are $\log_2 n$ loops for a given v. The complete transform thus requires operations of the order of $n \log_2 n$ operations. (For a rigorous proofs of efficiency, see the cited literature §11.8.)

Several aspects of implementing the FFT are important for us.

- The above constructions assume that n is a power of 2. When it is not, we add zero valued coefficients to construct a polynomial with such a property.

- The general FFT applies to complex valued coefficients. In our case, we deal with real valued coefficients. Then the FFT may be improved by packing two polynomials into a single DFT, thereby reducing the number of FFT calls by a half. (See the cited literature in §11.8 for details.)

****

The FFT is among the most widely used numerical algorithms today and has a very extensive literature. Above, we have only given the bare bones of its broad scope and theory. Its utility in many fields has been established beyond doubt. In the context of exact analysis for discrete data, however, the FFT method, as used thus far, has shown serious shortcomings. Unless these are remedied, its utility remains in question. In summary, among these problems are:

- In applications relating to exact analysis, the FFT projects real coefficients to the complex plane and back. Under finite precision arithmetic, the final coefficients contain an element of error. The error is increased when the ratio of the largest to the smallest coefficient (or, the dynamic range) is large. With integer coefficients, such an error is avoided by the use of Fourier primes and modular arithmetic, especially when the dynamic range is less than 2^{20}. The range of the coefficients found in discrete data analysis often exceeds this limit even when the data are sparse. It can thereby produce substantial transform generated errors.

- Such errors produce inaccurate exact distributions which in turn lead to seriously inaccurate p-values and CIs for the sparse data sets of the type encountered in practice.

  For example, for a case-control study data given in Breslow and Day (1980), the FFT method gave an exact CI for the COR equal to (1.37,1.47). The correct CI, generated by the EE and RPM methods, however, was (3.57,7.76) (Vollset, Hirji and Elashoff 1991).

  For the test of conditional independence in a variety of data sets, the FFT method produced incorrect negative probability values in the sample space as well as erroneous p-values. A gain in accuracy under the FFT is offset by a major compromise of efficiency (Hirji 1996).

  For exact analysis of a $2 \times K$ table, the FFT was shown to be less time efficient and less accurate than the RPM algorithm (Hirji and Johnson 1996).

- The relevance of the exponent and other checks developed for the FFT to exact discrete data analysis has not as yet been established.

- The RPM method can readily handle multidimensional polynomials. The use of a multidimensional FFT, on the other hand, adds more concerns about its accuracy and efficiency.

- For the details and specific examples of the problems of the FFT approach in the discrete data context, see the cited literature in §11.8.

For application to exact analysis of discrete data, we argue that the FFT approach is the case of a theoretically efficient method which attains optimal efficiency at sample space sizes at which there is hardly any difference between exact and large sample methods. It sacrifices accuracy at the expense of efficiency, and leads to unacceptable errors in inference. Further research may indicate how such problems can be minimized.

11.8 Relevant Literature

There is an extensive literature on polynomial multiplication algorithms in general and the FFT in particular. Popular texts on algorithms usually cover these topics. See Aho, Hopcroft and Ullman (1974), Horowitz and Sahni (1984), Kronsjö (1987), Press et al. (1992) and Sedgewick (1988). A comprehensive though specialized approach to the subject is given by Nussbaumer (1982). Duhamel and Vetterli (1990) has a thorough review of the FFT. Some of these works also contain computer programs for the FFT.

Edgington (1995) and Manly (1997) are applications oriented books on simulation, Monte–Carlo and randomization methods which also cover discrete data analysis. Ross (2002) is a fine introduction to the subject. Many papers have applied refined MC methods like importance sampling and the Markov chain Monte–Carlo method to discrete data. They include Mehta, Patel and Senchaudhuri (1988), Forster, McDonald and Smith (1996), McDonald, De Roure and Michaelides (1998), McDonald and Smith (1995), McDonald, Smith and Forster (1999), Smith, Forster and McDonald (1996) and Smith and McDonald (1995). Skovlund and Bølviken (1996) give a general MC method for constructing randomization based CIs. Sampling techniques like the bootstrap are applicable in this context as well. See Efron and Tibshirani (1993).

The basic idea equivalent to exponent checks was first employed in the context of network algorithms by Mehta and Patel (1980), and was extended to multivariate analysis by Hirji, Mehta and Patel (1987) and Hirji, Mehta and Patel (1988).

Thomas (1975) and Thomas and Gart (1992) dealt with the EE approach for the analysis of the common odds ratio model for several 2×2 tables. Pagano and Tritchler (1983a) applied the FFT to it, and Pagano and Tritchler (1983b) extended its use to a wider class of exact analysis problems. Mehta, Patel and Gray (1985) applied the network algorithm to the COR model. Vollset, Hirji and Elashoff (1991) showed that their algorithm was equivalent to the RPM algorithm and that the FFT method can yield highly inaccurate CIs for the COR model. Martin and Austin (1991) also used the polynomial formulation and multiplication in this context. The issue was further developed by Vollset and Hirji (1991). The last two papers together with Hirji and Vollset (1994a) give noncommercial software or computer programs.

Yao and Tritchler (1993) applied the FFT for testing conditional independence in several 2×2 tables and compared their analytic approach with a two step approach of Zelen (1971). Hirji (1996) gave an RPM algorithm, and showed that the FFT algorithm compromised both accuracy and efficiency. The RPM approach was shown to be equivalent to the network approach.

General reviews and evaluation of the FFT approach for exact analysis of discrete data are given by Baglivo, Olivier and Pagano (1993) and Hirji (1997a, 1998). The RPM approach is not covered in works on algorithms or polynomials. See Hirji (1997a, 1998) for a broad introduction. van de Wiel, Di Bucchianico and van der Laan (1999) provide a general introduction to generating functions in statistical analysis.

11.9 Exercises

11.1. Consider the set of 2×2 tables with an overall total equal to 10. Generate all the tables from this set in which the unconditional mle, the conditional mle or the conditional mue

EXERCISES

for the log-odds ratio are not finite. What is the relationship between these subsets? Generalize your results to a table with the total equal to n.

Extend this investigation to analysis of the logit model interaction term in a $2 \times 2 \times 2$ table under a product binomial design and with equal and unequal group sizes.

11.2. For Ganciclovir data in Table 6.3, show that the number of points in the unconditional sample space is $1,228,122$.

11.3. Write computer programs for the EE and MC algorithms to generate the conditional distribution of $\mathcal{T}_3$ in relation to model (6.1). Implement the relevant exponent checks for both approaches. (i) Use data examples in Chapter 6 to study the error in estimation of the null probabilities. (ii) Document the impact of these errors on p-values, cmues, cmles and exact confidence limits. (iii) For both these algorithms, use the data examples in Chapter to 6 compare the computing times needed with and without the use of exponent checks. (iv) In the randomization method, investigate the efficiency associated with use of sampling from the uniform discrete distribution instead of the binomial distribution.

11.4. Write a computer program to implement the EE approach for the COR model and for tests for conditional independence in several 2×2 tables. For several values of K and for different strata sizes and structures, determine the number of multiplications and the cpu times used.

11.5. Write a computer program to implement the RPM approach for the COR model and for tests for conditional independence in several 2×2 tables. For several K values and different strata sizes and data structures, determine the number of multiplications and the cpu times used.

11.6. Investigate the use of a more efficient method of multiplying two polynomials at each stage of the RPM method. Can it be adapted to include exponent checks?

11.7. In the case of tests for conditional independence, does the use of a hash array instead of a regular array provide savings in terms of memory and cpu time? What hash function(s) would you use here?

11.8. Extend the above exercise for computation of conditional power for various test statistics for the COR model and for tests for conditional independence in several 2×2 tables.

11.9. Implement the exponent checks for the tail segments, and point and tail probabilities in the EE approach. Give empirical results showing improvement of efficiency resulting from their use in datasets with several values K and variety of strata sizes and structures.

11.10. Implement the exponent checks for tail segments, and point and tail probabilities in the RPM approach. Give empirical results showing improvement of efficiency resulting from their use in datasets with several values K and variety of strata sizes and structures.

11.11. Suppose the multiplication of four polynomials, each of degree n, is done as follows: $\{f_1(\phi) * f_2(\phi)\} * \{f_3(\phi) * f_4(\phi)\}$. Show that the number of multiplications involved here is $2(n+1)^2 + (2n+1)^2$. How does that compare with the RPM approach?

Also consider a divide and conquer approach given in Sedgewick (1988) for this problem. In this, the polynomials to be multiplied are successively divided into right and left portions until polynomials of degree one are obtained. The product is then obtained by an upward recursion. Show how you would implement this on a computer when n is a power of 2. How would you proceed when n is not a power of 2?

11.12. Derive the formulas for $\nu_{EE.3}$ (11.4) and $\nu_{RPM.1}$ (11.13).

11.13. Give formal justifications for the exponent checks used in EC 11.01 up to EC 11.07.

11.14. Suppose before multiplying K polynomials in the RPM method, we order them in terms of the values of n_k. Does an ascending or a descending ordering minimize the total number of multiplications? Is the reduction of practical consequence? Does an ascending or a descending ordering work better with some or all of the exponent checks? If so, specify the applicable conditions.

11.15. Derive the exponent checks to directly compute the exact p-value for the COR model using the conditional score statistic and implement it in the RPM method. Here we need the minimum and maximum values of $(N_k u - n_k s_k)^2$ subject to $u \in \Omega_k$.

11.16. Modify the COR model exponent checks to directly compute the conditional p-value for the (un)conditional score or the (un)conditional LR statistics. Undertake a similar exercise for the tests for conditional independence. Where you cannot explicitly determine the exact minimum of the test statistic components, use an appropriate bound.

11.17. Implement the RPM algorithm in which both a tail and a point probability are simultaneously generated directly. Use it to directly compute the TST mid-p-value. Compare the efficiency of this method with computation of the mid-p-value from the complete distribution. For this purpose, use real and hypothetical datasets. Are the differences relevant in practical terms?

11.18. Suppose we want to compute the TST p-value for the COR model using the RPM method. In addition to the point probability, we need the left or the right tail probability. Use real and hypothetical datasets to find when it is more efficient to compute the left tail probability and when to compute the right tail probability. If possible, derive a rule to determine in advance which tail we should compute.

11.19. Consider the Mantel–Haenszel statistic exact based p-value for the COR in several 2×2 tables. Can it be computed using the RPM method? If yes, show how. Otherwise discuss why not. Write a computer program for this task using the EE method. If possible, develop exponent checks for this statistic.

11.20. Investigate and implement other polynomial algorithms that can be used to compute exact distributions for the COR model. Which of them permit the use of exponent checks?

11.21. Write an efficient computer program to directly compute the conditional mle and exact CI for the COR model.

11.22. Implement the efficiency enhancing techniques of Liao and Hall (1995) for computing the exact CI for the COR. Compare them with the use of exponent checks in the RPM method.

11.23. Design and conduct simulation studies in order to compare various exact and asymptotic p-values and CIs for the COR model and for testing conditional independence in several 2×2 tables. (Virtually no work on the size and power of exact and mid-p tests for the latter problem has been done.)

11.24. Derive the properties of the roots of unity stated in §11.7. Use this to prove the validity of the inverse DFT.

11.25. Give a detailed derivation of the recursive and iterative methods for the FFT, and give a formal proof of their efficiency. Prove the bit reversal method. Also show how the FFT may be adapted for the inverse DFT.

11.26. Show how the FFT for two sets of real coefficients may be made more efficient by packing the two polynomials into a single DFT.

11.27. Write an efficient computer program to analyze the COR model and perform tests for conditional independence in several 2×2 tables using the FFT. Compare the efficiency of the FFT method to that of the EE and RPM methods. Can the FFT method be modified to incorporate exponent checks?

11.28. Give details of the problems of applying the FFT to discrete data analysis that have been reported in the literature. Investigate the issue using real and hypothetical data sets.

EXERCISES

11.29. Prove Lemma 11.1, Lemma 11.2, and give a detailed proof of Theorem 11.1.

11.30. Suppose, as in Lemma 11.2, $\mathcal{Y} \in \{0, 1\}$ and $\mathcal{X}$ are discrete random variables and $P[\mathcal{Y} = 1]$ does not depend on the value of $\mathcal{X}$. Then show that

$$\mathbf{E}[\mathcal{Y}^2 \mathcal{X}^2] = P[\mathcal{Y} = 1]\mathbf{E}[\mathcal{X}^2 \mid \mathcal{Y} = 1] \qquad (11.52)$$

Use this to compute the variance of the polynomial coefficients obtained from Algorithm 11.MC.2. Show how this can be used to obtain confidence limits for the analytic quantities obtained from this algorithm.

11.31. Show how you would compute, from the MC approach, the 95% confidence limits for exact TST *p*-value and mid-*p*-value, mue and CI for the common odds ratio in several 2×2 tables. Extend this to the test for conditional independence.

CHAPTER 12

Polynomial Algorithms: II

12.1 Introduction

This chapter extends the recursive polynomial multiplication (RPM) algorithms of Chapter 11 to multivariate polynomials and describes their application to exact analysis of discrete data. Its specific aims are:

- To present the RPM algorithms for bivariate, trivariate and higher dimensional polynomials and extend the idea of exponent checks.

- To describe the technique of backward induction, and use it for constructing conditional exponent checks and evaluating a conditional polynomial.

- To apply the bivariate and trivariate RPM algorithms to models for a series of 2×2 tables, a $2 \times K$ table and other discrete data problems.

- To introduce the network algorithm and demonstrate a one-to-one correspondence between them and respective RPM algorithms.

- To consider aspects of computing the exact power of discrete data tests.

- To address computer implementation of the extended RPM algorithm.

12.2 Bivariate Polynomials

For $k = 1, \ldots, K$, consider the polynomials

$$f_k(\theta, \phi) = \sum_{u \in \Omega_k} c_k(u) \theta^u \phi^{w_k(u)} \tag{12.1}$$

where $c_k(u) > 0$ and $\Omega_k = \{0, 1, \ldots, e_k\}$. The exponents of ϕ, $w_k(u)$, are called scores. Some of the common scores, noted in Chapter 10, are linear, quadratic and exponential scores. $w_k(u)$ may also be the kth component of a score, LR or some other statistic, or the value of a covariate.

The product of the K polynomials is

$$f(\theta, \phi) = \prod_{k=1}^{K} f_k(\theta, \phi) = \sum_{(u,v) \in \Omega} c(u, v) \theta^u \phi^v \tag{12.2}$$

where Ω is the set of exponents of θ and ϕ. This set and the set $\{c(u, v) : (u, v) \in \Omega\}$ suffice to specify the generic polynomial $f(\theta, \phi)$. This product is readily obtained on a computer using nested loops or an equivalent implementation. This EE method, described in §11.2, multiplies the

K polynomials in a term by term manner. As before, the number of multiplications it performs is

$$(K-1)(1+e_1)(1+e_2)\cdots(1+e_K) \tag{12.3}$$

A more efficient alternative multiplies the polynomials in a cumulative manner. In this RPM approach, we define $F_1(\theta,\phi) = f_1(\theta,\phi)$, and for $k = 2, \ldots, K$, implement the recursion

$$F_k(\theta,\phi) = F_{k-1}(\theta,\phi)f_k(\theta,\phi) \tag{12.4}$$

Obviously, $F_K(\theta,\phi) = f(\theta,\phi)$. An explicit formula for the number of multiplications involved in the RPM approach is not known. For integer scores from a finite set, the efficiency of the RPM method over EE can be established by theoretical arguments. For other settings, this is done in an empirical fashion. The efficiency of RPM is a result of the aggregation of terms at each step. It is more pronounced with linear scores and is markedly reduced with exponential scores, or with scores that are not integer valued.

For the purpose of computer implementation, for $k = 1, \ldots, K$, let

$$e_{+,k} = e_1 + \cdots + e_k \tag{12.5}$$

When the exponent of θ in $F_k(\theta,\phi)$ is equal to $u \in \{0, 1, \ldots, e_{+,k}\}$, let the number of the distinct exponents of ϕ be $N_k(u)$. And further let these distinct exponents be ordered in some fashion, and denoted as

$$\{w_{+,k}(1,u), \ldots, w_{+,k}(N_k(u), u)\}$$

Then we write

$$F_k(\theta,\phi) = \sum_{u=0}^{e_{+,k}} \theta^u \sum_{j=1}^{N_k(u)} d_k(j,u)\phi^{w_{+,k}(j,u)} \tag{12.6}$$

where $d_k(j,u)$ is the coefficient associated with the jth exponent of ϕ when the exponent of θ is u. Under our usual shorthand notation, we also write (12.6) as

$$F_k(\theta,\phi) = \sum_{u=0}^{e_{+,k}} \theta^u F_k(1,\phi;u,.) \tag{12.7}$$

Algorithm 12.RPM.1 is a scheme for a computer implementation of (12.4). It has three main arrays, $N(0:1, 0:e_+)$, $W(0:1, 0:e_+, 1:M)$ and $C(0:1, 0:e_+, 1:M)$. The first dimension in each array is an indicator for the current ($j = 0$), or the following ($j = 1$) stage in the recursion. The second dimension denotes the exponents of θ. The array N stores, for each stage and exponent of θ, the number of distinct exponents of ϕ. The mth places along third dimension in W and C store, respectively, the value of the mth distinct exponent of ϕ, and its coefficient. M is the maximum possible number of such exponents.

BIVARIATE POLYNOMIALS

Algorithm 12.RPM.1

Step 01.	Specify M and $e_+ \Leftarrow e_1 + \cdots + e_K$
Step 02.	For $u = 0, 1, \ldots, e_+$ do
Step 03.	$\quad N(0, u) = 0;\ N(1, u) = 0$
Step 04.	$\quad$ For $v = 1, \ldots, M$ do
Step 05.	$\quad\quad W(0, u, v) = -1;\ W(1, u, v) = -1$
Step 06.	$\quad\quad C(0, u, v) = 0;\ C(1, u, v) = 0$
Step 07.	End do; End do
Step 08.	Set $j = 0;\ k = 1;\ n = e_1$
Step 09.	For $u = 0, 1, \ldots, n$ do
Step 10.	$\quad N(j, u) = 1$
Step 11.	$\quad$ Compute $w_1(u);\ W(j, u, 1) \Leftarrow w_1(u)$
Step 12.	$\quad$ Compute $c_1(u);\ C(j, u, 1) \Leftarrow c_1(u)$
Step 13.	End do
Step 14.	$k \Leftarrow k + 1$
Step 15.	For $u = 0, 1, \ldots, e_k$ do
Step 16.	$\quad$ Compute $w_k(u)$ and $c_k(u)$
Step 17.	End do
Step 18.	For $v = 0, 1, \ldots, n$ do
Step 19.	$\quad$ For $u = 0, 1, \ldots, e_k$ do
Step 20.	$\quad\quad i = u + v$
Step 21.	$\quad\quad$ For $jj = 1, \ldots, N(j, v)$ do
Comment:	Perform exponent checks here
Step 22.	$\quad\quad\quad w = w_k(u) + W(j, v, jj)$
Step 23.	$\quad\quad\quad x = c_k(u) * C(j, v, jj)$
Step 24.	$\quad\quad\quad$ Determine $h(w)$
Comment:	Identical or empty record
Step 25.	$\quad\quad\quad$ If $(W(1-j, i, h(w)) = -1)$ then
Step 26.	$\quad\quad\quad\quad W(1-j, i, h(w)) = w$
Step 27.	$\quad\quad\quad\quad C(1-j, i, h(w)) = x$
Step 28.	$\quad\quad\quad\quad N(1-j, i) \Leftarrow N(1-j, i) + 1$
Step 29.	$\quad\quad\quad$ Else
Step 30.	$\quad\quad\quad\quad C(1-j, i, h(w)) \Leftarrow C(1-j, i, h(w)) + x$
Step 31.	$\quad\quad\quad$ End If
Step 32.	End do; End do; End do
Step 33.	For $v = 0, 1, \ldots, n$ do
Step 34.	$\quad W(j, v, 1) = -1;\ C(j, v, 1) = 0;\ N(j, v) = 0$
Step 35.	End do
Step 36.	$n \Leftarrow n + e_k$
Step 37.	$j \Leftarrow 1 - j$
Step 38.	If $(k < K)$ go to Step 14

□□□□□□□

At each exponent of θ, $h(w)$ is a function assigning a storage location for the term with exponent of ϕ at w. This may be done using chaining or a hash function. Our implementation assumes that records are stored in a contiguous manner and the value -1 in W indicates an empty record. The final result is in the jth segment of N, W and C. Note, this is a skeleton program that has to be optimized and fine tuned for actual use.

12.3 A Conditional Polynomial

Often, we do not need the complete polynomial $f(\theta, \phi)$ but only some terms, or a computed quantity based on some terms. Let (12.2) be written as

$$f(\theta, \phi) = \sum_{u=0}^{e_+} \theta^u f(1, \phi; u, .) \tag{12.8}$$

where $e_+ = e_{+,K} = e_1 + \cdots + e_K$. Now suppose we need only the terms in which the exponent of θ is equal to t, or $f(\theta, \phi; t, .)$. Instead of extracting this from the complete polynomial, we obtain it more efficiently by proceeding as in Chapter 11 and applying exponent check EC 11.01 to the exponents of θ. For the purposes of this chapter, EC 11.01 needs to be reformulated. As before, let

$$t_{+,k} = t_1 + \cdots + t_k \tag{12.9}$$

Now, consider values of nonnegative integers $t_k, k = 1, \ldots, K$, that satisfy

$$0 \leq t_k \leq e_k \tag{12.10}$$

and

$$t_1 + \cdots + t_K = t \tag{12.11}$$

Let $\Lambda_k(t)$ be the set of values of $t_{+,k}$ consistent with (12.10) and (12.11). This set is derived as follows. First, we have that $0 \leq t_{+,1} = t_1 \leq e_1$ and

$$0 \leq t - t_1 = t_2 + \cdots + t_K \leq e_+ - e_{+,1}$$

which implies that

$$\max(0, t - e_+ + e_{+,1}) \leq t_{+,1} \leq \min(e_1, t)$$

In general, the kth cumulative sum, $t_{+,k}$, satisfies

$$\max(0, t - e_+ + e_{+,k}) \leq t_{+,k} \leq \min(e_{+,k}, t) \tag{12.12}$$

or that

$$\Lambda_k(t) = \{M_{1k}(t), M_{1k}(t) + 1, \cdots, M_{2k}(t)\} \tag{12.13}$$

where

$$M_{1k}(t) = \max(0, t - e_+ + e_{+,k}) \tag{12.14}$$

and

$$M_{2k}(t) = \min(e_{+,k}, t) \tag{12.15}$$

A CONDITIONAL POLYNOMIAL

It is readily seen that

$$\Lambda_K(t) = \{t\}$$

as required. Further, we also set

$$\Lambda_0(t) = \{0\}$$

Suppose we have the set $\Lambda_{k-1}(t)$ and, for each $v \in \Lambda_{k-1}(t)$, we need the values of $0 \leq t_k \leq e_k$ that also satisfy $v + t_k \in \Lambda_k(t)$. This means that

$$M_{1k}(t) \leq v + t_k \leq M_{2k}(t)$$

which implies that

$$\max(0, M_{1k}(t) - v) \leq t_k \leq \min(e_k, M_{2k}(t) - v)$$

which can then be written as

$$t_k \in \{m_{1k}(t,v), m_{1k}(t,v) + 1, \cdots, m_{2k}(t,v)\} \quad (12.16)$$

where

$$m_{1k}(t,v) = \max(0, t - e_+, e_{+,k} - v) \quad (12.17)$$

and

$$m_{2k}(t,v) = \min(e_k, e_{+,k} - v, t - v) \quad (12.18)$$

These quantities are adapted to the recursive process (12.4) as follows.

Modification 1: At each stage k, retain only those terms of $F_k(\theta, \phi)$ in which the exponent of θ is included in the set $\Lambda_k(t)$. Based on (12.6), we write this modified polynomial as:

$$F_{k,t}(\theta, \phi) = \sum_{u=M_{1k}(t)}^{M_{2k}(t)} \theta^u \sum_{j=1}^{N_k(u)} d_k(j,u) \phi^{w_{+,k}(j,u)} \quad (12.19)$$

Modification 2: When performing multiplications at the kth stage, we multiply terms of the so restricted polynomial $F_{k-1,t}(\theta, \phi)$ with only the corresponding terms of $f_k(\theta, \phi)$ in which the exponent of θ satisfy (12.16).

This selective multiplication scheme ensures that at the final stage, we only have terms of θ with exponent equal to t. This scheme is equivalent to applying the exponent check EC 11.01.

Algorithm 12.RPM.2

Step 01.   Specify t, M and $e_+ \Leftarrow e_1 + \cdots + e_K$
Step 02.   For $u = 0, 1, \ldots, e_+$ do
Step 03. $N(0, u) = 0$; $N(1, u) = 0$
Step 04.   For $v = 1, \ldots, M$ do
Step 05. $W(0, u, v) = -1$; $W(1, u, v) = -1$
Step 06. $C(0, u, v) = 0$; $C(1, u, v) = 0$
Step 07.   End do; End do
Step 08.   For $k = 1, \ldots, K$ do
Step 09.   Compute $M_{1k}(t)$ and $M_{2k}(t)$
Step 10.   End do
Step 11.   Set $j = 0$; $k = 1$
Step 12.   For $u = M_{11}(t), \ldots, M_{21}(t)$ do
Step 13. $N(j, u) = 1$
Step 14.   Compute $w_1(u)$; $W(j, u, 1) \Leftarrow w_1(u)$
Step 15.   Compute $c_1(u)$; $C(j, u, 1) \Leftarrow c_1(u)$
Step 16.   End do
Step 17. $k \Leftarrow k + 1$
Step 18.   For $u = 0, 1, \ldots, e_k$ do
Step 19.   Compute $w_k(u)$ and $c_k(u)$
Step 20.   End do
Step 21.   For $v = M_{1,k-1}(t), \ldots, M_{2,k-1}(t)$ do
Step 22.   Compute $m_{1k}(t, v)$ and $m_{2k}(t, v)$
Step 23.   For $u = m_{1k}(t, v), \ldots, m_{2k}(t, v)$ do
Step 24. $i = u + v$
Step 25.   For $jj = 1, \ldots, N(j, v)$ do
Step 26. $w = w_k(u) + W(j, v, jj)$
Comment: Perform exponent checks here
Step 27. $x = c_k(u) * C(j, v, jj)$
Step 28.   Determine $h(w)$
Comment: Identical or empty record
Step 29.   If $(W(1 - j, i, h(w)) = -1)$ then
Step 30. $W(1 - j, i, h(w)) = w$
Step 31. $C(1 - j, i, h(w)) = x$
Step 32. $N(1 - j, i) \Leftarrow N(1 - j, i) + 1$
Step 33.   Else
Step 34. $C(1 - j, i, h(w)) \Leftarrow C(1 - j, i, h(w)) + x$
Step 35.   End If
Step 36.   End do; End do; End do
Step 37.   For $v = M_{1,k-1}(t), \ldots, M_{2,k-1}(t)$ do
Step 38. $W(j, v, 1) = -1$; $C(j, v, 1) = 0$; $N(j, v) = 0$
Step 39.   End do
Step 40. $j \Leftarrow 1 - j$
Step 41.   If $(k < K)$ go to Step 17

□□□□□□□

Algorithm 12.RPM.2 is an adaptation of 12.RPM.1 under these modifications. The main arrays N, W and C and the storage function $h(w)$ remain the same. This algorithm also needs to be fine tuned and optimized for actual use.

12.4 Backward Induction

Backward induction is an elegant computational technique we can use to derive exponent checks in multidimensional situations. It is part of a widely applied general technique called dynamic programming, and is suitable for settings where the problem either is in the form of a series of finite, consecutive stages, or can be formulated as such. The **Principle of Optimality** is invoked to resolve the problem.

> The Principle of Optimality
>
> An optimal policy has the property that whatever the
> initial state and initial decision are, the remaining
> decisions must constitute an optimal policy with regard
> to the state resulting from the first decision. Bellman
> and Dreyfus (1962), page 15.

This simple principle provides a powerful and efficient device for solving an extensive array of problems in networks, engineering and planning as well as in the theory of optimization, calculus and stochastic processes.

The logic behind it is clarified by considering a travel problem. Suppose you need to go from one point to another in such a way that you have to pass through K regions during the trip in a sequential manner. That is, you must pass through the $(k-1)$th region before entering the kth region. Suppose you are taking the shortest path from the origin to the destination and you have reached a point in the k region. Since the sequence of regions to be passed is fixed, and jumping of regions is disallowed, it follows that the path from your current position to the final destination must also be the shortest path from that point to the final destination.

Below we apply the optimality principle to two specific problems. In these problems, we proceed in a backward manner from stage k to $k-1$, $k = K, K-1, \ldots, 1$. At stage K, we set initial values. Then at stage $K-1$ we derive optimal values from all positions in that stage which are consistent with the overall constraints in the problem to the position in stage K. Next, from these points, we step back to all possible consistent positions in the stage $K-2$ to the consistent positions in stage $K-1$. And so the process is continued until the sole position in the stage 0 is reached. This process has a major advantage in that it not only gives the overall optimal value but, when needed and without further computation, the optimal values from all consistent intermediate positions up to the final stage. The latter are stored as they are computed for subsequent use.

First we apply this scheme to the computation of the sum of a conditional product. Suppose, for some $h_k(t_k) \geq 0$, we need the value of the sum

$$\Delta = \sum \left[\prod_{j=1}^{K} h_k(t_k) \right] \tag{12.20}$$

where the summation is taken over all values of $t_k, k = 1, \ldots, K$ that satisfy (12.10) and (12.11). The following backward induction scheme is relevant for this task.

Step 0: With $\Lambda_K(t) = \{t\}$, let $\Delta_K(t) = 1$.

Step 1: For each $v \in \Lambda_{K-1}(t)$, all feasible values of t_K are given by $t_K = t - v$. For each such v, we then compute

$$\Delta_{K-1}(v) = h_K(t-v)\Delta_K(t) = h_K(t-v)$$

Step 2: For each $v \in \Lambda_{K-2}(t)$, the feasible values of t_{K-1} must, by (12.16), (12.17) and (12.18), satisfy

$$m_{1,K-1}(t,v) \leq t_{K-1} \leq m_{2,K-1}(t,v)$$

For each $v \in \Lambda_{K-2}(t)$, we then compute

$$\Delta_{K-2}(v) = \sum_{u=m_{1,K-1}(v,t)}^{m_{2,K-1}(v,t)} h_{K-1}(u)\Delta_{K-1}(u+v)$$

..... And so until

Step k: For each $v \in \Lambda_{K-k}(t)$, we compute

$$\Delta_{K-k}(v) = \sum_{u=m_{1,K-k+1}(v,t)}^{m_{2,K-k+1}(v,t)} h_{K-k+1}(u)\Delta_{K-k+1}(u+v) \qquad (12.21)$$

.... And so on until

Step K: And finally, since $\Lambda_0(t) = \{0\}$, we compute

$$\Delta_0(0) = \Delta = \sum_{u=m_{1,1}(0,t)}^{m_{2,1}(0,t)} h_1(u)\Delta_1(u) \qquad (12.22)$$

****

Next we apply the above scheme to determining, for $k = 1, \ldots, K$, the following:

$$\eta_{1,k}(v) = \min\left[\sum_{j \geq k} h_j(t_j)\right] \qquad (12.23)$$

$$\eta_{2,k}(v) = \max\left[\sum_{j \geq k} h_j(t_j)\right] \qquad (12.24)$$

where the maximization and minimization are subject to

$$\sum_{j \geq k} t_j = v, \quad \text{and} \quad 0 \leq t_j \leq e_j, \; j \geq k$$

and further where v ranges over values consistent with the overall conditions (12.10) and (12.11).

CONDITIONAL VALUES

In this case, we modify the above scheme so that at each step, instead of computing sums, we compute maxima or minima, as needed. Accordingly, for maximization, the kth step computation becomes

Step k: For each $v \in \Lambda_{K-k}(t)$, we compute

$$\eta_{2,K-k}(v) = \max\left[h_{K-k+1}(u)\Delta_{K-k+1}(u+v) \right] \qquad (12.25)$$

over $u \in \{m_{1,K-k+1}(v,t), m_{1,K-k+1}(v,t)+1, \cdots, m_{2,K-k+1}(v,t)\}$.

At the end, we get $\eta_{2,t}$, or $\eta_{1,t}$, respectively, the maxima, or the minima of

$$\sum_{k=1}^{K} h_k(t_k)$$

subject to

$$\sum_{k=1}^{K} t_k = t, \quad \text{and} \quad 0 \leq t_k \leq e_k, \quad k \geq 1$$

Additionally, we get the optimal values of the intermediate sums which may be stored for use when needed.

****

We note the following:

- The overall sum (12.20), minimum (12.23) and minimum (12.24) can be computed within a single backward induction pass.
- Computing the partial sums, maxima and minima does not require any further computational effort as they are generated within the overall computational process.

12.5 Conditional Values

Suppose that for the polynomial (12.2), we need one or more of the following quantities:

- The coefficient of the term in which the exponent of θ is t and the exponent of ϕ is equal to w_*, that is, $f(1, 1; t, w_*)$.
- The terms for which the exponent of θ is t and the exponents of ϕ are greater than or equal to w_*, that is, $f(\theta, \phi; t, \geq w_*)$.
- The sum of the coefficients of all terms for which the exponent of θ is t, that is, $f(1, 1; t, .)$.
- The value of that part of the polynomial in which the exponent of θ is t and the exponents of ϕ are greater than or equal to w_* when $\theta = 1$ and $\phi = \phi_0$, that is, $f(1, \phi_0; t, \geq w_*)$. For the special case $\phi_0 = 1$, $f(1, 1; t, \geq w_*)$ is the sum of the coefficients of the terms for which the exponent of θ is t and the exponents of ϕ are greater than or equal to w_*.

Algorithm 12.RPM.2 gives that segment of the polynomial $f(\theta, \phi)$ in which the exponent of θ is fixed. As this requirement is common to the above four cases, we may use the output of 12.RPM.2 to obtain any of these quantities. This may be a good strategy when such quantities are needed for different values of w_* and ϕ_0. With a more focused goal, or when the number of terms in the intermediate stages is excessive, it is, however, generally more efficient to apply additional exponent criteria on the exponents of ϕ. These criteria relate to but are somewhat different from those developed in Chapter 11. The principal difference is the need to take into consideration the fixed value of the final exponent of θ.

To develop such criteria, using (12.6), write the product at stage k of Algorithm 12.RPM.2, $F_k(\theta, \phi)$, as:

$$\sum_u \sum_v \theta^{(u+v)} \sum_w c_k(v) d_{k-1}(u, w) \phi^{(w + w_k(v))} \quad (12.26)$$

where the first summation is over $\{u \in \Lambda_{k-1}(t)\}$, the next summation is over

$$\{v = m_{1k}(t, u), m_{1k}(t, u) + 1, \cdots, m_{2k}(t, u)\}$$

and the final summation is over the $N_k(u)$ set of exponents of ϕ in $F_{k-1}(\theta, \phi)$ when the exponent of θ is u.

Now consider first $f(1, 1; t, w_*)$. Since (12.26) is processed using Algorithm 12.RPM.2, we only worry about the coefficient of ϕ. In particular, we check that the value of $w + w_k(v)$ is such that all later additions to it will produce a final value equal to w_*. These additions have the form

$$\sum_{j \geq k+1} w_j(u_j)$$

Accordingly, we need

$$\min/\max \left[\sum_{j \geq k+1} w_j(u_j) \right] \quad (12.27)$$

subject to

$$\sum_{j \geq k+1} u_j = t - (u+v), \quad \text{and} \quad 0 \leq u_j \leq e_j, \; j \geq k+1$$

These optimum values, respectively, $\eta_{1,k+1}(t - (u+v))$ and $\eta_{2,k+1}(t - (u+v))$, are available from the backward induction scheme of the previous section by putting $h_j(u_j) = w_j(u_j)$. Before implementing Algorithm 12.RPM.2, we thus use this scheme to compute and store these optimum values. To obtain $f(1, 1; t, w_*)$, we apply the following exponent checks on the coefficients of ϕ in (12.26).

CONDITIONAL VALUES

Check EC 12.01

$$\langle \text{ If } w + w_k(v) + \eta_{1,k+1}(t - (u+v)) > w_* \rangle$$

Or

$$\langle \text{ If } w + w_k(v) + \eta_{2,k+1}(t - (u+v)) < w_* \rangle$$

Do not perform the multiplication. Continue on to the next term.

$$\langle \text{ Otherwise } \rangle$$

Perform the multiplication $d_{k-1}(u, w) \times c_k(v)$, accumulate the result in $(u + v, w + w_k(v))$th position in $F_k(\theta, \phi)$ and continue.

□□□□□□□

These criteria are applied after Step 26 of Algorithm 12.RPM.2 on the quantity w. If one of the first two conditions hold, we jump to Step 36 and generate the next record. At the Kth stage, we get a single record which is $f(1, 1; t, w_*)$.

****

For the computation of $f(\theta, \phi; t, \geq w_*)$, the following checks are used:

Check EC 12.02

$$\langle \text{ If } w + w_k(v) + \eta_{2,k+1}(t - (u+v)) < w_* \rangle$$

Do not perform the multiplication. Continue on to the next term.

$$\langle \text{ Otherwise } \rangle$$

Perform the multiplication $d_{k-1}(u, w) \times c_k(v)$, accumulate the result in $(u + v, w + w_k(v))$th position in $F_k(\theta, \phi)$ and continue.

□□□□□□□

To compute $f(1, 1; t, .)$, we note that

$$f(\theta, 1) = \sum_u \theta^u \sum_v c(u, v) = \prod_{k=1}^{K} f_k(\theta, 1) \qquad (12.28)$$

In this case, we treat the problem as one of multiplying K univariate polynomials and use EC 11.01 to obtain the result. $f(1, 1; t, .)$ may also be obtained by a single backward induction pass using $h_j(u_j) = c_j(u_j)$ and computing the conditional product (12.20). In some applications, as we shall see in the next section, an explicit formula for $f(1, 1; t, .)$ is known.

****

Now consider the computation of $f(1, \phi_0; t, \geq w_*)$. We first use a backward induction scheme to compute the conditional product (12.20) with

$$h_j(u_j) = c_{*j}(u_j) = c_j(u_j)\phi_0^{w_j(u_j)}$$

We store the intermediate stage conditional product values, labeled $\eta_{3k}(.)$, as well. Further, within this backward induction pass, we simultaneously compute the minimum and maximum values (12.23) and (12.24). Then starting with $\Delta = 0$, we implement Algorithm 12.RPM.2 and apply the following checks.

Check EC 12.03

$\langle$ If $w + w_k(v) + \eta_{2,k+1}(t - (u+v)) < w_*$ $\rangle$

Do not perform the multiplication. Continue on to the next term.

$\langle$ If $w + w_k(v) + \eta_{1,k+1}(t - (u+v)) \geq w_*$ $\rangle$

Perform the multiplication $\rho = d_{k-1}(u,w) \times c_{*k}(v) \times \eta_{3,k+1}(t-(u+v))$ and set $\Delta = \Delta + \rho$. Do not store anything in $F_k(\theta, \phi)$ and continue on to the next term.

$\langle$ Otherwise $\rangle$

Perform the multiplication $d_{k-1}(u,w) \times c_{*k}(v)$, accumulate the result in $(u+v, w+w_k(v))$th position in $F_k(\theta, \phi)$ and continue.

□□□□□□

The value of Δ at the final stage is the desired quantity, $f(1, \phi_0; t, \geq w_*)$.

12.6 Applications

The algorithms and other techniques developed above are applicable to models for several 2×2 tables, and a single $2 \times K$ table. We outline these applications below.

12.6.1 Application I

First consider the HOR model (8.2) for several 2×2 tables given in Chapter 8. Here,

$$c_k(u) = \binom{n_k}{u}\binom{m_k}{s_k - u} \qquad (12.29)$$

and

$$\Omega_k = \{l_{1k}, l_{1k}+1, \ldots, l_{2k}\} \qquad (12.30)$$

where

$$l_{1k} = \max(0, s_k - m_k) \quad \text{and} \quad l_{2k} = \min(s_k, n_k) \qquad (12.31)$$

Then, with $w_k(v) = vz_k$, we have

APPLICATIONS

$$f_k(\phi, \psi) = \sum_{v \in \Omega_k} c_k(v) \phi^v \psi^{vz_k} \qquad (12.32)$$

For this application, we note that

$$f(1, t; 1, .) = \binom{N+M}{s}$$

12.6.2 Application II

Next consider the problem of testing the homogeneity of the odds ratios in several 2×2 tables addressed in §9.3. With Ω_k and $c_k(u)$ defined as in the previous application, we have

$$f_k(\phi, \psi) = \sum_{u \in \Omega_k} c_{*k}(u) \phi^u \psi^{w_k(u)} \qquad (12.33)$$

where

$$c_{*k}(u) = c_k(u) \rho_k^u$$

and further where the values of $\rho_k, k = 1, \ldots, K$, are specified and fixed. In this case, $w_k(u)$ is obtained from the statistics for testing homogeneity given in §9.4:

12.6.3 Application III

Next consider the analysis of a single $2 \times K$ table. First we look at the dose-response model (10.6). Here $\Omega_k = \{0, 1, \ldots, n_k\}$ and

$$c_k(u) = \binom{n_k}{u} \qquad (12.34)$$

Then we have

$$f_k(\theta, \phi) = \sum_{u \in \Omega_k} c_k(u) \theta^u \phi^{ux_k} \qquad (12.35)$$

For this application, we note from (10.15) that

$$f(1, t; 1, .) = \binom{N}{t} \qquad (12.36)$$

12.6.4 Application IV

In the case of the models for an unordered $2 \times K$ table given in §10.3, the same definitions of $c_k(u)$ and Ω_k apply. Then

$$f_k(\theta, \psi) = \sum_{u \in \Omega_k} c_{*k}(u) \theta^u \psi^{w_k(u)} \qquad (12.37)$$

where

$$c_{*k}(u) = c_k(u) \phi_k^u$$

and further where the values ϕ_k are fixed and specified. The kth component of test statistic for this case, $w_k(u)$, is obtained from the statistics for testing association given in §8.4.

12.6.5 Other Applications

Efficient methods of multiplying several bivariate polynomials are also applicable to the exact analysis of the incidence density data model in §9.7 and the inverse sampling design of §9.8. Among the relevant problems are the analysis of an incidence density study with a linear stratum effect; of a stratified inverse sampling design with a quantitative interaction effect; the homogeneity testing of relative risk; and checking linearity of relative risk in this model under inverse sampling.

****

These applications also need to be refined in terms of computing exact distributions, p-values, CIs, power and estimates. These are detailed matters of combining the relevant material in the previous chapters with the computational aspects given here.

12.7 Trivariate Polynomials

The algorithms and exponent checks developed for bivariate polynomials readily extend to trivariate polynomials of the following form.

$$f_k(\theta, \phi, \psi) = \sum_{u \in \Omega_k} c_{*k}(u) \theta^u \phi^{uz_k} \psi^{q_k(u)} \qquad (12.38)$$

where, for fixed and specified values of ρ_k,

$$c_{*k}(u) = c_k(u) \rho_k^u$$

Here Ω_k is a finite set of integers, $z_k \geq 0$, $q_k(u) \geq 0$, $\rho_k \geq 0$ and $c_k(u) > 0$. Consider two specific examples.

(i) Let $c_k(u)$ and Ω_k be the entities defined in §12.6.1. Polynomials of this form occur when testing the assumption of linearity in the TOR model for several 2×2 tables considered in §9.6. The quantity $q_k(u)$ can be a component of a relevant score or a likelihood ratio statistic.

(ii) Let $c_k(u)$ and Ω_k be the entities defined in §12.6.3. Polynomials of this form occur when testing the assumption of linearity in the dose-response model for a $2 \times K$ table of §10.6. The forms of the statistic component $q_k(u)$ are given by Theorem 10.6.

TRIVARIATE POLYNOMIALS

A specific instance of this is the quadratic dose-response model (10.123) for an ordered $2 \times K$ table.

$$f_k(\theta, \phi, \psi) = \sum_{u=0}^{n_k} \binom{n_k}{u} \theta^u \phi^{uz_k} \psi^{uk^2} \qquad (12.39)$$

****

For such trivariate polynomials, we consider the efficient determination of the product

$$f(\theta, \phi, \psi) = \prod_{k=1}^{K} f_k(\theta, \phi, \psi) = \sum_{(u,v,w) \in \Omega} c(u, v, w) \theta^u \phi^v \psi^w \qquad (12.40)$$

In applications, we need components of the product when (i) the exponent of θ is equal to t, namely, $f(1, \phi, \psi; t, ., .)$, (ii) the exponent of θ is equal to t, and that of ϕ is z, namely, $f(1, 1, \psi; t, z, .)$ and (iii) when the exponent of θ is equal to t, that of ϕ is z, and the exponent of ψ is within a certain range, for example, $f(1, 1, \psi; t, z, \geq q)$. The cases in (iii) follow the pattern given for bivariate polynomials in §12.5.

****

The general RPM method readily extends to such problems. Algorithm 12.RPM.2, in particular, applies to give $f(\theta, \phi, \psi; t, ., .)$, provided the dimension of the main storage arrays is increased by one to accommodate the exponents of ψ.

When we have other restrictions on the exponents of ϕ and/or ψ, additional exponent criteria are needed. Though these are, in principle, constructed in a way similar to that for bivariate polynomials, they have to be formulated and implemented with care.

For example, consider the determination of $f(1, 1, \psi; t, z, \geq q)$. To allow for the restrictions on the final exponents of ϕ, we compute, as before, $\eta_{4,k+1}(a)$ and $\eta_{5,k+1}(a)$ respectively as

$$\min/\max \left[\sum_{j \geq k+1} u_j z_j \right] \qquad (12.41)$$

subject to

$$\sum_{j \geq k+1} u_j = a, \quad \text{and} \quad 0 \leq u_j \leq e_j, \ j \geq k+1$$

To allow for the restrictions on the final exponents of ψ, we consider two exponent checks conditional on the final values of the exponents of θ and ϕ. For this, we compute $\eta_{6,k+1}(a)$ given by

$$\max \left[\sum_{j \geq k+1} q_j(u_j) \right] \qquad (12.42)$$

subject to

$$\sum_{j \geq k+1} u_j = a, \quad \text{and} \quad 0 \leq u_j \leq e_j, \quad j \geq k+1$$

and $\eta_{7,k+1}(b)$ given by

$$\max \left[\sum_{j \geq k+1} q_j(u_j) \right] \tag{12.43}$$

subject to

$$\sum_{j \geq k+1} u_j z_j = b, \quad \text{and} \quad 0 \leq u_j \leq e_j, \quad j \geq k+1$$

The values of a and b are consistent with the overall constraints. The quantities $\eta_{4,k}(a)$, $\eta_{5,k}(a)$ and $\eta_{6,k}(a)$ emerge from the backward induction scheme of §12.4. $\eta_{7,k}(b)$ is also computed with backward induction but the recursion is of a slightly more involved form. We apply these as follows. Consider the product

$$F_k(\theta, \phi, \psi) = F_{k-1}(\theta, \phi, \psi) f_k(\theta, \phi, \psi) \tag{12.44}$$

Suppose we are multiplying the term

$$d_{k-1}(u, v, w) \theta^u \phi^v \psi^w$$

of $F_{k-1}(\theta, \phi, \psi)$ with the term

$$c_{*k}(v_*) \theta^{v_*} \phi^{v_* z_k} \psi^{v_*} q_k(v_*)$$

of $f_k(\theta, \phi, \psi)$ in the context of Algorithm 12.RPM.2. Then we apply EC 12.04 given below.

Check EC 12.04

$$\langle \text{ If } \quad v + v_* z_k + \eta_{4,k+1}(t - (u + v_*)) > z \ \rangle$$

Or

$$\langle \text{ If } \quad v + v_* z_k + \eta_{5,k+1}(t - (u + v_*)) < z \ \rangle$$

Or

$$\langle \text{ If } \quad w + q_k(v_*) + \eta_{6,k+1}(t - (u + v_*)) < q \ \rangle$$

Or

$$\langle \text{ If } \quad w + q_k(v_*) + \eta_{7,k+1}(z - (v + v_* z_k)) < q \ \rangle$$

Do not perform the multiplication. Continue on to the next term.

$$\langle \text{ Otherwise } \rangle$$

Perform the multiplication $d_{k-1}(u, v, w) \times c_k(v_*)$, accumulate the result in $(u + v_*, v + z_k v_*, w + q_k(v_*))$th position in $F_k(\theta, \phi, \psi)$ and continue.

□□□□□□

AN EXTENSION

These exponent criteria can be modified further to directly compute the conditional sum of the coefficients $f(1, 1, 1; t, z, \geq q)$. In principle, the method is similar to that given in EC 12.03 but constructing the actual computation scheme is a rather involved process.

The details of applying the results of such polynomial based computations to the task of calculating exact distributions, p-values and other inferential entities for the applications noted above are left to the reader.

12.8 An Extension

We now consider a polynomial that occurs in the analysis of a dose-response model for several $2 \times K$ tables, namely, model (10.125). For $j = 1, \ldots, J$ and $k = 1, \ldots, K$, let

$$f_{jk}(\theta_j, \phi) = \sum_{u \in \Omega_{jk}} c_{jk}(u) \theta_j^{u_j} \phi^{u_j z_k} \qquad (12.45)$$

where $\Omega_{jk} = \{0, 1, \ldots, n_{jk}\}$, $z_k \geq 0$, and

$$c_{jk}(u) = \binom{n_{jk}}{u} \qquad (12.46)$$

Let $\boldsymbol{\theta} = (\theta_1, \ldots, \theta_J)$. We are interested in the product

$$f(\boldsymbol{\theta}, \phi) = \prod_{j=1}^{J} \prod_{k=1}^{K} f_{jk}(\theta_j, \phi) \qquad (12.47)$$

Suppose we need terms of (12.47) in which the exponent of θ_j is t_j, $j = 1, \ldots, J$, and the exponent of ϕ is greater than or equal to z. One method of getting this is to apply, for $j = 1, \ldots, J$, Algorithm 12.RPM.2 to the product

$$F_{*j}(\theta_j, \phi) = \prod_{k=1}^{K} f_{jk}(\theta_j, \phi) \qquad (12.48)$$

to obtain $F_{*j}(\theta_j, \phi; t_j, .)$. Then we multiply the resultant J univariate polynomials using Algorithm 11.RPM.1 to obtain

$$F_*(\mathbf{1}, \phi; \boldsymbol{t}, .) = \prod_{j=1}^{J} F_{*j}(1, \phi; t_j, .) \qquad (12.49)$$

Incorporating exponent criteria EC 11.03 upon the exponents of ϕ in this process directly generates $F_*(\mathbf{1}, \phi; \boldsymbol{t}, \geq z)$. A more efficient way is to combine the two separate recursions by incorporating checks on the exponent of ϕ in the process of determining each of the J polynomials $F_{*j}(1, \phi; t_j, .)$. This will reduce the number of redundant multiplications. Other relevant parts of the final product, like $F_*(\mathbf{1}, \phi; \boldsymbol{t}, \leq z)$, are computed in a similar fashion.

The polynomial (12.45) occurs in a statistical model that has a common trend parameter across the J strata. More generally, we consider the polynomial

$$f_{jk}(\theta_j, \phi_j) = \sum_{u \in \Omega_{jk}} c_{jk}(u) \theta_j^{u_j} \phi_j^{u_j z_k} \qquad (12.50)$$

In that case, we may test, for the relevant model, the hypothesis $\phi_j = \phi$ for all j. For this task, after selecting an appropriate statistic, we consider the multiplication of conditional J trivariate polynomials and select the terms which satisfy additional conditions. The approach combines relevant RPM algorithms for bivariate and univariate polynomials with appropriate exponent criteria. It also follows and extends the approach used in the problem of testing the homogeneity of the odds ratio in several 2×2 tables.

12.9 Network Algorithms

Efficient recursive algorithms for exact analysis of discrete data have often appeared in the literature in the form of network algorithms. A network consists of a set of nodes, and of arcs that connect the nodes to one another. The networks of relevance to us have nodes grouped into a series of $K+1$ stages, with a single node at Stage 0. For $k = 1, \ldots, K$, the arcs from a node in stage $k-1$ connect only to one or more nodes in stage k. A path in the network has K arcs and sequentially goes from the single node in Stage 0 to a node in Stage K. Often, the final stage also has a single node.

A network is helpful for representing polynomial multiplication algorithms in a pictorial form. Take a network constructed from the product of five polynomials. Consider the multiplication of bivariate polynomials described in §12.2 and §12.3. With $K = 4$, let $e_1 = e_2 = e_3 = e_4 = e_5 = 2$, and $w_k(u) = ku$. Then, for $k = 1, 2, 3, 4, 5$, let

$$f_k(\theta, \phi) = c_k(0) + c_k(1)\theta\phi^k + c_k(2)\theta^2\phi^{2k}$$

and

$$f(\theta, \phi) = \prod_{k=1}^{5} f_k(\theta, \phi)$$

Suppose we need the terms of this product represented by $f(1, \phi; t, .)$ when $t = 2$. We set up the nodes of the kth stage of the network using the elements of Λ_k, using the formulas from §12.3. Each node has the label (k, u), where k is the stage and $u \in \Lambda_k(t)$.

For the given example, $\Lambda_0(2) = \{0\}$; $\Lambda_1(2) = \{0, 1, 2\}$; $\Lambda_2(2) = \{0, 1, 2\}$; $\Lambda_3(2) = \{0, 1, 2\}$; $\Lambda_4(2) = \{1, 2\}$; and $\Lambda_5(2) = \{2\}$. The network with 5 stages is then set up as follows:

Arcs from a node $(k-1, u)$ in stage $k-1$ to nodes in stage k are constructed with the following rule: There is an arc from a node $(k-1, u)$ to a node (k, v) if and only if $v \in \Lambda_k(t)$ and $m_{1k}(t, u) \le v - u \le m_{2k}(t, u)$.

For the example in question, this network is given in Figure 12.1.

Further, for such a network, we define, for an arc connecting a node $(k-1, u)$ to a node (k, v), the arc length as $w_k(v-u)$ and the arc weight as $c_k(v-u)$. For any path in the network, we define the path length as the sum of its K arc lengths and the path weight as the product of its K path weights.

The operation of processing this network starts from the single node in Stage 0 and goes sequentially from one stage to the next along the connecting nodes. There is a single record in

NETWORK ALGORITHMS

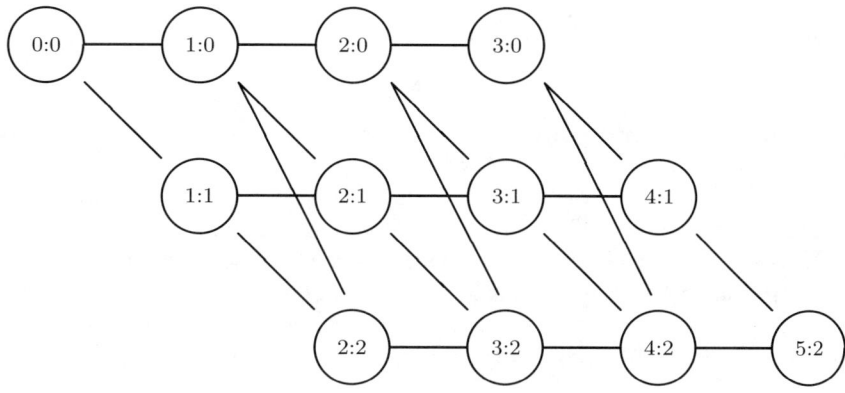

Figure 12.1 *A Five Stage Network.*

the starting node $(0,0)$ with cumulative length equal to 0 and cumulative weight equal to 1. The situation in stage 0 is written as

$$(0,0); [0,1]$$

When we step from one node to next, we add to all the records in origin node the associated arc length. We also multiply the arc weight with their respective cumulative values. These values are stored as records in the destination node. Records with the same cumulative path length are combined by adding their cumulative path weights. All paths leading to any node in stage k are processed before going to stage $k+1$. As we process the network from one stage to the next, each node accumulates a collection of paired values with distinct cumulative arc length and respective cumulative arc weight. For a node (k, a), these paired values are denoted as

$$(k, a); [(d_k(1, a), w_{+,k}(1, a)), \ldots, (d_k(1, a), w_{+,k}(N_k(a), a))]$$

At the final stage K, the set of values at node (K, t) has the paired values of the exponents and coefficients we need.

****

From how we construct the network, the manner of processing it is seen as a graphical version of the polynomial multiplication algorithm 12.RPM.2. Their essence is identical; only the terminology has changed. Instead of sums of exponents, we refer to sums of arc lengths; instead of products of coefficients, we have the products of arc weights. The various exponent checks we have developed are equivalent to looking at shortest and longest paths within the network. This construction leads to two fundamental results:

Theorem 12.1: All RPM algorithms given in this text are network algorithms. □

Theorem 12.2: All RPM algorithms for exact analysis of discrete data described in the statistical literature to date can be represented as network algorithms. Conversely, all network algorithms for exact analysis of discrete data given in the statistical literature to date are pictorial representation of equivalent RPM algorithms. □

Using the constructive arguments given above, these theorems are straightforward to prove; see the relevant papers mentioned in §12.12.

12.10 Power Computation

We need the exact power of an exact or asymptotic conditional test for some parameter(s) of interest in a model with PBD variates. The number of nuisance parameters in the model sets the dimension of the unconditional sample space, and is a major determinant of the effort involved. If it is more than two, to get the exact power even at small sample sizes can involve a prohibitive amount of computation. Taking this into account, we give two strategies for computing exact power. This is done in the context of two specific situations.

```
                   Algorithm 12.PW.1
```

Step 01. Specify null value ϕ_0, test statistic $\mathcal{W}$, and α_*.
Step 02. Specify K, and scores x_k, $k = 1, \ldots, K$.
Step 03. Specify n_k, $k = 1, \ldots, K$, and set $N = \Sigma_k n_k$.
Step 04. Specify nonnull values θ_1 and ϕ_1.
Step 05. Use Algorithm 12.RPM.1 to generate the generic gp $F_K(\theta, \phi)$, and store it on a term by term basis.
Step 06. Generate or compute the value of each term of $F_K(\theta_1, \phi_1)$, and store it on a term by term basis.
Step 07. Set $t = -1$ and $\Delta_* = 0$.
Step 08. $t \Leftarrow t + 1$
Step 09. Extract the gp $F_K(1, \phi; t, .)$ from Step 05.
Step 10. Construct the null distribution $P[\mathcal{W} = w \mid \mathcal{T} = t; \phi_0]$
Step 11. Perform the α_* level test for each $w \in \Omega(t)$.
Step 12. Identify the critical region, $\Omega_c(t; \phi_0, \alpha_*)$.
Step 13. Use values from Step 06 to get the quantity

$$\eta = \theta_1^t \sum_{v \in \Omega_c} c(t, v)\phi_1^v$$

Step 14. $\Delta_* \Leftarrow \Delta_* + \eta$
Step 15. If $t < N$ go to Step 08
Step 16. Compute $\Delta = F_K(\theta_1, \phi_1)$ from

$$\ln(\Delta) = \sum_{k=1}^{K} n_k \ln[1 + \theta_1 \phi_1^{x_k}]$$

Step 17. Compute the exact power as

$$\Pi = \frac{\Delta_*}{\Delta}$$

□□□□□□□

POWER COMPUTATION   369

12.10.1 A Dose-Response Model

Consider a $2 \times K$ table under the product binomial design and model (10.6) for ordered data. Suppose we test $H_0 : \phi = \phi_0$ against the $H_1 : \phi \neq \phi_0$ at level $\alpha_* (0 < \alpha_* < 1)$ with a conditional test. The exact power for this problem was formulated in §10.10.2.

This problem has only one nuisance parameter. For this, we consider the **primarily unconditional approach** for power computation. This first generates the exact unconditional distribution, or its gp and then works within it. The latter is done by a systematic generation of the marginal sample space of the sufficient statistics for the nuisance parameters, and using the quantities computed at the initial step to perform the conditional test, identify the conditional critical region, and accumulating the power value until all the marginal possibilities are exhausted.

An implementation of this approach for the dose-response problem, based on equations (10.100) to (10.104), is given in Algorithm 12.PW.1.

As a point of clarification, note that in Step 06, the quantity $F_K(\theta_1, \phi_1)$ is a point by point representation of the bivariate polynomial values. But in Step 16, it is the actual value of this polynomial at $\theta = \theta_1$ and $\phi = \phi_1$. Using the former to get the latter can produce an inaccurate value. The following additional points are also relevant in this exercise.

- The formulation in §10.10.2 is more in line with the joint approach described next.
- Step 05 and Step 06 may be implemented in a single recursive process. This need an extra memory array. The sample points (exponents) are the same but the coefficients or terms are not.
- We may leave computation of the nonnull critical values till the end. Then, Step 06 is eliminated. At each t, we identify the conditional critical region and store the corresponding exponents of $F_K(\theta, \phi; t, .)$ in $F_{cK}(\theta, \phi)$. At the end, we also evaluate $F_{cK}(\theta, \phi)$ at the nonnull values, set it equal to Δ_* and take the ratio to compute exact power. If exact power is needed at several nonnull values, this tends to be a more efficient option.
- In Step 11, we need not perform the test at each conditional sample point. We may start at one boundary point, proceed inwards, stop when a critical point is reached and repeat the same process from the other boundary point.
- When N is large, we apply the δ precision approach described in §7.4.
- Alternatively, we use a Monte–Carlo approach and draw random samples from the marginal distribution of $\mathcal{T}$. In that case, this marginal distribution has to be computed first (see §10.10.2). The details of the MC approach are given in another context below.
- See also the points noted in §12.11.

12.10.2 The COR Model

Now suppose we need the exact power of a conditional test for the COR model in several 2×2 tables. Here, the number of nuisance parameters is equal to the number of strata. As this can be higher than two, we adopt the **joint conditional unconditional approach.** As above, this navigates similarly through the unconditional sample space, but generates the distributions as conditional slices. In the former, the distributional computations are done at the start, while in the latter, they are obtained as needed. To get a specific conditional distribution, the latter is more efficient; but to get all the conditional distributions, it is not necessarily so.

The joint approach computes the marginal distribution of $\mathcal{S}$, the sufficient statistics for the nuisance parameters and then proceeds through this sample space in a systematic way. At each distinct point, the conditional gp for $\mathcal{T}$ given $\mathcal{S}$ is computed, the test is performed, the critical region identified and power value accumulated until all the marginal possibilities are exhausted.

Consider the COR model (8.1) under the product binomial design and test $H_0 : \phi = \phi_0$ against the $H_1 : \phi \neq \phi_0$ at level $\alpha_* \ (0 < \alpha_* < 1)$. Suppose we use a conditional test based on fixing the margins of all the 2×2 tables. We continue with the notation of Chapter 8 here. The Algorithm 12.PW.2 implements the rhs of (9.59).

Algorithm 12.PW.2

Step 01. Specify null value ϕ_0, test statistic $\mathcal{W}$, and α_*.
Step 02. Specify K, and pairs (n_k, m_k), $k = 1, \cdots, K$.
Step 03. Specify nonnull values θ_k, $k = 1, \cdots, K$, and ϕ_1.
Step 04. Compute and store $f_k(\theta_k, \phi_1)$, $k = 1, \cdots, K$, at the nonnull values on a term by term basis.
Step 05. Specify R; $r \Leftarrow 0$ and $\Delta_* \Leftarrow 0$.
Step 06. $r \Leftarrow r + 1$
Step 07. Generate $s_r = (s_{1r}, \cdots, s_{Kr})$ from items in Step 04.
Step 08. Use Algorithm 11.RPM.1 to get the generic gp

$$f(\mathbf{1}, \phi; \mathbf{s}_r, .) = \prod_{k=1}^{K} f_k(1, \phi; s_{rk}, .)$$

Step 09. Perform the conditional test for all $t \in \Omega(s_r)$.
Step 10. Identify the conditional critical region, $\Omega_c(s_r; \phi_0, \alpha_*)$.
Step 11. Use values stored in Step 08 to compute the sum

$$\eta_r = \sum_{v \in \Omega_c} c(s_r, v) \phi_1^v$$

Step 12. Compute

$$\pi_r = \sum_{k=1}^{K} s_{rk} \ln(\theta_k)$$

Step 13. $\Delta_* \Leftarrow \Delta_* + \eta_r \exp(\pi_r)$
Step 14. If $r < R$ go to Step 06
Step 15. Compute $\Delta = F_K(\boldsymbol{\theta}, \phi_1)$ as

$$\ln(\Delta) = \sum_{k=1}^{K} m_k \ln(1 + \theta_k) + n_k \ln(1 + \theta_k \phi_1)$$

Step 16. Compute the exact power as

$$\Pi = \frac{\Delta_*}{\Delta}$$

□□□□□□

Note, the generation of distinct s_r vectors may be done within nested 'do loops' or by simulating such loops (see Chapter 13). For exact computation, the value of R is given by (9.60). If this is too high, then a δ-precision approach may be used. In a nested generation of the vectors, this method can be particularly efficient (see Hirji et al. (1994) for details). However, when there are many strata, a Monte–Carlo implementation is advisable.

POWER COMPUTATION

The MC approach randomly generates values from the marginal distribution of $\mathcal{S}$ under the alternative, computes the conditional power and, at the end, takes their mean value as the estimate of power. This is implemented in Algorithm 12.PW.3.

Algorithm 12.PW.3

Step 01. Specify null value ϕ_0, test statistic $\mathcal{W}$, and α_*.
Step 02. Specify K, and pairs (n_k, m_k), $k = 1, \ldots, K$.
Step 03. Specify nonnull values θ_k, $k = 1, \ldots, K$, and ϕ_1.
Step 04. Compute and store $\mathrm{P}[\mathcal{S}_k = s_k; \theta_k, \phi]$, $k = 1, \cdots, K$
at the nonnull values on a term by term basis.
Step 05. Specify R and set $r = 0$, and $\Delta_* = \Delta_{**} = 0$.
Step 06. $r \Leftarrow r + 1$
Step 07. From distributions in Step 04, randomly generate K
independent values in $\boldsymbol{s}_r = (s_{1r}, \cdots, s_{Kr})$
Step 08. Use Algorithm 11.RPM.1 to get the gp

$$f(\mathbf{1}, \phi; \boldsymbol{s}_r, .) = \prod_{k=1}^{K} f_k(1, \phi; s_{rk}, .)$$

Evaluate this product at $\phi = \phi_1$ and set it to π_r.
Step 09. Perform the conditional test for all $t \in \Omega(\boldsymbol{s}_r)$.
Step 10. Identify conditional critical region, $\Omega_c(\boldsymbol{s}_r; \phi_0, \alpha_*)$.
Step 11. Use values stored in Step 08 to compute the sum

$$\eta_r = \sum_{v \in \Omega_c} c(\boldsymbol{s}_r, v) \phi_1^v$$

Step 12. Compute conditional power.

$$h_r = \frac{\eta_r}{\pi_r}$$

Step 13. $\Delta_* \Leftarrow \Delta_* + h_r$
Step 14. $\Delta_{**} \Leftarrow \Delta_{**} + h_r^2$
Step 15. If $r < R$ go to Step 06
Step 16. Estimate the exact power as

$$\hat{\Pi} = \frac{\Delta_*}{R}$$

□□□□□□□

Under the MC approach, the estimate of the exact power is the mean of the values computed in Step 12. We can show that $\hat{\Pi}$ is an unbiased estimate. Then, a sample based unbiased estimate of its variance is

$$\hat{\sigma}^2 = \frac{1}{R-1}\left\{\Delta_{**} - \hat{\Pi}^2\right\} \tag{12.51}$$

Therefore, an approximate 95% CI for the exact power is

$$\hat{\Pi} \pm 1.96\hat{\sigma}/\sqrt{R} \tag{12.52}$$

Several of the additional points made after Algorithm 12.PW.1 are also relevant for Algorithm 12.PW.2 and Algorithm 12.PW.3. See also the next section.

Exact power algorithms for the tests for conditional independence, homogeneity of odds ratios, and trend in the odds ratios for $K \times 2 \times 2$ tables and for homogeneity tests and tests for linearity in a $2 \times K$ table are developed in a similar way. See Tang (1999, 2001a) and Tang and Hirji (2002) for the details. See the latter paper for ways of integrating the exponent checks within the process of exact power computation. Such efficient algorithms may also be developed for case-control, incidence density, multinomial, RMF and inverse sampling designs, both for ordered and unordered factors. There are more nuisance parameters in these cases, making the task more complex. But the principles of computation are the same.

12.11 Practical Implementation

When implementing the algorithms given in this chapter one needs to bear in mind issues noted in §11.4. These include (i) rounding the statistic $w_k(u)$ to three or more decimal digits of precision; (ii) prevention of numeric overflow and appropriate methods for retaining accuracy; and (iii) using an efficient storage method for array records. Exact power computation needs particular care on all these fronts. In all the cases, a modular, robust, object oriented and optimized implementation is advisable.

The exponent checks are applicable under an EE version of the basic algorithms. Applying such checks requires careful memory management to avoid duplicate records. The practical problems are greater with higher dimensional polynomials but the principles remain the same.

12.12 Relevant Literature

Before efficient algorithms appeared on the scene, exact tests were avaliable for only a restricted set of problems, and even in that case, one had to have published tables for specific situations. For example, Bennett and Nakamura (1963) give such tables for the Freeman–Halton test for a 2×3 table for a limited number of configurations.

Gentleman (1975) gave an algorithm to simulate nested do loops. This is used to multiply, in an EE approach, the bivariate and trivariate polynomials under certain conditions (see Chapter 13). Bayer and Cox (1979) gave a general EE algorithm for a binary regression model that is applicable to the problems of this chapter. RPM algorithms for exact analysis have appeared in the literature in a variety of other guises. Besides the network formulation, other equivalent formulations include multivariate shift algorithms and algorithms for addition of multisets (see Chapter 13).

The formulation of exponent checks for bivariate and multivariate settings and the application of backward induction in that context was first proposed for network algorithms. Initially, the criteria were applied in a modified exhaustive enumeration context, that is, by a depth-wise processing of the network and later in a recursive, stage-wise processing of the network (Mehta and Patel 1980, 1983).

Recursions equivalent to those in network and RPM algorithms also appeared in papers that formulated the problem as a convolution of characteristic functions and applied the fast Fourier transform. For a bivariate formulation, see Tritchler (1984). For general references, see Baglivo, Olivier and Pagano (1993), Hirji (1997a, 1998) and Agresti (2001).

The shift and RPM formulations of various problems began with Hirji, Mehta and Patel (1987) and Streitberg and Rommel (1984, 1986). Other relevant papers are Hirji and Vollset (1994b). The relationship between the various algorithmic formulation was explored in Hirji, Mehta and Patel (1987), Vollset, Hirji and Elashoff (1991), Vollset and Hirji (1991), Martin and Austin

EXERCISES

(1991), Hirji (1992), Vollset and Hirji (1992), Hirji (1996), Martin and Austin (1996), Hirji and Johnson (1996) and generally in Hirji (1997a, 1998).

Network algorithms for one and several ordered $2 \times K$ tables were given by Mehta, Patel and Tsiatis (1984) and Mehta, Patel and Senchaudhuri (1992), respectively. Hirji et al. (1996) presented an RPM algorithm for testing the homogeneity of the odds ratios across several 2×2 tables. Tang and Hirji (2002) gave a trivariate RPM algorithm for the analysis of dose-response data.

Bennett (1986) considered the problem of combining relative risk estimates under the negative binomial design. Martin and Austin (1996) examined a stratified rate ratio model for cohort data.

The references for algorithms for exact power computation were noted in Chapter 7.

12.13 Exercises

12.1. Write an EE method based computer program to obtain the polynomial $f(\theta, \phi)$ given by (12.2). Incorporate the natural and logarithmic scale options into the program. Implement exponent check EC 11.01 for the exponents of θ in this program.

12.2. Convert Algorithms 12.RPM.1 and 12.RPM.2 into computer programs with the natural and logarithmic scale options. Explore different array storage methods for use here.

12.3. Show that computing a conditional polynomial under the method of §12.3 is equivalent to applying exponent check EC 11.01 to the exponents of θ.

12.4. For the bivariate polynomial (12.1), let $c_k(u) = 1$ and $w_k(u) = ku$ for all k and u. Consider two cases: (i) $e_k = 5$ and (ii) $e_k = 10$ for all k. Further, let K, the number of such polynomials, vary in the range $5, \ldots, 10$.

For each combination of these cases, compare the number of multiplications, under the EE and RPM programs, required to multiply the K polynomials.

For selected values of t, determine the gain in efficiency by using Algorithm 12.RPM.2 instead of obtaining the conditional polynomial $f(\theta, \phi; t, .)$ from Algorithm 12.RPM.1.

For a variety t, compare the multiplications needed when the exponent checks are used in the EE and 12.RPM.2 algorithms.

Repeat all the computations done above for the cases with $w_k(u) = k^2 u$ and $w_k(u) = 2^k u$.

12.5. Provide theoretical arguments to show that with linear scores and $e_k = n$ for all k, recursion (12.4) tends to be more efficient than EE.

12.6. Consider, for some $a > 0$, the maximization and minimization of

$$\text{(i)} \sum_{k=1}^{K} t_k^a \quad \text{(ii)} \sum_{k=1}^{K} k^2 t_k \quad \text{(iii)} \sum_{k=1}^{K} k^2 t_k^a$$

and

$$\text{(iv)} \sum_{k=1}^{K} (K+1-k) t_k^2$$

subject to t_k integer and

$$\sum_{k=1}^{K} t_k = t, \quad \text{and} \quad 0 \leq t_k \leq e_k, \quad k \geq 1$$

By considering situations with either $e_1 \leq e_2 \leq \cdots \leq e_K$ or $e_1 \geq e_2 \geq \cdots \geq e_K$, develop, where possible, direct methods of computing these optimal values. Where not possible, develop a backward induction scheme for the task.

Repeat this exercise when the optimization is subject to the condition t_k integer and

$$\sum_{k=1}^{K} t_k^2 = t, \quad \text{and} \quad 0 \leq t_k \leq e_k, \quad k \geq 1$$

12.7. Write backward induction programs to compute the product (12.20) and optimal values (12.23) and (12.24) under the stated constraints.

12.8. Provide rigorous arguments for the validity of the exponent checks EC 12.01 to 12.04. Incorporate these checks into your EE and RPM programs for bivariate polynomials.

12.9. Write computer programs for methods based on EE, Algorithm 12.RPM.1 and Algorithm 12.RPM.2 methods for multiplying a series of trivariate polynomials of the form (12.38). Empirically compare the methods using the types of variations given in Exercise 12.5. Implement the natural and logarithmic scale option in the programs. Construct a program to directly obtain $f(1, 1, \psi; t, z, \geq q)$, $f(1, 1, \psi; t, z, q)$ and $f(1, 1, \psi; t, z, \leq q)$.

12.10. For $k = 1, \ldots, K$, consider possible values of intermediate sums of the form

$$z_{+,k} = \sum_{j \geq k} u_j z_j$$

which are subject to the overall conditions

$$\sum_{j \geq 1} u_j z_j = z, \quad \text{and} \quad 0 \leq u_j \leq e_j, \quad j \geq 1$$

Construct a scheme to enumerate and list all possible values of $z_{+,k}$, $k = 1, \ldots, K$. Refer to (12.43) and use this scheme to develop a backward induction scheme to compute, for $q_j(u_j) > 0$,

$$\max \left[\sum_{j \geq 1} q_j(u_j) \right]$$

subject to

$$\sum_{j \geq 1} u_j z_j = z, \quad \text{and} \quad 0 \leq u_j \leq e_j, \quad j \geq 1$$

Apply this scheme to compute the optimum values for the four functions in Exercise 12.7 subject to the above condition. Implement this scheme in the programs written in Exercise 12.9.

12.11. Develop a detailed backward induction scheme to compute

$$\max \left[\sum_{j \geq 1} q_j(u_j) \right]$$

EXERCISES

subject to
$$\sum_{j\geq 1} u_j = t, \quad \sum_{j\geq 1} u_j z_j = z, \quad \text{and} \quad 0 \leq u_j \leq e_j, \quad j \geq 1$$

12.12. Adapt the computer programs you have written thus far with exponent criteria to compute exact distributions, p-values, confidence intervals and estimates for various models for (i) several 2×2 tables and (ii) a single $2 \times K$ table under different study designs.

12.13. Write computer programs for large sample methods to compute approximate p-values, confidence intervals and estimates for various models for (i) several 2×2 tables and (ii) a single $2 \times K$ table under different study designs. Use these and previous programs to compare the properties of exact and large sample methods for a variety of situations.

12.14. Show that the number of distinct values of t satisfying the equations
$$t_1 + \cdots + t_k = t$$
$$0 \leq t_k \leq n_k \quad \text{and} \quad 0 \leq s_k - t_k \leq m_k$$
is $1 + \Sigma_k \min\{n_k, m_k, s_k, n_k + m_k - s_k\}$. This is the size of the conditional sample space under COR model and product binomial design. What is the number of observations in the unconditional sample space? Use numbers from real and hypothetical to show the implications of this for exact power computation.

12.15. Develop in detail a recursive algorithm, with appropriate exponent checks, for obtaining conditional polynomials from the products of the form (12.47). In particular develop exponent criteria to obtain $F_*(1, \phi; t, z)$, $F_*(1, \phi; t, \geq z)$, $F_*(1, \phi; t, \leq z)$ and $F_*(1, 1; t, \geq z)$. For this problem, combine the two separate recursions by incorporating checks on the exponent of ϕ in the process of determining each of the J polynomials.

Develop an appropriate backward induction scheme. Implement this on a computer and apply it to the analysis of models for several $2 \times K$ tables.

Construct statistics for testing the homogeneity of the trend parameter across the J strata and develop an efficient RPM algorithm to compute its exact distribution and right tail value.

12.16. Prove that the RPM algorithm for testing homogeneity of the proportions in an unordered $2 \times K$ table with exponent checks is a network algorithm. Extend this to the problems with several ordered and unordered $2 \times K$ tables.

12.17. Using $c_k(u) = 1$ for u and k, develop a backward induction scheme to determine the total number of paths in a network for an ordered $2 \times K$ table. Extend this to the case of relevant models for several 2×2 tables. Can you determine an explicit formula to get this number in each of these two cases?

12.18. For the RPM algorithm for performing exact tests for an unordered or ordered $2 \times K$ table, investigate the impact of changing the order in which the rows are processed on computational efficiency. Is there a column sorting strategy that will optimize efficiency? Investigate the same issue, in terms of sorting the strata, for the relevant models (like the TOR model) for several 2×2 tables.

12.19. Develop RPM algorithms and exponent checks for the incidence density data model in §9.7 and the inverse sampling designs of §9.8. Extend them to (i) a stratified incidence density study with a linear stratum effect; (ii) a stratified inverse sampling study with a quantitative interaction effect; (iii) the homogeneity testing of relative risk and checking linearity of relative risk in relevant models under inverse sampling and incidence density designs.

12.20. Prove that Algorithm 12.PW.3 gives an unbiased estimate of exact power. Also prove the variance formula (12.51).

12.21. Write computer programs for Algorithm 12.PW.1, Algorithm 12.PW.2 and Algorithm 12.PW.3 as applied to the exact and asymptotic conditional tests on the (i) COR model for several 2×2 tables, and (ii) a logit dose-response model for a $2 \times K$ table. Implement the efficiency enhancing points noted there, including the δ precision method. Use real and simulated data to compare the accuracy and performance of the exact, δ precision and randomization approaches to power computation.

12.22. Write a program to compute the asymptotic power of asymptotic conditional tests on (i) the COR model for several 2×2 tables and (ii) the logit dose-response model for a $2 \times K$ table. Use this program and the program of the previous exercise to compare exact and asymptotic approaches to computation of the power for various data configurations and model parameters.

12.23. Use the above programs to compare the exact power of various exact, mid-p and asymptotic tests for these two discrete data settings. Plot their respective exact power functions and comment on their shape.

12.24. Develop exact power algorithms for the tests for conditional independence, homogeneity of odds ratios, and trend in the odds ratios for $K \times 2 \times 2$ tables and for homogeneity tests and tests for linearity in a $2 \times K$ table.

12.25. Extend the above four exercises to logit models for case-control, incidence density, multinomial, RMF and inverse sampling designs, both with ordered and unordered factors. Compare the exact and randomization approaches in terms of feasibility and accuracy.

12.26. Develop a strategy for computing sample sizes based on exact power for the COR model, the dose-response model and other models for similar data.

12.27. Develop an integer partitioning algorithm (see Chapter 13) to compute the exact p-value for an exact test of association in an unordered $2 \times K$ table. Compare it with the RPM algorithm in terms of time efficiency and accuracy.

CHAPTER 13

Multinomial Models

13.1 Introduction

Multinomial models form the general foundation for analyzing multivariate or multicategory discrete data. We now consider a set of multinomial models with one and two discrete variates. After formulating their distributions in a polynomial form, we describe methods for exact inference and computation for these models. The specific aims of this chapter are:

- To show how to generate fixed size subsets (combinations) of a set, and the compositions and partitions of an integer.
- To present exact goodness-of-fit tests for a single multinomial. A related incidence density design is also examined.
- To study models for an unordered and ordered $3 \times K$ table, derive their exact conditional distributions and present computational methods for exact analysis of these models.
- To outline the exact analysis of unordered and ordered $J \times K$ tables.

The presentation in this chapter, which builds on that given previously, is in a more condensed form; the details have been left for the exercises.

13.2 Compositions and Partitions

Exact distributions for discrete data models, as we have seen, usually contain a combinatorial coefficient. Combinatorial algorithms are thus relevant for the exact analysis of such data. In this section, we introduce three combinatorial algorithms applied to exact analysis of multinomial data. These algorithms deal with generating combinations, compositions and partitions. We first define the idea of a dictionary or lexicographic order for a list of numbers.

Let $\{a_1, a_2, \cdots, a_n\}$ and $\{b_1, b_2, \cdots, b_n\}$ be two ordered sets of numbers. Suppose the set of a's is listed bfore the set of b's. This list is said to be in an (increasing) dictionary or lexicographic order if for the first j for which $a_j \neq b_j$, $a_j < b_j$. For example, the placing the set $\{2,3,1,4\}$ before $\{2,4,3,1\}$ is consistent with lexicographic ordering but placing the sets $\{1,2,4,3\}$ before $\{1,2,3,4\}$ is not.

If there are more than two sets in such a list, then the entire list is said to be in an increasing dictionary or lexicographic order if all the consecutive pairs of sets in the list are in that order.

13.2.1 Combinations

Suppose we need all subsets of size k of the set $\{1, 2, \cdots, n\}$, $(k \leq n)$. The number of such subsets, also called combinations of size k, is $n!/\{(n-k)!k!\}$. Suppose the values in each subset increase from left to right. In that case, the minimum value for the jth element is j and the maximum value is $n - k + j$. The dictionary order list of these subsets begins with the subset $\{1, 2, \cdots, k\}$, and ends with the subset $\{n + k - 1, \cdots, n - 1, n\}$.

For example, consider the set $\{1, 2, 3, 4, 5, 6, 7, 8, 9, 10\}$. There are 210 ordered subsets of size $k = 6$ of this set. Listed in a dictionary order, the first two subsets are $\{1, 2, 3, 4, 5, 6\}$ and $\{1, 2, 3, 4, 5, 7\}$, and the final two are $\{4, 6, 7, 8, 9, 10\}$ and $\{5, 6, 7, 8, 9, 10\}$.

The size k subsets of the set $\{1, 2, \cdots, n\}$ can be generated in a dictionary order with the following scheme. We begin with $\{1, 2, \cdots, k\}$. At any step, we first scrutinize the current set from left to right to find the right most element that has not attained its maximum value. This value is increased by 1. Conditional on that, we replace each value to its right by the lowest possible value, that is one higher than the previous value. We stop when the first element of the subset has attained its maximal value. Otherwise, we continue with scrutinization and replacement. An outline of a computer program for this scheme appears in Algorithm 13.SS.1 below.

Algorithm 13.SS.1

Step 01.	Input k and n $k \leq n$
Step 02.	For $u = 1, \cdots, k$ $d(u) \Leftarrow u$
Step 03.	Output $(d(1), d(2), \cdots, d(k))$
Step 04.	For $u = 1, \cdots, k$ do
Step 05.	$\quad v \Leftarrow u$
Step 06.	$\quad$ If $(d(v) = n - k + v)$ then
Step 07.	$\quad\quad$ If $(v = 1)$ go to Step 18
Step 08.	$\quad$ Else
Step 09.	$\quad\quad v \Leftarrow v - 1$
Step 10.	$\quad\quad$ Go to Step 13
Step 11.	$\quad$ End If
Step 12.	End do
Step 13.	$d(v) \Leftarrow d(v) + 1$
Step 14.	For $u = v + 1, \cdots, k$ do
Step 15.	$\quad d(u) \Leftarrow d(u - 1) + 1$
Step 16.	End do
Step 17.	Go to Step 03
Step 18.	Stop

□□□□□□□

If $k > n/2$, we generate subsets of size $n - k$ of the set $\{1, 2, \cdots, n\}$, and obtain corresponding subsets of size k from them.

13.2.2 Compositions

Consider nonnegative integers t_j, $j = 1, \cdots, K$, that satisfy

$$t_1 + \cdots + t_K = n \qquad (13.1)$$

A particular solution to (13.1), say $(t_{*1}, \cdots, t_{*K})$, is called a K way composition of n. For example, when $n = 5$ and $K = 3$, $(0, 2, 3)$ is a 3 way composition of 5.

COMPOSITIONS AND PARTITIONS

Suppose, for some K and n, we need to list or enumerate all the K way composition of n. This is easily done by formulating the problem in terms of nested sums and implementing it as K nested 'do loops.' This was done for the exhaustive enumeration algorithms in §11.2. For example, when $K = 4$, we implement the scheme represented by the summation

$$\sum_{t_1=0}^{n} \sum_{t_2=t_1}^{n} \sum_{t_3=t_1+t_2}^{n} \{t_4 = n - (u + v + w)\} \tag{13.2}$$

A more robust and general implementation makes use of the Algorithm 13.SS.1. Let $m = n + K - 1$ and $k = K - 1$ and consider the problem of generating subsets of size k from the set $\{1, 2, \cdots, m\}$. Let one such ordered subset be $\{j_1, j_2, \cdots, j_k\}$ written so that $j_i < j_{i+1}$ for $i \geq 1$. Obviously, $1 \leq j_i \leq m$. Let $j_0 = 1$ and, for $1 \leq i \leq k$, let

$$t_i = j_i - j_{i-1} \tag{13.3}$$

and

$$t_K = n - (t_1 + \cdots + t_{K-1}) \tag{13.4}$$

It is clear that $(t_1, \cdots, t_K)$ so obtained is a K way composition of n. It then follows that the total number of solutions (or K way compositions) that satisfy $0 \leq t_j \leq n$ for all j and $t_1 + \cdots + t_K = n$ is

$$\binom{m}{k} = \frac{(n+K-1)!}{(K-1)!\, n!} \tag{13.5}$$

This method of generating all K way compositions of n is implemented in the Algorithm 13.CO.1 below.

Algorithm 13.CO.1

Step 01.   Input k and n $k \leq n$
Step 02.   Set $m \Leftarrow n + K - 1$ and $k \Leftarrow K - 1$
Step 03.   Generate all subsets of $\{1, 2, \cdots, m\}$
           of size k sequentially using Algorithm 13.SS.1.
Step 04.   For each generated subset, $\{j_1, j_2, \cdots, j_k\}$,
           obtain the corresponding K way composition
           of n, $(t_1, \ldots, t_K)$, using (13.3) and (13.4).

□□□□□□

13.2.3 Partitions

Using the above definitions, we have that the 3 part compositions of 5 are:

$$(0, 0, 5), (0, 5, 0), (5, 0, 0),$$
$$(0, 1, 4), (0, 4, 1), (1, 0, 4), (1, 4, 0), (4, 0, 1), (4, 1, 0),$$
$$(1, 1, 3), (1, 3, 1), (3, 1, 1),$$
$$(0, 2, 3), (0, 3, 2), (2, 0, 3), (2, 3, 0), (3, 0, 2)(3, 2, 0),$$
$$(1, 2, 2), (2, 1, 2), (2, 2, 1)$$

If the order in which the numbers appear in each composition is not relevant, then the list reduces to

$$\{0,0,5\}, \{0,1,4\}, \{1,1,3\}, \{0,2,3\}, \{1,2,2\}$$

The latter is called a list of 3 part partitions of 5. In general a K part partition of n is a nonnegative solution $\{t_1, \cdots, t_K\}$ of (13.1) where the order of the t_j's is immaterial. We can write a K part partition of n as

$$\{m_0 * 0, m_1 * 1, \cdots, m_n * n\}$$

where $m_j * j$ denotes that the number j appears with frequency m_j in the partition. Obviously,

$$\sum_{j=0}^{n} m_j = K$$

and

$$\sum_{j=0}^{n} j m_j = n$$

The number of compositions associated with the partition $\{m_0 * 0, m_1 * 1, \cdots, m_n * n\}$ is

$$\frac{K!}{m_0! \, m_1! \, \cdots \, m_n!}$$

An exact formula for the number of partitions of an integer is not known. It, however, is smaller, usually considerably smaller, than the number of compositions. Thus, while there are $46,376$ 5 part compositions of 30, the corresponding partitions number only 674.

All the k part partitions of n are generated in a dictionary order as follows. We start with $\{0,0,\cdots,n\}$. At any step, we first scrutinize the current set from left to right to find the right most element that does not differ from the last value by two or more. Then this value, say x, and all values to its right except the last one are replaced by $x+1$. The kth value is set to n minus the sum of all other $k-1$ values. We stop when none of the first $k-1$ values differs from the kth by more than one. Otherwise, we continue with scrutinization and replacement. (We leave the formal proof for correctness of this scheme to the reader.) An outline of a computer program for this scheme appears in Algorithm 13.PA.1 below.

We note, without going into the details, that a more efficient method of generating partitions utilizes the multiset representation of a partition.

13.3 A Single Multinomial

Suppose n is a positive integer and let $\boldsymbol{T} = (T_1, \ldots, T_K)$ be a nonnegative random vector of multinomial outcomes with $\Sigma_k T_k = n$. Let the probability vector associated with these outcomes be $\boldsymbol{\pi} = (\pi_1, \ldots, \pi_K)$ with $0 < \pi_k < 1$, $1 \leq k \leq K$ and $\Sigma_k \pi_k = 1$. This means that

$$P[\boldsymbol{T} = \boldsymbol{t}] = n! \prod_{k=1}^{K} c_k(t_k) \tag{13.6}$$

A SINGLE MULTINOMIAL

Algorithm 13.PA.1

```
Step 01.   Input k and n,  k ≤ n
Step 02.   For u = 1, ···, k − 1   d(u) ⇐ 0
Step 03.   d(k) ⇐ n
Step 04.   Output (d(1), d(2), ···, d(k))
Step 05.   For u = 1, ···, k  do
Step 06.     If (d(k) − d(u) ≤ 1) then
Step 07.       v ⇐ u − 1
Step 08.       Go to 11
Step 09.     End If
Step 10.   End do
Step 11.   If (v = 0) go to 21
Step 12.   For u = v, ···, k − 1  do
Step 13.     d(u) ⇐ d(v) + 1
Step 14.   End do
Step 15.   S ⇐ 0
Step 16.   For u = 1, ···, k − 1 do
Step 17.     S ⇐ S + d(u)
Step 18.   End do
Step 19.   d(k) ⇐ n − S
Step 20.   Go to Step 04
Step 21.   Stop
```

□□□□□□□

where $t = (t_1, \ldots, t_K)$ and

$$c_k(u) = \frac{1}{u!}\pi_k^u \tag{13.7}$$

Multinomial data often occurs in practice. A frequent task for such data is to test the hypothesis

$$H_0 : \pi_k = \pi_{0k}, \quad 0 < \pi_{0k} < 1, \quad k = 1, \ldots, K$$

A statistical test for H_0 is called a goodness-of-fit test. Such tests were among the earliest techniques of statistical inference. One special case, the equiprobability case, occurs when π_k is postulated to be the same for all k. That is,

$$\pi_{0k} = K^{-1}$$

Example 13.1: An historic example relates to data on the orbital eccentricity of 116 binary stars (Jeans 1935; Inman 1994). Here, $n = 116$, $K = 9$ and

$$t = (11, 9, 14, 24, 25, 6, 13, 7, 7)$$

Theoretical computations done by astronomers implied the need to test whether the data were consistent with the null probability

$$\pi_0 = (.039, .052, .069, .091, .112, .129, .147, .172, .189)$$

13.3.1 Test Statistics

Many statistics for testing H_0 exist. Among the two commonly used ones are the score, or the Pearson chisquare statistic

$$W_s = \sum_{k=1}^{K} \frac{(t_k - n\pi_{0k})^2}{n\pi_{0k}} \qquad (13.8)$$

and the LR statistic

$$W_l = 2 \sum_{k=1}^{K} t_k \ln\left[\frac{t_k}{n\pi_{0k}}\right] \qquad (13.9)$$

The Pearson chisquare statistic is often denoted as X^2, and the LR statistic, G^2. In general, let $\mathcal{W}$ denote a test statistic for H_0 and assume that it is of the form

$$\mathcal{W} = \sum_{k=1}^{K} w_k(\mathcal{T}_k) \qquad (13.10)$$

If w_* is the observed value of $\mathcal{W}$, the p-value for H_0 is

$$\mathrm{P}[\mathcal{W} \geq w_* \mid H_0] \qquad (13.11)$$

When n is large relative to K, the Pearson chisquare, the LR and some other statistics for H_0 approximately follow a chisquare distribution with $K - 1$ degrees of freedom. We use this to compute p-values in large samples. When the data are sparse, this approach may compromise accuracy. In this case, an exact method is needed.

Example 13.2: Consider data from Hirji (1997b). Let $K = 10$, $n = 20$ and

$$\boldsymbol{t} = (4, 2, 1, 1, 4, 1, 1, 1, 1, 4)$$

We test $H_0 : \boldsymbol{\pi} = (.03, .04, .06, .08, .09, .10, .12, .14, .16, .18)$. For the above data, the large sample LR p-value is 0.0450; the exact LR p-value, on the other hand, is 0.0628.

13.3.2 Exact Distributions

The exact distribution of $\mathcal{W}$ is

$$\mathrm{P}[\mathcal{W} = w] = n! \sum \left[\prod_{k=1}^{K} c_k(t_k)\right] \qquad (13.12)$$

where the summation in (13.12) is over the set of values of $\boldsymbol{t} = (t_1, \ldots, t_K)$ which satisfy

$$0 \leq t_k \leq n, \quad k = 1, \ldots, K \qquad (13.13)$$

A SINGLE MULTINOMIAL

$$\sum_{k=1}^{K} t_k = n \quad (13.14)$$

and

$$\sum_{k=1}^{K} w_k(t_k) = w \quad (13.15)$$

A straightforward construction of the distribution enumerates all the possible outcomes (or K way compositions of n) with Algorithm 13.CO.1. For each outcome, the value and the probability of the test statistic are computed and aggregated to produce the distribution of $\mathcal{W}$. A recursive polynomial method is, however, a more efficient alternative. To formulate this distribution in polynomial terms, let, for $k = 1, \ldots, K$,

$$f_k(\theta, \phi) = \sum_{u=0}^{n} c_k(u) \theta^u \phi^{w_k(u)} \quad (13.16)$$

The product of these K polynomials is

$$F_K(\theta, \phi) = \prod_{k=1}^{K} f_k(\theta, \phi) = \sum_{(u,v) \in \Omega} c(u, v) \theta^u \phi^v \quad (13.17)$$

where Ω is the set of exponents of θ and ϕ. This polynomial has the following important property.

Theorem 13.1:

$$P[\mathcal{W} = w] = n! c(n, w) \quad (13.18)$$

Proof: Follows from (13.12), and algebraic expansion of the polynomial product in (13.17). □

Consider now the equiprobable case, and assume that $w_k(u) = h(u)$ for all k. Let $\theta_* = \theta/K$. In this case, all the K way compositions of n associated with a particular K way partition of n have the same value of the test statistic and also have the same null probability. For computing a p-value, generating all partitions (using, say, Algorithm 13.PA.3), is a more efficient enumerative method in the equiprobable case. Under the polynomial method, the equiprobable generating polynomial is

$$F_K(\theta, \phi) = \left[\sum_{u=0}^{n} (u!)^{-1} \theta_*^u \phi^{h(u)} \right]^K \quad (13.19)$$

We need terms of (13.19) in which the exponent of θ, and thus of θ_*, is n. After appropriate computation, we extract these terms and normalize the coefficients by multiplying them with K^{-n}.

****

As (13.16) and (13.17) are special cases of the polynomials considered in Chapter 12, the RPM algorithms developed there apply here. However, a simpler formulation is also available here. To develop this, define, as usual, $F_k(1, \phi; m, .)$ as the subpolynomial of terms from the product

$$\prod_{j=1}^{k} f_j(\theta, \phi)$$

in which the exponent of θ is equal to m, and for which the value of θ is set to 1. Then, using the notation of Chapter 12, and for $0 \leq m \leq n$,

$$F_k(1, \phi; m, .) = \sum_{j=1}^{N_k(m)} d_k(m, j) \phi^{w+k(m,j)} \qquad (13.20)$$

From the formulas in §12.3, we have that

$$M_{1k}(n) = 0, \ k < K \quad \text{and} \quad M_{1k}(n) = n, \ k = K$$

and

$$M_{2k}(n) = n$$

Further,

$$m_{1k}(n, v) = 0, \ k < K, \quad \text{and} \quad m_{1k}(n, v) = n - v, \ k = K$$

and

$$m_{2k}(n, v) = n - v$$

These imply that the recursion to obtain terms of the final product in which the exponents of θ are n is as follows. Start with $F_0(1, \phi; m, .) \equiv 1$ for all $0 \leq m \leq n$. Then, for $0 \leq m \leq n$. implement

$$F_k(1, \phi; m, .) = \sum_{u=0}^{m} c_k(u) \phi^{w_k(u)} F_{k-1}(1, \phi; m - u, .) \qquad (13.21)$$

when $k < K$ and, finally

$$F_K(1, \phi; n, .) = \sum_{u=0}^{n} c_K(u) \phi^{w_K(u)} F_{K-1}(1, \phi; n - u, .) \qquad (13.22)$$

This algorithm is equivalent to the Algorithm 12.RPM.2.

13.3.3 Tail Area

The above recursion gives us the complete distribution of W. For a p-value, we need the right tail area under H_0 (13.11). We obtain this in an efficient manner by applying exponent checks

A SINGLE MULTINOMIAL

EC 12.04. Some of the quantities needed to implement such checks can be explicitly specified. First, to get the sums of the product like in (12.20), we use the identity

$$\sum \left[\prod_{k=1}^{K} \frac{\pi_k^{t_k}}{t_k!} \right] = \frac{\{\pi_1 + \ldots + \pi_K\}^n}{n!}$$

where the summation is over values of $t_1, \ldots, t_K$ that satisfy (13.13) and (13.14). In this connection, for later use, define

$$\delta_{+,k}(m) = \frac{\{\pi_{0k} + \ldots + \pi_{0K}\}^m}{m!} \qquad (13.23)$$

Next, as in Chapter 12, for $1 \leq k \leq K$, and $0 \leq v \leq n$, let

$$\eta_{1,k}(v) = \min \left[\sum_{j \geq k} w_j(t_j) \right] \qquad (13.24)$$

$$\eta_{2,k}(v) = \max \left[\sum_{j \geq k} w_j(t_j) \right] \qquad (13.25)$$

where the maximization and minimization are subject to

$$\sum_{j \geq k} t_j = v, \quad \text{and} \quad 0 \leq t_j \leq v, \quad j \geq k$$

and further where the v is an integer such that $0 \leq v \leq n$. For the current problem, $\eta_{1,k}(v)$ and $\eta_{2,k}(v)$, or relevant bounds, can be specified explicitly for some commonly used statistics (Exercise 13.12). Otherwise, we use backward induction. Assume that one of that has been done. Let $\Delta = 0$ and suppose we implement recursion (13.21)–(13.22), and are about to multiply the term

$$c_k(u)\phi^{w_k(u)}$$

with the term

$$d_{k-1}(m - u, j)\phi^{w_{+,k-1}(m - u, j)}$$

Then we apply the following exponent checks.

Check EC 13.01

⟨ If $w_k(u) + w_{+,k-1}(m - u, j) + \eta_{2,k+1}(n - m) < w_*$ ⟩

Do not perform the multiplication. Continue on to the next term.

⟨ If $w_k(u) + w_{+,k-1}(m - u, j) + \eta_{1,k+1}(n - m) \geq w_*$ ⟩

Perform the multiplication

$$\rho = d_{k-1}(m - u, j) \times c_k(u) \times \delta_{+,k+1}(n - m)$$

Set $\Delta = \Delta + \rho$. Do not store anything in $F_k(1, \phi; m, .)$ and continue on to the next term.

⟨ Otherwise ⟩

Perform the multiplication $d_{k-1}(m - u, j) \times c_k(u)$, accumulate the result in the position for the exponent $(w_k(u) + w_{+,k-1}(m - u, j))$ of ϕ in $F_k(1, \phi; m, .)$ and continue.

❏❏❏❏❏❏❏

The value of Δ at the final stage is the desired quantity, $F_K(1, 1; n, \geq w_*)$.

****

The history of application of the FFT to exact inference for the single multinomial problem is a long one. Despite the early promise, later investigations showed that it can give unreliable results and is not as efficient as the RPM method (Hirji 1997b). For details, consult the literature listed in §13.8. For the equiprobable case with moderate n, the partition generating method is a good alternative. For larger n, the RPM method is more efficient. In the general probability case, the RPM is the method of choice. For details on the theoretical connections and empirical comparisons between various algorithmic approaches, see Hirji (1998).

13.3.4 Equivalent Statistics

The concept of equivalent statistics helps simplify the computation in the single multinomial case.

Theorem 13.2: For the single multinomial problem:

(i) Under equiprobability, the Pearson chisquare statistic is equivalent to setting

$$w_k(t_k) = t_k^2$$

and the LR statistic is equivalent to setting

$$w_k(t_k) = t_k \log t_k$$

(ii) In general, the Pearson chisquare statistic is equivalent to setting

A SINGLE MULTINOMIAL

$$w_k(t_k) = (\nu t_k^2)/\pi_k$$

and the LR statistic is equivalent to setting

$$w_k(t_k) = t_k \log(\nu t_k/\pi_k)$$

where $\nu = \max\{\pi_1, \ldots \pi_K\}$. (The factor ν in the last two expressions enhances computational stability.)

Proof: Left to the reader. □

The power divergence family of statistics is applicable to this problem as well (see Chapter 10, and Cressie and Read 1984). Another statistic applied to the single multinomial has $w_k(t_k) = t_k(t_k - 1)$ (Holtzman and Good 1986).

Example 13.1 (continued): For the data on the orbital eccentricity of binary stars and the postulated null hypothesis, total enumeration is not feasible. An RPM approach with exponent checks yields the result readily. The exact chisquare p-value is 0.1051×10^{-7} and the exact LR p-value is 0.4983×10^{-10}. The computing times for the RPM approach, compared to the FFT method, are lower by a factor of four to eight. This difference persists despite the improvements for the FFT method given by Hirji (1997b).

13.3.5 An Incidence Density Design

Now consider data from an incidence density study. Assume that over time, events from K categories of outcomes are observed. Let t_k be the number of observations and τ_k, the associated unit time (person time). And let λ_k denote the rate of occurrence per unit time of events in category k. Assuming a Poisson model for each category, we have

$$P[\boldsymbol{T} = \boldsymbol{t}] = \prod_{k=1}^{K} \exp(-\lambda_k \tau_k) \frac{(\tau_k \lambda_k)^{t_k}}{t_k!} \qquad (13.26)$$

Let the observed number of the total number of events be n. It follows then that

$$P[\boldsymbol{T} = \boldsymbol{t} \mid \sum_k T_k = n] = n! \prod_{k=1}^{K} c_k(t_k) \qquad (13.27)$$

where, as before,

$$c_k(u) = \pi_k^u (u!)^{-1} \qquad (13.28)$$

and further where

$$\pi_k = \frac{\lambda_k \tau_k}{\lambda_1 \tau_1 + \cdots + \lambda_K \tau_K} \qquad (13.29)$$

This conditional probability is used to perform exact inference on these data along the lines developed for a single multinomial. For example, we may test the hypothesis

Table 13.1 *ESR and Pulmonary Infection in Kwashiorkor*

	ESR Level			
	≤ 10	11–25	26–99	100+
No Infection	26	9	4	0
Pulmonary Tuberculosis	1	0	3	3
Other Pneumonia	7	0	1	0

Source: Hameer (1990); Used with permission.

$$\lambda_1 = \ldots = \lambda_K$$

by testing

$$\pi_k = \frac{\tau_k}{\tau_+}$$

where $\tau_+ = \tau_1 + \ldots + \tau_K$.

The Poisson model extends to the case with an ordering between the categories. Suppose that the kth category has a numerical score or associated dose equal to x_k, and

$$\lambda_k = \mu \lambda^{x_k}$$

and

$$\pi_k = \frac{\tau_k \lambda^{x_k}}{\tau_1 \lambda^{x_1} + \cdots + \tau_K \lambda^{x_K}}$$

The hypothesis of no dose effect is $H_0 : \lambda = 1$. A point estimate and CI for λ assess the magnitude of the effect.

An inverse sampling design for the K outcomes case may also be devised. For example, subjects are sampled until a fixed number of observations occur in category 1. Exact analysis is then developed accordingly.

13.4 Trinary Response Models

Multinomial response data often involve situations with one or more covariates. Now, we consider a special case, namely, trinary response models with one covariate.

Example 13.3: Table 13.1 shows the data from Table 1.2 with the age variable removed. These concern young children in a pediatrics unit who were afflicted with Kwashiorkor. One aim of the study was to see whether the erythrocyte sedimentation rate (ESR) level was related with the type of infection (pulmonary tuberculosis or other pneumonia) when such a condition is present.

The general notation for such data is in Table 13.2. $\mathcal{Y} \in \{0, 1, 2\}$ is the row (response) variable,

TRINARY RESPONSE MODELS

Table 13.2 *Notation for a $3 \times K$ Table*

Response	Factor Score				
$\mathcal{Y}$	x_1	$\cdots$	x_k	$\cdots$	x_K
0	t_{01}	$\cdots$	t_{0k}	$\cdots$	t_{0K}
1	t_{11}	$\cdots$	t_{1k}	$\cdots$	t_{1K}
2	t_{21}	$\cdots$	t_{2k}	$\cdots$	t_{2K}
Total	n_1	$\cdots$	n_k	$\cdots$	n_K

and x_k is the factor (column) variable. The study units are statistically independent of each other. The column totals, n_k, $k = 1, \ldots, K$, are deemed fixed by design with $\Sigma_k n_k = n$. The number of units at level k expressing the response j is denoted by $\mathcal{T}_{jk}$, with t_{jk} being its observed value. Of course,

$$t_{0k} + t_{1k} + t_{2k} = n_k$$

for all k. Denote the cell probabilities, $P[\mathcal{Y} = j \mid x_k]$, as π_{jk} with

$$\pi_{0k} + \pi_{1k} + \pi_{2k} = 1$$

The sampling distribution is a product of K independent trinomial distributions. We also ask if the response variable is ordered or not. Different statistical models then apply to nominal and ordinal responses. The outcome in a clinical trial, for example, may be worse, no change, or improved. When response levels are ordered, the label for each level may not reflect the numerical score attached to it.

In this section, we use four types of models based on four combinations of response and factor variables. These models generalize the binary response models of Chapter 8.

13.4.1. Nominal Response, Nominal Factor

Suppose the values of $\mathcal{Y}$ are arbitrary, and x_k is an indicator variable with $x_k = 1$ if the kth level occurs, and is 0 otherwise. Then, for $j = 1, 2$, and $k = 1, \ldots, K$, consider the model

$$\frac{\pi_{jk}}{\pi_{0k}} = \theta_{jk} \tag{13.30}$$

Level 0 is the reference level, though any other level can be used in its place without changing the essence of the model. We write the joint distribution of the data, $P[\mathcal{T}_{jk} = t_{jk}; j = 1, 2; k = 1, \ldots, K]$, in a polynomial form as

$$\prod_{k=1}^{K} \frac{c_k(t_{1k}, t_{2k}) \theta_{1k}^{t_{1k}} \theta_{2k}^{t_{2k}}}{(1 + \theta_{1k} + \theta_{2k})^{n_k}} \tag{13.31}$$

where

$$c_k(u,v) = \frac{n_k!}{u!\,v!\,(n_k - u - v)!} \tag{13.32}$$

The denominator in (13.31) is also written as

$$\sum_{(u,v)\in\Omega_k} c_k(u,v)\theta_{1k}^u \theta_{2k}^v \tag{13.33}$$

where

$$\Omega_k = \{(u,v) : 0 \le u,v \le n_k, \text{ and } 0 \le u+v \le n_k\} \tag{13.34}$$

****

Are the response probabilities identical at all factor values? This is often the question of interest. Hence we need to test, for $j = 1, 2$, the null hypothesis $H_0 : \theta_{jk} = \theta_j$ for all k. Let, for $j = 1, 2$,

$$\mathcal{T}_j = \sum_{k=1}^{K} \mathcal{T}_{jk} \tag{13.35}$$

Note that $\mathcal{T}_{0k} = n_k - (\mathcal{T}_{1k} + \mathcal{T}_{2k})$ and $\mathcal{T}_0 = n_k - (\mathcal{T}_1 + \mathcal{T}_2)$. It is easy to show that $\mathcal{T}_j$ is sufficient for θ_j. Consider a test statistic for H_0 of the form

$$\mathcal{W} = \sum_{k=1}^{K} w_k(\mathcal{T}_{1k}, \mathcal{T}_{2k}) \tag{13.36}$$

This, for example, may be a score or a LR statistic. The former is given by

$$w_k(\mathcal{T}_{1k}, \mathcal{T}_{2k}) = \sum_{j=0}^{2} \left(\mathcal{T}_{jk} - \frac{n_k t_j}{n}\right)^2 \frac{n}{n_k t_j} \tag{13.37}$$

and the latter is

$$w_k(\mathcal{T}_{1k}, \mathcal{T}_{2k}) = 2\sum_{j=0}^{2} \mathcal{T}_{jk}\ln\left(\frac{n\mathcal{T}_{jk}}{n_k t_j}\right) \tag{13.38}$$

The probability statistic is also used. In general, assume that $w_k \ge 0$ and that large values of $\mathcal{W}$ are unlikely under the null. To derive an exact conditional test based on $\mathcal{W}$, consider the polynomials

$$f_{1k}(\theta_1, \theta_2, \psi) = \sum_{(u,v)\in\Omega_k} c_k(u,v)\theta_1^u \theta_2^v \psi^{w_k(u,v)} \tag{13.39}$$

where ψ is a dummy variable. Also let

TRINARY RESPONSE MODELS

$$F_{1K}(\theta_1, \theta_2, \psi) = \prod_{k=1}^{K} f_{1k}(\theta_1, \theta_2, \psi) \tag{13.40}$$

If w_* is the observed value of $\mathcal{W}$, then the conditional p-value for H_0 is

$$P[\mathcal{W} \geq w_* \mid \mathcal{T}_j = t_j, j = 1, 2; H_0] = \frac{F_{1K}(1, 1, 1; t_1, t_2, \geq w_*)}{F_{1K}(1, 1, 1; t_1, t_2, .)} \tag{13.41}$$

where

$$F_{1K}(1, 1, 1; t_1, t_2, .) = \frac{n!}{t_1! \, t_2! \, (n - t_1 - t_2)!} \tag{13.42}$$

In large samples, we often use the chisquare distribution with $2(K-1)$ degrees of freedom to approximate this p-value.

13.4.2. Nominal Response, Scored Factor

Now consider the case with nominal $\mathcal{Y}$ and x_k, a numeric scored variable. Without loss of generality, we assume that $x_k \geq 0$ and in particular, $x_1 = 0$. Further, assume that $x_k < x_{k+1}$. For $j = 1, 2$, and $k = 1, \ldots, K$, consider the model

$$\ln\left(\frac{\pi_{jk}}{\pi_{0k}}\right) = \ln(\theta_j) + x_k \ln(\phi_j) \tag{13.43}$$

With $\mathcal{T}_j$, $j = 1, 2$, as defined above, let

$$\mathcal{W}_j = \sum_{k=1}^{K} t_{jk} x_k \tag{13.44}$$

Under the product trinomials model, $\mathcal{T}_j$ is sufficient for θ_j and $\mathcal{W}_j$ is so for ϕ_j. In data analysis, interest focuses on whether the factor affects response. ϕ_j, $j = 1, 2$, are then the parameters of interest. In that regard, we consider the k gps

$$f_{2k}(\theta_1, \theta_2, \phi_1, \phi_2) = (1 + \theta_1 \phi_1^{x_k} + \theta_2 \phi_2^{x_k})^{n_k} \tag{13.45}$$

and

$$F_{2K}(\theta_1, \theta_2, \phi_1, \phi_2) = \prod_{k=1}^{K} f_{2k}(\theta_1, \theta_2, \phi_1, \phi_2) \tag{13.46}$$

The sufficient statistics for the parameters of interest are $(\mathcal{W}_1, \mathcal{W}_2)$. The required conditional distribution, $P[\mathcal{W}_j = w_j, j = 1, 2 \mid \mathcal{T}_j = t_j, j = 1, 2]$, is then

$$\frac{F_{2K}(1, 1, \phi_1, \phi_2; t_1, t_2, w_1, w_2)}{F_{2K}(1, 1, \phi_1, \phi_2; t_1, t_2, ., .)} \tag{13.47}$$

This conditional distribution can be used to test for the presence of a covariate effect, that is, to test $H_0: \phi_1 = \phi_2 = 0$. To do this, we derive a score statistic from (13.47) and use it compute the exact or mid-p-value. If we want to estimate or analyze each ϕ_j separately, we further condition on the sufficient statistic for the other parameter. For example, to analyze ϕ_1, we consider the distribution

$$P[\mathcal{W}_1 = w_1 \mid \mathcal{T}_j = t_j, j = 1, 2; \mathcal{W}_2 = w_2] \tag{13.48}$$

We may also consider whether a simpler model applies to this situation. For example, we may test the hypothesis $H_0: \phi_1 = \phi_2 = \phi$. In this case, the sufficient statistic for ϕ is

$$\mathcal{W}_3 = \sum_{k=1}^{K} (t_{1k} + t_{2k}) x_k \tag{13.49}$$

and the generating product polynomial is

$$F_{*2K}(\theta_1, \theta_2, \phi_1, \phi_2) = \prod_{k=1}^{K} (1 + \{\theta_1 + \theta_2\}\phi^{x_k})^{n_k} \tag{13.50}$$

The conditional distribution of $\mathcal{W}_3$ has a form similar to that of (13.48) derived from (13.47).

13.4.3. Ordinal Response, Nominal Factor

Now let $\mathcal{Y}$ be an ordinal and x_k, a nominal variable. For $k = 1, \ldots, K$, consider the model

$$\frac{\pi_{1k}}{\pi_{0k}} = \delta_{1k} \tag{13.51}$$

$$\frac{\pi_{2k}}{\pi_{1k}} = \delta_{2k} \tag{13.52}$$

This model is an extension of the adjacent categories model for ordinal response (10.66). Define

$$\mathcal{R}_1 = \sum_{k=1}^{K} (\mathcal{T}_{1k} + \mathcal{T}_{2k}) \quad \text{and} \quad \mathcal{R}_2 = \sum_{k=1}^{K} \mathcal{T}_{2k} \tag{13.53}$$

Then $\mathcal{R}_j$ is sufficient for δ_j, $j = 1, 2$. The hypothesis of the absence of association between response and factor is then equivalent to $H_0: \delta_{jk} = \delta_j, j = 1, 2; k = 1, \ldots, K$. Consider a test statistic for H_0 of the form

$$\mathcal{Q} = \sum_{k=1}^{K} q_k(\mathcal{T}_{1k}, \mathcal{T}_{2k}) \tag{13.54}$$

For example, this may be an appropriate score or likelihood ratio statistic for model (13.51)–(13.52). Assume that $q_k \geq 0$ and that larger values of $\mathcal{Q}$ are improbable under the null. To derive its exact conditional distribution, consider the polynomials

TRINARY RESPONSE MODELS

$$f_{3k}(\delta_1, \delta_2, \psi) = \sum_{(u,v) \in \Omega_k} c_k(u,v)\theta_1^{(u+v)}\theta_2^v \psi^{q_k(u,v)} \tag{13.55}$$

where ψ is a dummy variable and $c_k(u,v)$ and Ω_k are as above. Also let

$$F_{3K}(\delta_1, \delta_2, \psi) = \prod_{k=1}^{K} f_{3k}(\delta_1, \delta_2, \psi) \tag{13.56}$$

If q_* is the observed value of $\mathcal{Q}$, then the conditional p-value for H_0 is

$$P[\mathcal{Q} \geq q_* \mid \mathcal{R}_j = r_j, j=1,2; H_0] = \frac{F_{3K}(1,1,1;r_1,r_2,\geq q_*)}{F_{3K}(1,1,1;r_1,r_2,.)} \tag{13.57}$$

where

$$F_{3K}(1,1,1;r_1,r_2,.) = \frac{n!}{(r_1 - r_2)!\, r_2!\,(n-r_1)!} \tag{13.58}$$

In large samples, we often use the chisquare distribution with $2(K-1)$ degrees of freedom to approximate this p-value.

13.4.4. Ordinal Response, Scored Factor

Now consider when $\mathcal{Y}$ is ordinal and x_k, a scored variable. Without loss of generality, we assume that $x_k \geq 0$ and in particular, $x_1 = 0$. Further, assume that $x_k < x_{k+1}$. Then, for $j = 1, 2$, and $k = 1, \ldots, K$, let

$$\ln\left(\frac{\pi_{1k}}{\pi_{0k}}\right) = \ln(\delta_1) + x_k \ln(\mu_1) \tag{13.59}$$

$$\ln\left(\frac{\pi_{2k}}{\pi_{1k}}\right) = \ln(\delta_2) + x_k \ln(\mu_2) \tag{13.60}$$

This is the adjacent categories logit model ordinal response and a scored factor. Let $\mathcal{R}_1$ and $\mathcal{R}_2$ be defined as above and further let

$$\mathcal{Q}_1 = \sum_{k=1}^{K}(\mathcal{T}_{1k} + \mathcal{T}_{2k})x_k \tag{13.61}$$

$$\mathcal{Q}_2 = \sum_{k=1}^{K}\mathcal{T}_{2k}x_k \tag{13.62}$$

Under the product trinomials design, $\mathcal{R}_j$ is sufficient for δ_j, and $\mathcal{Q}_j$ is so for μ_j. To check if the factor affects response, we focus on μ_j, $j = 1, 2$. Then consider the k gps

$$f_{4k}(\delta_1, \delta_2, \mu_1, \mu_2) = (1 + \delta_1 \mu_1^{x_k}\{1 + \delta_2 \mu_2^{x_k}\})^{n_k} \tag{13.63}$$

and

$$F_{4K}(\delta_1, \delta_2, \mu_1, \mu_2) = \prod_{k=1}^{K} f_{4k}(\delta_1, \delta_2, \mu_1, \mu_2) \qquad (13.64)$$

The required conditional distribution, $P[\mathcal{Q}_j = w_j, j = 1, 2 \mid \mathcal{R}_j = t_j, j = 1, 2;]$, is then

$$\frac{F_{4K}(1, 1, \mu_1, \mu_2; r_1, r_2, q_1, q_2)}{F_{4K}(1, 1, \mu_1, \mu_2; r_1, r_2, \cdot, \cdot)} \qquad (13.65)$$

This conditional distribution can be used to test for the presence of a covariate effect, that is, to test $H_0 : \mu_1 = \mu_2 = 0$. To do this, we derive a score statistic from (13.65) and use it to compute the exact or mid-p-value. To analyze each parameter separately, we further condition on the relevant sufficient statistic. For example, to analyze μ_1, consider the distribution

$$P[\mathcal{Q}_1 = q_1 \mid \mathcal{R}_j = t_j, j = 1, 2, \mathcal{Q}_2 = q_2] \qquad (13.66)$$

We consider whether a simpler uniform effect model applies here. We then test the hypothesis $H_0 : \mu_1 = \mu_2 = \mu$. In this case, the sufficient statistic for μ is

$$\mathcal{Q}_3 = \sum_{k=1}^{K} (t_{1k} + 2t_{2k}) x_k \qquad (13.67)$$

and the generating product polynomial is

$$F_{*4K}(\delta_1, \delta_2, \mu_1, \mu_2) = \prod_{k=1}^{K} (1 + \delta_1 \mu^{x_k} \{1 + \delta_2 \mu^{x_k}\})^{n_k} \qquad (13.68)$$

The conditional distribution of $\mathcal{Q}_3$ is similar in form to that of (13.66).

13.5 Conditional Polynomials

Now we outline construction of efficient computational algorithms for the above types of trinary response models. First observe that since $r_1 = t_1 + t_2$ and $r_2 = t_2$, then $T_j = t_j, j = 1, 2$ if and only if $R_j = r_j, j = 1, 2$. The conditional sample space for all the trinary response models given above thus obtains from the set of $3 \times K$ tables with both margins fixed, or from terms of respective generating polynomials in which the coefficients of θ_j (or equivalently, those of δ_j) are fixed. Another way of stating this is to note that if $\mathcal{W}$ given by (13.36) is identical to $\mathcal{Q}$ given by (13.54) then

$$F_{1K}(\theta_1, \theta_2, \phi; t_1, t_2, \cdot) = F_{3K}(\theta_1, \theta_2, \phi; r_1 - r_2, r_2, \cdot)$$

Similar relations also exist for other gps given in the previous section. This implies that, up to a point, we can adopt a common computational strategy for these models. To develop this, consider the product of polynomials of the form (13.39). Dropping the subscript 1, let, for $k = 1, \ldots, K$,

CONDITIONAL POLYNOMIALS

$$F_k(\theta_1, \theta_2, \psi) = \prod_{j=1}^{k} f_j(\theta_1, \theta_2, \psi) \tag{13.69}$$

Suppose we need $F_K(\theta_1, \theta_2, \psi; t_1, t_2, .)$, that is terms of the product in which the exponent of θ_j equals t_j. The EE approach generates K trinomial distributions, multiplies them overall in a term by term fashion and extracts the desired segment from the product. Since $f_k(\theta_1, \theta_2, \psi)$ is also written as

$$f_k(\theta_1, \theta_2, \psi) = \sum_{u=0}^{n_k} \theta_1^u \sum_{v=u}^{u} c_k(u,v) \theta_2^v \psi^{w_k(u,v)} \tag{13.70}$$

This method entails a computational effort of the order

$$2^{-K} \prod_{k=1}^{K} (n_k+1)(n_k+2) \tag{13.71}$$

The RPM approach is a more efficient alternative, especially with integer types of scores. Under this, for $k = 1, \cdots, K$, we implement

$$F_k(\theta_1, \theta_2, \psi) = F_{k-1}(\theta_1, \theta_2, \psi) f_k(\theta_1, \theta_2, \psi) \tag{13.72}$$

where $F_0(\theta_1, \theta_2, \psi) \equiv 1$. The required conditional distribution then obtains from selecting the terms of the final product in which the exponent of θ_j is t_j.

13.5.1. Exponent Checks

Efficiency is enhanced by applying exponent checks. A simple way here is to apply checks EC 9.01 separately to the exponents of θ_j. Suppose we are about to multiply a term in $F_{k-1}(\theta_1, \theta_2, \psi)$ with a term in $f_k(\theta_1, \theta_2, \psi)$. Let t_{+jk} be the resultant exponent of θ_j if that was to be done. Obviously,

$$t_{+jk} = t_{j1} + \ldots + t_{jk} \tag{13.73}$$

for some $(t_{j1}, \cdots, t_{jk})$. Then we apply the following checks.

Check EC 13.02

$$\langle \text{ If } t_{+1k} > t_1 \quad \text{or} \quad t_{+1k} + n_{+,k+1} < t_1 \ \rangle$$

Or

$$\langle \text{ If } t_{+2k} > t_2 \quad \text{or} \quad t_{+2k} + n_{+,k+1} < t_2 \ \rangle$$

Do not perform the multiplication. Continue on to the next term.

$$\langle \text{ Otherwise } \rangle$$

Perform the multiplication, accumulate the result in $F_k(\theta_1, \theta_2, \psi)$ and continue.

□□□□□□□

Let κ_k be the number of terms in the polynomial $F_k(\theta_1, \theta_2, \psi)$ as truncated with EC 13.02, with $\kappa_0 = 1$. Then the RPM approach used with EC 13.02 entails a computational effort of the order

$$2^{-(K-1)} \sum_{k=1}^{K} \kappa_{k-1}(n_k+1)(n_k+2)$$

Consider an empirical comparison of the basic EE approach with the RPM approach augmented with EC 13.02. Take the situation with $K = 3, 4, 5$ and the observed $t_{jk} = a$, where $a \in \{1, 2, 3, 4, 5\}$. Then $n_k = 3a$, $n = 3aK$ and $t_j = aK$ for all k and j. For comparison, we consider the model (13.41) for a nominal response, a scored factor with $x_k = k-1$ and $\phi_1 = \phi_2 = \phi$. We then generate the conditional distribution of $\mathcal{W}_3$ given $\mathcal{T}_1$ and $\mathcal{T}_2$. Table 13.3 gives the number of records processed to obtain the final distribution under the two approaches. It is clear from it that the crude EE approach soon runs out of steam. Note that these comparisons were done for balanced data; when the data are unbalanced the impact of the exponent checks is even more pronounced.

Table 13.3 *A Comparison of EE and RPM + EC 13.02*

		Number of Records Processed		
K	$3aK$	EE	RPM	RPM/EE
3	9	1000	410	0.41
	18	21952	4732	0.22
	27	166375	24255	0.15
	36	753571	82901	0.11
	45	2515456	222089	0.88×10^{-1}
4	12	10000	1240	0.12
	24	614656	17528	0.28×10^{-1}
	36	68575000	98230	0.14×10^{-2}
	48	342102000	353081	0.10×10^{-2}
5	15	100000	3231	0.32×10^{-1}
	30	17210300	50653	0.29×10^{-2}
	45	6240320000	295626	0.40×10^{-4}

13.5.2. Improved Exponent Checks

Check EC 13.02, applied above, considers each constraint separately. It is more efficient to consider them simultaneously. And that can also be used to formulate a backward induction scheme for directly computing tail areas. It is also helpful to embed the exponent checks within the recursive scheme, as done for Algorithm 12.RPM.2 of Chapter 12.

Consider then values of nonnegative integers (t_{1k}, t_{2k}), $k = 1, \ldots, K$, that satisfy, for $j = 1, 2$,

CONDITIONAL POLYNOMIALS

$$0 \leq t_{jk} \leq n_k, \text{ and } 0 \leq t_{1k} + t_{2k} \leq n_k \tag{13.74}$$

for $k = 1, \ldots, K$, and

$$t_{j1} + \cdots + t_{jK} = t_j \tag{13.75}$$

where $0 \leq t_j \leq n$ and $0 \leq t_1 + t_2 \leq n$.

Let $\Lambda_k(t_1, t_2)$ be the set of (t_{+1k}, t_{+2k}) that satisfy (13.73), (13.74) and (13.75). This is the set of exponents of θ_1 and θ_2 of the k cumulative product polynomial, (13.72), trimmed so that the final exponents of these parameters will respectively be t_1 and t_2. It is also the set of feasible values of the row margins of a $3 \times k$ table in which the cells of the first k columns have been filled in such a way that the final $3 \times K$ table can have row margin sums equal to $n - t_1 - t_2$, t_1 and t_2. We use this analogy to generate the required set. Define

$$M_{11k}(t_1) = \max(0, t_1 + n_{+k} - n) \tag{13.76}$$
$$M_{12k}(t_1) = \min(t_1, n_{+k}) \tag{13.77}$$

and

$$M_{21k}(t_1, t_2, t_{+1k}) = \max(0, t_1 + t_2 + (n_{+k} - t_{+1k}) - n) \tag{13.78}$$
$$M_{22k}(t_1, t_2, t_{+1k}) = \min(t_2, n_{+k} - t_{+1k}) \tag{13.79}$$

$\Lambda_k(t_1, t_2)$ then obtains by first letting t_{+1k} take integer values that satisfy

$$M_{11k}(t_1) \leq t_{+1k} \leq M_{12k}(t_1)$$

and for each such value of t_{+1k}, letting t_{+2k} take integer values that satisfy

$$M_{21k}(t_1, t_2, t_{+1k}) \leq t_{+2k} \leq M_{22k}(t_1, t_2, t_{+1k})$$

This set gives us the terms of the truncated polynomial $F_k(\theta_1, \theta_2, \psi)$ we need. When the exponent of θ_1 and θ_2 in this polynomial equals u and v respectively, let the number of the distinct exponents of ψ be $N_k(u, v)$. Further let the distinct values of the exponents be ordered in some fashion, and denoted as

$$\{w_{+k}(1, u, v), \ldots, w_{+k}(N_k(u, v), u, v)\}$$

Then $F_k(\theta_1, \theta_2, \psi)$ is written as

$$\sum_{u = M_{11k}(t_1)}^{M_{12k}(t_1)} \theta_1^u \left\{ \sum_{v = M_{21k}(t_1, t_2, u)}^{M_{22k}(t_1, t_2, u)} \theta_2^v \, F_k(1, 1, \psi; u, v, .) \right\} \tag{13.80}$$

where

$$F_k(1, 1, \psi; u, v, .) = \sum_{j=1}^{N_k(u,v)} d_k(j, u, v) \phi^{w_{+,k}(j, u, v)} \quad (13.81)$$

and further where $d_k(j, u, v)$ is the coefficient associated with the jth exponent of ϕ when the exponent of θ_1 is u, and of θ_2 is v.

****

Now we need to determine terms of $f_{k+1}(\theta_1, \theta_2, \psi)$ to use when multiplying it with $F_k(\theta_1, \theta_2, \psi)$ so that the final exponents will be as needed. Assume we have given values for (t_{+1k}, t_{+2k}) and we need feasible values of $(t_{1,k+1}, t_{2,k+1})$ so that the final values of the cumulative sums will be as desired. Then define,

$$m_{11k}(t_1, t_{+1k}) = \max(0, t_1 - t_{+1k} + n_{+,k+1} - n) \quad (13.82)$$
$$m_{12k}(t_1, t_{+1k}) = \min(n_{k+1}, t_1 - t_{+1k}) \quad (13.83)$$

Further let $m_{21k}(t_1, t_2, t_{+1k}, t_{+2k}, t_{1,k+1})$ equal

$$\max(0, t_2 - t_{+2k} + (t_1 - t_{+1k} - t_{1,k+1}) - (n - n_{+,k+1})) \quad (13.84)$$

and $m_{22k}(t_1, t_2, t_{+1k}, t_{+2k}, t_{1,k+1})$ equal

$$\min(t_2 - t_{+2k}, n_{k+1} - t_{k+1}) \quad (13.85)$$

The set of feasible values of $(t_{1,k+1}, t_{2,k+1})$ given (t_{+1k}, t_{+2k}) obtains by letting $t_{1,k+1}$ assume integer values that satisfy

$$m_{11k}(t_1, t_{+1k}) \leq t_{1,k+1} \leq m_{12k}(t_1, t_{+1k}) \quad (13.86)$$

And then, for each such value of $t_{1,k+1}$, we let $t_{2,k+1}$ assume integer values that satisfy

$$m_{21k}(t_1, t_2, t_{+1k}, t_{+2k}, t_{1,k+1}) \leq t_{2,k+1} \leq m_{22k}(t_1, t_2, t_{+1k}, t_{+2k}, t_{1,k+1}) \quad (13.87)$$

We use the above expressions to construct Algorithm 13.RPM.1 outlined below.

Algorithm 13.RPM.1

Step 01.	Input data and initialize arrays.
Step 02.	Set $F_0(\theta_1, \theta_2, \psi) \Leftarrow 1$
Step 03.	Set $k \Leftarrow 0$
Step 04.	Set $k \Leftarrow k + 1$
Step 05.	Select one term at a time of $F_k(\theta_1, \theta_2, \psi)$.
Step 06.	For each such term, use (13.85) and (13.86) to select terms of $f_{k+1}(\theta_1, \theta_2, \psi)$ and multiply them.
Step 07.	Store the result in $F_{k+1}(\theta_1, \theta_2, \psi)$.
Step 08.	Continue until all multiplications have been done.
Step 09.	If $k = K$ Stop; Else go to Step 4; End If

□□□□□□□

CONDITIONAL POLYNOMIALS

Algorithm 13.RPM.1 is an RPM algorithm with built in exponent checks on the exponents of θ_1 and θ_2 that produces final values of their exponents respectively as t_1 and t_2. We write a detailed computer program for it along the same lines as indicated for Algorithm 12.RPM.2 of Chapter 12.

13.5.3. Backward Induction

The exponent selection scheme in §13.5.2 is helpful in constructing a backward induction scheme for computing conditional tail sums and products. Suppose, for some $h_k(t_{1k}, t_{2k}) \geq 0$, we need the value of the sum of products

$$\Delta(t_1, t_2) = \sum \left[\prod_{k \geq 1} h_k(t_{1k}, t_{2k}) \right] \tag{13.88}$$

or the optimum values

$$\eta_1(t_1, t_2) = \min \left[\sum_{k \geq 1} h_k(t_{1k}, t_{2k}) \right] \tag{13.89}$$

$$\eta_2(t_1, t_2) = \max \left[\sum_{k \geq 1} h_k(t_{1k}, t_{2k}) \right] \tag{13.90}$$

where the summation and the optimizations are subject to the constraint

$$\sum_{k \geq 1} t_{jk} = t_j$$

for $j = 1, 2$, and where, for $k = 1, \cdots, K$, we further have that

$$0 \leq t_{jk} \leq n_k \quad \text{and} \quad 0 \leq t_{1k} + t_{2k} \leq n_k$$

Even though we now have two linear constraints instead of one, the method of constructing a backward induction scheme in principle remains the same as that used for a single constraint in Chapter 12. We leave the details as an exercise.

13.5.3. ESR and Infection in Kwashiorkor

Example 13.3 (continued): We analyzed the ESR data in Table 13.1 by assigning the scores $0, 1, 2, 3$ to the ESR levels ≤ 10, 11–25, 26–99 and ≥ 100, respectively, and response labels 0 = No Infection, 1 = Pulmonary Tuberculosis, and 2 = Other Pneumonia. Using the nominal response, quantitative score model (13.43), $t_1 = 7$, $t_2 = 8$, $w_1 = 15$ and $w_2 = 2$. With $j = 1, 2$, let $\beta_j = \phi_j$. Then the results of the conditional exact and asymptotic analyses are shown in Table 13.4. (The two conditional gps are not shown.) We thereby see that, essentially, ESR tends to be raised with pulmonary tuberculosis and not the other lung infections. The exact, mid-p and asymptotic conditional analyses point to a similar conclusion but the exact CIs are wider than their mid-p and asymptotic counterparts. For these data, and for the analysis of other models, we need to plot the respective evidence functions to get a full picture.

Table 13.4 *Analyses of Table 13.1 Data*

	β_1	β_2
Conditional mle	-0.441	1.90
Conditional Score Tests		
Exact p-value	0.610	0.000
Mid-p-value	0.514	0.000
Asymptotic p-value	0.510	0.000
95% Conditional Score CI		
Exact	$(-2.06, 0.783)$	$(0.772, 3.47)$
Mid-p	$(-1.76, 0.674)$	$(0.783, 3.18)$
Asymptotic	$(-1.64, 0.719)$	$(0.781, 3.05)$

13.6 Several $3 \times K$ Tables

The analysis of trinomial response data may also need to adjust for the effect of a stratification variable.

Example 13.4: In studying the relationship between ESR and pulmonary infection, we may adjust for the influence, if any, of age. Table 13.5 has data from the same study described in Table 1.2, but in this case, the children had marasmus, another form of malnutrition.

For such data, we use stratified versions of the models used for a single $3 \times K$ table. The response variable may be nominal or ordinal, and the factor variable may also be a nominal or scored variable. The factor used for stratification may also be nominal or scored. In this section, we only show how to construct one such model and leave the other types to the exercises.

Let the response variable $\mathcal{Y}$ be a nominal variable, and x_{ik}, the score for the kth column in the ith stratum be a numeric scored variable. Assume, as before that $x_{ik} \geq 0$, $x_1 = 0$ and $x_{ik} < x_{i,k+1}$. Let t_{ijk} denote the observed count in the jth row of the kth column in the ith stratum and n_{ik}, the total in the k column of the ith stratum. For $j = 1, 2$, consider the model

$$\ln\left(\frac{\pi_{ijk}}{\pi_{i0k}}\right) = \ln(\theta_{ij}) + x_{ik}\ln(\phi) \tag{13.91}$$

This model implies that the stratification factor is a nominal variable. Now, let, for $j = 1, 2$,

$$\mathcal{T}_{ij} = \sum_{k=1}^{K} t_{ijk} \tag{13.92}$$

and

$$\mathcal{W} = \sum_{i=1}^{I}\sum_{k=1}^{K}(t_{i1k} + t_{i2k})x_{ik} \tag{13.93}$$

Table 13.5 *ESR and Pulmonary Infection by Age in Marasmus*

ESR Level

Age ≤ 12 Months

	≤ 10	11–25	26–99	100+
No Infection	10	9	9	1
Pulmonary Tuberculosis	0	1	0	1
Other Pneumonia	7	4	6	4

12 Months < Age ≤ 24 Months

	≤ 10	11–25	26–99	100+
No Infection	10	5	9	0
Pulmonary Tuberculosis	0	0	1	3
Other Pneumonia	5	3	4	1

Age > 24 Months

	≤ 10	11–25	26–99	100+
No Infection	2	0	1	1
Pulmonary Tuberculosis	0	0	3	0
Other Pneumonia	1	3	2	0

Source: Hameer (1990); Used with permission.

Under the product trinomials model, $\mathcal{T}_{ij}$ is sufficient for θ_{ij} and $\mathcal{W}$ is so for ϕ. Consider the gp

$$F_K(\boldsymbol{\theta_1}, \boldsymbol{\theta_2}, \phi) = \prod_{i=1}^{I} \prod_{k=1}^{K} (1 + \{\theta_{i1} + \theta_{i2}\}\phi^{x_{ik}})^{n_{ik}} \qquad (13.94)$$

The required conditional distribution for the analysis of ϕ, $P[\mathcal{W} = w \mid \mathcal{T}_{ij} = t_{ij}, j = 1, 2; i = 1, \cdots, I]$, is then

$$\frac{F_K(\boldsymbol{1}, \boldsymbol{1}, \phi; \boldsymbol{t_1}, \boldsymbol{t_2}, w)}{F_K(\boldsymbol{1}, \boldsymbol{1}, \phi; \boldsymbol{t_1}, \boldsymbol{t_2}, .)} \qquad (13.95)$$

The gp $F_K(\boldsymbol{1}, \boldsymbol{1}, \phi; \boldsymbol{t_1}, \boldsymbol{t_2}, w)$ is used to obtain the required distributions. The two computational stages may also be integrated with one another, in a way similar to that for several $2 \times K$ tables. The basic scheme (for univariate and bivariate factor parameters) is shown in Algorithm 13.RPM.2.

```
                Algorithm 13.RPM.2
Step 01.   Input data and initialize arrays.
Step 02.   For i = 1, ··· , I do
Step 03.      Compute $F_{iK}(\theta_{i1}, \theta_{i2}, \phi; t_{i1}, t_{i2}, .)$
              Or
              Compute $F_{iK}(\theta_{i1}, \theta_{i2}, \phi_1, \phi_2; t_{i1}, t_{i2})$
              using Algorithm 13.RPM.1.
Step 04.   End do
Step 05.   For k = 1, ··· , K, recursively multiply
              $F_{ik}(\theta_{i1}, \theta_{i2}, \phi; t_{i1}, t_{i2}, .)$ using Algorithm 12.RPM.1
              Or
              $F_{ik}(\theta_{i1}, \theta_{i2}, \phi; t_{i1}, t_{i2}, .)$ using Algorithm 11.RPM.1.
```

□□□□□□□

We add other exponent checks if we need to directly compute tail areas or tail portions. Other models for a series of $3 \times K$ tables are constructed using stratified versions of the models for a single $3 \times K$ table. An additional concern is whether the stratification variable is ordered or not, as done for the case of several $2 \times K$ tables. The relevant RPM algorithms are developed accordingly by combining the strategies given in Chapter 12 with the one shown above.

****

Example 13.4 (continued): Consider now the analysis of the age stratified ESR data in Table 13.5 using model (13.91). Here, $t_{11} = 2$, $t_{12} = 21$, $t_{21} = 4$, $t_{22} = 13$, $t_{31} = 3$, $t_{32} = 6$ and $w_* = 70$. To perform a conditional analysis of $\beta = \ln(\phi)$, we need the conditional polynomial $F_K(1, 1, \phi; t_1, t_2, w)$. This is shown in Table 13.6.

Using the gp in Table 13.6, we obtain the analytic results shown in Table 13.7. For these data then, all the methods point to a marginal ESR effect. However, the exact score method is slightly more conservative than the asymptotic and mid-p methods. The three CIs cover a similar range of values. The respective evidence functions can be plotted using the gp in Table 13.6.

13.7 $J \times K$ Tables

This section outlines the generalization of some models for a $3 \times K$ table to a $J \times K$ table. Label its rows and columns as $0, 1, \cdots, J - 1$, and $1, \cdots, K$ respectively. Let the (j, k)th random entry be $\mathcal{T}_{jk}$ with the observed value, t_{jk}, and $\mathcal{T}_k = (\mathcal{T}_{0k}, \mathcal{T}_{1k}, \cdots, \mathcal{T}_{J-1,k})$ and $\boldsymbol{t}_k = (t_{0k}, t_{1k}, \cdots, t_{J-1,k})$.

13.7.1 Unordered Tables

First we deal with tables with both the rows and columns not ordered. Assume that the column totals, $n_k, k = 1, \cdots, K$, are fixed by design. Then for $k = 1, \cdots, K$, let

$$\mathrm{P}[\mathcal{T}_k = \boldsymbol{t}_k] = c_k(\boldsymbol{t}_k) \prod_{j=0}^{J-1} \pi_{jk}^{t_{jk}} \qquad (13.96)$$

where

$J \times K$ TABLES

Table 13.6 *Conditional Generating Polynomial for Example 13.4*

w	$c(2, 21, 4, 13, 3, 6, w)$	w	$c(2, 21, 4, 13, 3, 6, w)$	w	$c(2, 21, 4, 13, 3, 6, w)$
13	0.50630339760000D+13	42	0.12433389864397D+33	71	0.27025263791544D+32
14	0.32873556315600D+15	43	0.20157365655145D+33	72	0.14326813784356D+32
15	0.10154457106794D+17	44	0.31355509954177D+33	73	0.72672020474179D+31
16	0.20339559854914D+18	45	0.46833762306913D+33	74	0.35229940947374D+31
17	0.30014188965641D+19	46	0.67214614256992D+33	75	0.16301008866705D+31
18	0.34973533101551D+20	47	0.92744862029934D+33	76	0.71885349159327D+30
19	0.33607932760570D+21	48	0.12310315527210D+34	77	0.30163467328677D+30
20	0.27435887092792D+22	49	0.15725514366856D+34	78	0.12021118235420D+30
21	0.19441139393528D+23	50	0.19340739647680D+34	79	0.45409106035838D+29
22	0.12154601715772D+24	51	0.22909961442259D+34	80	0.16220959726700D+29
23	0.67910475809706D+24	52	0.26144758160383D+34	81	0.54653500454415D+28
24	0.34259945139659D+25	53	0.28751210406432D+34	82	0.17317639005124D+28
25	0.15738781748601D+26	54	0.30473205461564D+34	83	0.51431764701636D+27
26	0.66308366252406D+26	55	0.31133467505127D+34	84	0.14261832447299D+27
27	0.25774422045285D+27	56	0.30663307150335D+34	85	0.36761311236895D+26
28	0.92912224976866D+27	57	0.29114251136210D+34	86	0.87625839121868D+25
29	0.31200435402651D+28	58	0.26648859184589D+34	87	0.19198122748479D+25
30	0.97982206156318D+28	59	0.23512964812025D+34	88	0.38382451916117D+24
31	0.28875067481242D+29	60	0.19995822492157D+34	89	0.69418477528599D+23
32	0.80095162875851D+29	61	0.16386987151467D+34	90	0.11237352329347D+23
33	0.20968561251489D+30	62	0.12938642568019D+34	91	0.16067256651944D+22
34	0.51934442626422D+30	63	0.98297954000269D+33	92	0.19951530581140D+21
35	0.12195590289835D+31	64	0.72051726755148D+33	93	0.21046578940147D+20
36	0.27205143309487D+31	65	0.50779799197643D+33	94	0.18307183130964D+19
37	0.57750550513440D+31	66	0.34429108238675D+33	95	0.12594160112147D+18
38	0.11684215141009D+32	67	0.22445005471289D+33	96	0.64456823615954D+16
39	0.22562958159653D+32	68	0.14060888357996D+33	97	0.22341278096060D+15
40	0.41639031243784D+32	69	0.84588547638458D+32	98	0.44915669876000D+13
41	0.73521504062919D+32	70	0.48830008391805D+32	99	0.38356018000000D+11

Note: The observed value is $w_* = 70$.

$$c_k(\boldsymbol{t}_k) = \frac{n_k!}{t_{0k}! \cdots t_{J-1,k}!} \qquad (13.97)$$

and

$$\sum_{j=0}^{J-1} t_{jk} = n_k \quad \text{and} \quad \sum_{j=0}^{J-1} \pi_{jk} = 1 \qquad (13.98)$$

Now let, for $j \geq 1$,

Table 13.7 *Analyses of Table 13.5 Data*

	β
Conditional mle	0.384
Conditional Score Tests	
Exact p-value	0.054
Mid-p-value	0.049
Asymptotic p-value	0.050
95% Conditional Score CI	
Exact	$(-0.011, 0.777)$
Mid-p	$(0.008, 0.765)$
Asymptotic	$(0.001, 0.767)$

$$\theta_{jk} = \frac{\pi_{jk}}{\pi_{0k}} \tag{13.99}$$

The hypothesis of no association between row and column variables, $\pi_{jk} = \pi_j$; $k = 1, \cdots, K$, is equivalent to $H_0 : \theta_{jk} = \theta_j$ for all k. Let, for $j \geq 1$,

$$\mathcal{T}_j = \sum_{k=1}^{K} \mathcal{T}_{jk} \tag{13.100}$$

Then $\mathcal{T}_j$ is sufficient for θ_j. Test statistics for H_0 generally have the form

$$\mathcal{W} = \sum_{k=1}^{K} w_k(\mathcal{T}_k) \tag{13.101}$$

In particular, the score and LR statistics have the usual form as in (13.37) and (13.38) respectively. In large samples, the chisquare distribution with $(J-1)(K-1)$ degrees of freedom provides an asymptotic p-value for H_0. The probability (Freeman–Halton) statistic for this problem is given as

$$w_k(\mathcal{T}_k) = \sum_{j=0}^{J-1} \ln[c_k(t_k)] \tag{13.102}$$

For such statistics, we need the exact conditional distribution of $\mathcal{W}$. Consider then the gp

$$f_k(\boldsymbol{\theta}, \psi) = \sum_{\boldsymbol{u} \in \Omega_k} \left\{ c_k(\boldsymbol{u}) \psi^{w_k(\boldsymbol{u})} \prod_{j \geq 1} \theta_j^{u_j} \right\} \tag{13.103}$$

where $\boldsymbol{u} = (u_1, \cdots, u_{J-1})$, $\boldsymbol{\theta} = (\theta_1, \cdots, \theta_{J-1})$ and ψ is a dummy variable and also where

$$\Omega_k = \{\boldsymbol{u} : 0 \leq u_j \leq n_k, j \geq 1; \text{ and } 0 \leq \sum_{j \geq 1} u_j \leq n_k\} \quad (13.104)$$

Now define

$$F_K(\boldsymbol{\theta}, \psi) = \prod_{k=1}^{K} f_k(\boldsymbol{\theta}, \psi) \quad (13.105)$$

The exact conditional distribution of $\mathcal{W}$ obtains from the collection of terms of $F_K(\boldsymbol{\theta}, \psi)$ in which the exponent of θ_j equals to t_j for all j. That is,

$$P[\mathcal{W} \geq w_* \mid \boldsymbol{\mathcal{T}} = \boldsymbol{t}; H_0] = \frac{F_K(\boldsymbol{1}, 1; \boldsymbol{t}, \geq w_*)}{F_K(\boldsymbol{1}, 1; \boldsymbol{t}, .)} \quad (13.106)$$

where $\boldsymbol{\mathcal{T}} = (\mathcal{T}_1, \cdots, \mathcal{T}_{J-1})$ and $\boldsymbol{t} = (t_1, \cdots, t_{J-1})$. In (13.106), we can readily show that

$$F_K(\boldsymbol{1}, 1; \boldsymbol{t}, .) = \frac{n!}{t_0! \cdots t_{J-1}!} \quad (13.107)$$

13.7.2 Doubly Ordered Tables

Now consider a $J \times K$ table in which both the rows and columns are ordered. Assume that y_j is the score attached to row j and x_k, the score for column k, in a study design in which the cell count $\mathcal{T}_{jk}$ is an independent Poisson variate with mean λ_{jk}. Further, let $x_1 < x_2 < \cdots < x_K$ with $x_1 = 0$ and $y_0 < y_1 < \cdots < y_{J-1}$.

The **linear by linear association model** is commonly used for such data (Agresti 1990). One form of this model is

$$\ln(\lambda_{jk}) = \mu + \mu_j^Y + \mu_k^X + \lambda y_j x_k \quad (13.108)$$

with $\mu_0^Y = \mu_1^X = 0$. In an exponential form, this model becomes

$$\lambda_{jk} = \delta \delta_j^Y \delta_k^X \phi^{y_j x_k} \quad (13.109)$$

where $\phi = \exp(\lambda)$ is the parameter of interest. The row and column margins are sufficient for the other parameters and the sufficient statistic for ϕ is

$$\mathcal{W} = \sum_k \sum_j \mathcal{T}_{jk} y_j x_k \quad (13.110)$$

The conditional distribution of $\mathcal{W}$ with both margins of the table fixed is obtained in two steps. First we fix the column margins. Let n_k be the total in the kth column. The cell entries in this column then constitute a multinomial distribution with probabilities

$$\pi_{jk} = \frac{\lambda_{jk}}{\lambda_{0k} + \cdots + \lambda_{J-1,k}} = \frac{\delta_j \phi^{y_j x_k}}{\sum_j \delta_j \phi^{y_j x_k}} \quad (13.111)$$

Note, the superscript X in δ_j has been dropped. The conditional distribution of $\mathcal{W}$ obtains from multiplying the K multinomial gps

$$f_{*k}(\boldsymbol{\delta}, \phi) = \left(\phi^{y_0 x_k} + \sum_{j \geq 1} \delta_j \phi^{y_j x_k}\right)^{n_k} \tag{13.112}$$

Let $F_{*K}(\boldsymbol{\delta}, \phi)$ be the product of these K polynomials. Then

$$\mathrm{P}[\mathcal{W} = w \mid \mathcal{T} = \boldsymbol{t}; H_0] = \frac{F_{*K}(\mathbf{1}, \phi; \boldsymbol{t}, w)}{F_{*K}(\mathbf{1}, \phi; \boldsymbol{t}, .)} \tag{13.113}$$

The linear by linear association model is related to the adjacent categories model for ordinal response. For $j \geq 1$, let $\theta_j = \delta_j/\delta_{j-1}$ and $y_j = j$. Then the kth column-wise gp (13.112) becomes

$$f_{**k}(\boldsymbol{\theta}, \phi) = \left(1 + \theta_1 \phi^{x_k} + \theta_1 \theta_2 \phi^{2x_k} \cdots + \theta_1 \cdots \theta_{J-1} \phi^{(J-1)x_k}\right) \tag{13.114}$$

which is the same as the gp for the adjacent categories ordinal response model. This generalizes (13.59) and (13.60) to more than three categories.

Other models for $J \times K$ tables are developed in a similar manner. The case of a $J \times 2$ table, namely, data with J categories of response and a binary factor, is of special interest.

Theorem 13.3: Consider exact conditional tests of association in a situation with J response categories and a binary factor. (i) If the response variable is nominal, then the conditional exact score, LR or probability tests are identical to the respective exact conditional tests for a binary response model with a J level nominal factor. (ii) If the response variable is an ordinal one following the adjacent categories model, then the conditional exact analysis for the parameter embodying the relationship between response and factor is identical to the exact conditional analysis for the trend parameter in a binary response model with a J level ordered factor with integer scores.

Proof: Left to the reader. □

Similar results can also be shown to hold for the case of large sample analysis.

This result implies that for exact inference on a $J \times 2$ table, we can use the computational methods developed in Chapter 12. In other cases, more elaborate methods are needed.

13.7.3 Computation

There is a voluminous literature on algorithms for exact analysis of $J \times K$ table. Initial work in the field focused on complete or random enumeration of such tables with fixed marginal totals. Many recursive schemes, some tailored for a particular test statistic, were developed. Below we give one such scheme for a 3×3 table given by Mielke and Berry (1992). Let the data be in Table 13.8.

Let $L = (t_0 + t_1 + n_1 + n_2 - n)$. The following scheme generates all tables 3×3 with margins identical to the observed one.

Table 13.8 *A 3 × 3 Table*

Response	Factor Score			
y	x_1	x_2	x_3	Total
0	t_{01}	t_{02}	t_{03}	t_0
1	t_{11}	t_{12}	t_{13}	t_1
2	t_{21}	t_{22}	t_{23}	t_2
Total	n_1	n_2	n_3	n

- Generate values of t_{01} that satisfy

$$\max(0, L - t_1 - n_2) \leq t_{01} \leq \min(t_0, n_1)$$

- Given each value of t_{01}, generate values of t_{02} that satisfy

$$\max(0, L - t_1 - t_{01}) \leq t_{02} \leq \min(t_0 - t_{01}, n_2)$$

- Given each pair of values of (t_{01}, t_{02}), generate values of t_{11} that satisfy

$$\max(0, L - n_2 - t_{01}) \leq t_{11} \leq \min(t_1, n_1 - t_{01})$$

- Given each triplet of values of (t_{01}, t_{02}, t_{11}), generate values of t_{12} that satisfy

$$\max(0, L - t_{01} - t_{02} - t_{11}) \leq t_{12} \leq \min(t_1 - t_{11}, n_2 - t_{01})$$

- For each $(t_{01}, t_{02}, t_{11}., t_{12})$, perform the partial computation to get a *p*-value or exact conditional distribution.
- Stop when all tables have been enumerated and perform overall summary computation.

Mielke and Berry (1992) used this scheme to get an exact conditional significance level with the probability statistic. It can, however, be used for other statistics as well as to analyze situations with singly or doubly ordered tables. Pagano and Halvorsen (1981) gave a recursive scheme for a $J \times K$ table analyzed with the probability statistic.

With higher dimensional tables, more efficient methods are needed. The methods we developed for a $3 \times K$ table can readily be extended for a $J \times K$ table. When the data are sparse, and the smallest of J or K does not exceed 4, this is a reasonable approach.

With larger sample sizes and higher values of J or K, we need more elaborate computational methods. Many sophisticated exact and Monte–Carlo approaches to deal with such problems exists. Mehta, Patel and their colleagues have developed refined and elaborate network algorithms for such data. In principle, it is equivalent to an RPM method with added backward induction based simultaneous multiple exponent checks. They and others have produced efficient, accurate Monte Carlo algorithms as well. A venture into this field is beyond the scope of the present text; the articles mentioned below and related references in Chapter 11 explore that field.

13.8 Relevant Literature

There is a large literature on the theory and methods of generating combinations, compositions and partitions. We only presented the basic idea. Other methods for generating combinations and compositions use the backtrack approach, and the theory of binary codes. Consult Lehmer (1964) and Reingold, Nievergelt and Deo (1977). The latter also gives a multiset based method for generating partitions. The presentation in §13.2 utilized these two works. Other useful works are Sedgewick (1983) and Press et al. (1992). Gentleman (1975) has a basic computer code for generating combinations.

The single multinomial was one of the first discrete data problems for which the question of efficient exact computation was tackled. Katti (1973) and Smith et al. (1979) used integer partitioning algorithms for the case of equiprobability; I. J. Good and colleagues (Good, Gover and Mitchell 1970, 1971; Good and Crook 1978; Good 1981, 1982, 1983; Holtzman and Good 1986) used bivariate and univariate FFT algorithms for the same problem; Baglivo, Olivier and Pagano (1992) extended the FFT approach to the general probability case. Hirji (1997b) improved the FFT method and applied the RPM method to the equiprobable and general probability case. He also empirically compared the various approaches. For comparisons between exact and large sample p-values for the single multinomial problem, see, among others, Holtzman and Good (1986) and Greenwood and Nikulin (1996).

Methods for exact analysis of a single and several $3 \times K$ tables given here are special cases of the models and methods in Hirji (1992). A recursive scheme for enumerating 3×3 tables is in Mielke and Berry (1992). Zelterman, Chan and Mielke (1995) gave simple recursions for 2^3 and 2^4 tables. Boulton and Wallace (1973) gave methods for recursive enumeration of two way tables with fixed margins; see also Gail and Mantel (1977).

For general methods for analysis of $J \times K$ tables see Bishop, Fienberg and Holland (1975); Agresti (1992); Cressie and Read (1984); and Greenwood and Nikulin (1996).

Some early papers on exact methods and computational algorithms for $J \times K$ contingency tables were Freeman and Halton (1951), March (1972), Agresti and Wackerly (1977), Agresti, Wackerly and Boyett (1979), Patefield (1981) and Pagano and Halvorsen (1981). More early papers are listed in a comprehensive review by Verbeek and Kroonenberg (1985). The foundations for an efficient approach were laid by Mehta and Patel (1983, 1986a, 1986b), Mehta, Patel and Senchaudhuri (1988), Agresti, Mehta and Patel (1990) and Hirji (1992).

For an overview of exact methods for two way and higher dimensional contingency tables, see Kreiner (1987), Morgan and Blumenstein (1991), Agresti (1992), Baglivo, Olivier and Pagano (1992), Mehta (1994), Hirji (1997a) and Agresti (2001).

Exact analyses of log-linear and logistic models have been advanced by Forster, McDonald and Smith (1996) and McDonald, Smith and Forster (1999). This team of researchers has expanded exact conditional testing to various types of contingency tables. These include: (i) incomplete tables (Smith and McDonald 1995); (ii) triangular tables (McDonald and Smith 1995); (iii) square tables (Smith, Forster and McDonald 1996); (iv) symmetric tables (McDonald, De Roure, Michaelides 1998); and (v) log-linear models for multidimensional rates (McDonald, Smith and Forster 1999). See also the papers noted in Chapter 11.

13.9 Exercises

13.1. (i) What is the number of integer vectors $(t_1, \ldots, t_K)$ that satisfy $0 < t_k \leq n$ for all k, and $t_1 + \ldots + t_K = n$? (ii) What is the number of integer vectors $(t_1, \ldots, t_K)$ that satisfy $0 < t_k \leq n$ for all k and $t_1 + \ldots + t_K \leq n$?

EXERCISES

13.2. Provide formal proofs of the correctness of the methods of generating subsets, composition and partitions given in §13.2.

13.3. Implement Algorithms 13.SS.1, 13.CO.1 and 13.PA.1 on a computer. Compare Algorithm 13.SS.1 with Gentleman (1975).

13.4. Consider the problem of generating subsets of size k of $\{1, 2, \cdots, n\}$. If $k > n/2$, tackle the problem by first generating subsets of size $n - k$ of this set and then obtaining corresponding subsets of size k from them. Implement the method on a computer.

13.5. Investigate the backtrack and binary code methods of generating compositions and the multiset based method of generating partitions and compare them with the methods given in §13.2 (see Lehmer (1964) and Reingold, Nievergelt and Deo (1977)).

13.6. Derive the Pearson chisquare (13.8) and the LR (13.9) statistics for a single multinomial.

13.7. Prove Theorem 13.1.

13.8. Give details of the derivations of the formulas for $M_{1k}(n)$, $M_{2k}(n)$, $m_{1k}(n,v)$ and $m_{2k}(n,v)$ given in §13.3.

13.9. Show that the recursion (13.21)–(13.22) is equivalent to applying Algorithm 12.RPM.2 to (13.17).

13.10. Provide a rigorous justification for exponent checks EC 13.01.

13.11. Provide details of the proof of Theorem 13.2.

13.12. For the equivalent versions of the Pearson chisquare and the LR statistic for a single multinomial given in Theorem 13.2, let $a = n \bmod K$, $b = (n-a)/K$, $\pi_* = \min\{\pi_1, \ldots, \pi_K\}$ and $\nu \geq \max\{\pi_1, \ldots, \pi_K\}$. Then prove the following:

(i) Equiprobable chisquare:
$$\eta_{1,K}(n) = (n-a)(n+a)/K + a \geq n^2/K$$
$$\eta_{2,K}(n) = n^2$$

(ii) Equiprobable LR:
$$\eta_{1,K}(n) = a(b+1)\log(b+1) + (K-b)b\log b \geq n\log(n/K)$$
$$\eta_{2,K}(n) = n\log n$$

(iii) General chisquare:
$$\eta_{1,K}(n) \geq n^2/\Sigma_k \pi_k$$
$$\eta_{2,K}(n) = n^2/\pi_*$$

(iv) General LR:
$$\eta_{1,K}(n) \geq n\log(n\nu/\Sigma_k \pi_k)$$
$$\eta_{2,K}(n) = n\log(n\nu/\pi_*)$$

13.13. Derive explicit exponent checks for using the statistic $w_k(t_k) = t_k(t_k - 1)$ in testing goodness-of-fit in a single multinomial (Holtzman and Good 1986).

13.14. Develop all the details of the RPM method with exponent checks for the equiprobable single multinomial data case.

13.15. Develop in detail the integer partitioning algorithm for the equiprobable multinomial model. Can it also be used for the general probability case? Also develop FFT based algorithms for the general probability and equal probability cases.

13.16. Write EE, integer partitioning, FFT and RPM based computer programs with, where possible, exponent checks (specific and backward induction based) for testing goodness-of-fit in a single multinomial. Include the equiprobable case as a special case, and incorporate the logarithmic scale option. Can you incorporate the exponent checks for generating the tails of a distribution of a test statistic with the methods of generating compositions and partitions?

13.17. Use these programs to compute exact and large sample p-values for the data in Example 13.1.

13.18. Show that the network algorithm for a single multinomial is an RPM algorithm.

13.19. Compute and compare exact and asymptotic Pearson chisquare and LR p-values for testing the hypothesis of equiprobability for the following datasets:

Dataset (A): $(3, 3, 5, 5, 6, 9, 9)$
Dataset (B): $(4, 5, 8, 11, 12)$
Dataset (C): $(0, 0, 6, 6, 6, 11, 11)$
Dataset (D): $(2, 2, 2, 2, 2, 8, 8, 8, 8, 8)$
Dataset (E): $(2, 2, 4, 7, 7, 7, 7, 10, 12, 12)$

Compare your answers with Hirji (1997b).

13.20. Compute and compare exact and asymptotic Pearson chisquare and LR p-values for testing the stated respective hypotheses for datasets given below.

Dataset (A): $(4, 2, 5, 6, 8)$, and $\pi_0 = (.1, .2, .1, .2, .4)$
Dataset (B): $(4, 2, 2, 3, 3, 2, 9)$ and $\pi_0 = (.02, .1, .2, .1, .1, .2, .28)$
Dataset (C): $(8, 13, 3, 2, 19)$ and $\pi_0 = (.2, .4, .1, .05, .25)$
Dataset (D): $(3, 4, 3, 0, 5, 3, 2, 2, 4, 4)$ and
$\pi_0 = (0.5, .05, .05, .1, .1, .1, .1, .15, .15, .15)$
Dataset (E): $(4, 3, 4, 1, 5, 4, 3, 5, 5, 6)$ and
$\pi_0 = (0.5, .05, .05, .05, .05, .1, .1, .15, .20, .20)$

Compare your answers with Hirji (1997b).

13.21. Consider the single multinomial problem in which there is an ordering between the K categories. Suppose, for $k > 1$, we can write

$$\ln(\pi_k/\pi_1) = \beta(x_k - x_1)$$

Derive score and LR statistics for this model and develop a method for exact analysis. Develop an RPM algorithm with exponent checks for the purpose.

13.22. Consider a K category incidence density design with and without an ordering between the categories, as described in §13.3.5. Derive score and LR statistics for these models and develop methods for exact analyses. Also develop RPM algorithms with exponent checks for doing the analyses.

13.23. Consider a K category problem with data $(t_1, \ldots, t_K)$. Suppose the data were generated by inverse sampling. That is, suppose units were sampled until, say, a fixed t_1 number of observations occurred in category 1. What is the probability distribution for the data? Derive relevant test statistics, exact analysis and EE and RPM algorithms for the problem. Extend the inverse sampling approach to the case when there is a numeric score x_k associated with category k.

EXERCISES

13.24. For the nominal response model (13.30), show that $\mathcal{T}_j$, defined by (13.35), is sufficient for θ_j. Derive score and LR statistics for testing $H_0 : \theta_{jk} = \theta_j$ for all k in this model and produce simpler equivalent forms of the statistics. Prove that the right tail conditional probability of the distribution of such a statistic is given by expression (13.41).

13.25. For the nominal response model (13.43), show that $\mathcal{T}_j$, defined by (13.35), is sufficient for θ_j, and $\mathcal{W}_j$, defined by (13.44), is sufficient for ϕ_j. Show that the conditional distribution of $(\mathcal{W}_1, \mathcal{W}_2)$ is given by expression (13.47). How would you estimate and construct exact and large sample CIs for ϕ_1 and ϕ_2? How would you construct an exact joint confidence region for (ϕ_1, ϕ_2)?

Consider the hypothesis of no covariate effect, $\phi_1 = \phi_2 = 0$, in (13.43). Derive the score and LR statistics (and simpler equivalent forms) for testing it, and express their conditional distributions in a polynomial form.

Consider the uniform effect hypothesis, $\phi_1 = \phi_2 = \phi$, in (13.43). Derive the score and LR statistics (and simpler equivalent forms) for testing it and express their conditional distributions in a polynomial form.

Further show that under the condition $\phi_1 = \phi_2 = \phi$, $\mathcal{W}_3$, defined by (13.49), is sufficient for ϕ and derive its conditional distribution. How would you construct exact, mid-p and large sample CIs for ϕ?

13.26. For the ordinal response model (13.51)–(13.52), show that $\mathcal{R}_j$, defined by (13.53), is sufficient for δ_j. Derive the score and LR statistics (and simpler equivalent forms) for testing $H_0 : \delta_{jk} = \delta_j$ for all k in this model. Prove that the right tail conditional probability of the distribution of such a statistic is given by expression (13.57).

13.27. For the ordinal response model (13.59)–(13.60), show that $\mathcal{R}_j$, defined by (13.53), is sufficient for δ_j, and $\mathcal{Q}_j$, defined by (13.61)–(13.62), is sufficient for μ_j. Show that the conditional distribution of $(\mathcal{Q}_1, \mathcal{Q}_2)$ is given by (13.65). How would you estimate and construct exact and large sample CIs for μ_1 and μ_2? How would you construct an exact joint confidence region for (μ_1, μ_2)?

Consider the hypothesis of no covariate effect, $\mu_1 = \mu_2 = 0$, in this model. Derive the score and LR statistics (and simpler equivalent forms) for testing it, and express their conditional distributions in a polynomial form.

Consider the hypothesis of uniform covariate effect, $\mu_1 = \mu_2 = \mu$, in this model. Derive the score and LR statistics (and simpler equivalent forms) for testing it and express their conditional distributions in a polynomial form.

Further show that under the condition $\mu_1 = \mu_2 = \mu$, $\mathcal{Q}_3$, defined by (13.67), is sufficient for μ and derive its conditional distribution. How would you construct exact, mid-p and large sample CIs for μ?

13.28. Write computer programs for enumerative and RPM methods for the analysis of all the nominal and ordinal response models for $3 \times K$ described in §13.4. Incorporate exponent checks of the form EC 13.02 in these programs.

13.29. Give a detailed derivation of the improved exponent checks described in §13.5.2. Write the detailed version of Algorithm 13.RPM.1, and implement it on the computer.

13.30. Develop in detail the bivariate backward induction scheme of §13.5.3 and show how it may be used to compute quantities of the form (13.87), (13.88) and (13.89). Incorporate it within Algorithm 13.RPM.1 and write the relevant computer programs for the analysis of all the nominal and ordinal response models of §13.4.

13.31. Perform large sample and exact conditional analyses of the data in Table 13.1. Compute exact and mid-p values and TST based CIs and compare your results with those in Table 13.4. Plot the evidence functions.

13.32. Perform large sample and exact conditional analyses of the data in Table 13.1. assuming that the column factor is not ordered.

13.33. Show that in model (13.91), T_{ij} given by (13.92) and W given by (13.93) are respectively sufficient for θ_{ij} and ϕ and that the conditional distribution of W is given by (13.95).

13.34. For the situation with a nominal stratification factor, construct stratified versions of all models for a single $3 \times K$ table for nominal and ordinal response models of §13.4. Derive the exact conditional distributions for testing relevant hypotheses and analyzing relevant parameters. Develop exponent checks and backward induction schemes. Use Algorithm 13.RPM.2 as the template to implement each on the computer. Also develop large sample methods for analyzing these models.

13.35. For the situation with a scored stratification factor, construct stratified versions of all models for a single $3 \times K$ table for nominal and ordinal response models of §13.4. Derive exact conditional distributions for testing relevant hypotheses and analyzing the relevant parameters. Develop related exponent checks and backward induction schemes. Use Algorithm 13.RPM.2 as the template to implement each on the computer. Also develop large sample methods for analyzing these models.

13.36. Use the programs developed above to analyze the data in Table 13.5 under various scenarios and assumptions relating to the nature of response, factor and stratification variables. Perform large sample and exact conditional analyses of the data and compute exact and mid-p values and TST based CIs and compare your results with those in Table 13.7. Plot the evidence functions. Also apply the CT method, where appropriate.

13.37. Perform large sample and exact conditional analyses of the data in Table 13.5 assuming that (i) the column factor is not ordered, and/or (ii) the stratification factor is ordered. Plot the evidence functions. Where appropriate also use the CT method.

13.38. The data in Table 13.9 are also from the same study described for Table 1.1; but these data relate to children with marasmic kwashiorkor, a somewhat different form of malnutrition from that for Table 1.1 and Table 13.5. Analyze the data using all the relevant models described in this chapter and plot the relevant evidence functions.

13.39. Prove Theorem 13.3. Use it to develop computational methods for exact conditional analysis of $J \times 2$ tables. Develop large sample methods for the analysis of such data.

13.40. Show that for the model (13.99) for an unordered $J \times K$ table, the right tail of the conditional distribution of the test statistic W (13.101) is given (13.106). Extend the RPM algorithm for a $3 \times K$ table to compute this.

13.41. Show that for the model (13.108) for a doubly ordered $J \times K$ table, the conditional distribution of the test statistic W (13.110) is given by (13.113).

13.42. Extend all the models for a single $3 \times K$ table of §13.4 to a $J \times K$ table $(J \geq K)$. Derive appropriate test and analytic statistics and their equivalent forms. Develop expressions for relevant conditional distributions and construct related RPM computational methods with exponent checks and backward induction schemes. Implement them on a computer.

13.43. Using the enumeration scheme for a 3×3 table with fixed margins of Mielke and Berry (1992) as the starting point, investigate and develop in detail a set of complete enumeration methods for all $J \times K$ tables with fixed margins. Implement them on the computer and compare them with the RPM method.

13.44. Investigate and implement the network algorithm of Agresti, Mehta and Patel (1990) for the analysis of a doubly ordered $J \times K$ table. Show that this algorithm is also an RPM algorithm.

Table 13.9 *ESR and Infection by Age in Marasmic Kwashiorkor*

	\multicolumn{4}{c}{ESR Level}			
	\multicolumn{4}{c}{Age $\leq$ 12 Months}			
	≤ 10	11–25	26–99	100+
No Infection	12	3	3	0
Pulmonary Tuberculosis	0	0	1	0
Other Pneumonia	5	1	1	1
	\multicolumn{4}{c}{12 Months $<$ Age $\leq$ 24 Months}			
	≤ 10	11–25	26–99	100+
No Infection	10	0	0	0
Pulmonary Tuberculosis	0	0	0	1
Other Pneumonia	1	1	0	0
	\multicolumn{4}{c}{Age $>$ 24 Months}			
	≤ 10	11–25	26–99	100+
No Infection	2	3	1	0
Pulmonary Tuberculosis	0	0	1	0
Other Pneumonia	0	0	0	0

Source: Hameer (1990); Used with permission.

13.45. Investigate and implement the network algorithms of Mehta and Patel (1983, 1986a, 1986b), and Mehta, Patel and Senchaudhuri (1988) for the analysis of an unordered $J \times K$ table. Show that these algorithms are also RPM algorithms.

13.46. Suppose the count t_{jk} in a $J \times K$ table is associated with a person time value τ_{jk}. Assuming Poisson rates for the counts, develop the models and RPM based methods for exact conditional analysis of (i) unordered, (ii) singly ordered and (iii) doubly ordered settings. Apply these to the data in McDonald, Smith and Forster (1999) and other real and simulated data.

CHAPTER 14

Matched and Dependent Data

14.1 Introduction

In the discrete data models considered thus far, the outcomes or study units were statistically independent. Now this assumption is relaxed to examine models with some form of relationship or dependency between the outcomes or units. That may be induced by the study design, or may be intrinsic to the phenomena under study. We show that exact distributions from several studies of this type can be formulated in a polynomial form, and give methods for inference and efficient computation for them. The specific aims of this chapter are:

- To present the exact analysis of matched case-control studies with two unmatched variables.
- To present exact analysis of logit models for paired binary outcome data with covariates.
- To present exact analysis of data from a two state time homogeneous Markov chain.
- To introduce efficient computational methods for analyzing these models.

14.2 Matched Designs

We introduced the basic case-control design in §5.8 and developed a formal justification for using conditional methods of analysis for it in §6.10. Consider now the case-control studies in which subjects are sampled from subpopulations delineated by one or more matching variables. For example, cases and controls may be matched by gender, age at exposure and location. The aim, as before, is to assess existence and magnitude of the relationship between a given condition and some exposure variables. The main features of this design are:

- Clear specification of the outcome and exposure variable(s).
- Identification and justification of variables to be used for matching, and construction of a sampling scheme based on combinations of these variables.
- For each combination, random sampling a predetermined number of cases and controls from the respective subpopulation.
- For each subject sampled, confirming the outcome status, and assessing the values of the exposure variable(s) of interest.

Each group (of fixed numbers) of cases and controls sampled is called a matched set. Suppose we have J matched sets, and z_j is the vector of matching variables, m_j, the number of cases, and $n_j - m_j$, the number of controls for the jth set. Under prospective sampling, we let $\mathcal{Y}_{ij}$ denote the case (1) or control (0) status, and x_{1ij}, x_{2ij}, the exposure variables of interest, for the ith subject in the jth set. Also, let $\pi_{ij} = P[\mathcal{Y}_{ij} = 1 \mid z_j, x_{1ij}, x_{2ij}]$. We assume the logit model

$$\ln\left(\frac{\pi_{ij}}{1-\pi_{ij}}\right) = \alpha + \boldsymbol{\gamma}'\boldsymbol{z}_j + \beta_1 x_{1ij} + \beta_2 x_{2ij} \tag{14.1}$$

Write $\alpha_j = \alpha + \boldsymbol{\gamma}'\boldsymbol{z}_j$. Then (14.1) simplifies to

$$\ln\left(\frac{\pi_{ij}}{1-\pi_{ij}}\right) = \alpha_j + \beta_1 x_{1ij} + \beta_2 x_{2ij} \tag{14.2}$$

The sampling method used, however, is retrospective with known case and control information. Under the noted justifications, we can still use models (14.1) and (14.2), provided appropriate conditioning is employed, and the nonestimable parameters are factored out. Hence, we consider the conditional probability for the jth set,

$$P[\mathcal{Y}_{ij} = y_{ij}, i = 1, \cdots, n_j \mid \sum_{i=1}^{n_j} \mathcal{Y}_{ij} = m_j]$$

which, after some algebra, emerges as equal to

$$\frac{\phi_1^{(\Sigma_i y_{ij} x_{1ij})} \phi_2^{(\Sigma_i y_{ij} x_{2ij})}}{f_j(1, \phi_1, \phi_2; m_j, ., .)} \tag{14.3}$$

where $\phi_k = \exp(\beta_k)$ and $f_j(1, \phi_1, \phi_2; m_j, ., .)$ is the polynomial extracted from terms of

$$\prod_{i=1}^{n_j} \left(1 + \theta_j \phi_1^{x_{1ij}} \phi_2^{x_{2ij}}\right) \tag{14.4}$$

in which the exponent of θ_j equals m_j. Now, for $k = 1, 2$, define

$$\mathcal{T}_k = \sum_{j=1}^{J} \sum_{i=1}^{n_j} \mathcal{Y}_{ij} x_{kij} \tag{14.5}$$

$\mathcal{T}_k$ is (conditionally) sufficient for ϕ_k and, further,

$$P[\mathcal{T}_1 = t_1, \mathcal{T}_2 = t_2 \mid \sum_{i=1}^{n_j} \mathcal{Y}_{ij} m_j, j = 1, \cdots, J]$$

is equal to

$$\frac{c(t_1, t_2) \phi_1^{t_1} \phi_2^{t_2}}{F_J(\phi_1, \phi_2)} \tag{14.6}$$

where

$$F_J(\phi_1, \phi_2) = \prod_{j=1}^{J} f_j(1, \phi_1, \phi_2; m_j, ., .) \tag{14.7}$$

MATCHED DESIGNS

and $c(t_1, t_2)$ is the coefficient of the term in $F_J(\phi_1, \phi_2)$ in which the exponent of ϕ_1 is t_1 and that of ϕ_2 is t_2.

Two matched sets are said to be **permutationally equivalent** if they have same number of cases and controls and share a common exposure profile. In the situation at hand, the common covariate profile of vectors is $\{(x_{11j}, x_{21j}), \cdots, (x_{11n_j}, x_{21n_j})\}$. But they may differ in terms of the persons (case or control) associated with a particular covariate vector. The important point is that permutationally equivalent sets have identical conditional gps. This is clear from (14.4).

If there are K separate covariate profiles and for the kth profile, there are r_k permutationally equivalent matched sets, then (14.7), the conditional gp of (T_1, T_2), is also written as

$$F_K(\phi_1, \phi_2) = \prod_{k=1}^{K} \{f_k(1, \phi_1, \phi_2; m_k, ., .)\}^{r_k} \qquad (14.8)$$

where

$$\sum_k r_k = J$$

The following remarks relate to data analysis for the above situation.

- We can more rigorously show, as in Chapter 6, that under a reasonable set of assumptions, the conditional distribution (14.6) can be used for inference on the parameters ϕ_1 and ϕ_2.
- The gp (14.8) resembles the bivariate gps dealt with in the analysis of several 2×2, and one or several $2 \times K$ tables. The computational methods developed earlier can then be readily extended to the present problem.
- The hypothesis of no exposure effect, $H_0: \phi_1 = \phi_2 = 1$, is tested by deriving appropriate score and LR statistics, and their exact conditional distributions from (14.6) in the usual manner.
- We perform inference on each exposure parameter by further conditioning on the sufficient statistic for the other. The required univariate conditional distribution may be obtained using exponent checks of the form EC 14.01 given below.
- The large sample conditional analysis is based on the conditional likelihood derived from (14.6).

The exact conditional distributions in several specific forms of matched studies can be given in an explicit form. Below we consider two such special situations with a single exposure variable.

14.2.1 1:1 Matched Design

Commonly, one case is matched with one control. Then, $m_j = 1$ and $n_j = 2$ for all j. Suppose we have one (unmatched) ordinal factor taking three values: $0, 1, 2$. In each matched pair, the case and the control are at one of these levels, giving nine possible types of permutationally equivalent sets. The combined data from the study are typically summarized in Table 14.1, where t_{ij} denotes the number of pairs where the case is at level i and the control is at level j of exposure.

According to (14.4), the unconditional gp for the jth pair is

Table 14.1 *1:1 Case-Control Design*

Case Factor	Control Factor		
	0	1	2
0	t_{00}	t_{01}	t_{02}
1	t_{10}	t_{11}	t_{12}
2	t_{20}	t_{21}	t_{22}

$$(1+\theta_j)^{r_{0j}}(1+\theta_j\phi)^{r_{1j}}(1+\theta_j\phi^2)^{r_{2j}}$$

where r_{aj} is the number of individuals at level a in set j with the provision that

$$r_{0j} + r_{1j} + r_{2j} = n_j = 2$$

Extracting terms in which the coefficient of θ_j is equal to 1, we have

$$f_j(1, \phi; 1, .) = r_{0j} + r_{1j}\phi + r_{2j}\phi^2$$

The six possible values of (r_{0j}, r_{1j}, r_{2j}) are noted below. Their associated conditional gps are then:

$$(0,0,2) \Rightarrow f_j(1,\phi) = 2\phi^2; \quad (0,1,1) \Rightarrow f_j(1,\phi) = \phi + \phi^2$$
$$(0,2,0) \Rightarrow f_j(1,\phi) = 2\phi; \quad (1,0,1) \Rightarrow f_j(1,\phi) = 1 + \phi^2$$
$$(2,0,0) \Rightarrow f_j(1,\phi) = 2; \quad (1,1,0) \Rightarrow f_j(1,\phi) = 1 + \phi$$

For each possibility on the left side, both pair members have the same exposure level, giving a degenerate probability distribution. We thus ignore such concordant pairs in the data analysis. Now let

$$r_1 = t_{10} + t_{01}; \quad r_2 = t_{20} + t_{02}; \quad r_3 = t_{12} + t_{21}$$

and

$$T = T_{10} + T_{12} + 2(T_{20} + T_{21})$$

Ignoring the concordant pairs, the conditional gp, $F_K(\phi)$, for T is

$$\phi^{r_3}[1+\phi]^{(r_1+r_3)}[1+\phi^2]^{r_2} \tag{14.9}$$

which expands to

$$F_K(\phi) = \phi^{r_3} \sum_{u=0}^{r_1+r_3} \sum_{v=0}^{r_2} \binom{r_1+r_3}{u}\binom{r_2}{v}\phi^{(u+2v)} \tag{14.10}$$

MATCHED DESIGNS

Table 14.2 *1:n Case-Control Design*

Case Factor	No. of Exposed Controls			
	0	1	$\cdots$	n
0	t_{00}	t_{01}	$\cdots$	t_{0n}
1	t_{10}	t_{11}	$\cdots$	t_{1n}

This can then be used for conditional large sample and exact inference on ϕ or $\beta = \ln(\phi)$ in the usual way. The conditional log-likelihood, ℓ_c, is

$$\beta t - r_3\beta - (r_1 + r_3)\ln[1 + \exp(\beta)] - r_2\ln[1 + \exp(2\beta)] \tag{14.11}$$

The conditional mle for β obtains from solving the cubic equation

$$\phi^3 + \frac{(t - r_3 - 2r_2)}{(t - r_3)}\phi^2 + \frac{(t - r_1 - 2r_3)}{(t - r_3)}\phi + 1 = 0 \tag{14.12}$$

and the conditional score statistic for testing $\beta = 0$ is

$$\mathcal{W} = \frac{[2t - (r_1 + 2r_2 + 3r_3)]^2}{r_1 + 4r_2 + r_3} \tag{14.13}$$

14.2.2 1:n Matched Design

Now consider the situation where one case is matched with more than one control. Suppose $m_j = 1$ and $n_j = n + 1$, and we have one (unmatched) binary factor of interest taking values: $0, 1$. Data from such a study are typically summarized in Table 14.2 where t_{ij} denotes the number of sets with i exposed cases and j exposed controls.

According to (14.4), the unconditional gp for the jth set is

$$(1 + \theta_j)^{r_{0j}}(1 + \theta_j\phi)^{r_{1j}}$$

where r_{0j} is the number of unexposed cases and controls and r_{1j} is the exposed number of cases and controls in set j. Extracting terms in which the coefficient of θ_j is equal to 1, we have that

$$f_j(1, \phi; 1, .) = r_{0j} + r_{1j}\phi$$

Now since

$$r_{0j} + r_{1j} = n_j = 1 + n$$

we get that

$$f_j(1, \phi; 1, .) = (1 + n - r_{1j}) + r_{1j}\phi$$

with $0 \leq r_{1j} \leq 1 + n$. Now for the concordant sets let $r_0 = t_{00}$ and $r_{n+1} = t_{1n}$, and for the others, for $k = 1, \cdots, n$ let

$$r_k = t_{1,k-1} + t_{0k}$$

denote the number of sets with k subjects exposed, and

$$T = T_{10} + T_{11} + \cdots + T_{1,n-1}$$

denote the total (random) number of exposed cases. Ignoring the two forms of concordant pairs, the overall conditional gp for T is

$$\prod_{k=1}^{n} [(1 + n - k) + k\phi]^{r_k} \tag{14.14}$$

which can be used to perform conditional large sample and exact inference on ϕ or $\beta = \ln(\phi)$ in the usual style. The conditional log-likelihood, ℓ_c, is

$$\beta t - \sum_{k=1}^{n} r_k \ln[(1 + n - k) + k \exp(\beta)] \tag{14.15}$$

The conditional mle obtains from maximizing this likelihood. The conditional score statistic for testing $\beta = 0$ is

$$W = \frac{[t(n+1) - \sum_{k=1}^{n} k r_k]^2}{\sum_{k=1}^{n} k(1 + n - k) r_k} \tag{14.16}$$

****

The above formulation readily extends to the situation where the number of controls is variable. Suppose, for example, that for some sets, one case is matched with n_1 controls and for others, one case is matched with n_2 controls, and so on. In this instance, we obtain the conditional gp for each $1:n_j$ level of matching, and multiply them to get the overall conditional gp.

Example 14.1: The hypothetical data in Table 14.3 show two types of matched sets. In some, the matching ratio was 1:2 matching and in others, it was 1:3. For the sets with 1:2 matching,

$$F_a(\phi) = (2 + \phi)^5 (1 + 2\phi)^2$$

and for the sets with 1:3 matching,

$$F_b(\phi) = (3 + \phi)^4 (2 + 2\phi)^2 (1 + 3\phi)^5$$

and overall

$$F(\phi) = F_a(\phi) F_b(\phi)$$

MATCHED DESIGNS

Table 14.3 *1:n_j Case-Control Design*

1:2 Sets	Exposed Controls		
Case Factor	0	1	2
0	3	1	2
1	4	0	5

1:3 Sets	Exposed Controls			
Case Factor	0	1	2	3
0	4	0	2	3
1	4	0	2	1

with the observed value of T equal to $4 + 6 = 10$.

Other situations where we can specify the exact conditional distribution include the 1:1 design with two binary covariates with and without an interaction term, and the $m:(n-m)$ design with a single binary covariate (Peritz 1992).

14.2.3 The General Matched Design

In general, when several cases have been matched to several controls, we need to use recursive polynomial multiplication to compute the conditional gp. When there are two exposure variables, the relevant RPM methods are similar to the RPM methods for exact analysis of a $2 \times K$ table developed in Chapter 10 and Chapter 12.

Consider the instance of a matched set with m cases and $n-m$ controls and one scored unmatched covariate. Suppose the covariate x has I distinct levels, with $0 \leq x_1 < \cdots < x_I$ and let e_i be the total number of cases and controls at level x_i. Then consider the polynomial

$$F_I(\theta, \phi) = \prod_{i=1}^{I} \{1 + \theta \phi^{x_i}\}^{e_i} \tag{14.17}$$

Using arguments from Chapters 12, we have that $F_I(1, \phi; m, .)$ is the conditional gp for the matched set. Now define, as in §12.3, $e_+ = e_1 + \cdots + e_I$, $e_{+,i} = e_1 + \cdots + e_i$, and

$$\begin{aligned} M_{1i}(m) &= \max(0, m - e_+ + e_{+,i}) \\ M_{2i}(m) &= \min(e_{+,i}, m) \end{aligned} \tag{14.18}$$

and

$$\begin{aligned} m_{1i}(m, s) &= \max(0, m - e_+, e_{+,i} - s) \\ m_{2i}(m, s) &= \min(e_i, e_{+,i} - s, m - s) \end{aligned} \tag{14.19}$$

Then, in a way similar to (13.21), we write the recursion to obtain $F_I(1, \phi; m, .)$ as follows. For

$$s \in \{M_{1i}(m), M_{1i}(m) + 1, \cdots, M_{2i}(m)\}$$

we implement

$$F_i(1, \phi; s, .) = \sum_{u=m_{1i}(m,s)}^{m_{2i}(m,s)} \binom{e_i}{u} \phi^{ux_i} F_{i-1}(1, \phi; s - u, .) \qquad (14.20)$$

Implementing this process from $i = 2, \cdots, J$, we get the required conditional gp for this matched set. If there is only one matched set, the conditional distribution of T has the form

$$P[T = t \mid m] = \frac{c(t)\phi^t}{F_I(1, \phi; m, .)} \qquad (14.21)$$

- If there are K matched sets, then we obtain the overall conditional gp by multiplying the conditional gps obtained from the above recursive scheme implemented K times.
- If there are two (or more) covariates, this procedure, in principle, extends in a straightforward way. In practice, we need better exponent checks and possible recourse to advanced Monte–Carlo methods.

In the general case of m to $n - m$ matching, even the large sample conditional analysis requires recursive computational methods. Assume, for example, that we have one matched set with one unmatched exposure factor. The conditional log-likelihood then is

$$\ell_c = \beta t - \ln[F_I(1, \exp(\beta); m, .)] \qquad (14.22)$$

We require first and second derivatives of the conditional likelihood. For the former,

$$\frac{\partial \ell_c}{\partial \beta} = t - \frac{F_I'(1, \exp(\beta); m, .)}{F_I(1, \exp(\beta); m, .)} \qquad (14.23)$$

$$-\frac{\partial^2 \ell_c}{\partial \beta^2} = \frac{F_I''(1, \exp(\beta); m, .) F_I(1, \exp(\beta); m, .) - \{F_I'(1, \exp(\beta); m, .)\}^2}{\{F_I(1, \exp(\beta); m, .)\}^2} \qquad (14.24)$$

From a computational point of view, these derivatives may obtained in one of two ways:

- Use recursion (14.20) to obtain the generic polynomial $F_I(1, \phi; m, .)$. Then for any specific value of β, use the distribution (14.21) to compute the values of the log-likelihood and its first and second derivatives.
- Instead of computing the generic polynomial $F_I(1, \phi; m, .)$, use the recursion (14.20) to compute the value of the log-likelihood at specific values of β. Further, differentiate this recursion to obtain the recursions for computing the values of the first and second derivative at specific values of β. These recursions are given in (14.25) and (14.26) below.

Then $F_i'(1, \phi; s, .)$ is given by

$$\sum_{u=m_{1i}(m,s)}^{m_{2i}(m,s)} \binom{e_i}{u} \left\{ \phi^{ux_i} F_{i-1}'(1, \phi; s - u, .) + x_i u \phi^{(ux_i - 1)} F_{i-1}'(1, \phi; s - u, .) \right\} \qquad (14.25)$$

and $F_i''(1,\phi;s,.)$ is given by

$$\sum_{u=M_{1i}(s)}^{M_{2i}(s)} \binom{e_i}{u} \left\{ \phi^{ux_i} F_{i-1}''(1,\phi;s-u,.) + 2x_i u \phi^{(ux_i-1)} F_{i-1}'(1,\phi;s-u,.) \right.$$
$$\left. + x_i u(x_i u - 1)\phi^{(ux_i-2)} F_{i-1}(1,\phi;s-u,.) \right\} \quad (14.26)$$

- When there is more than one exposure variable, and thus more than one parameter of interest, the recursions are readily extended to obtain all the partial first and second derivatives.
- When there are two or more matched sets, we apply the above strategies to the log-likelihood summed over all the matched sets and the overall conditional gp.

Of the two computational approaches given above, which is better? That depends on how often the log-likelihood needs to be maximized, the number of parameters of interest and the number of intermediate stage records generated in the process of computing the generic conditional gp. We leave it to the reader to investigate details.

14.2.4 Exponent Checks

When there are two or more matched sets, we compute the product (14.7) in a recursive fashion to obtain the overall conditional gp. At times, we analyze each parameter in this overall conditional distribution separately. Suppose we need to compute an exact CI for ϕ_2. Then we need only the portion of $F_J(\phi_1,\phi_2)$ in which the exponent of ϕ_1 is fixed at t_1. Let η_{1j} and μ_{1j} respectively be the largest and smallest exponents of ϕ_1 in $f_j(1,\phi_1,\phi_2;m_j,.,.)$. Further, let

$$\eta_{+1,j} = \sum_{i=j}^{J} \eta_{1j} \quad \text{and} \quad \mu_{+1,j} = \sum_{i=j}^{J} \mu_{1j}$$

Now assume we are about to multiply a term in $F_{j-1}(\phi_1,\phi_2)$ with a term in $f_j(1,\phi_1,\phi_2;m_j,.,.)$. Let (t_{+1j}, t_{+2j}) be the resultant exponent of (ϕ_1,ϕ_2). Then we apply the following checks.

Check EC 14.01

$\langle$ If $\; t_{+1j} + \mu_{+1,j+1} > t_1 \; \rangle$

Or

$\langle$ If $\; t_{+1j} + \eta_{+1,j+1} < t_1 \; \rangle$

Do not perform the multiplication. Continue on to the next term.

$\langle$ Otherwise $\rangle$

Perform the multiplication, accumulate the result in $F_j(\phi_1,\phi_2)$ and continue.

□□□□□□

The required conditional distribution is obtained from the resultant conditional gp and analyzed in the usual manner. Plotting respective evidence functions is a key part of this exercise.

14.3 Paired Binary Outcomes

At times, each study unit expresses a pair of binary outcomes. This may happen in a variety of ways. In a two period cross over clinical trial, subjects are randomly allocated to one of two treatments. After their responses have been gauged, they are crossed over (allowing for a possible wash out period) to the other treatment. The responses are again noted, giving a pair of responses for each subject. In some trials, two subjects at a time are matched in terms of prognostic factors, and randomized to one of two treatments. Pairs of identical twins or subjects selected by family are followed up to monitor the occurrence of a particular outcome. In a study of eye conditions, a reaction is elicited in each eye; in an investigation of ear disease, the left and right ears of each person are evaluated individually. And so on.

Unlike the matched pairs case-control study, these are prospective designs. To analyze data from them, it is advisable to invoke models that allow for dependence between the paired responses.

Suppose, with n pairs, $\mathcal{Y}_{1i}$ and $\mathcal{Y}_{2i}$ are respectively the random binary responses of the first and second member of the ith pair. Further, for $a, b \in \{0, 1\}$, let

$$\pi_i(a, b) = P[\mathcal{Y}_{1i} = a, \mathcal{Y}_{2i} = b] \qquad (14.27)$$

Questions of interest in paired studies include:

- Are the responses within the pairs independent?
- Are the marginal probabilities of response for each pair member equal?
- Is $\pi_i(a, b)$ affected by some covariate?
- Is the covariate effect the same for each pair member?

14.3.1 Models Without Covariates

First consider the situation without covariates under the logit model

$$\ln\left(\frac{\pi_i(y_{1i}, y_{2i})}{1 - \pi_k(y_{1i}, y_{2i})}\right) = \alpha_1 y_{1i} + \alpha_2 y_{2i} + \alpha_{12} y_{1i} y_{2i} \qquad (14.28)$$

Table 14.4 *Number of Pairs by Response*

First Pair Member	Second Pair Member		
	0	1	Total
0	$n - r_1 - r_2 + r_{12}$	$r_2 - r_{12}$	$n - r_1$
1	$r_1 - r_{12}$	r_{12}	r_1
Total	$n - r_2$	r_2	n

This extends the logit model for a single response. Its parameters can also be interpreted in terms of log-odds ratios. In fact, if, for $j = 1$ or $j = 2$, we condition on $\mathcal{Y}_{ji} = y_{ji}$ in (14.28),

PAIRED BINARY OUTCOMES

we get a univariate logit model. Further, if $\alpha_{12} = 0$, the responses of the pair members are independent. To analyze the data, let

$$\mathcal{R}_1 = \sum_i \mathcal{Y}_{1i}, \quad \mathcal{R}_2 = \sum_i \mathcal{Y}_{2i}, \quad \mathcal{R}_{12} = \sum_i \mathcal{Y}_{1i}\mathcal{Y}_{2i} \tag{14.29}$$

with observed values of (14.29) denoted r_1, r_2 and r_{12}, respectively. The data from all the n pairs are summarized in Table 14.4, whose counts follow a quadrinomial distribution. $P[\mathcal{R}_1 = r_1, \mathcal{R}_2 = r_2, \mathcal{R}_{12} = r_{12} \mid n]$ is then given by

$$\frac{n! \, \theta_1^{r_1} \theta_2^{r_2} \theta_{12}^{r_{12}} (1 + \theta_1 + \theta_2 + \theta_1 \theta_2 \theta_{12})^{-n}}{(n - r_1 - r_2 + r_{12})! (r_1 - r_{12})! (r_2 - r_{12})! r_{12}!} \tag{14.30}$$

where $\theta_1 = \exp(\alpha_1)$, $\theta_2 = \exp(\alpha_2)$ and $\theta_{12} = \exp(\alpha_{12})$. It is apparent that $\mathcal{R} = (\mathcal{R}_1, \mathcal{R}_2, \mathcal{R}_{12})$ is sufficient for $(\theta_1, \theta_2, \theta_{12})$.

- The hypothesis $\theta_1 = \theta_2$ in (14.28) implies that the marginal probabilities are homogeneous. Testing it leads to a binomial test based on discordant pairs, namely, the McNemar's exact binomial test for paired data.
- The presence and magnitude of interaction between the paired responses is reflected in the parameter θ_{12}. The conditional distribution of its sufficient statistic, $P[\mathcal{R}_{12} = u \mid \mathcal{R}_1 = r_1, \mathcal{R}_2 = r_2]$, is a hypergeometric distribution whose gp is

$$\sum_{u = \min(r_1, r_2)}^{\max(0, r_1 + r_2 - n)} \binom{r_1}{u} \binom{n - r_1}{r_2 - u} \theta_{12}^u \tag{14.31}$$

The associated distribution is analyzed in the usual manner.

14.3.2 Models With Covariates

Now consider models with one or more explanatory covariates. In paired studies, we distinguish between two types of covariates.

- Subunit specific covariates are factors that pertain to and may have distinct values for each member of the pair.
- Common covariates are factors which are intrinsically identical for both pair members.

For example, in a study of ear infections, whether the ear drum is bulging or not is particular to each ear while the age of the patient is a common covariate. Suppose, for the ith pair, we have one common covariate, x_i, and one subunit specific covariate, z_{1i} and z_{2i}. Then define

$$\Delta_{1i} = \exp(\alpha_1 + \beta_1 x_i + \gamma_1 z_{1i}) = \theta_1 \phi_1^{x_i} \psi_1^{z_{1i}} \tag{14.32}$$

$$\Delta_{2i} = \exp(\alpha_2 + \beta_2 x_i + \gamma_2 z_{2i}) = \theta_2 \phi_2^{x_i} \psi_2^{z_{2i}} \tag{14.33}$$

$$\Delta_{12i} = \exp(\alpha_{12} + \beta_{12} x_i + \gamma_{12} z_{1i} z_{2i}) = \theta_{12} \phi_{12}^{x_i} \psi_{12}^{z_{1i} z_{2i}} \tag{14.34}$$

Consider the paired logit model

Table 14.5 *Paired Outcomes with Two Binary Covariates*

Group	Covariates			No. of (y_{1i}, y_{2i}) Pairs				Total
k	x_k	z_{1k}	z_{2k}	$(0,0)$	$(0,1)$	$(1,0)$	$(1,1)$	n_k
1	0	0	0	r_{001}	r_{011}	r_{101}	r_{111}	n_1
2	0	0	1	r_{002}	r_{012}	r_{102}	r_{112}	n_2
3	0	1	0	r_{003}	r_{013}	r_{103}	r_{113}	n_3
4	0	1	1	r_{004}	r_{014}	r_{104}	r_{114}	n_4
5	1	0	0	r_{005}	r_{015}	r_{105}	r_{115}	n_5
6	1	0	1	r_{006}	r_{016}	r_{106}	r_{116}	n_6
7	1	1	0	r_{007}	r_{017}	r_{107}	r_{117}	n_7
8	1	1	1	r_{008}	r_{018}	r_{108}	r_{118}	n_8

$$\ln\left(\frac{\pi_i(y_{1i}, y_{2i} \mid x_i, z_{1i}, z_{2i})}{1 - \pi_i(y_{1i}, y_{2i} \mid x_i, z_{1i}, z_{2i})}\right) = y_{1i}\ln(\Delta_{1i}) + y_{2i}\ln(\Delta_{2i}) + y_{1i}y_{2i}\ln(\Delta_{12i}) \quad (14.35)$$

with

$$\pi_i(0,0) + \pi_i(0,1) + \pi_i(1,0) + \pi_i(1,1) = 1$$

The parameters in this model have the usual interpretations in terms of constant effect and covariate mediated effect on the odds ratio scale. In particular, θ_{12} reflects the odds-ratio scale background or constant interaction between the pair members, ϕ_{12} reflects the odds ratio scale interaction mediated through the common covariate and ψ_{12}, the odds ratio scale interaction mediated through the subunit specific covariate.

For data with covariates, it is convenient, and, from a computational perspective, more efficient to use the data in a grouped format. The groups are formed by the distinct profiles of common and subunit specific covariates. Suppose we have K groups formed by K distinct covariate profiles. We then count, for each group, the numbers of each of the four types of paired responses. For example, Table 14.5 shows the layout with one common binary covariate and one binary subunit specific covariate. For such data, $\mathcal{R}_{ijk}$ and r_{ijk} are defined separately for each group in a manner similar to (14.29).

The gp for the data derives from the product of K quadrinomial gps. Let $\boldsymbol{\theta} = (\theta_1, \theta_2, \theta_{12})$, $\boldsymbol{\phi} = (\phi_1, \phi_2, \phi_{12})$ and $\boldsymbol{\psi} = (\psi_1, \psi_2, \psi_{12})$. Then the gp for grouped data is

$$F_K(\boldsymbol{\theta}, \boldsymbol{\phi}, \boldsymbol{\psi}) = \prod_{k=1}^{K} (1 + \Delta_{1k} + \Delta_{2k} + \Delta_{1k}\Delta_{2k}\Delta_{12k})^{n_k} \quad (14.36)$$

Define

PAIRED BINARY OUTCOMES

Table 14.6 *Paired Outcomes with One Binary Covariate*

Group	Covariate	No. of Pairs				Total
k	x_k	$(0,0)$	$(0,1)$	$(1,0)$	$(1,1)$	n_k
1	0	3	1	2	0	6
2	1	0	2	1	3	6

Note: Hypothetical data.

$$\mathcal{R}_1 = \sum_k (\mathcal{R}_{10k} + \mathcal{R}_{11k}), \qquad \mathcal{R}_2 = \sum_k (\mathcal{R}_{01k} + \mathcal{R}_{10k}),$$

$$\mathcal{R}_{12} = \sum_k \mathcal{R}_{11k} \tag{14.37}$$

$$\mathcal{T}_1 = \sum_k (\mathcal{R}_{10k} + \mathcal{R}_{11k})x_k, \qquad \mathcal{T}_2 = \sum_k (\mathcal{R}_{01k} + \mathcal{R}_{11k})x_k,$$

$$\mathcal{T}_{12} = \sum_k \mathcal{R}_{11k} x_k \tag{14.38}$$

$$\mathcal{S}_1 = \sum_k (\mathcal{R}_{10k} + \mathcal{R}_{11k})z_{1k}, \qquad \mathcal{S}_2 = \sum_k (\mathcal{R}_{01k} + \mathcal{R}_{11k})z_{2k},$$

$$\mathcal{S}_{12} = \sum_k \mathcal{R}_{11k} z_{1k} z_{2k} \tag{14.39}$$

Then

- $\mathcal{R} = (\mathcal{R}_1, \mathcal{R}_2, \mathcal{R}_{12})$, is sufficient for $\boldsymbol{\theta} = (\theta_1, \theta_2, \theta_{12})$.
- $\mathcal{T} = (\mathcal{T}_1, \mathcal{T}_2, \mathcal{T}_{12})$, is sufficient for $\boldsymbol{\phi} = (\phi_1, \phi_2, \phi_{12})$.
- $\mathcal{S} = (\mathcal{S}_1, \mathcal{S}_2, \mathcal{S}_{12})$, is sufficient for $\boldsymbol{\psi} = (\psi_1, \psi_2, \psi_{12})$.

Further

$$P[\mathcal{R} = r, \mathcal{T} = t, \mathcal{S} = s] = \frac{c(r, t, s)\boldsymbol{\theta}^r \boldsymbol{\phi}^t \boldsymbol{\psi}^s}{F_K(\boldsymbol{\theta}, \boldsymbol{\phi}, \boldsymbol{\psi})} \tag{14.40}$$

The parameters in this distribution are analyzed by appropriate conditioning. We illustrate this for a specific case below.

14.3.3 One Binary Common Covariate

Consider the problem where there is a single common binary covariate, x_k. Thus, $K = 2$, $x_1 = 0$ and $x_2 = 1$.

Example 14.2: A hypothetical data example appears in Table 14.6.

For these data, the gp (14.36) becomes

$$(1+\theta_1+\theta_2+\theta_1\theta_2\theta_{12})^{n_1}(1+\theta_1\phi_1+\theta_2\phi_2+\theta_1\theta_2\theta_{12}\phi_1\phi_2\phi_{12})^{n_2} \tag{14.41}$$

We analyze a simpler model in which the covariate effect is expressed through a single parameter ϕ and where a covariate mediated interaction effect is not present. In terms of model (14.32)–(14.34), this means that $\phi_1 = \phi_2 = \phi$, $\phi_{12} = 1$ and, for $k = 1, 2$,

$$\Delta_{1k} = \theta_1 \phi^{x_k}, \quad \Delta_{2k} = \theta_2 \phi^{x_k}, \quad \Delta_{12k} = \theta_{12} \phi^{2 x_k} \tag{14.42}$$

Now let

$$\mathcal{T} = \sum_{k=1}^{2}(\mathcal{R}_{10k} + \mathcal{R}_{01k} + 2\mathcal{R}_{11k})x_k$$

$\mathcal{T}$ is sufficient for ϕ. $F(\theta_1, \theta_2, \theta_{12}, \phi)$, the gp of the joint distribution of $(\mathcal{R}_1, \mathcal{R}_2, \mathcal{R}_{12}, \mathcal{T})$, is

$$[1+\theta_1+\theta_2+\theta_1\theta_2\theta_{12}]^{n_1}[1+\theta_1\phi+\theta_2\phi+\theta_1\theta_2\theta_{12}\phi^2]^{n_2} \tag{14.43}$$

To analyze the parameter ϕ, we need terms of this polynomial in which the exponent of θ_1 is r_1, that of θ_2 is r_2 and that of θ_{12} is r_{12}. Expanding each term in (14.43) as a binomial and rearranging terms, it becomes

$$\sum_{u=0}^{n_1+n_2} [\theta_1\theta_2\theta_{12}]^{(n_1+n_2-u)} \Bigg\{ \sum_{v=\max(0,u-n_1)}^{\min(u,n_2)} \binom{n_1}{u-v}\binom{n_2}{v} \\ \times \phi^{2(n_2-v)}[1+(\theta_1+\theta_2)]^{(u-v)}[1+\phi(\theta_1+\theta_2)]^{v} \Bigg\} \tag{14.44}$$

From this, we select terms in which the exponent θ_{12} is r_{12}. These are terms within the first summation in which $n_1 + n_2 - u = r_{12}$ or $u = n_1 + n_2 - r_{12}$. Hence, $F(\theta_1, \theta_2, \theta_{12}, \phi; ., ., r_{12}, .)$ is

$$[\theta_1\theta_2\theta_{12}]^{r_{12}} \sum_{v=\max(0,n_2-r_{12})}^{\min(n_1+n_2-r_{12},n_2)} \binom{n_1}{n_1+n_2-r_{12}-v}\binom{n_2}{v} \\ \times \phi^{2(n_2-v)}[1+(\theta_1+\theta_2)]^{(n_1+n_2-r_{12}-v)}[1+\phi(\theta_1+\theta_2)]^{v} \tag{14.45}$$

Now examine products of the form

$$[1+\theta_1+\theta_2]^a [1+\phi(\theta_1+\theta_2)]^b \tag{14.46}$$

Expanding each part as a binomial and rearranging terms, it becomes

$$\sum_{x=0}^{a}\sum_{y=0}^{b}\sum_{w=0}^{x+y} \binom{a}{x}\binom{b}{y}\binom{x+y}{w} \theta_1^w \theta_2^{(x+y-w)} \phi^y \tag{14.47}$$

We need terms in (14.47) in which the exponent of θ_1 is $r_1 - r_{12}$ and that of θ_2 is $r_2 - r_{12}$.

PAIRED BINARY OUTCOMES

These are terms with $w = r_1 - r_{12}$, $x + y = r_1 + r_2 - 2r_{12}$ and $x + y - w = r_2 - r_{12}$. With $x + y = r_1 + r_2 - 2r_{12}$, the relevant terms satisfy

$$\max(0, r_1 + r_2 - 2r_{12} - a) \le y \le \min(b, r_1 + r_2 - 2r_{12})$$

The required terms then are

$$\binom{r_1 + r_2 - 2r_{12}}{r_1 - r_{12}} \theta_1^{(r_1 - r_{12})} \theta_2^{(r_2 - r_{12})}$$

$$\times \sum_{y=\max(0, r_1+r_2-2r_{12}-a)}^{\min(b, r_1+r_2-2r_{12})} \binom{a}{r_1 + r_2 - 2r_{12} - y}\binom{b}{y}\phi^y \quad (14.48)$$

With $a = n_1 + n_2 - r_{12} - v$, and $b = v$, combining (14.45) with (14.48), and omitting the factor in (14.48) outside the summation sign, the required conditional gp for the conditional distribution of T, $F(1,1,1,\phi; r_1, r_2, r_{12}, .)$ equals

$$\sum_{v=\xi_1}^{\xi_2} \binom{n_1}{n_1 + n_2 - r_{12} - v}\binom{n_2}{v}\phi^{[2(n_2 - v)]} \left\{ \sum_{w=\xi_3(v)}^{\xi_4(v)} \binom{n_1 + n_2 - r_{12} - v}{r_1 + r_2 - 2r_{12} - w}\binom{v}{w}\phi^w \right\}$$

(14.49)

where

$$\begin{aligned}
\xi_1 &= \max(0, n_2 - r_{12}) & (14.50) \\
\xi_2 &= \min(n_1 + n_2 - r_{12}, n_2) & (14.51) \\
\xi_3(v) &= \max(0, r_1 + r_2 - r_{12} - n_1 - n_2 + v) & (14.52) \\
\xi_4(v) &= \min(r_1 + r_2 - 2r_{12}, v) & (14.53)
\end{aligned}$$

Note that since $x_1 = 0$ and $x_2 = 1$, the observed value of T is given by

$$t = r_{102} + r_{012} + 2r_{112}$$

****

Example 14.2 (continued): For the data in Table 14.6,

$$n_1 = n_2 = 6, r_1 = r_2 = 6, r_{12} = 3$$

Then $n_1 - r_{12} = 3$, $n_1 + n_2 - r_{12} = 9$, $r_1 + r_2 - 2r_{12} = 6$ and $r_1 + r_2 - r_{12} - n_1 - n_2 + v = v - 3$. The gp $F(1,1,1,\phi; 6,6,3,.)$ is

$$\sum_{v=\max(0,3)}^{\min(9,6)} \binom{6}{9-v}\binom{6}{v}\phi^{[2(6-v)]} \left\{ \sum_{w=\max(0,v-3)}^{\min(6,v)} \binom{9-v}{6-w}\binom{v}{w}\phi^w \right\}$$

which equals

$$40\phi^3(10 + 45\phi + 108\phi^2 + 136\phi^3 + 108\phi^4 + 45\phi^5 + 10\phi^6)$$

Further

$$t = 2 + 1 + 2 \times 3 = 9$$

The TST mid-p-value for the hypothesis $\phi = 1$ is

$$2 \times \frac{10}{462} - \frac{10}{462} = 0.0216$$

The conditional gp is then used to plot the mid-p evidence function, and obtain the mue and associated CI.

14.3.4 Computation

Explicit formulation of the conditional distribution becomes too onerous in a more complex model involving a common or a subunit specific binary or nonbinary covariate. With more than one covariate, recourse to an efficient computation method is all but essential. We apply the recursive polynomial multiplication with appropriate exponent checks to such problems. Under this, we implement, for $k = 2, \cdots, K$

$$F_k(\boldsymbol{\theta}, \boldsymbol{\phi}, \boldsymbol{\psi}) = F_{k-1}(\boldsymbol{\theta}, \boldsymbol{\phi}, \boldsymbol{\psi})\left(1 + \Delta_{1k} + \Delta_{2k} + \Delta_{1k}\Delta_{2k}\Delta_{12k}\right)^{n_k} \qquad (14.54)$$

The entities Δ_{1k}, Δ_{2k} and Δ_{12k} depend on the type and number of covariates and the type of model used. As the number of parameters in the polynomials may easily exceed five, good memory management is a key aspect of the computer implementation. The basic exponent checks are those that ensure fixed final values of the exponents of θ_1, θ_2 and θ_{12}. These are formulated in a way similar to that done in EC 11.02. Suppose we are about to multiply a term in $F_{k-1}(\boldsymbol{\theta}, \boldsymbol{\phi}, \boldsymbol{\psi})$ with a term in

$$\left(1 + \Delta_{1k} + \Delta_{2k} + \Delta_{1k}\Delta_{2k}\Delta_{12k}\right)^{n_k}$$

Let $(r_{+1k}, r_{+2k}, r_{+12k})$ be the resultant exponent of $(\theta_1, \theta_2, \theta_{12})$. Then we apply the following checks.

PAIRED BINARY OUTCOMES

Checks EC 14.02

$$\left\langle \text{If} \quad r_{+1k} > r_1 \quad \text{or} \quad r_{+1k} + \sum_{j>k} n_j < r_1 \right\rangle$$

Or

$$\left\langle \text{If} \quad r_{+2k} > r_2 \quad \text{or} \quad r_{+2k} + \sum_{j>k} n_j < r_2 \right\rangle$$

Or

$$\left\langle \text{If} \quad r_{+12k} > r_{12} \quad \text{or} \quad r_{+12k} + \sum_{j>k} n_j < r_{12} \right\rangle$$

Do not perform the multiplication. Continue on to the next term.

$$\langle \text{Otherwise} \rangle$$

Perform the multiplication, accumulate the result in $F_k(\boldsymbol{\theta}, \boldsymbol{\phi}, \boldsymbol{\psi})$ and continue.

□□□□□□□

Other exponent checks may be added either to directly compute tail areas of some statistic, or to condition on the values of the sufficient statistics of one or more of common or subunit specific covariates.

Suppose we need to condition on the final values $\boldsymbol{T}$. At stage k, let $(t_{+1k}, t_{+2k}, t_{+12k})$ be a possible exponent of $(\phi_1, \phi_2, \phi_{12})$. Then apply the following checks.

Checks EC 14.03

$$\left\langle \text{If} \quad t_{+1k} > t_1 \quad \text{or} \quad t_{+1k} + \sum_{j>k} x_j n_j < t_1 \right\rangle$$

Or

$$\left\langle \text{If} \quad t_{+2k} > t_2 \quad \text{or} \quad t_{+2k} + \sum_{j>k} x_j n_j < t_2 \right\rangle$$

Or

$$\left\langle \text{If} \quad t_{+12k} > t_{12} \quad \text{or} \quad t_{+12k} + \sum_{j>k} x_j n_j < t_{12} \right\rangle$$

Do not perform the multiplication. Continue on to the next term.

$$\langle \text{Otherwise} \rangle$$

Perform the multiplication, accumulate the result in $F_k(\boldsymbol{\theta}, \boldsymbol{\phi}, \boldsymbol{\psi})$ and continue.

□□□□□□□

Suppose we need to condition on the final values $\boldsymbol{S}$. At stage k, let $(s_{+1k}, s_{+2k}, s_{+12k})$ be a possible exponent of $(\psi_1, \psi_2, \psi_{12})$. Then apply the following checks.

Check EC 14.04

$$\left\langle \text{If} \quad s_{+1k} > s_1 \quad \text{or} \quad s_{+1k} + \sum_{j>k} z_{1j} n_j < s_1 \right\rangle$$

Or

$$\left\langle \text{If} \quad s_{+2k} > s_2 \quad \text{or} \quad s_{+2k} + \sum_{j>k} z_{2j} n_j < s_2 \right\rangle$$

Or

$$\left\langle \text{If} \quad s_{+12k} > s_{12} \quad \text{or} \quad s_{+12k} + \sum_{j>k} z_{1j} z_{2j} n_j < s_{12} \right\rangle$$

Do not perform the multiplication. Continue on to the next term.

$$\langle \text{Otherwise} \rangle$$

Perform the multiplication, accumulate the result in $F_k(\boldsymbol{\theta}, \boldsymbol{\phi}, \boldsymbol{\psi})$ and continue.

□□□□□□□

The following are important to note:

- The number of exponent checks in EC 14.03 and EC 14.04 are adjusted depending, respectively, on how many of the exponents of the parameters in $(\phi_1, \phi_2, \phi_{12})$ and $(\psi_1, \psi_2, \psi_{12})$ are fixed.
- These checks vary according to the model under study.
- EC 14.02, EC 14.03 and EC 14.04 are crude checks which work reasonably well with moderately sized samples. They are improved by checks in which a particular check depends on one or more of the previous checks (see Hirji 1994).

14.4 Markov Chain Models

Consider three examples depicting one or more of a series of Bernoulli events.

Example 14.3: The status of a business telephone line was checked every 30 seconds over a period of 25 minutes. It was labeled 0 if the line was free, and 1 if busy. The result, from Bhat (1972), page 44, was:

1101110101110011010111100111110100111011010010111010

Example 14.4: The blood glucose levels of three healthy 50 year old men of similar height and weight were measured for seven consecutive days. With 0 indicating a normal level, and 1 indicating an abnormal level, the results were:

(I): 0000001; (II): 0000100; (III): 0011101

Example 14.5: Persons with arthritis were entered into a clinical trial and randomized to treatment A or B. Once a week for five weeks, their pain level was assessed and recorded as 1, if it was severe, and 0 otherwise. With five patients in each arm, the data collected were:

MARKOV CHAIN MODELS

Treatment A: 10010; 01001; 01010; 00001; 10000

Treatment B: 11101; 10011; 10111; 11011; 01110

Are the events in these sequences independent? Perhaps not. In arthritis, for example, a painful day may be more likely to lead to another painful day. If independence cannot be assumed, the usual Binomial model does not apply.

****

Dependent discrete data regularly occur in practice. Many models for dependent events are available and are usually analyzed with large sample methods. Exact methods have been developed for a few discrete dependent data models. One of these is the **Markov chain model**.

Consider $\mathcal{Y}_k$, $k = 0, 1, \cdots, n$, a series of $n+1$ discrete events or trials. They are said to constitute an mth order **Markov chain** if the outcome or the **state** of a given trial depends only on the outcomes or states in the previous m trials. In a **first order** Markov chain, the outcome of the kth trial depends only on the outcome of the $(k-1)$th trial. That is:

$$P[\mathcal{Y}_k = y_k \mid \mathcal{Y}_l = y_l, l < k] = P[\mathcal{Y}_k = y_k \mid \mathcal{Y}_{k-1} = y_{k-1}] \quad (14.55)$$

Using this property, the joint probability of a first order chain is written as:

$$P[\mathcal{Y}_k = y_k, k = 0, 1, \cdots, n] = P[\mathcal{Y}_0 = y_0] \prod_{k=1}^{n} P[\mathcal{Y}_k = y_k \mid \mathcal{Y}_{k-1} = y_{k-1}] \quad (14.56)$$

Let $\pi_{ij}(k)$ be the probability of going from state i in trial $(k-1)$ to state j in trial k. This is called a **transition probability.** Note that, for any i, k,

$$\sum_j \pi_{ij}(k) = 1 \quad (14.57)$$

In a **homogeneous chain,** the transition probabilities remain constant over the trials, that is, $\pi_{ij}(k) = \pi_{ij}$. In this section, we only deal with **homogeneous first order two state Markov chains** with states 0 and 1. Further, as shown in Table 14.7, we write $\pi_{01} = \pi_0$ and $\pi_{11} = \pi_1$.

Table 14.7 *Two State Transition Probabilities*

Initial State	Subsequent State	
	0	1
0	$1 - \pi_0$	π_0
1	$1 - \pi_1$	π_1

Suppose M persons are randomly selected from a population of persons with a chronic disease. The disease is classified into one of two states, 0 for nonsevere disease and 1 for severe disease.

All subjects are followed up at regular intervals. For person m, let $\mathcal{Y}_{km}$ denote his or her state at trial k, and $1 + n_m$ the number of follow up visits.

For now suppose we have data from a single chain, labeled chain m. Let n_{ijm} denote the total number of transitions from state i to state j. These are shown in Table 14.8. For this chain, (14.56) becomes

Table 14.8 *Transition Counts for mth Chain*

Initial State	Subsequent State		
	0	1	Total
0	n_{00m}	n_{01m}	$n_m - s_m$
1	n_{10m}	n_{11m}	s_m
Total	$n_m - r_m$	r_m	n_m

$$P[\mathcal{Y}_{0m} = y_{0m}](1-\pi_0)^{n_{00m}}(\pi_0)^{n_{01m}}(1-\pi_1)^{n_{10m}}(\pi_1)^{n_{11m}} \qquad (14.58)$$

Now let

$$\theta = \frac{\pi_0}{(1-\pi_0)} \quad \text{and} \quad \phi = \frac{\pi_1/(1-\pi_1)}{\pi_0/(1-\pi_0)} \quad \text{and} \quad \rho = \frac{(1-\pi_1)}{(1-\pi_0)} \qquad (14.59)$$

Under this reparametrization, and conditioning on the initial state, (14.58) becomes

$$\frac{\theta^{r_m}\phi^{t_m}}{(1+\theta)^{(n_m - s_m)}(1+\theta\phi)^{s_m}} = \theta^{r_m}\phi^{t_m}\rho^{s_m}(1-\pi_0)^{n_m} \qquad (14.60)$$

where

$$t_m = n_{11m} \qquad (14.61)$$

The dimension of distribution (14.60) is **three**. But it has only **two** free parameters. In the distributions we have encountered thus far, these two quantities were identical.

14.4.1 Large Sample Analysis

The likelihood from (14.58) is identical to the likelihood obtained for a one margin fixed 2×2 table of counts. This also holds in the case of M chains. To see this, let

MARKOV CHAIN MODELS

$$\mathcal{R} = \sum_{m=1}^{M} (\mathcal{N}_{01m} + \mathcal{N}_{11m}) = \sum_{m=1}^{M} \mathcal{R}_m \tag{14.62}$$

$$\mathcal{S} = \sum_{m=1}^{M} (\mathcal{N}_{10m} + \mathcal{N}_{11m}) = \sum_{m=1}^{M} \mathcal{S}_m \tag{14.63}$$

$$\mathcal{T} = \sum_{m=1}^{M} \mathcal{N}_{11m} = \sum_{m=1}^{M} \mathcal{T}_m \tag{14.64}$$

Assume a homogeneous first order Markov dependency with the same transition probabilities for each person. Fixing the initial states, and with (r, s, t), the realized value of $(\mathcal{R}, \mathcal{S}, \mathcal{T})$, the joint likelihood for the data is the product of the individual chain likelihoods. This equals

$$\ell(\theta, \phi) = \frac{\theta^r \phi^t}{(1+\theta)^{(n-s)}(1+\theta\phi)^s} \tag{14.65}$$

where $n = \Sigma_m n_m$.

Suppose we test the hypothesis that the trials are independent trials as opposed to a first order homogeneous Markov chain. Then we test $\pi_1 = \pi_0$ ($\phi = 1$) versus $\pi_1 \neq \pi_0$ ($\phi_1 \neq 1$).

When the total number of trials is large, we can treat the likelihood (14.63) as if it is from a conventional 2×2 table of counts, and apply the asymptotic analysis given in Chapter 5. Thus the chisquare and LR tests developed there also apply here.

Maximum likelihood estimates and large sample CIs for π_0 and π_1 (or for θ and ϕ) are computed in a corresponding manner.

14.4.2 Exact Distributions

Exact analysis of Markov chain models requires the exact distribution of transition counts. Consider first a single chain. We drop the subscript m for now. For some n, y_0 and (r, s, t), let $c(r, s, t)$ be the number of binary sequences $\{y_0, y_1, \cdots, y_n\}$ with $y_m \in \{0, 1\}$ that satisfy

$$\sum_{k=0}^{n-1} y_k y_{k+1} = n_{11} = t \tag{14.66}$$

$$\sum_{k=0}^{n-1} y_k (1 - y_{k+1}) = n_{10} = s - t \tag{14.67}$$

$$\sum_{k=0}^{n-1} (1 - y_k) y_{k+1} = n_{01} = r - t \tag{14.68}$$

$$\sum_{k=0}^{n-1} (1 - y_k)(1 - y_{k+1}) = n_{00} = n - r - s + t \tag{14.69}$$

and for given n and y_0, let

$$\Omega = \{(r, s, t) : c(r, s, t) > 0\} \tag{14.70}$$

Note that although not made explicit by the notation, both $c(r,s,t)$ and Ω depend on y_0 and n. Then we can prove the following:

Theorem 14.1: For a single two state first order Markov chain with transition counts in Table 14.8, the exact probability for the transition matrix given the initial state, $P[(\mathcal{R},\mathcal{S},\mathcal{T})=(r,s,t) \mid y_0; \theta, \phi]$, is equal to

$$\frac{c(r,s,t)\theta^r \rho^s \phi^t}{\sum_{(u,v,w)\in\Omega} c(u,v,w)\theta^u \rho^v \phi^w} \tag{14.71}$$

Proof: Linking the counts in Table 14.8 to the sums (14.66), (14.67), (14.68) and (14.69), then using expression (14.60) and noting that the probabilities add up to one yields the desired result. □

Consider the multichain situation. Adding the subscript m in each of the above quantities, and for the given y_{0m}, let us define

$$f_m(\theta, \rho, \phi) = \sum_{(u,v,w)\in\Omega_m} c_m(u,v,w)\theta^u \rho^v \phi^w \tag{14.72}$$

and

$$f(\theta, \rho, \phi) = \prod_{m=1}^{M} f_m(\theta, \rho, \phi) \tag{14.73}$$

which we also expand as

$$f(\theta, \rho, \phi) = \sum_{(u,v,w)\in\Omega} c(u,v,w)\theta^u \rho^v \phi^w \tag{14.74}$$

where $c(u,v,w)$ is the coefficient of the term in (14.73) in which the exponent of θ is u, of ρ is v and of ϕ is w.

Further, let

$$\mathcal{T}_m = (\mathcal{R}_m, \mathcal{S}_m, \mathcal{T}_m) \quad \text{and} \quad t_m = (r_m, s_m, t_m) \tag{14.75}$$

and

$$\mathcal{T} = (\mathcal{R}, \mathcal{S}, \mathcal{T}) \quad \text{and} \quad t = (r, s, t) \tag{14.76}$$

where

$$\mathcal{T} = \sum_{m=1}^{M} \mathcal{T}_m \quad \text{and} \quad t = \sum_{m=1}^{M} t_m \tag{14.77}$$

with similar definitions for the components in (14.76).

MARKOV CHAIN MODELS

Theorem 14.2: Consider M independent two state homogeneous first order Markov chains with common transition probabilities as in Theorem 14.1. Then $(\mathcal{R}, \mathcal{S}, \mathcal{T})$ is jointly sufficient for (θ, ϕ), and the exact distribution of $\mathcal{T}$ given the initial states is

$$P[\mathcal{R} = r, \mathcal{S} = s, \mathcal{T} = t \mid \boldsymbol{y}_0] = \frac{c(r,s,t)\theta^r \rho^s \phi^t}{f(\theta, \rho, \phi)} \qquad (14.78)$$

where $\boldsymbol{y}_0 = (y_{01}, \cdots, y_{0M})$.

Proof: Follows from Theorem 14.1 and independence between the chains.□

To complete the specification of these exact distributions, we need the formulas for the coefficients and support set for a single chain. These are given below.

Theorem 14.3: The coefficients of the gp for the single chain specified in Theorem 14.1 are

$$c(r,s,t) = \binom{n-r-y_0}{n-r-s+t}\binom{r+y_0-1}{t} \qquad (14.79)$$

Further the set Ω is obtained as follows. First define

$$\begin{aligned}
l_1(r, y_0) &= (1-y_0)\max(0, r-1) + y_0 r & (14.80) \\
l_2(r, y_0) &= (1-y_0)r + y_0 \min(n, r+1) & (14.81) \\
l_{11}(r, s, y_0) &= \max(0, r+s-n) & (14.82) \\
l_{12}(r, s, y_0) &= \min(r+y_0-1, s-y_0) & (14.83)
\end{aligned}$$

with the conditions that (i) if $y_0 = r = s = 0$, then $l_{11}(r, s, y_0) = l_{12}(r, s, y_0) = 0$, (ii) if $y_0 = 1$ and $r = s = n$, then $l_{11}(r, s, y_0) = l_{12}(r, s, y_0) = n$, (iii) when $y_0 = 0$, the case with $r = s = n$ cannot occur and (iv) when $y_0 = 1$, the case with $r = s = 0$ cannot occur.

Then we implement the following scheme:

- Select r from the set $\{0, 1, \cdots, n\}$.
- For this r, select s from the set $\{s : l_1(r, y_0) \leq s \leq l_2(r, y_0)\}$.
- For this pair (r, s), select t from the set $\{t : l_{11}(r, s, y_0) \leq t \leq l_{12}(r, s, y_0)\}$.
- Continue until all feasible configurations of (r, s, t) are exhausted.

Proof: Given y_0, of the 2^n sequences $\{y_1, \cdots, y_n\}$ with $y_k \in \{0, 1\}$, we need the number which satisfy the equations (14.66), (14.67), (14.68) and (14.69). These equations first imply that

$$\sum_{k=1}^{n} y_k = r \quad \text{and} \quad r - s = y_n - y_0 \qquad (14.84)$$

The sequence $\{y_0, y_1, \cdots, y_n\}$ has alternating blocks (or runs) of zeros and ones. With fixed y_0, the number of blocks of zeros is $a = 1 - y_0 + n_{10} = 1 - y_0 + s - t$ and the number of blocks of ones is $b = y_0 + n_{01} = y_0 + r - t$.

Table 14.9 *Conditional GP for Telephone Data*

Record No.	u	$c(r,s,u)$	Record No.	u	$c(r,s,u)$
1	15	0.565723E+09	10	24	0.255700E+12
2	16	0.102184E+11	11	25	0.654592E+11
3	17	0.769383E+11	12	26	0.112150E+11
4	18	0.320576E+12	13	27	0.124611E+10
5	19	0.826749E+12	14	28	0.855848E+08
6	20	0.139721E+13	15	29	0.337280E+07
7	21	0.159681E+13	16	30	0.674560E+05
8	22	0.125463E+13	17	31	0.544000E+03
9	23	0.681866E+12	18	32	0.100000E+01

Note: $y_0 = 1, r = 32, s = 33, t = 18$.

Let $z_{01}, \cdots, z_{0a}$ be the numbers 0 to 0 transitions in these blocks of zeros and $z_{11}, \cdots, z_{1b}$, numbers 1 to 1 transitions in these blocks of ones. Then

$$z_{01} + \cdots + z_{0a} = n - r - s + t \quad \text{and} \quad z_{11} + \cdots + z_{1b} = t \tag{14.85}$$

with $0 \leq z_{0k}, z_{1k}$ for all k. Using the results of §13.2.2, the number of solutions to the first system in (14.85) is

$$\binom{a+n-r-s+t-1}{n-r-s+t} = \binom{n-r-y_0}{n-r-s+t} \tag{14.86}$$

and to the second system is

$$\binom{b+t-1}{t} = \binom{r+y_0-1}{t} \tag{14.87}$$

The generation of the set Ω starts with (14.84) which leads to identification of the feasible values of r and s. Given a feasible pair (r,s), the possible values of t are identified by considering those values of t for $c(r,s,t) > 0$ and detailing all situations for which a feasible chain exists. The details are left to the reader. □

14.4.3 Exact Analysis

Consider first an exact test for independence for first order chains. To test $\phi = 1$ versus $\phi \neq 1$ in the context of (14.78), the relevant conditional distribution is

$$P[\mathcal{T} = t \mid \mathcal{R} = r, \mathcal{S} = s, \boldsymbol{y}_0] = \frac{c(r,s,t)\phi^t}{f(1,1,\phi;r,s,.)} \tag{14.88}$$

Example 14.3 (continued): For the telephone line data, the initial state is 1 and the transition matrix is

MARKOV CHAIN MODELS   439

$$\begin{bmatrix} n_{00} & n_{01} \\ n_{10} & n_{11} \end{bmatrix} = \begin{bmatrix} 3 & 14 \\ 15 & 18 \end{bmatrix} \quad (14.89)$$

Applying Theorem 14.3, and selecting terms with $r = 32$, and $s = 33$, gives the required conditional gp, whose exponents and related coefficients are shown in Table 14.9. The exact TST p-value for $\phi = 1$ versus $\phi_1 \neq 1$ then is 0.1257 while the exact probability statistic based p-value is 0.1148. We may use this gp to plot an evidence function for ϕ.

****

Example 14.4 (continued): For the blood glucose level data, the three transition counts matrices are

$$\begin{bmatrix} 5 & 1 \\ 0 & 0 \end{bmatrix} \quad \begin{bmatrix} 4 & 1 \\ 1 & 0 \end{bmatrix} \quad \begin{bmatrix} 1 & 2 \\ 1 & 2 \end{bmatrix}$$

Assume that the counts are from homogeneous first order chains with common underlying transition probabilities. Since $y_0 = 0$ and $n = 6$ for all three persons, they have the same gp. Using Theorem 14.3, for $m = 1, 2, 3$, this is equal to

$$\begin{aligned} f_m(\theta, \rho, \phi) = \; & 1 + \theta + 5\theta\rho + 4\theta^2\rho + \theta^2\rho\phi + 6\theta^2\rho^2 + 4\theta^2\rho^2\phi + \\ & 3\theta^3\rho^2 + 6\theta^3\rho^2\phi + \theta^3\rho^2\phi^2 + \theta^3\rho^3 + 6\theta^3\rho^3\phi + \\ & 3\theta^3\rho^3\phi^2 + 3\theta^4\rho^3\phi + 6\theta^4\rho^3\phi^2 + \theta^4\rho^3\phi^3 + 3\theta^4\rho^4\phi^2 + \\ & 2\theta^4\rho^4\phi^3 + 4\theta^5\rho^4\phi^3 + \theta^5\rho^4\phi^4 + \theta^5\rho^5\phi^4 + \theta^6\rho^5\phi^5 \end{aligned}$$

Combining the three chains, the overall transition counts are

$$\begin{bmatrix} 10 & 4 \\ 2 & 2 \end{bmatrix}$$

The gp for the joint distribution of $(\mathcal{R}, \mathcal{S}, \mathcal{T})$ is given by

$$f(\theta, \rho, \phi) = \{f_1(\theta, \rho, \phi)\}^3$$

This polynomial has 280 terms and is not shown. Suppose we need the mues for the transition probabilities π_0 and π_1. At observed value $(r, s, t) = (6, 4, 2)$, the gp for the distribution of $\mathcal{R}$ given $\mathcal{S} = 4$ and $\mathcal{T} = 2$ is extracted from this polynomial and is

$$f(\theta, 1, 1;., 4, 2) = 147\theta^4 + 819\theta^5 + 1179\theta^6 + 489\theta^7$$

Similarly, the gp for the distribution of $\mathcal{T}$ given $\mathcal{R} = 6$ and $\mathcal{S} = 4$ is

$$f(1, 1, \phi; 6, 4, .) = 807 + 1854\phi + 1179\phi^2 + 240\phi^3 + 15\phi^4$$

The mues obtained from these univariate gps are equal to

$$\hat{\theta} = 1.38 \quad \text{and} \quad \hat{\phi} = 2.76$$

Using (14.59), these are solved to give the mues for π_0 and π_1. Evidence functions for the parameters are plotted in the usual fashion.

****

To analyze the clinical trial data of the sort given in Example 14.5, we assume a first order homogeneous chain. A key question is whether the two treatments have equivalent transition probabilities or not. Let $x = 0$ for treatment A and $x = 1$ for treatment B. Then for a person on treatment x, let

$$\pi(y, x) = P[\mathcal{Y}_k = 1 \mid \mathcal{Y}_{k-1} = y; x] \qquad (14.90)$$

We consider the logit model for the transition probabilities

$$\ln\left(\frac{\pi(y, x)}{1 - \pi(y, x)}\right) = \alpha_x + \beta_x y \qquad (14.91)$$

In this model, the two treatments are equal if $\alpha_0 = \alpha_1$ and $\beta_0 = \beta_1$. Let us write these parameters as

$$\theta_x = \exp(\alpha_x) = \theta\psi^x \quad \text{and} \quad \phi_x = \exp(\beta_x) = \phi\delta^x \qquad (14.92)$$

Treatment equivalence means that $\psi = \delta = 1$. Asymptotic tests of this hypothesis are constructed in the usual way. This theory shows that the data may be analyzed with logistic regression software commonly used for analysis of binary data.

To develop an exact test, we consider, for treatment x, the distribution of the combined transition counts given the initial states and the total number of transitions from state 1. That is,

$$P[\mathcal{R}_x = r_x, \mathcal{T}_x = t_x \mid \mathcal{S}_x = s_x, \boldsymbol{y}_{0x}] \qquad (14.93)$$

For given treatment x, this is obtained as follows:

- Determine the gp for all the individual chains in treatment x.
- Multiply these gps.
- From the product polynomial, select the terms in which the coefficient of ρ_x is equal to s_x.
- Setting $\rho_x = 1$, we write this subpolynomial as $f_x(\theta_x, 1, \phi_x; ., s_x, .)$.

To combine the data from both treatments, we consider the joint conditional distribution, $P[\mathcal{R}_x = r_x, \mathcal{T}_x = t_x, x = 0, 1 \mid \mathcal{S}_x = s_x, \boldsymbol{y}_{0x}, x = 0, 1]$. This distribution is a PBD from which it is readily seen that $\mathcal{R}_0 + \mathcal{R}_1$ is conditionally sufficient for θ, $\mathcal{T}_0 + \mathcal{T}_1$ is so for ϕ, $\mathcal{R}_1$ is so for ψ and $\mathcal{T}_1$ is so for δ.

To test the hypothesis of treatment equivalence, the distribution of $(\mathcal{R}_1, \mathcal{T}_1)$ upon further fixing the values of $\mathcal{R}_0 + \mathcal{R}_1$ and $\mathcal{T}_0 + \mathcal{T}_1$ is appropriate. The gp of this distribution is obtained by computing the product

$$f_0(\theta, 1, \phi; ., s_0, .) \times f_1(\theta\psi, 1, \phi\delta; ., s_1, .) \qquad (14.94)$$

and selecting terms in which the exponent of θ is $r_0 + r_1$ and that of ϕ is $t_0 + t_1$. With the PBD constructed from this gp, the analysis of ψ and δ is then done in the usual way.

MARKOV CHAIN MODELS

Table 14.10 *Conditional GP for Example 14.5*

Record No.	u	v	$c_*(u,v)$
1	11	5	0.355680000E+05
2	11	6	0.894880000E+06
3	11	7	0.234446400E+07
4	11	8	0.997936000E+06
5	11	9	0.687120000E+05
6	12	6	0.176070400E+07
7	12	7	0.154697760E+08
8	12	8	0.154873920E+08
9	12	9	0.223788000E+07
10	13	7	0.961497600E+07
11	13	8	0.306961200E+08
12	13	9	0.102233280E+08
13	14	8	0.748440000E+07
14	14	9	0.765489600E+07
15	15	9	0.587934000E+06

Note: $r_0 = 6$, $s_0 = 6$, $t_0 = 0$, $r_1 = 14$, $s_1 = 14$, $t_1 = 9$.

Example 14.5 (continued): For the two treatment groups, the overall transition matrices are

$$\text{A:} \begin{bmatrix} 8 & 6 \\ 6 & 0 \end{bmatrix} \qquad \text{B:} \begin{bmatrix} 1 & 5 \\ 5 & 9 \end{bmatrix}$$

Using the procedure outlined above, the bivariate conditional gp for the distribution of $(\mathcal{R}_1, \mathcal{T}_1)$ is determined as in Table 14.10. The observed values are $(r_1, t_1) = (14, 9)$. From this gp, and using a probability ordering based exact test, the *p*-value for the hypothesis of the equivalence of treatments in terms of the transition probabilities is found to be 0.2280. Evidence functions and exact CIs are obtained in the usual fashion.

14.4.4 Concluding Remarks

We conclude with some remarks on exact analysis of Markov chain data. First we address conceptual issues.

- Exact distributions for a single two and multistate chains were derived in the 1950s. A number of empirical studies done decades ago pointed to the limitations of the asymptotic methods for such data. Yet the exact analysis of Markovian data, especially in applied studies, has hardly begun.
- Earlier derivations of exact distributions for several chains were in error. A correct formulation was given by Hirji and Johnson (1999).
- For a single chain and several chains, the conditional distribution of the transition matrix upon fixing both margins of the matrix is NOT a hypergeometric distribution. Both the conditional sample space and the coefficients are different. This is sometimes not well understood in the applications oriented statistical works.

- Markov chain data represents a curious phenomenon: the conventional contingency table based (unconditional) large sample analysis is valid here, but the conventional contingency table based (conditional) exact analysis is NOT.
- Markov chain data demonstrate the versatility and utility of the polynomial formulation in the analysis of discrete data. For exact analysis of more general models for Markov chain data, see Hirji and Johnson (1999).

Now consider the relevant computational issues.

- Computing exact distributions for several two state chains uses the RPM method and applies relevant exponent checks as needed. These checks are derived in the usual manner. The details are left to the reader.
- Computing exact distributions for stratified two state chain data further applies the RPM method and relevant exponent checks for multiplying stratum specific polynomials. In the clinical trial example above, the number of strata was two.
- For general computational methods for exact analysis of Markov chain data, see Hirji and Johnson (1999).
- Exact analysis of a single and several multistate Markovian data is an open, unexplored research topic.

14.5 Relevant Literature

Rothman and Greenland (1998) provide an introduction to the design and analysis of matched case-control studies; Prentice (1976) gave a justification for the use of the conditional probability in retrospective studies. Other relevant books were noted in earlier chapters. Hirji (1991) compared exact, mid-p and asymptotic conditional score tests for $1:n_j$ matched designs with two binary covariates; Vollset, Hirji and Afifi (1991) evaluated exact and asymptotic CIs in logistic analyses of matched studies. In both cases, the mid-p method had better properties. See also Hirji et al. (1988)

Miettinen (1968) derived exact distributions for the matched pairs design, and Miettinen (1969) extended that to $1:n$ matching with one binary covariate. Peritz (1982) derived explicit formulas for exact distributions for matched pairs designs with one binary covariate and one possibly nonbinary covariate.

Efficient recursive algorithms for computing the conditional likelihood, and its first and second partial derivatives for matched case-control studies in which the number of cases and controls per set both exceed one, were given by Gail, Lubin and Rubinstein (1981) and Storer, Wacholder and Breslow (1983). For efficient algorithms for exact analysis of matched case-control studies in general see Martin and Austin (1991), Vollset, Hirji and Elashoff (1991) and Hirji, Mehta and Patel (1988).

Rosner (1984) and Glynn and Rosner (1992) provide a good introduction to, and a justification for, paired binary data models. Jones and Kenward (1989) apply paired data models to cross over clinical trials. See also Cox and Snell (1989) and Bonney (1987). The literature on exact analysis of paired binary data is scanty. The basic reference is Hirji (1994). Copas (1973) explored randomization models for paired binary data; Frisen (1980) cautioned against the use of conditional methods of inference for some forms of analysis for correlated categorical data.

For large sample analysis of Markov chain discrete data, see Hoel (1954), Andersen and Goodman (1957), Billingsley (1961), Chatfield (1976), Bishop, Fienberg and Holland (1975) and Agresti

EXERCISES   443

Table 14.11 *1:2 Case-Control Design*

Exposed Cases	Exposed Controls		
	0	1	2
0	2	1	2
1	5	4	3

Table 14.12 *A Paired Design with Two Binary Covariates*

x_{1j}	x_{2j}	Unconditional Generating Polynomial
0	0	$(1 + \theta_j)^{r_{00}}$
0	1	$(1 + \theta_j \phi_2)^{r_{01}}$
1	0	$(1 + \theta_j \phi_1)^{r_{10}}$
1	1	$(1 + \theta_j \phi_1 \phi_2 \phi 12)^{r_{11}}$

(1990), among many others. The exact distributions for a single two and multistate chain were derived by Whittle (1955) and Billingsley (1961). Further work was done by Klozt (1972) and Smith and Solow (1996). See also Cox and Snell (1989). The proof of Theorem 14.3 is based on Smith and Solow (1996). Exact distributions for multistate chains with a stratification factor were derived by Johnson (1997) and Hirji and Johnson (1999). They corrected some errors in the literature and dealt with efficient computation. Some of the material given above is derived from an unpublished paper, Hirji and Johnson (2000). The limitations of asymptotic methods for Markov chains were documented by Guthrie and Youseff (1970), Lissitz (1972), Johnson (1997) and Hirji and Johnson (1999).

14.6 Exercises

14.1. Starting with model (14.1), give a detailed derivation of the conditional distribution (14.6).

14.2. Carefully stating your assumptions, give a detailed justification for the use of the conditional distribution (14.6) in a matched case-control study.

14.3. Table 14.11 shows data from a 1:2 matched design with one binary covariate. Analyze them using exact and large sample conditional methods. Plot the evidence functions.

14.4. For the conditional likelihood (14.11), obtain an explicit formula for the conditional mle for β from the cubic equation (14.12) and derive the conditional score statistic formula given in (14.13). Also derive the related conditional LR statistic. Also give a formula for the large sample conditional CI for β.

14.5. In a 1:1 matched design with two binary covariates, first consider a model with a between the covariates interaction term. Using the unconditional gps in Table 14.12, derive explicit formulas for the univariate exact conditional distributions for each exposure parameter. Repeat the same exercise for the model without an interaction term (Peritz 1982).

Table 14.13 *A 3:3 Case-Control Design*

Exposed Cases	Exposed Controls			
	0	1	2	3
0	2	1	1	1
1	3	0	3	0
2	1	2	4	3
3	0	1	1	4

14.6. Consider the $1:n_j$ matched design with two exposure variables both of which are binary. Derive explicit formulas for the conditional mles of the exposure effect parameters and the associated bivariate and univariate score tests for them (Hirji 1991).

14.7. Consider the $m_j:n_j - m_j$ design with a single binary covariate. Derive explicit formulas for the conditional distribution of the exposure parameter when there is (i) one matched set and (ii) two matched sets. How would you obtain this when there are more than two matched sets?

14.8. Develop, in full detail, the RPM algorithm with exponent checks for a $1:n_j$ matched case-control study with two exposure variables. Implement it on the computer. How would you extend it to the situation with more than two exposure variables, possibly with interaction terms (Hirji, Mehta and Patel 1989).

14.9. Starting with the formulation in §14.2.3, develop, in full detail, the RPM algorithm with exponent checks for an $m_j:n_j$ matched case-control study with two exposure variables. Also justify use of the checks EC 14.01 in this context. Implement the RPM methods with the checks on the computer. How would you extend it to problems with more than two exposure variables, and possibly with interaction terms (Hirji, Mehta and Patel 1988)?

14.10. Lay out the details of performing large sample conditional analysis for matched case-control studies with (i) one exposure variable, (ii) two exposure variables and (iii) more than two exposure variables (Gail, Lubin and Rubinstein 1981; Storer, Wacholder and Breslow 1983). Give detailed derivations of the recursions for computing the conditional likelihood (14.22) and its first two derivatives given in §14.2.3. Implement and compare, with simulated data, the two strategies for performing the task when you have (i) one exposure variable, (ii) two exposure variables and (iii) more than two exposure variables.

14.11. The data in Table 14.13 are from an hypothetical case-control study with a single binary exposure variable and where, in each matched set, three cases were matched with three controls. Derive the exact conditional distribution for these data and compute exact, mid-p and large sample conditional exact CIs for β. Plot the evidence function.

14.12. For model (14.28), derive exact tests for (i) marginal homogeneity and (ii) the presence of interaction. Also, for (i) and (iii), derive large sample based unconditional and conditional score and LR tests.

14.13. Give a detailed derivation of the distribution (14.30).

14.14. Construct the portion of the polynomial below in which

$$(1 + \theta_1 + \theta_2 + \theta_1\theta_2\theta_{12})^{n_1}(1 + \theta_1\phi + \theta_2\phi + \theta_1\theta_2\theta_{12}\phi^2\phi_{12})^{n_2}$$

in which the exponents of θ_1, θ_2 and θ_{12} are set at fixed values.

EXERCISES 445

Table 14.14 *Eye Data*

Group	Age (years)	No. of Pairs				Total
k	x_k	(0,0)	(0,1)	(1,0)	(1,1)	n_k
1	≤ 30	2	0	1	1	4
2	31–40	0	4	0	3	7
3	41–50	0	3	1	1	5
4	≥ 51	0	1	0	0	1

Source: Berson et al. (1991); Used with permission.
©1991, American Medical Association. All rights reserved.

14.15. How would you derive the portion of the polynomial below in which

$$(1 + \theta_1 + \theta_2 + \theta_1\theta_2\theta_{12})^{n_1} (1 + \theta_1\phi_1 + \theta_2\phi_2 + \theta_1\theta_2\theta_{12}\phi_1\phi_2\phi_{12})^{n_2}$$

the exponents of θ_1, θ_2 and θ_{12} are set at fixed values? Apply these to the exact analysis of relevant paired binary data models.

14.16. Derive appropriate unconditional and conditional LR and score statistics for testing relevant hypotheses for paired binary outcome data models with one binary common covariate. Consider models with and without covariate mediated interaction effects.

14.17. Consider a paired binary outcomes study with one subunit specific binary covariate and consider several models based on (14.32)–(14.34). How would you compute exact conditional distributions for such models?

14.18. Consider a paired binary outcomes study with one common nonbinary covariate and consider several models based on (14.32)–(14.34). How would you compute exact conditional distributions for such data? Derive appropriate LR and score statistics for testing relevant hypotheses for the models.

14.19. Give a detailed derivation of the distribution (14.40).

14.20. The data in Table 14.14 give ocular findings from 17 unrelated patients with a particular genetic defect (Berson et al. 1991). The two binary eye conditions are lens opacity and presence of pigment. The findings for both eyes were identical in each patient; so the paired analysis here concerns the two outcomes from each person. The covariate of interest is age.

(i) Ignoring the covariate, test for the presence of interaction between the two outcomes.
(ii) Using Age ≤ 40 as a cut point, fit the binary covariate model (14.42) to these data.
(iii) Extend this analysis using a model with a covariate mediated interaction effect. (iv) Using scores 0,1,2,3 for the age group use a scored covariate version of model (14.42) to analyze these data. (v) Derive the LR and score statistics for testing relevant hypotheses for these models. Where appropriate, plot the evidence function.

14.21. Consider a paired binary outcomes study with one common binary covariate and one subunit specific binary covariate. For several models based on (14.32)–(14.34), show how you would compute exact conditional distributions for these models. Derive the LR and score statistics for testing relevant hypotheses for the models.

14.22. Give a detailed justification for exponent checks EC 14.02, EC 14.03 and EC 14.04.

Table 14.15 *BSE Maternal Cohort Study*

Group	Days Until Onset	No. of Pairs				Total
k	x_k	(0,0)	(0,1)	(1,0)	(1,1)	n_k
1	≤ 150 days	214	3	34	6	257
2	> 150 days	38	4	2	0	44

Source: Donnely et al. (1997), Table 5.
Used with permission of Blackwell Publishing Ltd.

14.23. Implement the RPM method for paired binary data with exponent checks on the computer. Start with models for two covariates and extend it to more than two covariates. Explore the use of stricter exponent checks (Hirji 1994).

14.24. Table 14.15 shows data from a cohort study of maternal transmission of Bovine Spongiform Encephalopathy (BSE) in cows. This was a paired design study in which an animal born *"to a dam diagnosed with BSE was paired with a control animal born on the same farm and at about the same time but to a dam that, with one exception, was at least 5 years old and at the time of recruitment had not been diagnosed with BSE."* (Curnow, Hodge and Wilesmith 1997). The data in the table are from Donnely et al. (1997).

The outcome we study is eventual diagnosis with BSE in the dam. One covariate of interest, denoted x_k, was the number of days from the date of birth of the maternally exposed animal until the onset of clinical signs of BSE.

Analyze these data using exact and large sample methods and relevant paired binary outcomes models. Plot the evidence functions.

14.25. Construct the exact distribution and test the hypothesis of a first order dependence in each of the following binary chains. Compare the exact and asymptotic test results.

(i): 1000100101001110010101000010000110101

(ii): 000101110111000000000000011010100000

(iii): 111110000000010000011111110001

14.26. Construct the exact distribution and test the hypothesis of a first order dependence using the combined data from the five binary chains below. Compare the exact and asymptotic test results.

(i): 0001101; (ii): 10011111; (iii): 0001111;

(iv): 01010; (v): 11011; (vi): 010111.

14.27. For the data in the previous exercise, compute the mue and mle of the transition probabilities. Also compute exact and large sample CIs for them. Plot the evidence functions for these parameters.

EXERCISES

14.28. Persons with arthritis were entered into a clinical trial and randomized to treatment A or B. Once a week for five weeks, their pain level was assessed and recorded as 1, if it was severe, and 0 otherwise. With five patients in each arm, the data collected were:

Treatment A: 11110; 01111; 01110; 00111; 11110

Treatment B: 10001; 10001; 10001; 10010; 01010

Analyze these data using exact and large sample methods. Plot the evidence functions.

14.29. Derive the expression (14.56).

14.30. Give a step by step formal justification of the use expression (14.94) to construct the exact distribution for two state Markov chain clinical trial data.

14.31. Formulate the likelihood for two state Markov chain clinical trial data, and derive large sample methods for analyzing them. Also derive asymptotic CIs for the logit model parameters.

14.32. Extend exact analysis of two state Markov chain data to the case with more than two strata (Hirji and Johnson 1999).

14.33. Implement an efficient method for computing the exact unconditional and conditional distributions for two state single chain, multichain, unstratified and stratified Markov chain data. Also derive the relevant exponent checks.

14.34. Derive the exact distribution for a three state homogeneous first order Markov chain. Extend this to multiple chains. Construct exact and asymptotic methods for analysis of relevant parameters.

14.35. Derive the exact distribution for a two state homogeneous second order Markov chain. Construct exact and asymptotic methods for analysis of relevant parameters. Construct an exact test for second order dependency.

14.36. For the examples given in this chapter and simulated data, investigate the differences between the conventional contingency table based 'exact' (or formally exact) analysis and the correct exact analysis for Markovian data. Also perform such comparisons for simulated data from three state chains.

CHAPTER 15

Reflections On Exactness

15.1 Introduction

This chapter addresses conceptual and controversial issues arising in the theory and practice of exact inference on discrete data. It has but a few formulas and no derivations. The particular topics of interest are:

- The descriptors associated with the term "exact" in statistical analysis, and their appropriateness in the discrete data context.
- Brief explanations of the Bayesian and frequentist framework for statistical inference.
- Elucidation of the relationship between study design and method of data analysis.
- Distinctive views on conditional analysis, external validity, sample size and power.
- The relations between the culture of *p*-values and exact analysis of discrete data.
- Formalistic application and misuse of exact methods for discrete data.
- Promoting a prudent and measured application of exact methods.

The issues covered in these pages have generated much discord in the statistical community. It is not possible to do justice to all those disparate views in this brief space. Unavoidably, the views of the author get a greater airing. Though, at the least, references to other views are provided. To maintain the flow of the argument, some key points are repeated when they arise in another context.

15.2 Inexact Terminology

The term "statistical method" refers not just to a test but to a conceptually unified set of tools of analysis. The label "exact" is often attached to a statistical method or its components. We find exact probability distributions, exact *p*-values, exact confidence intervals and other exact quantities in data analysis reports. What the label means is rarely made explicit, as if it is self evident, if not directly, then from the context. Yet, it does not carry the same implication for all. This becomes apparent when we now and then see it used interchangeably with or in place of terms like "nonparametric," "distribution free," "randomization," "resampling," "permutation" and even "conditional."

These terms refer to different underlying concepts. Their meaning for continuous data, moreover, is not the same as that for discrete data. In the latter context, not all of these terms are good descriptors for an **exact method,** especially in relation to the approach presented in this book. A related point is that arguments presented for exact tests, which are based on some of these concepts, are then at times erroneously taken to apply to the exact method as such.

The term **exact distribution**, with qualifiers unconditional and conditional, was defined in §6.7. The **design based sample space** was noted as the **frame of reference** for the study. Bear this in mind as we discuss the meanings assigned to the term exact method.

- An exact method, like an asymptotic method, can be a conditional, a partly conditional or an unconditional method. **Conditioning** and **exactness** are not inextricably tied to each other.

- The term **distribution free method** derives from the field of continuous data analysis. It refers to the analytic tools applicable to a general class of distributions and whose validity does not depend on a precise distributional specification. For example, it may as readily apply to a normal distribution as to a Weibull distribution (Krauth 1988).

    Of course, no statistical method is "distribution free." For continuous data, a wide spectrum of usable distributions exist. In that context, the implication of the term, namely "virtually distribution free," is generally clear. In the discrete data context, the analysis is, for the most part, based on the multinomial or Poisson distributions, or distributions related to them. This applies to exact and asymptotic methods for discrete data. Exact and asymptotic logistic regression, for example, derive from the same logistic model and the product binomial setup. None is more or less distribution free. In large samples, the latter is perhaps more so!

- The term **nonparametric** has a similar origin and connotation as the term "distribution free" (Pratt and Gibbons 1981). Its limitation for discrete data is similar. An exact confidence interval for the odds ratio is, parametrically and conceptually speaking, not that distinct from an asymptotic confidence interval. Both derive from the same parametric model formulation.

- The body of statistical methods known as **resampling methods** sample observations from empirically observed distribution functions to draw inference on data. They include **bootstrap methods,** which utilize sampling with replacement (Efron and Tibshirani 1993). **Permutation methods** utilize rearrangements of the observed data, or essentially sampling without replacement, to draw the inference. Some conditional exact discrete data tests are derived as permutational tests. The related exact confidence intervals or nonnull exact distributions have not been derived in a similar way.

- The term **randomization methods** covers both the two previously noted types of methods and more. Conceptually, it pertains as well to methods that invoke features of the study design, like random sampling or random allocation, for their justification.

    The term randomization in a computational sense refers to the practical technique of Monte–Carlo simulation. As noted in Chapter 11, this approximates a systematic or complete enumeration of the exact distribution. Complete randomization is then equivalent to drawing an infinite number of samples based on the study design, or aspects thereof (Manly 1997).

In our view, the terms "distribution free" and "nonparametric" have limited utility for discrete data in general, and do not accurately describe the exact methods for discrete data. The labels "resampling," "permutation" and "randomization" may be used when the meaning is clear. But they are not synonyms of the term "exact." In the discrete data setting, we define an exact method as follows:

An Exact Method

> An exact unconditional or conditional method is
> a method that respectively employs the exact
> unconditional or conditional distribution generated
> from the study design and postulated model.
> Particularly, it does not resort to a large sample
> approximation of any sort.

To put it simply, an exact method is a **nonasymptotic method**. Further, it retains its exactitude only if the model and assumptions upon which it is based are valid for the particular study, and if the study data have not been compromised through poor design and implementation.

To an extent, our views differ from Krauth (1988). In a comprehensive explanation of statistical terms, he classifies all contingency table tests as distribution free tests, defines exact tests as those tests for which one can specify the maximum type I error and regards conditional tests as a "*subset of exact tests*" (pages 1–39). Thereby, a TST mid-*p* test for a 2×2 table, though a conditional test, is not regarded as an exact test. On the other hand, an asymptotic chisquare test is a distribution free test. For Agresti (1996) as well, an exact test is a test for which one can bound the type I error rate (pages 39–45).

Our view is more in harmony with Storer and Kim (1990) and Newman (2001). For the latter, "*the term 'exact' means that the actual probability distribution is being used to perform calculations, as opposed to a normal approximation*" (page 12). As defined in §7.7, we call those exact methods which provide guaranteed coverage or type I error rates as **conventional exact methods**.

Terminological clarity is essential for conceptual clarity. Exactness pertains not just to tests, but more importantly, the whole panoply of tools of data analysis. These include evidence functions and other entities not dealt with in this text. The exact conditional method for a 2×2 table, for example, is based on the odds ratio model, and so is decidedly a parametric method. It is not applicable to a risk difference or a risk ratio parametric formulation. Yet, the Fisher–Irwin exact test for testing the equality of two proportions can, under strict assumptions, apply to the odds ratio as well as the other two formulations.

15.3 Bayesians and Frequentists

The **Bayesian** and the **frequentist** frameworks for statistical inference form the two main philosophies of statistical inference. In this section, we briefly present their basic tenets.

15.3.1 The Bayesian Framework

The crucial difference between Bayesians and frequentists lies in how they perceive probabilities. To the former, a probability expresses one's degree of belief about a proposition or state of nature. Consider π, the response rate to a particular therapy. The Bayesian world view posits π as an unknown random variable which lies between 0 and 1. Our knowledge of this parameter is encapsulated by its probability distribution, known as the **prior probability distribution** of π. If this is a uniform distribution, we call it a **vague or uninformative prior**. That is, we believe that the values of π values are equally spread out between 0 and 1. A scientific study of the therapy enables us to improve our knowledge of π in that the data from the study are used to transform the prior probability into a more informative **posterior probability**. This is done using the method shown below.

Suppose $f(\pi) = P[\pi]$ denotes the prior distribution of π. Let the treatment be tested on n patients, and the observed binary data given by $t = (t_1, \cdots, t_n)$. Then we update our prior beliefs using the **Bayes Rule**:

$$P[\pi \mid t] = \frac{P[t \mid \pi]f(\pi)}{P[t]} \qquad (15.1)$$

where $P[t]$ is the marginal distribution of the observations given by

$$P[t] = \int P[t \mid \pi] f(\pi) d\pi \qquad (15.2)$$

Suppose we need to test the hypothesis $\pi \geq \pi_0$ versus $\pi < \pi_0$. From the posterior distribution, we compute $P[\pi \geq \pi_0 \mid t]$. This is a Bayesian version of the *p*-value for the former hypothesis. A 95% **credible interval** for π is obtained by finding values π_l and π_u such that $P[\pi_l \leq \pi \leq \pi_u \mid t] = 0.95$. Another measure of evidence is the **Bayes Factor**, defined as the ratio of the posterior odds of the hypothesis to the prior odds of the hypothesis.

How to select the prior is a major concern; different priors can give substantially different results. A potential for injecting subjectivity into scientific inference is thus built into the framework. To overcome that hurdle, the **empirical Bayes** approach uses prior distributions derived from the data. This is a key transition in terms of the underlying philosophy, but not in terms of the technical aspects of the method, or conceptual inferential entities.

According to Stern (1998), adopting the Bayesian framework has the tangible benefits:

- It is a logically coherent approach.
- It employs the language of probability in an elegant manner to portray the state of scientific knowledge.
- It provides a conceptually simple and natural approach to data analysis that is also valid in finite samples.
- It has a built in mechanism to update scientific knowledge as more data are accumulated.
- It readily copes with complexities in study design and implementation, such as missing data.

Bayesian methods are particularly suited for unit level application, like those found in clinical decision making. Further, they have appeal in a situation where a large volume of information has been collected from a few individual units. This occurs, for example, in the analysis of gene microarray data or flow cytometry data (Efron 2005; Gill, Sabin and Schmid 2005).

Bayesian methods are based on complex assumptions. Even for simple problems, they are algebraically intricate and computationally intensive. Until recently, their appeal was thereby limited. Efficient computation methods such as Markov Chain Monte–Carlo algorithms have served to enhance their feasibility. Consequently, their use has been expanding at a fast pace since the early 1990s.

Woodworth (2004) gives a good, basic introduction to Bayesian methods for health and biomedical studies. Bayesian analyses of a 2×2 table are described in Altham (1969), Nurminen and Mutanen (1987) and Hashemi, Nandram and Goldberg (1997); the last paper cites other relevant papers. Interestingly, the first paper connects a Bayesian posterior probability to the Fisher–Irwin one-sided exact *p*-value. The second paper shows that under a general beta prior, a Bayesian analysis for the odds ratio gives results close to those from the corresponding mid-*p* analysis. For a concise and clear description of a Bayesian analysis of a 2×2 table, see Little (1989); a more general and simple exposition with practical examples is in Goodman (1999a, 1999b).

To date, the Bayesian method is not frequently applied to research data, though, as noted above, a change is underway. In fields like genetics, it has made substantial inroads. The area of meta-analysis of clinical trials has also seen an increasing number of papers advocating and applying this approach. Efron (2005) postulates an eventual integration of the two main branches of statistical inference, with the empirical Bayes approach serving as the essential bridge. Goodman (1999b, 2001), though sharply critical of *p*-values, also views both frameworks as necessary components of statistical inference. For details on the techniques, see the literature cited in §15.9.

15.3.2 The Frequentist Framework

In the **frequentist framework**, probabilities represent the proportions of events in a possible experiment with a very large number of trials. To Jerzy Neyman, one of the pioneers of this framework, a probability denoted *"an idealization of long run frequency in a long sequence of repetitions under constant conditions"* (as interpreted by Lehmann (1993)).

The parameters of a scientific model are fixed but unknown states of nature, or an objective reality. A scientific study is set up to add to our knowledge of the model. This includes questioning its form and estimating its parameters. The probability distribution applied to the study is formed by a conjunction of its design (which gives the sampling frame of this theoretically repeatable experiment), and the model taken to relate its variables. To develop our knowledge, we construct various measures of evidence, apply them to the data, and interpret the results in the long run frequentist framework.

The analytic tools given in this book are frequentist tools of evidence. *p*-values, estimates, confidence intervals and the evidence function capture the data driven uncertainty about the model parameters. As they are random entities, the conclusions we draw from them are also repeated sampling statements about the parameters.

There are two main divisions in the frequentist framework. The **Tail Area Based School** is illustrated by the general approach of this book. Historically, this school has had two subdivisions. The **Neyman–Pearson School** utilizes hypothesis tests, confidence intervals and power as its principal analytic tools. The **Fisherian School** stresses the null *p*-value (calling it a significance level) and questions the idea of power. Also, it promotes conditional tools while the former permits the use of unconditional and/or conditional tools. We discuss this matter in detail later, but note for now that there are valid reasons for setting that historic division aside, and promote a unified tail area based school of inference.

The **Likelihood Based School**, on the other hand, posits the likelihood, that is, the probability distribution seen as a function of the parameters, as the main measure of evidence. Consider two possible values, π_1 and π_2, for a Bernoulli parameter, π. This school employs the **likelihood ratio** as an evidentiary gauge to decide between one or the other value. Given the study data t, and the parameter π, we consider the ratio

$$\rho = \frac{\mathrm{P}[t \mid \pi_1]}{\mathrm{P}[t \mid \pi_2]} \qquad (15.3)$$

The higher this ratio, the greater the preference for π_1 over π_2, with various cut off levels used to quantify the level of evidence.

Blume (2002) provides a readable tutorial on the likelihood based method. In his view, it avoids the shortfalls of both the Bayesian and the classic Neyman–Pearson approaches, and is a logically sound method. Many prominent statisticians have championed it. Yet, it is hardly found in practical applications.

Ideas from the likelihood school, though, permeate other approaches to inference. In the tail area based school, the likelihood is one of the several devices used for constructing measures of evidence. For example, likelihood ratio tests are often applied in the classic analysis of log-linear models for contingency tables. The likelihood ratio also plays a key role in Bayesian analysis.

The frequentist tail area based framework is the framework generally adopted by practicing social, biomedical and other scientists today.

****

We illustrate a key difference between the approaches. Take the case of a binary response trial. In Chapter 2, we noted two types of trial designs. In the first, we fix the total number of subjects, giving us the binomial model, and in another, we fix the number of successes, resulting in the negative binomial model. Without going into the details, we then note:

Observation 15.1:

- Under any prior, the posterior probability of π is the same for both these designs.
- The likelihood and the likelihood ratio are also the same for the two designs.
- The large sample tail area based score test for π is also the same under both designs.
- The exact tail area based score test or confidence intervals for the two designs, however, are not the same.

The relationship between the study design and method of analyzing data is a primary issue in statistics. Results like these indicate that the relative import given to the study design in determination of the data analytic method is not the same for the different schools of inference. Some statisticians hold that methods of data analysis should not be constrained by the study design. Data are to be analyzed as they are without concern for how they arose. In part, this is because what is often said to be fixed by design is, in real studies, hardly fixed.

Consider, for example, the question of the sample size. This is generally a key part of the study design. But in studies of human subjects, the actual sample size is rarely the planned one. A study aiming for 40 patients may end up with 36 or 43 patients. As the data analysis invariably fixes or conditions on the sample size attained, is not the notion of a size fixed by design a fictional entity? A similar logic is also applied to the other aspects of study design. Is this not a reason to always condition on the data at hand and not worry about how they were obtained?

The Bayesian approach is a **fully conditional** approach. To update the prior, it conditions on the observed data. It is said that under this approach, how the data were collected plays little or no part in the inference drawn. (Though this is not really the case; see below.) In relation to sample size, one conditions on the size obtained; by what process that number was obtained, it is then claimed, is not particularly relevant.

Under the frequentist framework, one may argue that scientific studies are not like haphazard data collection. They have to be planned with care and conducted in a meticulous way. Yet, no study is perfect; some deviation from the plan always obtains, be it in measuring technique, or sample size. A major deviation indicates serious deficiencies at some stage in the process, and may generate data of such poor quality as to be not worth analyzing. While the methods for missing data have their legitimate place, it is not the task of statistics as a discipline to cover up the deficiencies of real research. Its role is to facilitate design, conduct, analysis and interpretation under the highest possible standards. Determining the required sample size and helping carry out the study in such way as to attain that size are a key part of that exercise. If how the data are collected, or if the planned size has no bearing on data analysis, why plan the study in the first place?

We hold the view that **no philosophy of statistical inference can entertain a total disconnect between design and analysis**. Consider two binary outcome clinical trials. In the first, three distinct groups of subjects ingest three doses of a drug; and in the second, all the subjects take the three doses at three time points. In both, the data may be given as a 2×3 table. In any data analysis, be it Bayesian or frequentist, a likelihood or p-value based, the former study is analyzed with methods and distributions appropriate for independent samples, and the latter with methods that permit possible dependency between responses over time. Or, to give a trivial case, when we select fixed numbers of cases and controls for a disease, we preclude the estimation of its prevalence.

DESIGN AND ANALYSIS

For more complex situations, and when we examine the issues at various levels of conditioning, the extent of invariance of the analysis to the study design for the Bayesian, likelihood and the tail area frequentist large sample approaches noted above does not necessarily hold.

Observation 15.2:

- The Bayesian analysis of the odds ratio in a two binomials setup is not necessarily the same as the Bayesian analysis of the cross product ratio in the 2×2 overall total fixed design.
- Some forms of frequentist analyses of the odds ratio in a two binomials setup is the same as the frequentist analysis of the cross product ratio in the 2×2 overall total fixed design.
- The unconditional likelihoods from a two group binary response trials for the binomial and negative binomial sampling schemes are the same. The conditional likelihoods, fixing the other margin, are distinct.
- The unconditional likelihood for a 2×2 transition matrix is identical to the likelihood for a regular 2×2 table of counts. The conditional likelihoods fixing both margins are, however, distinct.

Is the unconditional likelihood the better or more fundamental measure as opposed to the conditional likelihood? The issue of conditioning, thereby, also arises in the likelihood based school as well as the Bayesian framework.

How the data were collected does matter. To what extent it does so may differ under different approaches; but the link cannot be severed altogether. At times, the design calls for a certain type of analysis, and at times, it excludes some forms of analysis. Overall, the relation of the design to analysis is a complex issue that has to investigated along a number of avenues. Ultimately, the position one takes on this is a choice between the main philosophies of statistical inference.

15.4 Design and Analysis

We continue our discussion of the linkage between design and analysis in this and the following sections. But from now on, we mostly consider the differences within the frequentist framework. Even then the arena is vast; so our focus is selective. We look at four topics: (i) the validity of conditional analysis, (ii) random allocation as a basis for inference, (iii) the importance of random sampling and external validity and (iv) consideration of power and sample size. As elsewhere, we mainly attend to discrete data problems. Discussions on conditioning have too often occurred in the narrow context of a 2×2 table, and mainly with reference to exact tests. In that respect, we adopt a broader vision.

Among the queries of continued interest to us in this section are:

- To what extent and how do the study design and the assumed model impact data analysis?
- Are the tools of analysis predicated on a segment of the possible outcomes of the study valid measures of evidence? That is, is conditional analysis a justified mode of inference?

Confining ourselves to the tail area based frequentist framework, we conduct the initial part of our discussion on two separate but related planes: the pragmatic one and the conceptual one.

15.4.1 The Pragmatic Perspective

The findings of the previous chapters lead to a pragmatic approach on the connection between study design and data analysis method. These were:

Observation 15.3:

- For a 2×2 table, the mle of the cross product ratio as well as the null LR and score statistics are the same under the nothing fixed, total fixed and one margin fixed designs. The fully conditional exact distribution containing the odds ratio type of parameter is also the same irrespective of the starting design (Chapter 5).
- For some models for two 2×2 tables, the interaction term is analyzed in the same way at various levels of conditioning (Chapter 6).
- For an independent data $2 \times K$ table considered in Chapter 10:

    (i) In the unordered case, the null score and LR statistics for assessing association between the row and column variables are invariant under the column margin fixed, row margin fixed, total fixed and nothing fixed designs. Their large sample and exact conditional distributions are the same.

    (ii) Asymptotic and exact conditional analyses of trend in the ordered data models we examined for such designs are, to an extent, equivalent, in that they give identical estimates, confidence intervals and p-values.

Such results extend to many general contingency table models (Bishop, Fienberg and Holland 1975). Not only do they simplify exact and large sample analysis of discrete data, but also tell us that, in a real sense, the analysis method is design invariant. Exact or large sample analysis for discrete data can then proceed without concern, to a degree, for how the data came to be because **the same analytic method often derives from, and produces the same results in a broad class of designs.**

Practical studies have complex designs. Their models often have many parameters. As such, they may not be analyzed, or even be analyzable, in terms of the design based unconditional likelihood, or the exact probability. A degree of conditioning is almost always implicit in any analysis. Results of the type noted above give us a wide latitude as to consider what is "fixed by design," and analyze the data at a conditional level that has fewer parameters. Provided we look at some specific parameters, and often they are precisely the ones usually of interest, a conditional analysis regularly gives a similar, if not the same, result as an unconditional analysis.

It was in this spirit, and for a general class of contingency table data models, that Piegorsch (1997) commented:

```
                  Identical Inference

   [T]he identical-inference feature [of these models]
   .... provides a luxury of choice among an integrated
   set of statistical methods ....., unencumbered by
   subtle specifics such as which or whether certain
   marginal totals are fixed or not. Piegorsch (1997),
   page 416.
```

DESIGN AND ANALYSIS

A similar rationale is also given for the widespread application of binary response **logistic regression models.** The explanatory factors in such models are deemed fixed even when they are random entities. The likelihood is derived from a product of binomial probabilities even while the total number of successes for each binomial segment cannot be regarded as fixed. Upon exploring a diversity of models for such data, Cox and Snell (1989) state:

Binary Response Data

> [I]n studying the relation between a binary response
> and explanatory variables it is unnecessary and
> potentially confusing to model anything other than
> the conditional distribution of the response for fixed
> values of the explanatory variables. Cox and Snell
> (1989), page 147.

Practically useful as such observations are, overextending their import is fraught with danger. Exceptions exist even in the cases given.

Observation 15.4:

- While the mle for the cross product ratio based parameters in a 2×2 table does not vary as we successively condition from the nothing fixed to the one margin fixed case, it changes upon conditioning on the last margin.
- The score statistic for testing the odds ratio when we fix both the margins is not the same as that from the one margin fixed design and the score based confidence intervals are distinct.
- Similar results hold for the analysis of the interaction in two 2×2 tables, and for ordered and unordered $2 \times K$ tables.

Further we note that

- Exact unconditional analysis is much more dependent on the design.
- The noted invariance of analysis method to design holds only for a certain class of models. Risk difference and relative risk models, thus, do not share that property. For ordered outcomes $2 \times K$ tables, it holds for models consistent with the adjacent categories model but not, say, for the proportional odds model.

To use a common conditional method for a multiplicity of designs is at times a valid strategy, and at times, it is not. This holds for large sample and for exact methods. Each situation requires a careful scrutiny.

15.4.2 Conceptual Perspectives

At a conceptual level, and among the frequentist statisticians, there are three principal schools of thought on the role of conditioning in data analysis:

The **Firmly Unconditional School (FUS)** declares that data are to be analyzed in consonance with how they were obtained. Quantities not fixed by design ought not be fixed in the analysis. The validity of the conditional approach is doubted as a matter of principle.

Theoretical and empirical results demonstrating that conditional analysis entails a loss of accuracy and precision are invoked to support this view. Exact conditional tests are at times very conservative, and reduce power. When conditioning improves inference, it is minor in nature, or not meaningful for practice. Granted that unconditional tests are biased, the degree of bias is minimal and the gain in power, much larger. Recourse to a conditional test for a 2×2 table, or in general, is advisable only when the relevant margins are actually fixed by the design (Barnard 1945a, 1945b, 1947, 1949; Berkson 1978; Suissa and Shuster 1984, 1985; Rice 1988).

The **Firmly Conditional School (FCS)** postulates that the conditional approach is the only valid form of inference. Conditioning identifies the "relevant subset" of outcomes, the only ones worthy of consideration. A formal justification of this view argues that (i) sufficient statistics enable control of nuisance parameters; (ii) the principle of ancillarity identifies the quantities of no import to the question at hand; and (iii) fixing these quantities in the analysis involves little or no loss of information regarding the parameter(s) of interest.

Notions like "fixed by design" or "unconditional evaluation" are thereby deemed vague, and not reflective of real scientific practice. Some also query the very idea of generalizing study findings to a broad population (see below).

The two group binary response randomized clinical trial is often invoked to support this view. Under treatment equivalence, the total number of responders is said to be fixed. The conditional probability of the number of responses in a given treatment is then derived by permutation based arguments, and is used to compute an exact *p*-value. The fixed margin is called ancillary with respect to the odds ratio, the parameter of interest (Yates 1984; Barnard 1979, 1989, 1990; Camilli 1990; Upton 1992; Edgington 1995).

The **Frame of Reference School (FRS)**, like the FUS school, posits the set of possible outcomes given by the idealized study design as the primary contextual entity. Unlike the latter, however, it deems both the conditional and unconditional methods as valid methods in principle. Both have to be assessed in a uniform manner over the design space. Under this school, conditioning is an analytic technique, and not a philosophy. It is used to handle nuisance parameters in the models and designs for which that is possible, and where empirical or theoretical studies show that it has a good performance over the frame of reference (Haber 1990; D'Agostino, Chase and Belanger 1990; Hirji, Tan and Elashoff 1991; Lehmann 1993).

****

Consider various critical aspects of these schools of thought, starting with the first, the FUS approach to data analysis:

- In principle (but rarely in practice), it excludes the use of regression models in which quantities not fixed by design are treated as fixed.
- Thereby, if advocated strictly, it may end up with unwieldy models and an extensive number of parameters. This poses a challenge in terms of computation and interpretation, and may reduce the precision of the estimates.
- It ignores the known relative invariance of the study design to the level of conditioning for the cases we saw earlier.
- It ignores theoretical results showing that conditional *p*-values, tests, estimates and confidence intervals at times have optimal large sample properties, and the empirical studies showing, for some types of data, an equivalent if not a superior small sample performance of the conditional methods, even when assessed over the full unconditional sample space.
- A rigid exclusion of conditional analysis is hard to justify in terms of basic principles.
- Currently exact unconditional analysis is feasible for a very limited type of discrete data models and designs.

DESIGN AND ANALYSIS 459

- For a number of problems, when unconditional analysis (exact or asymptotic) is shown to be better than any conditional alternative, the gain is minimal while the added computational burden may be extensive.
- In complex settings, optimization over a multiparametric space of nuisance parameters may make exact unconditional analysis more conservative than its conditional counterpart.
- Exact unconditional confidence intervals for complex problems are difficult to compute.

Yet, to jump from this type of critique of the firmly unconditional school to adopting the firmly conditional perspective is problematic for several reasons.

- To assume, as is argued for a randomized clinical trial, that the number of responders under treatment equivalence is a deterministic quantity, and random allocation provides the sole source of stochastic variability is shaky on scientific grounds. Biological or social processes are characterized by an intrinsic element of randomness. Other factors (noted in Chapter 6) also inject randomness into clinical studies (more on this later).
- By taking the sample as the universe, the FCS also sidesteps the fundamental issue of generalization of the study findings to a broader population (more on this later).
- Under the permutation justification, the usual null conditional *p*-value or test becomes the main if not the sole index of evidence (more later). Note that the noncentral hypergeometric distribution has not been derived using permutation arguments (Copas 1973). This approach then faces a problem when a null hypothesis other than the usual is tested, as is done, for example, in testing for equivalence or inferiority.
- Under the strictly conditional viewpoint, the unconditional mle of the odds ratio is not a permissible estimate. One should only use the conditional mle. A similar case holds for confidence intervals. Such a point, though, is rarely (or not firmly) advanced by the adherents of the conditional view.
- The conditional school decries the use of the two binomials setup for computing power and actual type I error. Yet, it seems to embrace this setup by using the noncentral hypergeometric distribution which is derived from the two binomials setup. R. A. Fisher also used the two binomials setup to compute values of the odds ratio not contradicted by the data at 0.01 and 0.05 levels of significance (Fisher 1935b).
- Explicitly or otherwise, the strict fully conditional school limits itself to the logistic or log-linear types of models. Models like the risk difference and risk ratio models, or the rate difference and rate ratio models do not permit conditional analyses. Analytic convenience rather than scientific considerations then seems to drive model choice.
- The margins of a 2×2 table are **not ancillary** for the odds ratio; they are **approximately ancillary** (Chapter 5). On noting this, Little (1989) aptly portrays the arguments put forward for conditioning on the margins as a *"slippery slope"* towards a fully conditional Bayesian analysis.
- The idea of ancillary statistic was developed by R. A. Fisher. But his usage of the term was not precise or consistent. See, for example, his effusive response to the comments of J. Irwin in the discussion of his seminal paper, Fisher (1935). Note that Ludbrook and Dudley (1998), who ascribe to Fisher's main ideas, characterize his writings as *"enigmatic."*
- This lack of clarity unfortunately is also manifested in the current statistical literature. Hence, upon reviewing justifications given for exact conditional analysis, an author who does not necessarily reject conditional inference was led to declare that:

Vague Exactness

> .. arguments in support of conditional tests .... [a]s presented, ... are simply not logically compelling, perhaps because they appeal to rather vague and abstract concepts of ancillarity, conditionality and marginal information. Greenland (1991).

****

This type of critique of the FUS and FCS approaches inclines us to favor the Frame of Reference School in which conditioning is a useful device but not an inviolable requirement. We recommend conditional analysis when it performs well over the design based frame of reference. Forms of such analyses are to be compared among themselves, and with unconditional analyses in terms of unconditional ASL, exact power, coverage, width of confidence intervals, the distribution of *p*-values, features of evidence and power functions, and so on. Such criteria assist us in selecting the methods from among those advocated by the FCS and FUS schools.

At a theoretical level, Helland (1995) adopted such a frame of reference approach, and Lehmann (1993) gave a detailed rationale for an analytic synthesis constructed along these lines. Accordingly, he says that

A Unified Approach

> ... *p* values, fixed-level significance statements, conditioning and power considerations can be combined into a unified approach. When long-term power and conditioning are in conflict, specification of the appropriate frame of reference takes priority, because it determines the meaning of probability statements. Lehmann (1993).

A broad, unified approach (which may accumulate a large collection of specific methods) is at times critiqued for transforming the discipline of statistics into an eclectic collection of ad hoc recipes. Like any prime scientific discipline, statistics needs a coherent foundation. Thus, in his contribution to a debate on the 2×2 table, a fine statistician was led to conclude that:

Statistical Principles

> [P]rinciples of inference matter. The excellent trend in statistics towards applications should not replace the need for us to think carefully about the foundations of our methods. Little (1989).

This is a crucial observation. Modern statistical practice has techniques derived from different approaches to inference (Efron 1986). Their usage is inconsistent in a way; we use one method one day, and the next, quite the other, without concern or thought that they may reflect disparate basic principles.

Principles do matter. But what are these principles? That is an issue as well. Take the case of what goes under the name of the conditionality principle. Loosely speaking, it states that under certain conditions the evidence from a series of possible experiments is equivalent to that from a specific observed experiment (Birnbaum 1962). To some, this principle and its sequels provide the philosophical foundation for conditional analysis. Any analysis ought to always condition on the observed data, and use the likelihood as the measure of evidence. (Exploring the full ramification of this issue is beyond the scope of present discourse.)

However, the conditionality principle has been questioned by others. For example, Helland (1995) notes that the term evidence is used rather loosely in this context. Further, he demonstrates with counterexamples that *"the conditionality principle should not be taken to be of universal validity."*

The point we are making is that what to some is a basic principle is to others a matter of technique. In not giving a first priority to conditioning, Lehmann (1993) is then not discarding a basic principle but upholding a principled frequentist framework. The study design thereby provides the frame for repeated sampling and the meaning to the whole exercise, and takes priority. To allow a conditional analysis but then insist that it be assessed over this frame of reference then concurs with this principle.

In this text, we have focused on models for which conditional analysis is feasible. That is not because we consider it to be the sole mode of inference. Nor do we regard the models we study as superior to other models. We do note, however, that they (i) display practically useful flexibility under a variety designs, (ii) often allow partial and (almost) fully conditional analyses; (iii) readily permit adjustment for covariates; (iv) allow small sample analysis for a wide class of designs; and (v) permit valid analysis of case-control data. In line with Cox and Snell (1989), we employ binary response models and conditioning to analyze the data, but then we also note the limitations of such an approach.

We do not discount, in principle or practice, exact or large sample unconditional analysis, or models for which conditional analysis is not feasible. In writing a specialized work, we have focused on our area of expertise. The explication of the other models and methods is left to other, more qualified authors.

To conclude this section, we note that the ever present challenge for a data analyst is to distinguish when aspects of the design can be safely ignored and when they are to be carefully factored into the analysis. For some designs, a greater flexibility obtains, and for other designs, that is not so. For some class of designs, the large sample methods are less dependent on the design, and for others, exact methods are not as much influenced by it. A general claim that we can ignore the design, or how the data were collected, in our view, has the potential to mislead.

Design and Analysis

```
How the study design bears on the data analytic
technique is not reducible to some simple
pronouncements. A careful elucidation and evaluation
are needed for each class of designs, study model and
form of data analysis.
```

Primarily, the models used, and methods applied for data analysis have to depend on the scientific issues at hand. Mathematical convenience or tradition are not valid substitutes. Computational infeasibility at times drives us towards a compromise. When that is unavoidable, it should be clearly noted, and the implications explored. Too often convenience and tradition rule the day. Too often the communication between the analyst and subject matter expert is poor. And too often, the latter is simply expected to generate publishable p-values. In that instance, the analytic methods and the product do not reflect good science. If the label "exact" is then applied, it just gives the analysis more credence than it deserves, and deters a critical examination of the measures, models and methods used in the study.

To explore such and other issues, our discussion of conditional analysis continues below.

15.5 Status Quo Exactness

Cross sectional or observational studies in the social sciences often use or strive to use random samples from a defined population to select the study subjects. This is done, for example, in sociologic or demographic studies, surveys of opinions and marketing studies. Public health investigations also use random sampling. A large body of work on the issues associated with the design, implementation and analysis of such studies exists.

The utility of random sampling, particularly in the biomedical field and clinical trials, has been the subject of some disagreement among statisticians. In this section, we compare three perspectives on random sampling in that specific context.

In part, these perspectives relate to two distinctive visualizations of the frame of reference. These views go under the names of the **population model** and the **sample based model.** To an extent, these differences reflect the division in the frequentist camp noted in §15.3.2 and the related difference between the FCS and the other schools described in §15.4.2.

The population model derives from the classic frequentist vision. To it, the data in a study are a realization from a large external population. The latter is the basis for repeated (long run) sampling; the probability model used in the analysis and its results ultimately pertain to this population. Generalizability from the sample to the population is a crucial concern. Selecting random samples is a good device to attain this goal. **Random selection,** implemented in many different ways, then provides **external validity** to the study.

In contrast, under the **sample based model,** random selection is of little or no import. At best, it is unrealistic, and at worst, irrelevant for scientific inference. Internal features of the study, in this view, can form the complete basis for making valid probability statements. The relevant frame of reference is defined by possible rearrangements of the observed data under the given experimental conditions.

This reasoning is mostly applied to studies which allocate study units in a random fashion. The sample based model is thus also called the **randomization model.** Random allocation then provides the basis for analyzing the repeated or long run behavior of the arrangement actually observed. The Fisher–Irwin exact test and the permutation tests for continuous data are, for example, derived under such sample based permutation arguments. The formal validity of such tools of evidence is then restricted to the observed sample. Extending their conclusions further is at best an informal, and at worst, a logically questionable venture.

15.5.1 Paper I

We start our discussion with a survey paper that explored how aspects of the study design related to the analysis method used. Though this paper looked at analytic methods for continuous data, its arguments and conclusions apply equally to discrete data.

Ludbrook and Dudley (1998) surveyed five prime journals in surgery, physiology and pharmacology, and extracted information from 252 prospective studies with comparison groups. These were laboratory based studies (with animals, cells etc.) and clinical investigations with human subjects. The design and analytic method in each study was noted. The former was categorized in terms of whether or not random sampling from a defined population had been used, and whether or not allocation to comparison groups was by randomization. The analysis method was classified as a classical test based (t or F test), or a permutation test based. (The other details they looked at are not that relevant here).

Table 15.1 *Study Type and Design*

Design Feature	Study Type		Total
	Laboratory	Clinical	
Random Sampling	11	0	11
Random Allocation	216	25	241
Total	227	25	252

Constructed from Ludbrook and Dudley (1998).
Used with permission from *The American Statistician*.
©1998 by the American Statistical Association. All rights reserved.

Of the 252 studies examined, about 10% were clinical studies; the rest were done in the laboratory. 11 of the studies (4%) selected study units by random sampling; the rest (96%) employed random allocation but not random sampling. No study combined random sampling with random allocation. One remarkable finding was that there was not a single comparative study which had not used some type of random mechanism.

The sample sizes in the laboratory and clinical studies were generally small. The group size in the 216 laboratory studies with random allocation ranged from 2 to 77 (median = 6). The group size in the 25 clinical studies with random allocation ranged from 4 to 345 (median = 25).

More details on the studies are provided in Table 15.1 and Table 15.2. Usage of random sampling was totally absent in the 25 clinical studies. In the rest, it was clearly on the low side. Random allocation was, however, conspicuously prominent in both types of investigations. Permutation tests were deployed in 27% of the studies with random selection, and in 15% of those with random allocation.

Table 15.2 *Analysis Type and Design*

Design Feature	Analysis Type			Total
	None	Classical	Permutation	
Random Sampling	0	8	3	11
Random Allocation	16	189	36	241
Total	16	197	39	252

Constructed from Ludbrook and Dudley (1998).
Used with permission from *The American Statistician*.
©1998 by The American Statistical Association. All rights reserved.

This paper underscores the well known general observation that in a vast majority of clinical trials, study subjects are not selected at random from a delineated target population. At times, effort is made to make the study sample representative in a systematic manner (locations, timing and so on). Often, it is implicitly assumed that representativeness or demarcation of the target population is enhanced or assured by using clear and appropriate eligibility criteria for the study and by describing the sample characteristics.

From their pioneering investigation, Ludbrook and Dudley (1998) draw a commonly endorsed conclusion:

> Follow the Norm
>
> Because randomization rather than random sampling is
> the norm in biomedical research and because group sizes
> are usually small, exact permutation or randomization
> tests for the differences in location should be
> preferred to t or F tests. Ludbrook and Dudley (1998).

A similar view is advanced for the use of exact tests in trials with binary outcomes. Randomization then not only serves to reduce selection bias, promotes comparability of the treatment groups on average but also provides a basis for the use of an exact test. See, thus, Mehta, Patel and Wei (1988). Small samples combined with random allocation form the empirical and conceptual underpinning for the Fisher–Irwin exact test in a binary outcome clinical trial, in particular. Selecting the study units in a random manner is thereby transformed into a secondary matter. This view typifies the sample based randomization model.

15.5.2 Paper II

Random sampling is considered from more of a philosophical angle by **Rothman and Poole (1996)**. These authors begin with a lucid synopsis of the evolution and ramifications of the concept of causal or scientific inference. With a focus on cancer epidemiology, they argue that scientific inference entails careful elucidation of causal models, testing them on the data, estimating the parameters, and so on. A deductive and probabilistic logic of falsifiability is a central facet of the growth of knowledge. Models that cannot be falsified are not scientific models. In particular, the traditional view of science as an inductive process, namely, that of going from the particular to the general, is suspect.

Even though science involves probabilities, they stress that statistical inference is distinct from scientific inference. The former pertains to generalizability and representativeness. But the latter is not reducible to mechanical inference from the sample to the population. True science is not like taking a poll and making broad projections. Rather, it is a creative intellectual endeavor that involves *"the integration of scientific findings from a particular study into the larger fabric of scientific knowledge."* (Rothman and Poole 1996, page 9). It is thereby concluded that:

> Mindless Statistical Inference?
>
> Scientific generalizations are generated by the minds
> of the scientists who interpret the data. Statistical
> generalization from a sample to the population from
> which the sample was drawn does not merit the dignity
> of being called scientific inference. Rothman and Poole
> (1996).

15.5.3 Paper III

Finally, we consider a rather different view on generalizability. **Rothwell (2005a)** is the first in a series of papers dealing with clinical trials, and systematic reviews and clinical practice guidelines based on them. It contains the most comprehensive and in depth discussion of the external validity of such investigations to date. The following points are extensively documented:

- Research on external validity has been and is miniscule in comparison to that on internal validity;
- Guidelines on performance, reporting and assessment of clinical trials avoid or marginalize external validity;
- The pharmaceutical industry, governmental regulatory and other bodies, major funding agencies and ethics committees do not adequately, if at all, deal with representativeness of subjects or clinical practice;
- Even medical journals have neglected covering the issue.

Because of this, clinical and other researchers at the moment lack clear directions on how to enhance, protect and judge the external validity of their studies.

Rothwell (2005a) gives a detailed breakdown of the circumstances and practices that limit the external validity of clinical studies. These include trial setting, manner of selection of patients, the features of randomized patients, trial management, outcome measures and the adverse effects of treatment. He critiques the proposition that routinely collected data are more generalizable than clinical trial data, or that the characteristics of study subjects necessarily indicate well the type of subjects to whom its results will apply. Pioneers in the field like A. L. Cochrane and A. B. Hill, it is noted, were aware of the limitations of generalizing from randomized trials.

That historic and continued neglect of external validity in clinical trials (and other studies) has had untoward consequences for medical practice.

> External Validity
>
> Lack of consideration of external validity is the most frequent criticism by clinicians of RCTs [randomized controlled trials], systematic reviews, and guidelines, and is one explanation for the widespread under use in routine practice of treatments that were beneficial in trials and that are recommended in guidelines. Rothwell (2005a) (Reference numbers omitted).

Yet, Rothwell (2005a) does not minimize the centrality of internal validity. Progress in developing and implementing the mechanisms for protecting internal validity has helped make "*[r]andomized clinical trials (RCTs) and systematic reviews ... the most reliable methods of determining the effects of treatment.*" But the applicability of clinical trial results to general clinical practice remains a major hurdle to be surmounted. Towards that end, this paper provides four recommendations to improve external validity. These pertain to research on the topic, integrating it into the regulatory process and the inclusion of external validity in reporting guidelines and publication requirements.

15.5.4 Comments

What are we to make of such disparate views on the import of random sampling and external validity, especially in relation to clinical trials? The basic aim of a clinical trial is to assess the safety and efficacy of the treatments under study. The data in a trial are not to be collected in a casual way. First comes the design. The literature is reviewed, and the goals are clearly defined. The relevant instruments, variables, eligibility, diagnostic and outcome measures, schedule, location, timing, quality control and required sample size are dealt with in a thorough and systematic fashion. These tasks may take months. A key part of the effort is to pose and address the question, 'To what group or population will the findings apply?' and plan accordingly. What kind of subjects are to be studied, and how they are to be obtained deserves a major role in planning, conducting and reporting of the study.

Reproducibility is a primary tenet of scientific endeavor. A study with clear and specific aims, plan, materials, methods and subjects reduces bias and other errors, and allows other investigators to assess, check and critique its results. Haphazard data collection, or grab samples are not science. Internal and external validity as well as rigor and relevance are essential parts of this process.

A trial with a hundred subjects may form the basis to promote a treatment used by hundreds of thousands of patients. A treatment tested on patients with severe disease is recommended for the larger population of those with mild or moderate disease. If it has strong adverse effects, that may do more harm than good. Among the children with ear infections, for example, those under two years of age are, for good reason, considered a separate group. If an antibiotic for acute otitis media is tested on the older children only, its results may not apply to the younger group.

All the prognostic factors for a disease are rarely known. A random sample tends to produce a profile of these factors reflective of that in the population at large. A delineation of that population and a detailed process for drawing representative, or better, random samples are central to enhancing the practical relevance of biomedical and other research studies.

External validity is a key aspect of the frequentist frame of reference approach to inference. R. A. Fisher also strove to formulate a method for drawing valid inference from the sample to the population. The sample based school, which takes Fisher as its authority, virtually denies that possibility. The views of Ludbrook and Dudley (1998), who dealt with hypothesis tests, are typical in that regard. They categorically state that under this perspective, "*confidence intervals cannot be used.*" Consideration of power is also basically not relevant. Lack of random samples, or the fact of small sample sizes appear not as serious problems that need to be resolved, but as facts of life that can be adjusted for by permutations tests. See also Kempthorne (1955).

We do not agree with such a view. We also question the rationale for sample based inference used by the randomization school. Take the case of a randomized trial of two treatments used for high blood pressure. Assume that the treatments have the same efficacy. If the outcome is analyzed on a continuous scale, can we claim that the combined sum of the blood pressure values or the reductions in these values in the two treatments are fixed under the null? But if the outcome is dichotomized in some way and the data given in a 2×2 table, why is that said to be so? What if the dichotomization was done in some other way? Which is the truly fixed margin?

A general permutation test for a continuous outcome conditions on the observed data. Does this mean that these values are in essence fixed entities, not random variables? Its discrete equivalent is to condition on all the cells of the table, not just the margins. It is one thing to posit, for the sake of computing an index of evidence, that the margins are to be fixed, and quite another to claim that they are actually fixed if the null hypothesis holds. The latter is not a scientifically tenable proposition.

In contrast to the attention accorded to them by Ludbrook and Dudley (1998), both K. Rothman and C. Poole have, in their other writings, firmly critiqued hypothesis tests and the obsession with *p*-values. Both have advocated the use of confidence intervals and, also, the confidence interval (evidence) function (Poole 1987a, 1987b, 1987c; Rothman 2002) The editorial Rothman (1986a) had a major influence in advancing the use of confidence intervals in the reports of clinical and other health studies. Yet, on the issue of random sampling, generalizability and external validity, or on the practical implications thereof, their views hardly differ from Ludbrook and Dudley (1998).

What is particularly troubling is the presentation of the two basic pillars of validity in terms that make them appear as mutually exclusive dichotomies. It is as if one has either random sampling or random allocation, either external validity or internal validity, either statistical inference or scientific inference, and either deduction or induction, but not both. Ludbrook and Dudley (1998) mention the latter possibility in a cursory way but do not pursue it. Rothman and Poole (1996), in our view, essentially use a strawman analogy to dismiss statistical inference.

We are inclined towards the vision presented in Rothwell (2005a). This is also in line with the frame of reference school as advocated by Lehmann (1993) and the conclusions of Helland (1995). The latter, derived from a critical examination of the conditionality principle, are summed up as:

The Target Population

> Above all: Every statistical experiment and every
> statistical analysis must be related to a target
> population, either real or conceptual. There are many
> open questions concerning the exact way in which such
> a population point of view may help us in choosing
> the right way to condition in concrete cases. Helland
> (1995).

To scientific practice, such dichotomies are artificial and not useful. A practicing clinician and scientist such as Rothwell (2005a, 2005b) values both internal and external validity, celebrates the major achievements in promoting the former and yet decries the neglect of the latter. In that regard, statistical inference also pertains to scientific inference. Good science needs both these components of validity. Thus, to what Rothman and Poole (1996) say, we add

External Validity and Scientific Inference

> Ignoring the issue of generalization from the sample to
> the population from which it was drawn does not as well
> merit the dignity of being called scientific inference.

Deleterious deficiencies like the lack of representative samples, poor external validity and small sample size need to be remedied and not to be pushed under the rug by statistical or other contrivances that ignore the basic purpose of the study altogether. In particular, the use of exact or nonparametric methods as such, and in the absence of other remedies, is not a magic solution to the problem. The oft cited maxim - Garbage In, Garbage Out - applies here as well.

15.6 Practical Inexactness

This section examines how three oft proscribed statistical practices bear on exact analysis of discrete data. These are (i) an unwarranted use of one-sided tests, (ii) dredging the data for statistical significance and (iii) favoring *p*-values at the expense of estimation and confidence intervals.

We start with one-sided (or one-tailed) tests. For now we do not ask whether or not they should be used. A synopsis of the current consensus on that topic was stated in §2.7. Granting a place for them in analysis, we now ask whether, when used, they are used in a transparent and scientifically sound manner or not.

Table 15.3 *One- and Two-Tailed Exact Tests*

Test Type Specified?	Actual Test Type				
	2-Tail	1-Tail	1&2-Tail	Unclear	Total
Yes	17	4	2	0	23
No	5	12	0	16	33
Total	22	16	2	16	56

Constructed from the text of McKinney at al. (1989).

A landmark study by **McKinney at al. (1989)** provides an insight into how practical exact tests on discrete data relate to the above issues. They surveyed the use of the Fisher–Irwin exact test for a 2×2 table in six influential U.S. and British medical journals from the years 1983 to 1987. Their findings derive from 56 articles in which the use of this test was unequivocal and which also contained the data that could be used for verification.

Overall they found that 33 (59%) of the 56 articles did not clearly state whether a one or two-tailed test was used; 23 (41%) did. There was a significant difference between the journals in terms of clarity: The actual test type in articles that clearly noted the type and those that did not is shown in Table 15.3. An issue not explored was whether the one-sided tests were scientifically warranted. The authors, however, identified cases where they seemed to have been deployed to attain statistical significance. They also noted cases of computational errors.

From Table 15.3, we assess an observed proportion of use of the one-tailed Fisher–Irwin exact test per article that possibly ranges from 32% (18/56) to 61% (34/56). General information on the rate of one-sided testing in those journals, or the other medical literature of that period, does not seem to exist. Nonetheless, it is our impression that the rate of one-sided testing for exact tests they found is higher compared to that then prevailing for other types of tests. We reflect on the possible reasons after stating two other concerns about exact analysis.

****

Overanalysis of the data, or the strenuous practice of what is called a "religion of *p*-values," has a tendency to find superfluous associations. Such results are, not surprisingly, often contradicted by later studies. Repeated cycles of wide initial publicity of an apparently remarkable result from an unsound study or flawed analysis, and its subsequent debunking can hardly serve to enhance respect for science in the public mind.

The 56 articles examined by McKinney at al. (1989) employed the Fisher–Irwin exact test a total of 247 times, with the median usage at 3 times per article. At least one used it 25 times. Andersen (1990) noted a study in which a grand total of 850 pair-wise comparative tests were done, mainly with the Fisher–Irwin exact test and Student's t test (page 230). In the author's experience, and as others have noted, hunting for statistical significance is not uncommon in biomedical and other studies. Though, we do not know if such practices are more or less prevalent with the exact or randomization methods as compared to other methods. But that they occur as well in conjunction with exact tests is not in dispute. And, because in the latter instance, it may be more in the context of small samples, the implications are worse.

Consider now the reporting of confidence intervals, a key aspect of assessing the clinical or practical significance of the findings. This is required by many scientific journals now; some software programs also give exact confidence intervals. Yet, our impression is that in comparison to other methods, exact and randomization methods as used in practice are more biased towards testing and p-values, that is, more biased towards statistical significance than towards clinical or practical significance. No hard and fast comparative data on this issue have, however, been published.

****

Statisticians periodically raise concern about the misuses and abuse of statistical ideas and methods in research, and work with the scientific community to remedy them. (For a basic impression of the scope of the problem, see Chapter 1 of Andersen 1990.) One thing not sufficiently stressed, however, is that their own theories and practice may in part be an inadvertent contributing factor to this unhappy state of affairs. Consider this in relation to exact methods for discrete data.

First, we note once again that while R. A. Fisher computed two-sided significance levels for other tests, for the exact test he reported one-sided levels only. In fact, when performing exact tests, he explicitly said, without giving a sound rationale, that he favored one-sided tests (Upton 1992). As research practice has come to advocate two-sided levels as the default, and even as extensive research on two-sided entities has been done, this curious bias towards one-sided levels for exact tests persists even in some of the influential, review and key papers in the statistical literature on exact methods to this day. Inexplicably, exact types of p-values are shown in a one-sided form, and confidence intervals given as inversions of two one-sided tests, rarely as inversions of a two-sided test. Two-sided tests are at times given as a cursory afterthought, and occasionally dismissed by a few lines of unclear terminology (Barnard 1990; Upton 1992; Agresti 2000, 2001).

There is also a problem with how a one-sided p-value is described in books and software output, even on exact methods. It is often defined as half the two-sided value. Yet, this is not a correct statement, even for a symmetric distribution. Not only does it imply that a one-sided p-value is always less than 0.5 but also leads even critically minded statisticians to make declarations of the following sort:

```
                   A One-Tailed Test

     The advantage of a one-tailed test is that a
     significant outcome is easier to obtain. Andersen
     (1990), page 235.
```

Consider the example of testing $H_0 : \pi \leq 0.5$ versus $H_1 : \pi > 0.5$ under a binomial design with $n = 10$. If 2 successes are observed, the one-sided mid-p-value is 0.9673. On the other hand, the TST mid-p-value for the two-sided hypothesis setup $H_0 : \pi = 0.5$ versus $H_1 : \pi \neq 0.5$ is 0.0654. The former is larger than the latter! A more precise version of the above statement then would be:

A Convenient One-Tailed Test

> The advantage of a conveniently or opportunistically framed one-tailed test is that a significant outcome is easier to obtain.

Without *a priori* null and alternative hypotheses, one-sided *p*-values do not mean much. When the software packages specializing in exact methods define and give the exact one-sided *p*-value as half the two-sided one, or as the smaller of the two tails, and do not give directionality of the hypotheses, it can hardly facilitate the appropriate use of exact or other forms of one-sided *p*-values and tests.

****

To some, the randomization rationale for exact tests and the associated sample mode of inference has, as a matter of principle, no room for confidence intervals or unconditional power computations. A comprehensive work on randomization in clinical trials, mainly based on this approach, however, does cover these entities, though in a minimal fashion (Rosenberger and Lachin 2002).

While the randomization approach is associated with R. A. Fisher, his views on the role of significance tests in research was more nuanced. In an assessment in Rao (2001), his influential work, *Statistical Methods and Scientific Inference,* is summed up as promoting such tests in the following light.

- They are of limited value in general.
- They are more useful when the alternative to the null is unspecified.
- They should not be related to fixed cut points like 0.05.
- They should be combined with other relevant evidence.

These points appear in a sympathetic and laudatory rendition of Fisher's novel and extensive contributions to statistical science. (More critical portraits of Fisher's views also exist.) Yet, even C. R. Rao points out that

Testing the Null

> ... Fisher's emphasis on testing of null hypotheses in his earlier writings has probably misled the statistical practitioners in the interpretation of significance tests in research work ... Rao (2000).

While Fisher popularized fixed levels of significance, he also trenchantly argued against them in his theoretical works. Some of those who claim to carry on Fisher's legacy decry fixed significance levels (Upton 1992). But they pay undue attention to significance testing. And, while the followers of the Neyman–Pearson school are associated with usage of fixed levels, they also promote confidence intervals. As Barnard (1990) notes, p-values are an aid to judgment and not an end unto themselves. Clearly, in the spirit of Lehmann (1993), a valid middle ground exists here.

Some, like Goodman (1999a) and Hubbard and Bayarri (2003a, 2003b), say that the stand of Fisher on the one hand, and of Neyman and Pearson, on the other, cannot be reconciled. In particular, one cannot simultaneously have p's with the α's as measures of evidence. Discussing the latter paper, Berk (2003) and Carlton (2003) agree that p-values are often erroneously taken as type I error rates (see Chapter 2). At the same time, they do not consider the two concepts as irreconcilable. To them, they are complementary entities, related to but not identical with, one another. We are of the same opinion.

While the statisticians continue to argue among themselves, the fact is that the world of research has moved to the middle arena. In this day and age, the major biomedical journals discourage, in a Fisherian spirit, the crude reporting of fixed cut off values ($p < 0.05$ or $p < 0.01$). The actual p-value is desired. And, in the Neyman–Pearson spirit, they require a confidence interval as well. Despite its advantages, the latter indeed utilizes a fixed cut point (Gardner and Altman 1986; Weinberg 2001).

Another reason for positing a common ground relates to the mid-p. As noted in Chapter 2, statisticians coming from a wide variety of perspectives have advocated its use for discrete data analysis. Accordingly, the following conclusion has an even broader implication.

```
                  Towards Mid-p Inference

   Inference based on the mid-P-value seems to be a
   sensible compromise between conservativeness of
   [traditional] exact methods and the uncertain
   adequacy of large-sample methods. Agresti (2001).
```

The over, and even exclusive, reliance on the p-value is a cause for concern. For whatever one's persuasion, that fixed significance levels and p-values have been and are often misused in practice cannot be wished away. In our view, this is a function of the sociology of research than of the measures as such. One journal, *Epidemiology,* had until recently virtually banished the p-value from its pages. In general, the consensus now is to report both p-values and confidence intervals, with the stress on the latter. It is our view then that those who research on, write about and employ exact methods have to firmly incorporate that in their publications. Exact analysis should go hand in hand with full reporting of results, that is, including p-values, estimates, confidence intervals, and power, and also, where indicated, evidence functions. Further, these indices of inference need to be computed in an integrated fashion, as advocated in this text. Thus, the concept of mid-p as well needs to be implemented in relation to the related evidence function and derivative indices of inference.

15.7 Formal Exactness

A **formally exact analysis** is an analysis that takes superficial similarities in data presentation as its implicit basis for applying the exact method. Essentially, it amounts to formal manipulation of numbers that ignores key aspects of the study design. These include repeated measures, time to event outcomes, dependent study units and misspecification of the unit of analysis. That the data in question are sparse is then given as the explicit basis for the exact method.

As an illustration, take the case of a trial of an antibiotic for ear infection in which eligible

Table 15.4 *A Matched Pairs Trial*

Treatment	No CR	CR	Total
Control	4	11	15
Antibiotic	2	13	15
Total	6	24	30

Note: CR = Complete Response.

children were matched, prior to randomization, by the number (one or two) of ears affected. Hypothetical data from this study are shown in Table 15.4. The analysis was done as for a regular 2×2 table. As two of the expected values (computed for data in such a form) are less than 5, the Fisher–Irwin exact test was used to assess statistical significance. An exact confidence interval for the odds ratio was also computed.

Students in elementary classes learn that this is not an advisable form of analysis; the assumption of statistical independence is in question. Whether the analysis is in the form of the Fisher–Irwin test and a 95% exact confidence interval for the odds ratio, or the chisquare test and a 95% confidence interval for the difference of proportions, that basic concern is not mitigated. These are paired data, and should be analyzed as such.

The presentation for such data can also be improved. Table 15.4 may thus refer to any of the three matched pairs trials shown in Table 15.5. The exact confidence intervals for the odds ratio from these are distinct from that obtained in the regular analysis of Table 15.4.

Using an exact procedure instead of an asymptotic one in such a case may actually worsen the problem. By giving an impression of statistical sophistication, it may mask the critical flaw in the analysis. For actual cases of such inappropriate uses of the Fisher–Irwin and chisquare tests, see Andersen (1990), pages 217–219, 230. Below, we give more cases of formal exact analysis. The data, though hypothetical, relate to actual examples gleaned from applied and theoretical papers.

15.6.1 Dependent or Time to Event Data

A company hires six female and six male applicants, all equally qualified, for the same kind of position. At the end of the year, each employee is evaluated for, and granted or denied promotion. Gender and other forms of discrimination litigation present such data, often as several 2×2 tables. For our example, they are given in Table 15.6. As is the case here, such data tend to be sparse. Exact conditional tests of Chapter 8 are then applied to assess if the observed variation between genders can be ascribed to chance factors alone.

It is possible, and perhaps it is generally so, that the same individuals appeared in the pool of persons eligible for promotion more than once. In that instance, the key assumption of independent tables does not hold. The exact analysis in that instance is an example of a formally exact but otherwise flawed analysis.

Assume that the number of years to promotion for the male and female employees respectively were $\{1, 1, 1, 2, 2, 3\}$ and $\{1, 2, 2, 3, 4, 5\}$. A representation of these data in the form of a series of 2×2 is given in Table 15.7. Note that this is an extended form of Table 15.6. Such data are

Table 15.5 *Three Trials for Table 15.4*

Trial I	Antibiotic		
Control	No CR	CR	Total
No CR	2	2	4
CR	0	11	11
Total	2	13	15

Trial II	Antibiotic		
Control	No CR	CR	Total
No CR	1	3	4
CR	1	10	11
Total	2	13	15

Trial III	Antibiotic		
Control	No CR	CR	Total
No CR	0	4	4
CR	2	9	11
Total	2	13	15

Note: CR = Complete Response.

Table 15.6 *Promotions by Year*

Year	Outcome	Female	Male
1990	Not Promoted	5	3
	Promoted	1	3
	Eligible	6	6
1991	Not Promoted	3	1
	Promoted	2	2
	Eligible	5	3
1993	Not Promoted	2	0
	Promoted	1	1
	Eligible	3	1

Note: Hypothetical data.

known as time to event, or survival data, the latter name deriving from health studies where the event under study often is death.

Usually, such employment and promotion data are complex; individuals are hired at different times, some are not promoted even by the end of the study. Some leave the firm before being promoted. With such eventualities, the arguments we give are even more critical.

For survival data, each stratum or 2×2 table, as in Table 15.7, is called a risk set. Unlike the strata dealt with in this book, these are not statistically independent. Subjects are counted several times over and the overall sample size is inflated. Nevertheless, when the sample size becomes large, and under suitable conditions for dropouts and missing data, the asymptotic theory of partial likelihood shows that one can analyze these data in the same way as for regular, independent 2×2 tables of counts. A Mantel–Haenszel type of test (formally a conditional score test for a logit model) in this context is called a **log-rank test** (Newman 2001).

Exact analysis done along the same lines, however, and even if the data are sparse, is a formally exact but flawed analysis. It is a misnomer to call such a test, as was done in some earlier literature on exact tests, an exact log-rank test. Even when done in a supposedly conservative way, it does not guarantee nonexceedance of the nominal type I error rate (Soper and Tonkonoh 1993; Heinze, Gnant and Schemper 2003).

Table 15.7 *Promotions by Years of Service*

Years of Service	Outcome	Female	Male
1	Not Promoted	5	3
	Promoted	1	3
	Eligible	6	6
2	Not Promoted	3	1
	Promoted	2	2
	Eligible	5	3
3	Not Promoted	2	0
	Promoted	1	1
	Eligible	3	1
4	Not Promoted	1	0
	Promoted	1	0
	Eligible	2	0
5	Not Promoted	0	0
	Promoted	1	0
	Eligible	1	0

Note: Hypothetical data.

Suppose there are no dropouts, every hiree is eventually promoted and complete data are available. An appropriate representation of the data is shown in Table 15.8. This can form the basis for an appropriate exact or nonparametric analysis. Exact analysis for such data is not covered in this text. It bears on nonparametric analysis of continuous data and in some computational aspects, it resembles that done for a $2 \times K$ table in Chapter 10.

More generally, large sample analysis of multivariate survival data is typically done with the **Cox regression** method. This is, in computational terms, identical to applying the method of conditional logistic regression on the corresponding risk sets. In large samples, this is a valid

FORMAL EXACTNESS 475

approach. A small sample version, however, is formally exact but otherwise invalid. It would be a misnomer to call it "exact Cox regression."

Sound formulation of exact analysis of survival data is a complex issue, beyond the scope of this work. See Jennrich (1984), Soper and Tonkonoh (1993) and Heinze, Gnant and Schemper (2003) for details.

Table 15.8 *Years to Promotions by Gender*

Gender	Years to Promotion				
	1	2	3	4	5
Male	3	2	1	0	0
Female	1	2	1	1	1

Note: Same data as in Table 15.7.

15.6.2 Other Data Forms

Consider a study in which subjects are followed up over time for the occurrence of some event. When many subjects have been followed but the number of events is relatively few, a Poisson model (with a person time type of formulation) is often applied. This is based on the assumptions noted in §1.6, and extended as follows:

- The probability of an event in a small time interval is low, and of more than two events, is vanishingly low (the **rare event assumption**).
- The event probabilities for disjoint time intervals are statistically independent.
- The event probabilities for persons of a particular covariate combination (or cell) are homogeneous.
- For the purpose of analysis, the total follow up time in each cell is deemed a fixed quantity.
- In some studies, a person may contribute person time in more than one cell. This is also ignored in the analysis.

See Chapters 1, 2 and 10 for references on this. Under asymptotic considerations, the application of a Poisson model may be justified. When the number of events is low, or the data are sparse, exact Poisson analysis is generally recommended. See, for example, Gart et al. (1986) and McDonald, Smith and Forster (1999). In this book as well, we presented exact methods for analysis of cohort data for several designs under the same line of thought.

Yet, we are of the opinion that exact Poisson analysis of cohort data needs deeper scrutiny. Consider, for example, the analysis of the arsenic data in Chapter 6 (Table 6.6 and Example 6.4). In the original dataset, factor y is a time related variable. Also, in this type of study, a subject may contribute person time value to more than one cell of the table. Under such circumstances applying exact methods as is usually done is, in our view, going too far! Too many assumptions are needed to justify that analysis, and so even calling it exact is to stretch the meaning of the term beyond recognition.

We thus note that the authors of one comparative study of exact Poisson methods for cohort data found them only marginally attractive (Samuels, Beaumont and Breslow 1991). Their words of caution, made in the same spirit, are worth bearing in mind:

Analysis of Cohort Data

> Perhaps the virtues of ''exact'' probability calculations, or approximations to them should not be overemphasized .... the Poisson assumption itself is artificial and often unchecked. Samuels, Beaumont and Breslow (1991).

Exact conditional analysis for cohort data is thereby in need of more empirical and theoretical studies. The conditions under which it may be a valid form of analysis have to be clearly identified. Without such studies conducted for a wide variety of designs and data, and published in peer reviewed journals, a cloud of doubt will continue to hang over the use of exact Poisson methods for cohort data. At this stage, their inclusion in statistical software thereby needs to be accompanied with strong words of caution.

Next, consider the 2×2 inverse sampling design. For this, the unconditional likelihood is the same as the two binomials design. Some forms of large sample analysis are also the same. But the conditional likelihoods and exact conditional distributions for the two situations are different. Even the parameters retained by conditioning are distinct. Exact conditional analysis of inverse sampling data as if they were from a regular 2×2 design is another instance of a formally exact but otherwise questionable analysis.

Finally, consider the two state, homogeneous Markov chain. As shown in Chapter 14, the unconditional likelihood here is identical to that for a 2×2 table of counts. Large sample analysis of Markov chain data then uses the same methods as large sample methods for corresponding contingency tables. On the other hand, the conditional likelihood and exact conditional distributions are different. Applying a hypergeometric distribution based analysis here is also an instance of formally exact but actually flawed analysis.

****

We note an exchange in the medical literature which brought a number of points we have raised so far to the fore. That it did so in a peculiar manner makes it even more instructive:

Lens and Dawes (2002a) systematically reviewed eight clinical trials for interferon therapy for malignant melanoma. Its conclusions were later questioned by several correspondents. The critique by Kirkwood et al. (2002) is apropos here. They point out that the trials under review had time to relapse and time to death as the main outcomes, and were designed to be analyzed with the log-rank test for survival data. Raising a concern about possible abuse of statistical methods, they write:

A Flawed Analysis?

> Lens and Dawes base their analysis on the Fisher's exact test, using only the numbers of events for each treatment, although these outcomes are time-dependent. Kirkwood et al. (2002).

Reading Lens and Dawes (2002a), one notes that it does not mention the Fisher's exact test at all! These authors then point out that they had, as indicated in the methods section of their paper, analyzed the main outcomes in terms of odds ratios, and had used the Mantel–Haenszel method (Lens and Dawes 2002b). Accordingly, they state in their response:

IN PRAISE OF EXACTNESS   477

Not Really A Flawed Analysis?

> The log-rank test may be preferable, but in the absence
> of the raw data, we were not able to use it. However,
> we must be precise that we did not run Fisher's exact
> test on the published data. We calculated odds ratios
> with 95% confidence intervals, ... Lens and Dawes
> (2002b).

This exchange bears on important aspects of the design, conduct and analysis of clinical trials. The following peculiar features of this debate relating to statistical methods are of interest here:

- While Lens and Dawes (2002b) correctly say that they did not use the Fisher exact test, the underlying flaw of analyzing survival data as binary response data is not satisfactorily acknowledged.
- Lens and Dawes (2002b), for their part, rightly critique Kirkwood et al. (2002) for what appears to be a convenient reliance on one-sided *p*-values.
- Lens and Dawes (2002b) state that Kirkwood et al. (2002) place too much emphasis on *p*-values, But they fail to acknowledge that the latter did report a confidence interval.
- Though Kirkwood et al. (2002) did provide a confidence interval, they do not explain why it was a 90% interval. It appears that the 95% interval would have included the null value.
- While Kirkwood et al. (2002) mainly focus on statistical concerns relating to the systematic review of Lens and Dawes (2002a), they seem to sideline the key issue of the methodologic quality of the melanoma trials raised by the latter.

Each side thus makes relevant points and gives valid criticisms. In the process, each side falls prey to key statistical and methodologic errors. And, for a strange reason, the Fisher–Irwin exact test is sandwiched within this conflict, even though it was never applied anywhere by anyone!

15.8 In Praise of Exactness

Why did the Fisher–Irwin exact test feature in the above controversy? Is it because it has too often been embroiled in a controversy? Is it because in some minds it has come to embody the very idea of a controversy? Perhaps an illustration will clarify the matter.

Daniel (2005) is a popular introductory statistics text for biomedical and health science courses. Now it is into its 8th edition. Like other books at this level, it contains a well illustrated section on the Fisher–Irwin exact test. The reader is also informed that the test is the *"subject of some controversy."* (page 361). In a similar vein, it is not uncommon to find statistical books and papers portraying exact methods for discrete data as controversial. The irony is that these methods are called "exact" and the supposedly noncontroversial alternative is denoted as "approximate"!

Daniel (2005) correctly notes that assuming the two margins are fixed is not realistic for most 2×2 table data. And if they are not, he opines that the Fisher–Irwin test may not be appropriate. What makes exact methods in general controversial, accordingly, is that they fix what is not fixed by design. Or at least that is what we are told.

Now consider, for comparison, the Mantel–Haenszel method for stratified 2×2 tables. In Daniel (2005), it is introduced a few pages after the Fisher–Irwin test. This method also fixes both margins in each table and is regularly applied to data in which the margins were not set in advance. Thus, on page 642 of Daniel (2005), we find that the variance used is the sum of conditional hypergeometric variances. Yet, now there is no hint here that because the Mantel–Haenszel method fixes both margins of each table, it too should be branded "controversial."

Consider the multiplicity of nonparametric methods for continuous data. They are derived by conditioning arguments, even when the design does not condition in the manner assumed. Popular regression methods condition on the explanatory variables not fixed by design. Are they called controversial for that reason? Hardly ever. Why the disconnect with discrete data, and the Fisher–Irwin test in particular?

The historic schism in the statistical community, and the nature of the debates on the subject in part explain why exact analysis of discrete data is viewed differently. But is it not the time for a change, and to present the matter in more consistent terms?

Let us imagine a different history of statistics. Suppose that when its edifice was under construction in the 1930s, statisticians had access to good computing facilities. Instead of slide rulers, tables of logarithms and rudimentary calculating machines, they used electronic calculators and even mainframe computers of the 1970s generation.

Many techniques and practices from those days were developed as clever, often ingenuous, approximations to exact computation. Ease of calculation often drove selection of the method. Methods based on stricter assumptions were easier to implement. Large sample theory was assumed to apply unless proven otherwise. The Yates continuity correction was an easier version of the Fisher–Irwin test, and ample effort was devoted to better the correction. Many exact methods were formulated but then deemed infeasible. The Freeman–Halton exact test for an $r \times c$ table was formulated in 1951 but did not become practical until the 1980s. Exact analysis of logistic models was put forward in 1958, but only became feasible in the late 1980s.

With an early advent of computing power, the tables would perhaps be turned. In his historic debate with Fisher, Barnard would have given more striking cases of exact unconditional analysis (Barnard 1945a, 1945b, 1947). Others would have expanded on his work sooner. The case-control studies of the 1950s would have analyzed several 2×2 tables with exact methods, not the Cornfield–Mantel–Haenszel method. Advanced Monte–Carlo techniques and the bootstrap may have bloomed earlier. And, who is to say, Bayesian methods may have become the popular methods in research!

Under this hypothetical version of history, the Cochran rule for $r \times c$ tables would be: "Do not use the chisquare test unless all expected values exceed 10." Joseph Berkson may have penned a paper with the title "In Dispraise of Approximate Tests" (Berkson 1978). Each approximate (thereby controversial) method would be thoroughly checked on the basis of an exact (conditional or unconditional) standard.

Not that statistics as a discipline would be free of controversy. The selection of 95% level two-sided Bayesian credible intervals, especially in asymmetric posterior distributions, or fixed cut points based on the likelihood ratio would be hot topics. Debates on selecting the frame of reference, the importance of the likelihood and the Frequentist versus Bayesian approaches would, indeed, go on. But equating conditioning with exactness, and the narrow context of the 2×2 table would happily be absent.

What is acceptable and what is not, what is controversial and what is not derives in part from what was computable and what was not in an earlier era. A method that has gained common currency, and is present in all teaching materials, becomes the accepted standard. That it was initially adopted as a convenience is forgotten. This was why saying $p < 0.05$ instead of giving the p-value came to dominate research reports. Even after computers have made it as easy to report the latter, changing habits, even to a small degree, took considerable time and effort. In fact, the practice continues to this day.

The past cannot be easily conditioned away. Yet, the future need not be strictly conditioned on history. It is time for the statistical community to put aside historic acrimonies that are of little import now. To use a continuity correction or not? What is a two-tailed *p*-value? These are matters of detail, not principle. Key issues have to be tackled but in a broader context. Restricting their discussion to exact tests, and especially to the 2×2 table, sustains the confusion, fosters the obsession with *p*-values and hardly points the way ahead. The same arguments are instead repeated time and again.

****

Another misperception about exact methods for discrete and other data is about their computability. Let us give two cases. Consider the view of Brad Efron, the pioneer of a new field of computer intensive methods. After giving an exact 95% confidence interval for the odds ratio in the 2×2 table with $a = 1; b = 13; c = 15; d = 13$, he writes:

Prohibitive Calculation?

> However, the conditional distribution is not easy to
> calculate, even in this simple case, and it becomes
> prohibitive in more complicated situations. Efron
> (2000).

Brian Manly is an authority on randomization methods. In his work on the subject, exact computation, or complete enumeration, is characterized as follows. It is stated that in many practical problems:

Enumeration or Randomization?

> ... the large number of permutations makes a complete
> enumeration extremely difficult (if not impossible)
> with present-day computers. Manly (1997), page 14.

Such is the assessment of the difficulties involved in exact computation of notable experts in computational statistics. The ordinary statistician may well have a more extreme opinion. These opinions have some truth, but do not faithfully reflect the current state of affairs. Perhaps they are not at fault. The unnecessarily complex presentations of the computing algorithms by researchers specializing in exact methods make it hard for anyone not to form that kind of an impression.

Yet, such views also overlook the tremendous progress in making a wide array of discrete data problems amenable to exact analysis that transpired in the last two decades of the twentieth century. And now it is not just a matter of theory. The user friendly software products like StatXact and LogXact (Cytel Software Corporation 2004a, 2004b) and other statistical software have efficiently implemented exact methods for discrete and continuous data but a mouse click away from researchers and analysts. For cases where exact computation is more demanding, they incorporate efficient Monte–Carlo options that quickly generate results that are quite close to the exact results.

An analyst, when performing Cox regression on survival data, is not too concerned with the computational technique used to maximize the partial likelihood. If he or she needs a bootstrap confidence interval for dose-response data, the intricacies of the technique are not an overriding matter. That also holds when an exact confidence interval for the odds ratio in stratified binary data is required. If the software gives him or her what is required in a fast and accurate way, how that is done is, for daily applications, a secondary issue.

And, the story does not stop there. Lack of familiarity should not be confounded with complexity. We have aimed to demonstrate in detail that **highly efficient computational methods for exact analysis of discrete data are neither intricate nor complex in any comparative sense of the term.** They basically involve simple methods of multiplying a series of polynomials in a step by step way, and simple rules for selecting the terms to retain or discard. The logic involved in the process is a few steps beyond that learned in a good high school algebra course. A few techniques for efficient implementation are needed. But they are not more complex than, say, the methods used for matrix inversion when maximizing multiparametric likelihoods.

Adopting the motto: **compute simply, that others may simply compute,** we hope to encourage statisticians and users of statistical methods to **give exact methods a chance.** For researchers, teachers, analysts, statisticians and software developers, there is now an added reason for learning, implementing, teaching and using exact methods.

The scope of exact methods for discrete data goes beyond what has been covered in this introductory work. Besides what has been noted elsewhere, models that are amenable to exact analysis include models for exact measures of association and agreement, exact analysis of marginal homogeneity in an $R \times R$ table, data in triangular tables, genetic data including the Hardy–Weinberg model, general $R \times C$ and $R \times C \times K$ tables, tables with structural zeroes, binary and polytomous regression models, and three state and stratified Markov chains. Research extending the scope of exact methods to other models, and expanding the purview of exact unconditional analysis, is also underway. See the references noted in §15.9.

****

We end with a brief sojourn into the question sample size in biomedical research, a field of research that has expanded phenomenally since the 1950s. Hundreds of thousands of clinical trials have been conducted, and tens of thousands have been published. Yet, in many medical specialties, this quantitative expansion of research has been short on quality. In fact, it has been accompanied by the proliferation of small sample studies with too low a power to detect clinically meaningful effects. Poor quality and small sample size have partnered in creating an atmosphere of confusion. Not surprisingly, some researchers have called for simple, well designed, large sample size studies. To obtain a sampling of the views on this issue, and of the actual state of affairs, see Freiman et al. (1978), Michael, Boyce and Wilcox (1984), Yusuf, Collins and Peto (1984), Claessen et al. (1992), Moher, Dulberg and Wells (1994), Thornley and Adams (1998), Altman (2002), Vail and Gardner (2003), Ottenbacher et al. (2204) and Soares et al. (2004).

While recognizing this problem, some have nevertheless argued for the continued need for small sample studies that are at the same time well designed and of good quality. In an age of systematic reviews and meta-analysis, there is a basis for conducting good quality studies that may at times not have adequate power when seen in isolation. It becomes a matter of balancing bias against precision. Even here, there is a diversity of opinions, and continued discussion in the medical and biostatistical literature; see, for example, Barnard (1990), Matthews (1995), Halpern, Karlawish and Berlin (2002) and Schulz and Grimes (2005). On exact methods in meta-analysis, see Mosteller and Chalmers (1992).

Schulz and Grimes (2005), in particular, favor small sample trials with the stipulation that they are of high quality, that their methods and results are reported clearly and in full and that the publication of all the studies, be they negative or positive, is the norm. They go on to state:

```
              Small Sample But High Quality Trials

    Some shift of emphasis from a fixation on sample size
    to a focus on methodological quality would yield more
    trials with less bias. Unbiased trials with imprecise
    results trump no results at all. Schulz and Grimes
    (2005).
```

Exact methods are required for small sample and sparse data. The theoreticians and practitioners of exact methods thereby need to be aware of the concerns raised in such debates, and to emphasize the issue of the quality of the data to which these methods are applied. They must embrace the ongoing challenge for the statistical profession to promote the usage of its methods to conduct rigorous as well as ethically sound research. In particular, exact methods should not be, or should not be seen as, in theory or practice, a cloak for unsound science. Abuse of the p-value, overemphasis on statistical significance and neglect of power, confidence intervals, sample size and external validity are not practices with which "exactness" ought to be associated. For example, grant proposals written for funding institutions and scientific papers or approval reports for regulatory agencies like the FDA which deal with small sample size studies may deploy exact p-values. Funding and regulatory agencies, however, tend to rely excessively on the p-value. In consequence, the clinical or practical significance of the research does not get the attention it deserves. Use of exact methods in these settings should thereby be done in such a way as to pay due attention to the practical significance of the study. In our view, if the label "exact" is to denote a thoughtful and appropriate form of statistical analysis, its gurus and practitioners have to promote sound and relevant research methods, design, practice, analysis and reporting at all times and in an unequivocal way .

To sum it all: exact methods for discrete data are not necessarily controversial, or too complex to understand and implement in practice. In many models and designs, there are sound reasons for applying them. If thereby used appropriately, and in conjuction with studies performed under exacting scientific and ethical standards, their position as an essential component of the science of statistics can only be cemented further.

15.9 Relevant Literature

This chapter has barely touched the surface of a very deep arena, namely, the elucidation and discussion of the fundamental approaches to statistical inference. It is also biased towards the author's views. We encourage the reader to explore it further. The references below augment what has been noted earlier.

Abrams, Ashby and Errington (1994), Freedman (1996) and Stern (1998) give readable and sympathetic expositions of Bayesian methods for clinical studies. A critical view is provided by Fisher (1996). The stimulating piece by Efron (1986) and the ensuing discussion introduce the differing perspectives. Berry (1993) presents arguments for employing Bayesian methods in clinical trials. Congdon (2005) is a key reference work for Bayesian methods for categorical data.

Lehmann (1985) is the standard work on theory of the Neyman–Pearson approach. Efron (1986) describes its essential features and contrasts it with the Fisherian approach. Whitehead (1993) and Fisher (1996) make a solid case for frequentist methods in clinical trials. For a critique of the frequentist *p*-value, and the alternatives to it, see Goodman and Royall (1988) and Goodman (1992, 1999a, 1999b, 2001).

A general introduction to the likelihood approach is in Birnbaum (1962) and Blume (2002). See also Edwards (1972) and Royall (1997). Little (1989) is a discrete data related paper that outlines with illustrations the main issues from a comparative perspective. The references in Goodman (1992, 1999a, 1999b, 2001) and Helland (1995) are also of help.

For assessments of R. A. Fisher's contribution to statistics, see Lehmann (1993), Rao (2000) and Efron (2000). The references they provide are a basis for pursuing the topic further.

Definitions and discussion of ancillarity are found in Kalbfleisch (1982), Little (1989). See also Basu (1977). On the two binomials design for a randomized trial and related arguments, see Copas (1973) and Haber (1990).

For varied perspectives on the analysis of the 2×2 table, see the following papers and the comments they generated: Berkson (1978), Upton (1982), Yates (1984), D'Agostino, Chase and Belanger (1988), Rice (1988) and Haviland (1990). Other relevant papers are Barnard (1989), Little (1989), Greenland (1991), Hirji, Elashoff and Tan (1991) and Upton (1992). For a quick overview, consult the dialogue between Camilli (1990), who outlines the various approaches, and the succinct response of Haber (1990). Our views are close to those of Haber (1990).

Some papers on exact methods for discrete data models that are not covered, or only covered in brief in this text were noted in Chapters 10 to 14. Other papers in this regard are: multidimensional tables (Kreiner 1987, Morgan and Blumenstein 1991); measures of agreement (Sprott and Farewell 2000); general testing for marginal homogeneity (Patil 1975); outlier detection and model checking (Bedrick and Hill 1990); the Hardy–Weinberg model (Chapco 1976, Elston and Forthofer 1977, Haber 1981, Louis and Dempster 1987, Guo and Thompson 1992 and other papers referenced therein); and general genetic models (Lange 2002). See the papers on exact unconditional analysis noted in Chapter 7 as well.

15.10 Exercises

15.1. Discuss further the meaning of the descriptors of the term 'exact' noted in §15.2. Survey their usage in the statistical and applied literature, and write a report on your findings.

15.2. For the study designs of Chapter 5 and Chapter 6, develop a bootstrap based analyses of a single and two 2×2 tables. In practical data, how do they differ from the respective exact conditional and unconditional analyses? Conduct simulation studies to investigate this question as well.

15.3. Consider a Bayesian analysis of the binomial using the formulation based on odds. Suppose $\phi = \pi/(1 - \pi)$ and π has a uniform prior distribution. What is the corresponding distribution of ϕ? Apply this to equation (15.2) and use that to develop a Bayesian analysis of the single binomial. What other prior distributions can be used here?

15.4. Develop a Bayesian analysis of the odds ratio in a two binomials experiment, and a Bayesian analysis of the cross product ratio in a 2×2 design with the overall total fixed. Are the two analyses necessarily equivalent?

15.5. Consider a 2×2 table with both margins fixed. Assume the noncentral hypergeometric model for this problem and develop a Bayesian analysis of the parameter ϕ. Compare this with the Bayesian analysis of the two binomials problem. What are the practical and theoretical implications of the differences you found?

15.6. Develop a consistent likelihood based analysis for the single, two and several independent binomials problems. In the latter, include the case of ordered data. Extend your methods

EXERCISES

to several 2×2 tables. Compare and contrast them with the approach and methods used in the book.

15.7. Prove the assertions in Observation 15.1.

15.8. Prove the assertions in Observation 15.2.

15.9. Prove the assertions in Observation 15.3.

15.10. Prove the assertions in Observation 15.4.

15.11. Suppose we plant n seeds. Of these, m germinate. Further, the color of the flowers in t of the m plants is purple, and in the rest it is not ($t \leq m \leq n$). Let ψ denote the true proportion of seeds of this type that germinate, and π, the proportion of plants yielding purple flowers. Write a probability model for this situation. If we analyze π, is n an ancillary statistic? Compare and contrast the conditional and unconditional methods for analyzing π.

15.12. Suppose we recruit n eligible patients for a clinical study. m of them agree to participate, and are given the new treatment. The number of cures seen is t ($t \leq m \leq n$). Let ψ denote the proportion of patients that tend to accept participation and π, the underlying cure rate of the treatment. Write a model for this situation. If we analyze π, is n an ancillary statistic? What if the tendency to participate is related to the severity of the condition and hence the efficacy of the treatment? Compare and contrast the conditional and unconditional methods for analyzing π. What are the implications for external validity?

15.13. Consider the previous problem with J treatment groups. There are n_j subjects in group j. Of these m_j complete the study and t_j are deemed as cured. In terms of comparing the treatments, what is ancillary and what is not? And under what conditions?

15.14. Compare exact analysis of the data in Table 15.4 with the three paired data possibilities in Table 15.5. Conduct a general study of this issue using other real and simulated data.

15.15. Compare the formal "exact" analysis of the data in Table 15.7 with the appropriate analysis based on Table 15.8. Conduct a general study of this issue using other real and simulated data.

15.16. Explore the biomedical and other literature to uncover the instances of formal exact analysis. Summarize your findings and give suggestions how such practices can be avoided.

15.17. Develop, in a rigorous fashion and from first principles, exact methods for the analysis of cohort data under the designs shown in Chapters 5, 6, 9 and 10. Do we need to restrict the type of covariates in the models and the design for the exact analysis to be valid?

15.18. Investigate the actual use of Poisson distribution based exact methods for cohort data in biomedical and other literature. Are there instances where one may doubt the validity of their use?

15.19. Summarize the Bayesian critique of the *p*-value, and of the classic frequentist approach. How would a frequentist reply?

15.20. Summarize the frequentist critique of the Bayesian approach. How would a Bayesian reply?

15.21. Conduct a survey of practicing statisticians on the use of exact methods for discrete data. How frequently and under what conditions are they used?

15.22. Investigate the relationship between sample size and study quality in health and other types of research. What are the implications of your findings for the use of exact methods of analysis?

15.23. What are the arguments for and against the erection of a unified frequentist of inference that combines the Fisherian and Neyman–Pearson schools of inference. How do exact and asymptotic methods for discrete data fit into this divide? Give your views on the subject.

15.24. Develop an empirical Bayes analysis of the single and two binomials model. How does this method fit into the frequentist versus Bayesian discourse? Is the likelihood based approach a good alternative to these two approaches? Why or why not?

15.25. From the medical and other literature, explore the growth of exact methods in practical data analysis. Is there an association between study quality and the use of such methods? Are the methods used appropriately?

15.26. Survey the medical and other literature to assess whether or not one-sided tests for discrete data are used more frequently with exact tests as compared to other tests.

15.27. Investigate the importance attached to external and internal validity in actual biomedical and other studies and expand on our discussion of their relationship to the use of exact statistical methods.

15.28. Are evidence functions valid entities for deriving statistical measures of evidence?

15.29. Are evidence (or confidence interval) functions used in biomedical and other studies? Explore why or why not? Conduct a survey of statistics and epidemiology textbooks to ascertain the extent to which they cover the topic. What has to be done to promote their use?

15.30. Explore the use of exact methods in clinical trials and meta-analysis. Where possible, reanalyze the data from a variety of binary response trials and systematic reviews using exact methods. What are the implications of your findings?

References

Abrams K, Ashby D and Errington D (1994) Simple Bayesian analysis in clinical trials: a tutorial, *Controlled Clinical Trials*, **15**: 349–359.

Adams WT and Skopek TR (1987) Statistical tests for the comparison of samples from mutational spectra, *Journal of Molecular Biology*, **194**: 391–396.

Agresti A (1990) *Categorical Data Analysis*, John Wiley & Sons, New York.

Agresti A (1992) A survey of exact inference for contingency tables (with discussion), *Statistical Science*, **7**: 131–177.

Agresti A (1996) *An Introduction to Categorical Data Analysis*, John Wiley & Sons, New York.

Agresti A (1999) On logit confidence intervals for the odds ratio with small samples, *Biometrics*, **55**: 597–602.

Agresti A (2000) Challenges for categorical data analysis in the twenty-first century, in CR Rao and GJ Szekely (editors) (2000) *Statistics for the 21st Century: Methodologies for Applications of the Future*, pages 1–19, Marcel Dekker, New York.

Agresti A (2001) Exact inference for categorical data: recent advances and continuing controversies, *Statistics in Medicine*, **20**: 2709–2722.

Agresti A and Caffo B (2000) Simple and effective confidence intervals for proportions and differences of proportions: results from adding two successes and two failures, *The American Statistician*, **54**: 280–288.

Agresti A and Coull BA (1998) Approximate is better than "exact" for interval estimation of binomial proportions, *The American Statistician*, **52**: 119–126.

Agresti A, Lang JB and Mehta CR (1993) Some empirical comparisons of exact, modified exact, and higher-order asymptotic tests of independence for ordered categorical variables, *Communications in Statistics - Simulation and Computation*, **22**: 1–18.

Agresti A, Mehta CR and Patel NR (1990) Exact inference for contingency tables with ordered categories, *Journal of the American Statistical Association*, **85**: 453–458.

Agresti A and Min Y (2001) On small-sample confidence intervals for parameters in discrete distributions, *Biometrics*, **57**: 963–971.

Agresti A and Min Y (2002) Unconditional small-sample confidence intervals for the odds ratio, *Biostatistics*, **3**: 379–386.

Agresti A and Wackerly D (1977) Some exact conditional tests for independence for $R \times C$ cross-classification tables, *Psychometrika*, **42**: 111–125.

Agresti A, Wackerly D and Boyett J (1979) Exact conditional tests for independence for cross-classifications: approximation of attained significance levels, *Psychometrika*, **44**: 75–83.

Ahlbom A (1993) *Biostatistics for Epidemiologists*, Lewis Publishers, Boca Raton, Florida.

Aho AV, Hopcroft JE and Ullman JD (1974) *The Design and Analysis of Computer Algorithms*, Addison–Wesley Publishing Company, Reading, Massachusetts.

Ali MW (1990) Exact versus asymptotic tests of trend of tumor prevalence in tumorigenicity experiments: a comparison of *p*-values for small frequency of tumors, *Drug Information Journal*, **24**:727–737.

Altham PME (1969) Exact Bayesian analysis of a 2×2 table and Fisher's "exact" significance test, *Journal of the Royal Statistical Society, Series B*, **31**: 261–269.

Altman DG (2000) Statistics in medical journals: some recent trends, *Statistics in Medicine*, **19**: 3275–3289.

Altman DG (2002) Poor-quality medical research: what can journals do? *Journal of the American Medical Association*, **287**: 2765–2767.

Altman DG and Bland JM (2003) Interaction revisited: the difference between two estimates, *British Medical Journal*, **326**: 219.

Altman DG and Matthews JNS (1996) Interaction 1: heterogeneity of effects, *British Medical Journal*, **313**: 486.

Alvo M and Cabilio P (2000) Calculation of hypergeometric probabilities using Chebychev polynomials, *The American Statistician*, **54**: 141–144.

Andersen B (1990) *Methodological Errors in Medical Research*, Blackwell Scientific Publications, Oxford.

Andersen EB (1982) *Discrete Statistical Models with Social Science Applications*, North Holland–Publishing Company, Amsterdam.

Anderson I (1989) *A First Course in Combinatorial Mathematics*, Oxford University Press, Oxford.

Anderson TW and Goodman LA (1957) Statistical inference about Markov chains, *Annals of Mathematical Statistics*, **28**: 89–110.

Anonymous (1988) Supreme Court ruling on death penalty, *Chance*, **1**: 7–8.

Anscombe FJ (1981) *Computing in Statistical Science through APL*, Springer–Verlag, New York.

Appleton DR, French JM and Vanderpump MPJ (1996) Ignoring a covariate: an example of Simpson's paradox, *The American Statistician*, **50**: 340–341.

Armitage P (1955) Tests for linear trend in proportions and frequencies, *Biometrics*, **11**: 375–386.

Armitage P and Berry G (1987) *Statistical Methods in Medical Research*, 2nd edition, Blackwell Scientific Publications, Oxford.

Baglivo J, Olivier D and Pagano M (1992) Methods for exact goodness of fit tests, *Journal of the American Statistical Association*, **87**: 464–469.

Baglivo J, Olivier D and Pagano M (1993) Analysis of discrete data: rerandomization methods and complexity, *Computational Statistics and Data Analysis*, **16**: 175–184.

Baglivo J, Pagano M and Spino C (1996) Permutation distributions via generating functions with applications to sensitivity analysis of discrete data, *Journal of the American Statistical Association*, **91**: 1037–1046.

Bailar JC and Ederer F (1964) Significance factors for the ratio of a Poisson variable to its expectation, *Biometrics*, **20**: 639–643.

Baptista J and Pike MC (1997) Exact two-sided confidence limits for the odds ratio in a 2×2 table, *Applied Statistics*, **26**: 216–220.

Barnard GA (1945a) A new test for 2×2 tables, *Nature*, **156**: 177.

Barnard GA (1945b) A new test for 2×2 tables, *Nature*, **156**: 783.

Barnard GA (1947) Significance tests for 2×2 tables, *Biometrika*, **34**: 123–138.

Barnard GA (1949) Statistical inference, *Journal of the Royal Statistical Society, Series B*, **11**: 115–139.

Barnard GA (1979) In contradiction to J. Berkson's dispraise: conditional tests can be more efficient, *Journal of Statistical Planning and Inference*, **3**: 181–187.

Barnard GA (1989) On alleged gains in power from lower *p*-values, *Statistics in Medicine*, **8**: 1469–1477.

Barnard GA (1990) Must clinical trials be large? The interpretation of *p*-values and the combination of test results, *Statistics in Medicine*, **9**: 601–614.

Barnard GA, Jenkins GM and Winsten CB (1962) Likelihood inference and time series, *Journal of the Royal Statistical Society, Series A*, **125**: 321–372.

Bartlett MS (1935) Contingency table interactions, *Journal of the Royal Statistical Society, Supplement*, **2**: 248–252.

Basu D (1959) The family of ancillary statistics, *Sankhya*, **21**: 247–246.

Basu D (1977) On the elimination of nuisance parameters, *Journal of the American Statistical Association*, **72**: 355–366.

Baxt WG (1991) Use of an artificial neural network for the diagnosis of myocardial infarction (correction: 92V116 p94), *Annals of Internal Medicine*, **115**: 843–848.

Bayer L and Cox C (1979) Algorithm AS142: Exact tests of significance in binary regression models, *Applied Statistics*, **28**: 319–324.

Bedrick EJ and Hill JR (1990) Outlier tests for logistic regression: a conditional approach, *Biometrika*, **77**: 815–827.

Bellman RE and Dreyfus SE (1962) *Applied Dynamic Programming*, Princeton University Press, Princeton, New Jersey.

Bender R (2001) Calculating confidence intervals for the number needed to treat, *Controlled Clinical Trials*, **22**: 102–110.

Bennett BM (1981) On the use of the negative binomial in epidemiology, *Biometrical Journal*, **23**: 69–72.

Bennett BM (1986) On combining estimates of relative risk using the negative binomial model, *Biometrical Journal*, **28**: 859–862.

Bennett BM and Hsu P (1960) On the power function for the exact test for the 2×2 contingency table, *Biometrika*, **47**: 393–398.

Bennett BM and Nakamura E (1963) Tables for testing significance in a 2×3 contingency table, *Technometrics*, **5**: 501–511.

Bennett BM and Nakamura E (1964) The power function of the exact test for the 2×3 contingency table, *Technometrics*, **6**: 439–458.

Bennett JH (1990) *Statistical Inference and Analysis: Selected Correspondence of R A Fisher*, Oxford University Press, Oxford.

Berger RL (1996) More powerful test from confidence interval *p* values, *The American Statistician*, **50**: 314–318.

Berger JO and Boos DD (1994) *P* values maximized over a confidence set for the nuisance parameter, *Journal of the American Statistical Association*, **89**: 1012–1016.

Berger JO and Selke T (1987) Testing a point null hypothesis: the irreconcilability of *P* values and evidence, *Journal of the American Statistical Association*, **82**: 112–122.

Berk KN (2003) Discussion of Hubbard and Bayarri (2003), *The American Statistician*, **57**: 178–179.

Berkson J (1978) In dispraise of the exact test: do marginal totals of the 2×2 table contain relevant information respecting the table proportions, *Journal of Statistical Planning and Inference*, **2**: 27–42.

Berlin JA, Laird NM, Sacks HS and Chalmers TC (1989) A comparison of statistical methods for combining event rates from clinical trials, *Statistics in Medicine*, **8**: 141–151.

Berry DA (1993) A case for Bayesianism in clinical trials, *Statistics in Medicine*, **12**: 1377–1393.

Berry G and Armitage P (1995) Mid-*P* confidence intervals: a brief review, *The Statistician*, **44**: 417–423.

Berson EL, Rosner B, Sandberg MA and Dryja TP (1991) Ocular findings in patients with autosomal dominant retinitis pigmentosa and a rhodopsin gene defect (Pro-23-His), *Archives of Ophthalmology*, **109**: 92–101.

Bhat UB (1972) *Elements of Applied Stochastic Processes*, John Wiley & Sons, New York.

Billingsley P (1961) Statistical methods in Markov chains, *Annals of Mathematical Statistics*, **32**: 12–40.

Birch MW (1963) Maximum likelihood in three-way contingency tables, *Journal of the Royal Statistical Society, Series B*, **25**: 220–233.

Birnbaum A (1961) Confidence curves: an omnibus technique for estimation and testing statistical hypotheses, *Journal of the American Statistical Association*, **56**: 246–249.

Birnbaum A (1962) On the foundations of statistical inference (with discussion), *Journal of the American Statistical Association*, **57**: 269–326.

Birnbaum A (1964) Median unbiased estimators, *Bulletin of Mathematical Statistics*, **11**: 25–34.

Bishop YMM, Fienberg SE and Holland PW (1975) *Discrete Multivariate Analysis: Theory and Practice*, MIT Press, Cambridge, MA.

Blaker H (2000) Confidence curves and improved exact confidence intervals for discrete distributions, *The Canadian Journal of Statistics*, **28**: 783–798.

Bland JM and Altman DG (1994) One and two sided tests of significance, *British Medical Journal*, **309**: 248.

Blume JD (2002) Likelihood methods for measuring evidence, *Statistics in Medicine*, **21**: 2563–2599.

Blyth CR and Still HA (1983) Binomial confidence intervals, *Journal of the American Statistical Association*, **78**: 108–116.

Bonney GE (1987) Logistic regression with dependent binary observations, *Biometrics*, **43**: 951–973.

Borenstein M (1994) The case for confidence intervals in controlled clinical trials, *Controlled Clinical Trials*, **15**: 411–428.

Borse GJ (1985) *FORTRAN 77 and Numerical Methods for Engineers*, PWS–Kent Publishing Company, Boston.

Boschloo RD (1970) Raised conditional level of significance for the 2×2 table when testing for the equality of two proportions, *Statistica Neerlandica*, **21**: 1–35.

Boulton DM and Wallace CS (1973) Occupancy of a rectangular array, *The Computer Journal*, **16**: 57–63.

Brand R and Kragt H (1992) Importance of trends in the interpretation of an overall odds ratio in the meta-analysis of clinical trials, *Statistics in Medicine*, **11**: 2077–2082.

Breslow NE (1996) Statistics in epidemiology: the case-control study, *Journal of the American Statistical Association*, **91**: 14–28.

Breslow NE and Day NE (1980) *Statistical Methods in Cancer Research, Volume I: The Analysis of Case-Control Studies*, IARC Scientific Publications, Lyon.

Breslow NE and Day NE (1987) *Statistical Methods in Cancer Research, Volume II: The Design and Analysis of Cohort Studies*, IARC Scientific Publications, Lyon.

Brown CC and Fears TR (1981) Exact significance levels for multiple binomial testing with application to carcinogenicity screens, *Biometrics*, **37**: 763–774.

Burden RL and Faires JD (1993) *Numerical Analysis*, 5th edition, PWS–Kent Publishing Company, Boston.

Camilli G (1990) The test of homogeneity for 2×2 contingency tables: a review of and some personal opinions on the controversy, *Psychological Bulletin*, **108**: 135–145.

Caporaso N, Hayes RB, Dosemeci M, Hoover R, Ayesh R, Hetzel M and Idle J (1989) Lung cancer risk, occupational exposure, and the debrisoquine metabolic phenotype, *Cancer Research*, **49**: 3675–3679.

Carlton MA (2003) Discussion of Hubbard and Bayarri (2003), *The American Statistician*, **57**: 179–181.

Casagrande JT, Pike MC and Smith PG (1978a) The power function of the "exact" test for comparing two binomial proportions, *Applied Statistics*, **27**: 176–180.

Casagrande JT, Pike MC and Smith PG (1978b) Algorithm AS129: The power function of the "exact" test for comparing two binomial proportions, *Applied Statistics*, **27**: 212–219.

Casella G and Berger R (1999) *Statistical Inference*, Wadsworth, Pacific Grove, CA.

Chan ISF (1998) Exact tests of equivalence and efficacy with a non-zero lower bound for comparative studies, *Statistics in Medicine*, **17**: 1403–1413.

Chan ISF and Zhang Z (1999) Test-based exact confidence intervals for the difference of two binomial proportions, *Biometrics*, **55**: 1202–1209.

Chapco W (1976) An exact test of the Hardy–Weinberg law, *Biometrics*, **32**: 183–189.

Chapman DG and Nam JM (1968) Asymptotic power of chi square tests for linear trend in proportions, *Biometrics*, **24**: 315–327.

Chatfield C (1976) Statistical inference regarding Markov chains, *Applied Statistics*, **82**: 7–20.

Chen X (2002) A quasi-exact method for the confidence intervals of the difference of two independent binomial proportions in small sample cases, *Statistics in Medicine*, **21**: 943–956.

Claessen JQPJ, Appelman CLM, Touw–Otten FWMM, De Melker RA and Hordijk GJ (1992) A review of clinical trials regarding treatment of acute otitis media, *Clinical Otolaryngology*, **17**: 251–257.

Clayton DC and Hills M (1993) *Statistical Models in Epidemiology*, Oxford University Press, Oxford.

Clayton DG (1982) The analysis of prospective studies of disease aetiolgy, *Communications in Statistics - Theory and Methods*, **11**: 2129–2155.

Clemans KC (1959) Confidence limits in the case of the geometric distribution, *Biometrika*, **46**: 260–264.

Clopper CJ and Pearson ES (1934) The use of confidence or fiducial limits as illustrated in the case of the binomial, *Biometrika*, **26**: 404–413.

Cochran WG (1954a) The combination of estimates from different experiments, *Biometrics*, **10**: 101–129.

Cochran WG (1954b) Some methods for strengthening the common χ^2 tests, *Biometrics*, **10**: 417–451.

Coggon DIW and Martyn CN (2005) Time and chance: the stochastic nature of disease causation, *The Lancet*, **365**: 1434–1437.

Cohen A and Sackrowitz HB (1992) An evaluation of some tests of trend in contingency tables, *Journal of the American Statistical Association*, **87**: 470–475.

Cohen GR and Yang SY (1994) Mid-*P* confidence intervals for the Poisson expectation, *Statistics in Medicine*, **13**: 2189–2203.

Collet D (1991) *Modeling Binary Data*, Chapman and Hall, London.

Congdon P (2005) *Bayesian Models for Categorical Data*, John Wiley & Sons, New York.

Cook RJ and Sackett DL (1995) The number needed to treat: a clinically useful measure of treatment effect, *British Medical Journal*, **310**: 452–454.

Cook RJ and Walter SD (1997) A logistic model for trend in $2 \times 2 \times K$ tables with application to meta-analyses, *Biometrics*, **53**: 352–357.

Copas JB (1973) Randomization models for the matched and unmatched 2×2 tables, *Biometrika*, **60**: 467–476.

Cormack RS (1986) The meaning of probability in relation to Fisher's exact test, *Metron*, **44**: 1–30.

Cornfield J (1956) A statistical problem arising from retrospective studies, in J Neyman (editor), *Proceedings of the Third Berkeley Symposium*, **IV**: pages 135–148, University of California Press, Berkeley.

Cox DR (1958) The regression analysis of binary sequences (with discussion), *Journal of the Royal Statistical Society, Series B*, **20**: 215–242.

Cox DR (1970) *The Analysis of Binary Data*, Methuen, London.

Cox CR and Hinkley DV (1974) *Theoretical Statistics*, Chapman and Hall, London.

Cox DR and Snell EJ (1989) *Analysis of Binary Data*, 2nd edition, Chapman and Hall, London.

Cressie N and Read TRC (1984) Multinomial goodness-of-fit tests, *Journal of the Royal Statistical Society, Series B*, **46**: 440–464.

Crow EL (1956) Confidence intervals for a proportion, *Biometrika*, **43**: 423–435.

Crow EL and Gardner RS (1959) Confidence intervals for the expectation of a Poisson variable, *Biometrika*, **46**: 441–453.

Curnow RN, Hodge A and Wilesmith JW (1997) Analysis of the Bovine Spongiform Encephalopathy Maternal Cohort Study: the discordant case-control pairs, *Applied Statistics*, **46**: 345–349.

Cytel Software Corporation (2004a) *StatXact for Windows*, version 6, Cytel Software Corporation, Cambridge, MA.

Cytel Software Corporation (2004b) *LogXact for Windows*, version 5, Cytel Software Corporation, Cambridge, MA.

D'Agostino RB, Chase W and Belanger A (1988) The appropriateness of some common procedures for testing the equality of two binomial proportions, *The American Statistician*, **42**: 198–202.

Daniel WW (2005) *Biostatistics: A Foundation for Analysis in the Health Sciences*, 8th edition, John Wiley & Sons, New York.

Darroch JN and Borkent M (1994) Synergism, attributable risks and interaction for two binary exposure factors, *Biometrika*, **81**: 259–270.

Davey Smith G, Egger M and Phillips AN (1997) Meta-analysis: beyond the grand mean, *British Medical Journal*, **315**: 1610–1614.

Davis AB (1993) Power of testing proportions in small two-sample studies when sample sizes are equal, *Statistics in Medicine*, **12**: 777–787.

Davis LJ (1986) Exact tests for 2×2 contingency tables, *The American Statistician*, **40**: 139–141.

Dawson–Saunders B and Trapp RG (1994) *Basic & Clinical Biostatistics*, Appleton & Lange, Norwalk, Connecticut.

DerSimonian R and Laird N (1986) Meta-analysis in clinical trials, *Controlled Clinical Trials*, **7**: 177–188.

Dewar RA and Armstrong B (1992) A simple confidence interval for the odds ratio in a two-by-two table with one zero cell, *Epidemiology*, **3**: 385–387.

Donahue RMJ (1999) A note on information seldom reported via the *P* value, *The American Statistician*, **53**: 303–306.

Donnely CA, Ghani AC, Ferguson NM, Wilesmith JW and Anderson RM (1997) Analysis of the Bovine Spongiform Encephalopathy Maternal Cohort Study: evidence for direct maternal transmission, *Applied Statistics*, **46**: 321–344.

Dubey SD (1991) Some thoughts on the one-sided and two-sided tests, *Journal of the Biopharmaceutical Statistics*, **1**: 139–150.

Duhamel P and Vetterli M (1990) Fast Fourier transforms: a tutorial review and a state of the art, *Signal Processing*, **19**: 259–299.

Dunne A, Pawitan Y and Doody L (1996) Two-sided *P*-values from discrete asymmetric distributions based on uniformly most powerful unbiased tests, *The Statistician*, **45**: 397–405.

Dunnet CW (1955) A multiple comparison procedure for comparing several treatments with a control, *Journal of the American Statistical Association*, **50**: 1096–1121.

Dunnet CW and Gent M (1996) An alternative to the use of two-sided tests in clinical trials, *Statistics in Medicine*, **15**: 1729–1738.

Dupont WD (1986) Sensitivity of Fisher's exact test to minor perturbations in 2×2 contingency tables, *Statistics in Medicine*, **5**: 629–635.

Edgington ES (1995) *Randomization Tests*, 3rd edition, Marcel Dekker, New York.

Edwardes MD (1998) The evaluation of confidence sets with application to binomial intervals, *Statistica Sinica*, **8**: 393–409.

Edwards A (1972) *Likelihood*, Cambridge University Press, Cambridge, U.K.

Efron B (1986) Why isn't everyone a Bayesian? (with discussion), *The American Statistician*, **40**: 1–11.

Efron B (2000) R. A. Fisher in the 21st century, in CR Rao and GJ Szekely (editors) (2000) *Statistics for the 21st Century: Methodologies for Applications of the Future*, pages 109–144, Marcel Dekker, New York.

Efron B (2005) Bayesians, frequentists, and scientists, *Journal of the American Statistical Association*, **100**: 1–5.

Efron B and Tibshirani RJ (1993) *An Introduction to the Bootstrap*, Chapman and Hall, London.

Egger M, Davey Smith G and Phillips AN (1997) Meta-analysis: principles and procedures, *British Medical Journal*, **315**: 1533–1537.

Elmore JG, Barton MB, Moceri VM, Polk S, Arena PJ and Fletcher SW (1998) Ten-year risk of false positive screening mammograms and clinical breast examinations, *The New England Journal of Medicine*, **338**: 1089–1096.

Elston RC and Forthofer R (1977) Testing for Hardy–Weinberg equilibrium in small samples, *Biometrics*, **33**: 536–542.

Emerson JD (1994) Combining estimates of the odds ratio: the state of the art, *Statistical Methods in Medical Research*, **3**: 157–178.

Engels EA, Schmid CH, Terrin N, Olkin I and Lau J (2000) Heterogeneity and statistical significance in meta-analysis: an empirical study of 125 meta-analyses, *Statistics in Medicine*, **19**: 1707–1728.

Essex–Sorlie D (1995) *Medical Biostatistics & Epidemiology*, Appleton & Lange, Norwalk, Connecticut.

Farewell VT (1979) Some results on the estimation of logistic models based on retrospective data, *Biometrika*, **66**: 27–32.

Farrar DB and Crump KS (1988) Exact tests for any carcinogenic effect in animal bioassay, *Fundamental and Applied Toxicology*, **11**: 652–663.

Fernandez H, Baton C, Benifla JL, Frydman R and Lelaidier C (1993) Methotrexate treatment of ectopic pregnancy: 100 cases treated by primary transvaginal injection under sonographic control, *Fertility and Sterility*, **59**: 773–777.

Fienberg SE (1977) *The Analysis of Cross-Classification Categorical Data*, MIT Press, Cambridge, MA.

Finney DJ (1947) Errors of estimation in inverse sampling, *Nature*, **160**: 195–196.

Finney DJ (1949) On a method of estimating frequencies, *Biometrika*, **36**: 233–234.

Fisher LD (1991) The use of one-sided tests in drug trials: an FDA Advisory Committee member's perspective, *Journal of the Biopharmaceutical Statistics*, **1**: 151–156.

Fisher LD (1996) Comments on Bayesian and frequentist analysis and interpretation of clinical trials, *Controlled Clinical Trials*, **17**: 423–434.

Fisher LD and van Belle G (1993) *Biostatistics: A Methodology for the Health Sciences*, John Wiley & Sons, New York.

Fisher RA (1934) *Statistical Methods for Research Workers*, 5th edition, Oliver and Boyd, Edinburgh.

Fisher RA (1935a) *The Design of Experiments*, Oliver and Boyd, Edinburgh.

Fisher RA (1935b) The logic of inductive inference (with discussion), *Journal of the Royal Statistical Society*, **98, Part I**: 35–82.

Fisher et al. (1998) Tamoxifen for prevention of breast cancer: report on the National Surgical Adjuvant Breast and Bowel Project P-1 Study, *Journal of the National Cancer Institute*, **90**: 1371–1388.

Fleiss JL (1981) *Statistical Methods for Rates and Proportions*, 2nd edition, John Wiley & Sons, New York.

Folks JF (1981) *Ideas of Statistics*, John Wiley & Sons, New York.

Forster JJ, McDonald JW and Smith PWF (1996) Monte–Carlo exact conditional tests for log-linear and logistic models, *Journal of the Royal Statistical Society, Series B*, **58**: 445–453.

Foster DA and Sullivan DM (1987) Computer program produces *p*-value graphics (letter), *American Journal of Public Health*, **77**: 879–880.

Franck WE (1986) *P*-values for discrete test statistics, *Biometrical Journal*, **28**: 403–406.

Freeman GH and Halton JH (1951) Note on an exact treatment of contingency, goodness of fit and other problems of significance, *Biometrika*, **38**: 141–149.

Freedman L (1996) Bayesian statistical methods: a natural way to assess clinical evidence, *British Medical Journal*, **313**: 569–570.

Freeman PR (1993) The role of *P*-values in analyzing trial results, *Statistics in Medicine*, **12**: 1443–1452.

Freiman JA, Chalmers TC, Smith H Jr. and Kuebler RR (1978) The importance of beta, the type II error and sample size in the design and interpretation of the randomized controlled trial, *New England Journal of Medicine,* **299**: 690–694.

Frisen M (1980) Consequences of the uses of conditional inference in the analysis of a correlated contingency table, *Biometrika,* **67**: 23–30.

Fu YX and Arnold J (1992) A table of exact sample sizes for use with Fisher's exact test for 2×2 tables, *Biometrics,* **48**: 1103–1112.

Fuchs C (2001) UMPU and alternative tests for association in 2×2 tables, *Biometrics,* **57**: 535–538.

Gail M and Gart JJ (1973) The determination of sample sizes for use with the exact conditional test 2×2 comparative trials, *Biometrics,* **29**: 441–448.

Gail MH, Lubin JH and Rubinstein LV (1981) Likelihood calculations for matched case-control studies and survival studies with tied death times, *Biometrika,* **68**: 703–707.

Gail M and Mantel N (1977) Counting the number of $r \times c$ contingency tables with fixed margins, *Journal of the American Statistical Association,* **72**: 859–862.

Gardner MJ and Altman DG (1986) Confidence intervals rather than *P* values: estimation rather than hypothesis testing, *British Medical Journal,* **292**: 746–750.

Gardner MJ and Altman DG (1989) *Statistics with Confidence - Confidence Intervals and Statistical Guidelines,* British Medical Journal, London.

Garside GR (1971) An accurate correction for the X^2 test in the homogeneity case of 2×2 contingency tables, *New Journal of Statistical Operations Research,* **7**: 1–26.

Garside GR and Mack C (1967) Correct confidence limits for the 2×2 homogeneity contingency table with small frequencies, *New Journal of Statistical Operations Research,* **3**(2): 1–25.

Garside GR and Mack C (1968a) Alternative confidence limits for the 2×2 homogeneity contingency table with small frequencies, *New Journal of Statistical Operations Research,* **4**(1): 2–18.

Garside GR and Mack C (1968b) Further tables of confidence limits for the 2×2 homogeneity contingency table, *New Journal of Statistical Operations Research,* **4**(3): 9–34.

Garside GR and Mack C (1971) Actual type I error probabilities for the various tests in the homogeneity case of the 2×2 contingency table, *The America Statistician,* **30**: 18–21.

Gart JJ (1962) On the combination of relative risks, *Biometrics,* **18**: 601–610.

Gart JJ (1970) Point and interval estimation of the common odds ratio in the combination 2×2 tables with fixed marginals, *Biometrika,* **57**: 471–475.

Gart JJ (1971) The comparison of proportions: a review of significance tests, confidence intervals and adjustments for stratification, *Review of the International Statistical Institute,* **39**: 148–169.

Gart JJ (1972) Interaction tests for $2 \times s \times t$ contingency tables, *Biometrika,* **59**: 309–316.

Gart JJ (1977) Exact and approximate tests for relative potency, *Bulletin of the International Statistical Institute,* **47**: 172–175.

Gart JJ (1978) The analysis of ratios and cross-product ratios of Poisson variates with application to incidence rates, *Communications in Statistics - Theory and Methods,* **7**: 917–937.

Gart JJ, Krewski D, Lee PN, Tarone RE and Wahrendorf J (1986) *Statistical Methods in Cancer Research: Volume III: The Design and Analysis of Long-Term Animal Experiments,* IARC Scientific Publications, Lyon.

Garwood F (1936) Fiducial confidence limits for the Poisson distribution, *Biometrika,* **46**: 441–453.

Gatswirth JL (1988a) *Statistical Reasoning in Law and Public Policy*, Volume 1, Academic Press, San Diego, CA.

Gatswirth JL (1988b) *Statistical Reasoning in Law and Public Policy*, Volume 2, Academic Press, San Diego, CA.

Gautam S (2002) Analysis of mixed categorical data in $2 \times K$ contingency tables, *Statistics in Medicine*, **21**: 1471–1484.

Geddes J, Freemantle N, Harrison P and Bebbington P (for the National Schizophrenia Guideline Development Group) (2000) Atypical antipsychotics in the treatment of schizophrenia: systematic overview and meta-regression analysis, *British Medical Journal*, **321**: 1371–1376.

Gentleman, JF (1975) Algorithm AS88: Generation of all $_NC_R$ combinations by simulating nested Fortran Do loops, *Applied Statistics*, **24**: 374–376.

George VT and Elston RC (1993) Confidence limits based on the first occurrence of an event, *Statistics in Medicine*, **12**: 685–690.

Gibbons JD and Pratt JW (1975) P-values: interpretation and methodology, *The American Statistician*, **29**: 20–25.

Gill CJ, Sabin L and Schmid CH (2005) Why clinicians are natural Bayesians, *British Medical Journal*, **330**: 1080–1083.

Glasziou P, Irwig L, Bain C and Colditz G (2001) *Systematic Reviews in Health Care: A Practical Guide*, Cambridge University Press, Cambridge.

Glynn RJ and Rosner B (1992) Accounting for correlation between fellow eyes in regression analysis, *Archives of Opthalmology*, **110**: 381–387.

Good IJ (1981) The fast calculation of the exact distribution of Pearson's chi-squared and of the number of repeats within the cells of a multinomial by using a fast Fourier transform, *Journal of Statistical Computation and Simulation*, **14**: 71–78.

Good IJ (1982) The fast calculation of the exact distribution of Pearson's chi-squared and of the number of repeats within the cells of a multinomial by using a fast Fourier transform: an addendum, *Journal of Statistical Computation and Simulation*, **15**: 336–337.

Good IJ (1983) An improved algorithm for the fast calculation of the exact distribution of Pearson's chi-squared, *Journal of Statistical Computation and Simulation*, **17**: 236–242.

Good IJ and Crook JF (1978) Correction to "Exact distribution for X^2 and for the likelihood-ratio statistic for the equiprobable multinomial distribution," *Journal of the American Statistical Association*, **73**: 900.

Good IJ, Gover TN and Mitchell GJ (1970) Exact distribution for X^2 and for the likelihood-ratio statistic for the equiprobable multinomial distribution, *Journal of the American Statistical Association*, **65**: 267–283.

Good IJ, Gover TN and Mitchell GJ (1971) Correction to "Exact distribution for X^2 and for the likelihood-ratio statistic for the equiprobable multinomial distribution," *Journal of the American Statistical Association*, **66**: 229.

Goodman LA (1964) Simple methods for analyzing three-factor interaction in contingency tables, *Journal of the American Statistical Association*, **59**, 319–332.

Goodman R (1988) *Introduction to Stochastic Models*, Benjamin/Cummings, Menlo Park, CA.

Goodman SN (1992) A comment on replication, p-values and evidence, *Statistics in Medicine*, **11**: 875–879.

Goodman SN (1999a) Towards evidence-based medical statistics, 1: the P-value fallacy, *Annals of Internal Medicine*, **130**: 995–1004.

Goodman SN (1999b) Towards evidence-based medical statistics, 2: the Bayes factor, *Annals of Internal Medicine*, **130**: 1005–1013.

Goodman SN (2001) Of *P*-values and Bayes: a modest proposal, *Epidemiology*, **12**: 295–297.

Goodman SN and Royall R (1988) Evidence and scientific research, *American Journal of Epidemiology*, **78**: 1568–1574.

Gordon H (1997) *Discrete Probability*, Springer–Verlag, New York.

Gordon I (1994) Sample size for two independent proportions: a review, *Australian Journal of Statistics*, **36**: 199–209.

Gosh BK (1979) A comparison of some approximate confidence intervals for the binomial parameter, *Journal of the American Statistical Association*, **74**: 894–900.

Granville V and Schifflers E (1993) Efficient algorithms for exact inference in 2×2 contingency tables, *Statistics and Computing*, **3**: 83–87.

Graubard BI and Korn E (1987) Choice of column scores for testing independence in ordered $2 \times k$ contingency tables, *Biometrics*, **43**: 471–476.

Greenland S (1991) On the logical justification of conditional tests for two-by-two contingency tables, *The American Statistician*, **45**: 248–251.

Greenland S (1994) Invited commentary: a critical look at some popular meta-analytic methods (with comment by I Olkin and response), *American Journal of Epidemiology*, **140**: 290–301.

Greenland S (1995a) Dose-response and trend analysis in epidemiology: alternatives to categorical analysis, *Epidemiology*, **6**: 356–365.

Greenland S (1995b) Problems in the average-risk interpretation of categorical dose-response analyses, *Epidemiology*, **6**: 563–565.

Greenwood PE and Nikulin MS (1996) *A Guide to Chi-Squared Testing*, John Wiley & Sons, New York.

Guess HA, Lydick EG, Small RD and Miller LP (1987) Exact binomial confidence intervals for the relative risk in follow-up studies with incidence density data, *American Journal of Epidemiology*, **125**: 340–347.

Guess HA and Thomas JE (1990) A rapidly converging algorithm for exact binomial confidence intervals about the relative risk in follow-up studies with stratified incidence-density data, *Epidemiology*, **1**: 75–77.

Guo SW and Thompson EA (1992) Performing the exact test of Hardy–Weinberg proportion for multiple alleles, *Biometrics*, **48**: 361–372.

Guthrie D and Youssef MN (1970) Empirical evaluation of some chi-square tests for the order of a Markov chain, *Journal of the American Medical Association*, **65**: 631–634.

Haber M (1981) Exact significance levels of goodness-of-fit tests for the Hardy–Weinberg equilibrium, *Human Heredity*, **31**: 161–166.

Haber M (1983) Sample sizes for the exact test of "no interaction" in $2 \times 2 \times 2$ tables, *Biometrics*, **39**: 493–498.

Haber M (1986a) An exact unconditional test for the 2×2 comparative trial, *Psychological Bulletin*, **99**: 129–132.

Haber M (1986b) A modified exact test for 2×2 contingency tables, *Biometrical Journal*, **28**: 455–463.

Haber M (1987) A comparison of some conditional and unconditional exact tests for 2 by 2 tables, *Communications in Statistics - Simulation and Computation*, **16**: 999–1013.

Haber M (1990) Comments on "The test of homogeneity for 2×2 contingency tables: a review of and some personal opinions on the controversy" by G Camilli, *Psychological Bulletin*, **108**: 146–149.

Haberman SJ (1988) A warning on the use of chi-square statistics with frequency tables with small expected counts, *Journal of the American Statistical Association*, **83**: 555–560.

Haldane JBS (1945a) A labor-saving method of sampling, *Nature*, **155**: 49–50.

Haldane JBS (1945b) On a method of estimating frequencies, *Biometrika*, **33**: 222–224.

Hall GH and Round AP (1994) Logistic regression - explanation and use, *Journal of the Royal College of Physicians*, **28**: 242–246.

Halperin M, Ware JH, Byar DP, Mantel N, Brown CC, Koziol J, Gail M and Green SB (1977) Testing for interaction in an $I \times J \times K$ table, *Biometrika*, **64**: 271–275.

Halpern SD, Karlawish JH and Berlin JA (2002) The continuing unethical conduct of underpowered trials, *Journal of the American Medical Association*, **288**: 358–362.

Hameer MF (1990) *The Value of Erythrocyte Sedimentation Rate in the Diagnosis of Infection in Children with Protein Energy Malnutrition*, unpublished M.Med. dissertation, Faculty of Medicine, University of Dar es Salaam, Tanzania.

Hamilton MA (1979) Choosing a parameter for a 2×2 or a $2 \times 2 \times 2$ table analysis, *American Journal of Epidemiology*, **109**: 362–375.

Hanley JA and Lippman–Hand A (1983) If nothing goes wrong, is everything all right? Interpreting zero numerators, *Journal of the American Medical Association*, **249**: 1743–1745.

Harkness WL (1965) Properties of the extended hypergeometric distribution, *The Annals of Mathematical Statistics*, **36**: 938–945.

Haseman JK (1978) Exact sample sizes for use with the Fisher–Irwin test for 2×2 tables, *Biometrics*, **34**: 106–109.

Hashemi L, Nandram B and Goldberg R (1997) Bayesian analysis for a single 2×2 table, *Statistics in Medicine*, **16**: 1311–1328.

Haviland MG (1990) Yates's correction for continuity and the analysis of 2×2 contingency tables (with comments), *Statistics in Medicine*, **9**: 363–383.

Heinze G, Gnant M and Schemper M (2003) Exact log-rank tests for unequal follow-up, *Biometrics*, **59**: 1151–1157.

Helland IS (1995) Simple counterexamples against the conditionality principle, *The American Statistician*, **49**: 351–356.

Hill AB (1962) *Statistical Methods in Clinical and Preventive Medicine*, E & S Livingstone, Ltd., Edinburgh.

Hilton JF and Mehta CR (1993) Power and sample size calculations for exact conditional tests with ordered categorical data, *Biometrics*, **49**: 609–616.

Hirji KF (1991) A comparison of exact, mid-*P*, and score tests for matched case-control studies, *Biometrics*, **47**: 487–496.

Hirji KF (1992) Computing exact distributions for polytomous response data, *Journal of the American Statistical Association*, **87**: 487–492.

Hirji KF (1994) Exact analysis for paired binary data, *Biometrics*, **50**: 964–974.

Hirji KF (1996) A note on exact analysis of several 2×2 tables, *Biometrics*, **52**: 1018–1025.

Hirji KF (1997a) A review and a synthesis of the fast Fourier transform algorithms for exact analysis of discrete data, *Computational Statistics & Data Analysis*, **25**: 321–336.

Hirji KF (1997b) A comparison of algorithms for exact goodness-of-fit tests for multinomial data, *Communications in Statistics - Simulations and Computations*, **26**: 1197–1227.

Hirji KF (1998) Assessing fast Fourier transform algorithms, *Computational Statistics & Data Analysis*, **27**: 1–9.

Hirji KF, Elashoff RM, Moore DH II and Bennett DE (1988) Exact versus asymptotic analysis for a matched case-control study, *Statistics in Medicine*, **7**: 765–772.

Hirji KF and Johnson TD (1996) A comparison of algorithms for exact analysis of unordered $2 \times K$ contingency tables, *Computational Statistics & Data Analysis*, **21**: 419–429.

Hirji KF and Johnson TD (1999) Exact inference on stratified two-stage Markov chain models, *Computational Statistics & Data Analysis*, **31**: 159–186.

Hirji KF and Johnson TD (2000) Is the Fisher exact test appropriate for a transition matrix? (*unpublished paper*).

Hirji KF, Mehta CR and Patel NR (1987) Computing distributions for exact logistic regression, *Journal of the American Statistical Association*, **82**: 1110–1117.

Hirji KF, Mehta CR and Patel NR (1988) Exact inference for matched case-control studies, *Biometrics*, **44**: 803–814.

Hirji KF, Tan SJ and Elashoff RM (1991) A quasi-exact test for comparing two binomial proportions, *Statistics in Medicine*, **10**: 1137–1153.

Hirji KF and Tang ML (1998) A comparison of tests for trend, *Communications in Statistics - Theory and Methods*, **27**: 943–963.

Hirji KF, Tang ML, Vollset SE and Elashoff RM (1994) Efficient power computation for exact and mid-*p* tests for the common odds ratio in several 2×2 tables, *Statistics in Medicine*, **13**: 1539–1549.

Hirji KF, Tsiatis AA and Mehta CR (1989) Median unbiased estimation for binary data, *The American Statistician*, **43**: 7–11.

Hirji KF and Vollset SE (1994a) Algorithm AS 289: Computing exact distributions for several 2×2 tables, *Applied Statistics*, **43**: 270–274.

Hirji KF and Vollset SE (1994b) Algorithm AS 293: Computing exact distributions for several ordered $2 \times K$ tables, *Applied Statistics*, **43**: 541–548.

Hirji KF and Vollset SE (1994c) *Single Parameter Exact Inference on Discrete Data*, UCLA Statistics Series Report No. 148, University of California, Los Angeles, available at www.ucla.stat.edu.

Hirji KF, Vollset SE, Reis IM and Afifi AA (1996) Exact tests for interaction in several 2×2 tables, *Journal of Computational and Graphical Statistics*, **5**: 209–224.

Hlatky MA and Whittemore AS (1991) The importance of models in the assessment of synergy, *Journal of Clinical Epidemiology*, **44**: 1287–1288.

Hoel PG (1954) A test for Markoff chains, *Biometrika*, **41**: 430–433.

Holtzman GI and Good IJ (1986) The Poisson and chi-squared approximations as compared with the true upper-tail probability of Pearson's X^2 for equiprobable multinomials, *Journal of Statistical Planning and Inference*, **13**: 283–295.

Hooiveld M, Heederick DJJ, Kogevinas M, Boffetta P, Needham LJ, Patterson DG, Bueno-de-Mesquita HB (1998) Second follow-up of a Dutch cohort occupationally exposed to phenoxy herbicides, chlorophenols and contaminants, *American Journal of Epidemiology*, **147**: 891–901.

Horowitz E and Sahni S (1984) *Fundamentals of Computer Algorithms*, Computer Science Press, Rockville, Maryland.

Hosmer DW and Lemeshow S (1989) *Applied Logistic Regression*, John Wiley & Sons, New York.

Hsueh HM, Liu JP and Chen JJ (2001) Unconditional exact tests for equivalence or noninferiority for paired binary end points, *Biometrics*, **57**: 478–483.

Hubbard R and Bayarri MJ (2003a) Confusion over measures of evidence (*p*'s) versus errors (α's) in classical testing, *The American Statistician*, **57**: 171–178.

Hubbard R and Bayarri MJ (2003b) Rejoinder, *The American Statistician*, **57**: 181–182.

Hull R, Raskob G, Pineo G et al. (1993) A comparison of subcutaneous low-molecular-weight heparin with warfarin sodium for prophylaxis against deep vein thrombosis after hip or knee implantation. *New England Journal of Medicine*, **329**: 1370–1376.

Hulley SB and Cummings SR (editors) (1988) *Designing Clinical Research*, Williams & Wilkins, Baltimore.

Hwang JT and Yang M (2001) An optimality theory for mid *p*-values in 2×2 contingency tables, *Statistica Sinica*, **11**: 807–826.

Infante PF (1976) Oncogenic and mutagenic risks in communities with polyvinyl chloride production facilities, *Annals of the New York Academy of Sciences*, **271**: 49–57.

Inman HF (1994) Karl Pearson and R. A. Fisher on statistical tests: a 1935 exchange from Nature, *The American Statistician*, **48**: 2–11.

Irgens A, Krüger K, Skorve AH and Irgens LM (1997) Male proportion in offspring of parents exposed to strong static and extremely low-frequency electromagnetic fields in Norway, *American Journal of Industrial Medicine*, **32**: 557–561.

Irwin JO (1935) Tests of significance for differences between percentages based on small numbers, *Metron*, **12**: 83–94.

Jeans JH (1935) Age of the universe, *Nature*, **136**: 432.

Jekel JF, Elmore JG and Katz DL (1996) *Epidemiology, Biostatistics and Preventive Medicine*, W B Saunders Company, Philadelphia.

Jennrich RI (1984) Some exact tests for comparing survival curves in the presence of unequal right censoring, *Biometrika*, **71**: 57–64.

Johnson NL and Kotz S (1969) *Distributions in Statistics: Discrete Distributions*, John Wiley & Sons, New York.

Johnson NL, Kotz S and Kemp AW (1992) *Univariate Discrete Distributions*, John Wiley & Sons, New York.

Johnson TD (1997) *Exact Statistical Inference for Markov Chains*, unpublished Ph.D. Dissertation, Department of Biostatistics, University of California, Los Angeles.

Johnston AM, Valentine ME, Ottinger J, Baydo R, Gryszowka V, Vavro C, Weinhold K, St. Clair M and McKinney RE (2001) Immune reconstitution in human immunodeficiency virus-infected children receiving highly active antiretroviral therapy: a cohort study, *Pediatric Infectious Diseases Journal*, **20**: 941–946.

Jollife IT (1995) Sample size and the central limit theorem: the Poisson distribution as an illustration, *The American Statistician*, **49**: 3.

Jones B and Kenward MG (1989) *Design and Analysis of Cross-Over Trials*, London, Chapman and Hall.

Jones MP, O'Gorman TW, Lemke JH and Woolson RF (1989) A Monte Carlo investigation of homogeneity tests of the odds ratio under various sample size configurations, *Biometrics*, **45**: 171–181.

Kabaila P and Byrne J (2001) Exact short confidence intervals from discrete data, *Australia New Zealand Journal of Statistics*, **43**: 303–309.

Kaiser L, Lew D, Hirschel B, Aukenthaler R, Morabia A, Heald A, Benedict P, Terrier F, Wunderli W, Matter L, Germann D, Voegeli J, and Stalder H (1996) Effects of antibiotic treatment in the subset of common-cold patients who have bacteria in their nasopharyngeal secretions, *The Lancet*, **347**: 1507–1510.

Kalbfleisch JD (1982) Ancillary statistic, in S Kotz and NL Johnson (editors) *Encyclopedia of Statistical Sciences*, **1**, John Wiley & Son, New York, pages 77–81.

Katti SK (1973) Exact distribution for the chi-square test in the one way table, *Communications in Statistics*, **2**: 435–447.

Katz JN, Daltroy LH, Brennan TA and Liang MH (1992) Informed consent and the prescription on nonsteroidal antiinflammatory drugs, *Arthritis and Rheumatism*, **35**: 1257–1263.

Kempthorne O (1955) The randomization theory of experimental inference, *Journal of the American Statistical Association*, **50**: 946–967.

Kennedy WJ and Gentle JE (1980) *Statistical Computing*, Marcel Dekker, New York.

Kessler DA (1993) Introducing MedWatch: a new approach to reporting medication and device adverse effects and product problems, *Journal of the American Medical Association*, **269**: 2765–2768.

Kikuchi DA (1987) Inverse sampling in case control studies involving a rare exposure, *Biometrical Journal*, **29**: 243–246.

Kim D and Agresti A (1995) Improved exact inference about conditional association in three-way contingency tables, *Journal of the American Statistical Association*, **90**: 632–639.

Kirkwood JM, Ibrahim J, Sondack VK, Ernstoff MS, Flaherty L and Haluska FJ (2002) Use and abuse of statistics in evidence-based medicine (letter), *Journal of Clinical Oncology*, **20**: 4122–4123.

Kiviluoto T, Sirén J, Luukkonen P and Kivilaakso E (1998) Randomized trial of laparoscopic versus open cholesystectomy for acute and gangrenous cholecystitis, *The Lancet*, **351**: 321–325.

Klotz J (1972) Markov chain clustering of births by sex, in LM LeCam, J Neyman and EL Scott (editors) (1972) *Proceedings of the Sixth Berkeley Symposium on Mathematical Statistics and Probability*, **4**: pages 173–185, University of California Press, Berkeley.

Koch GG (1991) One-sided and two-sided tests and p values, *Journal of the Biopharmaceutical Statistics*, **1**: 161–170.

Kocherlakota S and Kocherlakota K (1992) *Bivariate Discrete Distributions*, Marcel Dekker, New York.

Korfhage RR (1984) *Discrete Computational Structures*, 2nd edition, Academic Press, New York.

Krauth J (1988) *Distribution-Free Statistics: An Applications-Oriented Approach*, Elsevier Science, Amsterdam.

Kreiner S (1987) Analysis of multidimensional contingency tables by exact conditional tests: techniques and strategies, *Scandinavian Journal of Statistics*, **14**: 97–112.

Kronsjö L (1987) *Algorithms: Their Complexity and Efficiency*, 2nd edition, John Wiley & Sons, Chichester, U.K.

Kudô A and Tarumi T (1978) 2×2 tables emerging out of different chance mechanisms, *Communications in Statistics - Theory and Methods*, **7**: 977–986.

Kulkarni PM, Tripathi RC and Michalek JE (1998) Maximum (max) and mid-p confidence intervals and p values for the standardized mortality and incidence ratios, *American Journal of Epidemiology*, **147**: 83–86.

Lancaster HO (1949) The combination of probabilities arising from data in discrete distributions, *Biometrika*, **36**: 370–382.

Lancaster HO (1952) Statistical control of counting experiments, *Biometrika*, **39**: 419–422.

Lancaster HO (1961) Significance tests in discrete distributions, *Journal of the American Statistical Association*, **56**: 223–234.

Lancaster HO (1969) *The Chisquared Distribution*, John Wiley & Sons, New York.

Lange K (2002) *Mathematical and Statistical Methods for Genetic Analysis*, 2nd edition, Springer–Verlag, New York.

Laupacis A, Sackett DL and Roberts RS (1988) An assessment of clinically useful measures of the consequence of treatment, *The New England Journal of Medicine*, **318**: 1728–1733.

Lazarou J, Pomeranz BH and Corey PN (1998) Incidence of adverse drug reactions in hospitalized patients: a meta-analysis of prospective studies, *Journal of the American Medical Association*, **279**: 1200–1204.

Lee YJ (1980) Test of trend in count data: multinomial distribution case, *Journal of the American Statistical Association*, **75**: 1010–1014.

Lehmann EL (1986) *Testing Statistical Hypotheses*, 2nd edition, John Wiley & Sons, New York.

Lehmann EL (1993) The Fisher, Neyman–Pearson theories of testing hypotheses: one theory or two? *Journal of the American Statistical Association*, **88**: 1242–1249.

Lehmann EL (2001) *Theory of Point Estimation*, 2nd edition, Springer, New York.

Lehmer DH (1964) The machine tools of combinatorics, in EF Beckenbach (editor), *Applied Combinatorial Mathematics*, John Wiley and Sons, New York, pages 5–31.

Lemen RA, Johnson WM, Wagoner JK, Archer VE and Saccomanno G (1976) Cytologic observations and cancer incidence following exposure to BCME, *Annals of the New York Academy of Sciences*, **271**: 71–80.

Lens MB and Dawes M (2002a) Interferon alfa therapy for malignant melanoma: a systematic review of the randomized controlled trials, *Journal of Clinical Oncology*, **20**: 1818–1825.

Lens MB and Dawes M (2002b) In reply: use and abuse of statistics in evidence-based medicine (letter), *Journal of Clinical Oncology*, **20**: 4123–4124.

Levin JL, McLarty JW, Hurst GA, Smith AN and Frank AL (1998) Tyler asbestos workers: mortality experience in a cohort exposed to amosite, *Occupational and Environmental Medicine*, **55**: 155–160.

Lewontin RC and Felsenstein J (1965) The robustness of homogeneity tests in $2 \times N$ tables, *Biometrics*, **21**: 19–33.

Liang KY and Self SG (1985) Tests for homogeneity of odds ratio when the data are sparse, *Biometrika*, **72**: 353–358.

Liao J (1992) An algorithm for the mean and variance of the noncentral hypergeometric distribution, *Biometrics*, **48**: 889–892.

Liao JG and Hall CB (1995) Fast and accurate computation of the exact confidence limits for the common odds ratio in several 2×2 tables, *Journal of Computational and Graphical Statistics*, **4**: 173–179.

Liddell D (1978) Practical tests of 2×2 contingency tables, *The Statistician*, **25**: 295–304.

Liddell FDK (1984) Simple exact analysis of the standardized mortality ratio, *Journal of Epidemiology and Community Health*, **38**: 85–88.

Ling RF (1992) Just say no to binomial (and other discrete distributions) tables, *The American Statistician*, **46**: 53–54.

Lissitz RW (1972) Comparison of the small sample power of the chi-square and likelihood ratio tests of the assumptions for stochastic models, *Journal of the American Statistical Association*, **67**: 574–577.

Little RJA (1989) Testing the equality of two independent binomial proportions, *The American Statistician*, **43**: 283–288.

Liu Q, Li Y and Boyett JM (1997) Controlling false positive rates in prognostic analyses with small samples, *Statistics in Medicine*, **16**: 2095–2101.

Lloyd CJ (1988) Doubling the one-sided P-value in testing independence in 2×2 tables against a two-sided alternative, *Statistics in Medicine*, **5**: 629–635.

Louis EJ and Dempster ER (1987) An exact test of Hardy–Weinberg and multiple alleles, *Biometrics*, **43**: 805–811.

Ludbrook J and Dudley H (1998) Why permutation tests are superior to t and F tests in biomedical research, *The American Statistician*, **52**: 127–132.

Lui KJ (1995a) Notes on conditional confidence intervals under inverse sampling, *Statistics in Medicine*, **14**: 2051–2056.

Lui KJ (1995b) Confidence limits for the population prevalence rate based on the negative binomial distribution, *Statistics in Medicine*, **14**: 1471–1477.

Lui KJ (1996a) Samples size for the exact conditional test under inverse sampling, *Statistics in Medicine*, **15**: 671–678.

Lui KJ (1996b) Point estimation of relative risk under inverse sampling, *Biometrical Journal*, **38**: 669–680.

Lui KJ (1997) Exact equivalence test for risk ratio and its sample size determination under inverse sampling, *Statistics in Medicine*, **16**: 1777–1786.

Lui KJ (2004) *Statistical Estimation of Epidemiological Risk*, John Wiley & Sons, New York.

Lydersen S and Laake P (2003) Power comparison of two-sided exact tests for association in 2×2 contingency tables using standard, mid-p, and randomized test versions, *Statistics in Medicine*, **22**: 3859–3871.

Maclure M and Greenland S (1992) Tests for trend and dose response: misinterpretations and alternatives, *American Journal of Epidemiology*, **135**: 96–104.

Mainous AG, Hueston WJ and Clark JR (1996) Antibiotics and upper respiratory infections: do some folks think there is a cure for the common cold? *The Journal of Family Practice*, **42**: 357–361.

Malone KE, Daling JR, Thompson JD, O'Brien CA, Francisco LV and Ostrander EA (1998) BRCA1 mutations and breast cancer in the general population, *Journal of the American Medical Association*, **279**: 922–929.

Manly BFJ (1997) *Randomization, Bootstrap and Monte Carlo Methods in Biology*, 2nd edition, Chapman and Hall, London.

Mantel N (1963) Chi-square tests with one degree of freedom: extensions of the Mantel–Haenszel procedure, *Journal of the American Statistical Association*, **58**: 690–700.

Mantel N (1992) An invalid comparison of several point estimators of the odds ratio in a single 2×2 contingency table, *Biometrics*, **48**: 1289–1292.

Mantel N, Brown C and Byar DP (1977) Tests for homogeneity of effect in epidemiological investigation, *American Journal of Epidemiology*, **106**: 125–129.

Mantel N and Haenszel W (1959) Statistical aspects of analysis of data from retrospective studies of disease, *Journal of the National Cancer Institute*, **22**: 719–748.

Mantel N and Hankey BF (1975) The odds ratio of a 2×2 contingency table, *The American Statistician*, **29**: 143–145.

March DL (1972) Exact probabilities for $R \times C$ contingency tables, *Communications of the Association of Computing Machinery*, **15**: 991–992.

Martin Andrés AA (1991) A review of classic non-asymptotic methods for two proportions by means of independent samples, *Communications in Statistics - Simulation and Computation*, **20**: 551–583.

Martin Andrés AA and Silva Mato A (1994) Choosing the optimal unconditional test for comparing two independent proportions, *Computational Statistics and Data Analysis*, **17**: 555–574.

Martin DO and Austin H (1991) An efficient algorithm for computing conditional maximum likelihood estimates and exact confidence limits for a common odds ratio, *Epidemiology*, **2**: 359–362.

Martin DO and Austin H (1996) Exact estimates for a rate ratio, *Epidemiology*, **7**: 29–33.

Matthews JN (1995) Small clinical trials: are they all bad?, *Statistics in Medicine*, **14**: 115–126.

Matthews JNS and Altman DG (1996a) Interaction 2: compare effect sizes not P values, *British Medical Journal*, **313**: 808.

Matthews JNS and Altman DG (1996b) Interaction 3: how to examine heterogeneity, *British Medical Journal*, **313**: 862.

May WL and Johnson WD (1997) The validity and power of tests for equality of two correlated proportions, *Statistics in Medicine*, **16**: 1081–1096.

McDonald LL, Davis BM and Milliken GA (1977) A nonrandomized unconditional test for comparing two proportions in 2×2 contingency tables, *Technometrics*, **19**: 145–157.

McDonald JW, De Roure DC and Michaelides DT (1998) Exact tests for two-way symmetric contingency tables, *Statistics and Computing*, **8**: 391–399.

McDonald JW and Smith PWF (1995) Exact conditional tests of quasi-independence for triangular contingency tables: estimating attained significance levels, *Applied Statistics*, **44**: 143–151.

McDonald JW, Smith PWF and Forster JJ (1999) Exact tests of goodness of fit for log-linear models for rates, *Biometrics*, **55**: 620–624.

McKinney WP, Young MJ, Hartz A and Lee MBF (1989) The inexact use of Fisher's exact test in six major medical journals, *Journal of the American Medical Association*, **261**: 3430–3433.

Mehrotra DV, Chan ISF and Berger RL (2003) A cautionary note on exact unconditional inference for a difference between two independent binomial proportions, *Biometrics*, **59**: 441–450.

Mehta CR (1994) The exact analysis of contingency tables in medical research, *Statistical Methods in Medical Research*, **3**: 135–156.

Mehta CR and Hilton JF (1993) Exact power of conditional and unconditional tests: going beyond the 2×2 contingency table (correction: 94V48 p175–176), *The American Statistician*, **47**: 91–98.

Mehta CR and Patel NR (1980) A network algorithm for the exact treatment of the $2 \times k$ contingency table, *Communications in Statistics - Simulation and Computation*, **9**: 649–664.

Mehta CR and Patel NR (1983) A network algorithm for performing the Fisher's exact test in $r \times c$ contingency tables, *Journal of the American Statistical Association*, **78**: 427–434.

Mehta CR and Patel NR (1986a) FEXACT: A FORTRAN subroutine for Fisher's exact test on unordered $r \times c$ contingency tables, *ACM Transactions on Mathematical Software*, **12**: 154–161.

Mehta CR and Patel NR (1986b) A hybrid algorithm for Fisher's exact test in unordered $R \times C$ contingency tables, *Communications in Statistics - Theory and Methods*, **15**: 387–403.

Mehta CR and Patel NR (1995) Exact logistic regression: theory and examples, *Statistics in Medicine*, **14**: 2143–2160.

Mehta CR, Patel NR and Gray R (1985) Computing an exact confidence interval for the common odds ratio in several 2 by 2 contingency tables (correction: 85V81 p1132), *Journal of the American Statistical Association*, **80**: 969–973.

Mehta CR, Patel NR and Senchaudhuri P (1988) Importance sampling for estimating exact probabilities in permutational inference, *Journal of the American Statistical Association*, **83**: 999–1005.

Mehta CR, Patel NR and Senchaudhuri P (1992) Exact stratified linear rank tests for ordered categorical and binary data, *Journal of Computational and Graphical Statistics*, **1**: 21–40.

Mehta CR, Patel NR and Senchaudhuri P (1998) Exact power and sample-size computation for the Cochran–Armitage trend test, *Biometrics*, **54**: 1615–1621.

Mehta CR, Patel NR and Tsiatis AA (1984) Exact significance testing to establish treatment equivalence with ordered categorical data, *Biometrics*, **40**: 819–825.

Mehta CR, Patel NR and Wei LJ (1988) Constructing exact significance tests with restricted randomization rules, *Biometrika*, **75**: 295–302.

Mehta CR and Walsh SJ (1992) Comparison of exact, mid-p, and Mantel–Haenszel confidence intervals for the common odds ratio across several 2×2 contingency tables (correction: 93V47 p86–87), *The American Statistician*, **46**: 146–150.

Merigan TC, Renlund DG, Keay S, Bristow MR, Starnes V, O'Connell JB, Resta S, Dunn D, Gamberg P, Ratkovec RM, Richenbacher WE, Millar RC, DuMond C, DeAmond B, Sullivan V, Cheney T, Buhles W and Stinson EB (1992) A controlled trial of ganciclovir to prevent cytomegalovirus disease after heart transplantation, *New England Journal of Medicine*, **326**: 1182–1186.

Michael M, Boyce WT and Wilcox AJ (1984) *Biomedical Bestiary: An Epidemiologic Guide to Flaws and Fallacies in the Medical Literature*, Little, Brown and Company, Boston.

Mielke PW and Berry KJ (1992) Fisher's exact probability test for cross-classification tables, *Educational and Psychological Measurement*, **52**: 97–101.

Miettinen OS (1968) The matched pairs design in the case all-or-none responses, *Biometrics*, **24**: 339–352.

Miettinen OS (1969) Individual matching with multiple controls in the case all-or-none responses, *Biometrics*, **25**: 339–355.

Miettinen OS (1974) Comment on "Some reasons for not using the Yates' continuity correction on 2×2 contingency tables" by WJ Conover, *Journal of the American Statistical Association*, **69**: 380–382.

Miettinen OS (1985) *Theoretical Epidemiology*, John Wiley & Sons, New York.

Miettinen O and Nurminen M (1985) Comparative analysis of two rates, *Statistics in Medicine*, **4**: 213–226.

Moher D, Dulberg CS and Wells GA (1994) Statistical power, sample size and their reporting in randomized controlled trials, *Journal of the American Medical Association*, **272**: 122–124.

Morgan WM and Blumenstein BA (1991) Exact conditional tests for heirarchichal models in multidimensional contingency tables, *Applied Statistics*, **40**: 435–442.

Morton AP and Dobson AJ (1990) Analyzing ordered categorical data from two independent samples, *British Medical Journal*, **301**: 971–973.

Moses LE, Emerson JD and Hosseini H (1984) Analyzing data from ordered categories, *The New England Journal of Medicine*, **311**: 442–448.

Mosteller F and Chalmers TC (1992) Some progress and problems in meta-analysis of clinical trials, *Statistical Science*, **7**: 227–236.

Moyé LA (2000) *Statistical Reasoning in Medicine: The Intuitive P-Value Primer*, Springer–Verlag, New York.

Mulder PGH (1983) An exact method for calculating a confidence interval of a Poisson parameter (letter), *American Journal of Epidemiology*, **117**: 377.

Newcombe RG (1998a) Two-sided confidence intervals for the single proportion: comparison of seven methods, *Statistics in Medicine*, **17**: 857–872.

Newcombe RG (1998b) Interval estimates for the difference between two independent proportions: comparison of eleven methods, *Statistics in Medicine*, **17**: 873–890.

Newman SC (2001) *Biostatistical Methods in Epidemiology*, John Wiley & Sons, New York.

Newman TB (1995) If almost nothing goes wrong, is almost everything all right? Interpreting small numerators (letter), *Journal of the American Medical Association*, **274**: 1013.

Neyman J (1965) Certain chance mechanisms involving discrete distributions, in GP Patil (editor) *Classical and Contagious Discrete Distributions*, pages 4–14, Pergamon Press, Oxford.

Nordrehaug JE, Chronos NAF, Priestly KA, Buller NP, Foran J, Wainwright R, Vollset SE and Sigwart U (1996) Randomized evaluation of an inflatable femoral artery compression device after cardiac catheterization, *Journal of Interventional Cardiology*, **9**: 381–387.

Nordstrom DL, DeStefano F, Vierkant RA and Layde PM (1998) Incidence of diagnosed carpal tunnel syndrome in a general population, *Epidemiology*, **9**: 342–345.

Norton NH (1945) Calculation of chi-square for complex contingency tables, *Journal of the American Statistical Association*, **40**: 251–258.

Nurminen M and Mutanen P (1987) Exact Bayesian analysis of two proportions, *Scandinavian Journal of Statistics*, **14**: 67–77.

Nussbaumer HJ (1982) *Fast Fourier Transform and Convolution Algorithms*, 2nd edition, Springer–Verlag, Berlin.

O'Brien KF (1994) Concerning the analysis of 2×2 tables, *Computers and Biomedical Research*, **27**: 434–440.

Olsen AO, Dillner J, Skrondal A and Magnus P (1998) Combined effect of smoking and human papilloma virus type 16 infection in cervical carcinogenesis, *Epidemiology*, **9**: 346–349.

Orrett FA and Premanand N (1998) Postpartum surveillance of bacteriuria in term vaginal deliveries, *Journal of the National Medical Association*, **90**: 177–180.

Ottenbacher KJ, Ottenbacher HR, Tooth L and Ostir GV (2004) A review of the two journals found that articles using multivariable logistic regression frequently did not report commonly recommended assumptions, *Journal of Clinical Epidemiology*, **57**: 1147–1152.

Overall JE (1991) A comment concerning one-sided tests of significance in New Drug Applications, *Journal of the Biopharmaceutical Statistics*, **1**: 157–160.

Pagano M and Halvorsen K (1981) An algorithm for finding exact significance levels of $r \times c$ contingency tables, *Journal of the American Statistical Association*, **76**: 931–934.

Pagano M and Tritchler DL (1983a) Algorithms for the analysis of several 2×2 tables, *SIAM Journal of Scientific and Statistical Computing*, **4**: 302–309.

Pagano M and Tritchler DL (1983b) On obtaining permutation distributions in polynomial time, *Journal of the American Statistical Association*, **78**: 435–440.

Patefield WM (1981) Algorithm AS 159: An efficient method of generating random $R \times C$ tables with given row and column totals, *Applied Statistics*, **30**: 91–97.

Patil GP (1985) Multivariate power series distributions, in S Kotz, NL Johnson, CB Read (editors), *Encyclopedia of Statistical Sciences*, **6**: pages 104–108, John Wiley & Sons, New York.

Patil GP (1986) Power series distributions, in S Kotz, NL Johnson, CB Read (editors), *Encyclopedia of Statistical Sciences*, **7**: pages 130–134, John Wiley & Sons, New York.

Patil KD (1975) Cochran's Q test: exact distribution, *Journal of the American Statistical Association*, **70**: 186–189.

Paul SR and Donner A (1989) A comparison of tests of homogeneity of odds ratio in K 2×2 tables, *Statistics in Medicine*, **8**: 1455–1468.

Paul SR and Donner A (1992) Small sample performance of tests of homogeneity of odds ratio in K 2×2 tables, *Statistics in Medicine*, **11**: 159–165.

Peace KE (1991) One-sided or two-sided *p* values: which most appropriately address the question of drug efficacy? *Journal of the Biopharmaceutical Statistics*, **1**: 133–138.

Pearson ES (1947) The choice of statistical tests illustrated on the interpretation of data classified in 2×2 tables, *Biometrika*, **51**: 139–167.

Pérez–Abreu V (1991) Poisson approximation to power series distributions, *The American Statistician*, **45**: 42–44.

Peritz E (1982) Exact tests for matched pairs: studies with covariates, *Communications in Statistics - Theory and Methods*, **11**: 2157–2167.

Petitti DB (2000) *Meta-Analysis, Decision Analysis, and Cost-Effectiveness Analysis: Methods for Quantitative Synthesis in Medicine*, Oxford University Press, New York and Oxford.

Petitti DB (2001) Approaches to heterogeneity in meta-analysis, *Statistics in Medicine*, **20**: 3625–3633.

Piantadosi S (1997) *Clinical Trials: A Methodologic Perspective*, John Wiley & Sons, New York.

Piegorsch WW (1994) Statistical methods for genetic susceptibility in toxicological and epidemiological investigations, *Environmental Health Perspectives*, **102, Supplement 1**: 177–182.

Piegorsch WW and Bailer AJ (1994) Statistical approaches for analyzing mutational spectra: some recommendations for categorical data, *Genetics*, **136**: 403–416.

Piegorsch WW and Bailer AJ (1997) *Statistics for Environmental Biology and Toxicology*, Chapman and Hall, London.

Pierce DA and Peters D (1999) Improving on exact tests by approximate conditioning, *Biometrika*, **86**: 265–277.

Plackett RL (1984) Discussion of "Tests of significance for 2×2 contingency tables" by F Yates, *Journal of the Royal Statistical Society, Series A*, **147**: 458.

Pocock SJ (1983) *Clinical Trials: A Practical Approach*, John Wiley & Sons, Chichester.

Poole C (1987a) Beyond the confidence interval, *American Journal of Public Health*, **77**: 195–199.

Poole C (1987b) Mr. Poole's response (letter), *American Journal of Public Health*, **77**: 880.

Poole C (1987c) Confidence intervals exclude nothing, *American Journal of Public Health*, **77**: 492–493.

Poole C (2001) Low *P*-values or narrow confidence intervals: which are more durable, *Epidemiology*, **12**: 291–294.

Portier C and Hoel D (1984) Type I error of trend tests in proportions and the design of cancer screens, *Communications in Statistics - Theory and Methods*, **13**: 1–14.

Potter DM (2005) A permutation test for inference in logistic regression with small- and moderate-sized data sets, *Statistics in Medicine*, **24**: 693–708.

Pratt JW and Gibbons JD (1981) *Concepts of Nonparametric Theory*, Springer–Verlag, New York.

Prentice R (1976) Use of the logistic model in retrospective studies, *Biometrics*, **32**: 599–606.

Prentice RL (1978) Linear rank tests with right censored data, *Biometrika*, **65**: 167–179.

Prentice RL and Pyke R (1979) Logistic disease incidence models and case-control studies, *Biometrika*, **66**: 403–411.

Press WH, Flannery BP, Teukolsky SA and Vetterling WT (1992) *Numerical Recipes: The Art of Scientific Computing*, 2nd edition, Cambridge University Press, Cambridge, UK.

Prummel MF, Mourits MP, Blank L, Berghout A, Koornneef L and Wiersinga WM (1993) Randomized double-blind trial of prednisone versus radiotherapy in Graves' ophthalmopathy, *The Lancet*, **342**: 949–954.

Radhakrishna S (1965) Combination of results from several 2×2 contingency tables, *Biometrics*, **21**: 86–98.

Ramakrishnan V and Meeter D (1993) Negative binomial cross-tabulations with applications to abundance data, *Biometrics*, **49**: 195–207.

Rao CR (1973) *Linear Statistical Inference and Its Applications*, 2nd edition, John Wiley & Sons, New York.

Rao CR (2000) R. A. Fisher: the founder of modern statistics, in CR Rao and GJ Szekely (editors) (2000) *Statistics for the 21st Century: Methodologies for Applications of the Future*, pages 311–350, Marcel Dekker, New York.

Rao CR and Szekely GJ (editors) (2000) *Statistics for the 21st Century: Methodologies for Applications of the Future*, Marcel Dekker, New York.

Read CB (1985) Median unbiased estimators, in S Kotz and NL Johnson (editors), *Encyclopedia of Statistical Sciences*, **5**: pages 424–426, John Wiley & Sons, New York.

Reiczigel J (2003) Confidence intervals for the binomial parameter: some new considerations, *Statistics in Medicine*, **22**: 611–621.

Reingold EM, Nievergelt J and Deo N (1977) *Combinatorial Algorithms: Theory and Practice*, Prentice Hall, Englewood Cliffs, NJ.

Reis IM (1996) *Exact Tests for Homogeneity of Odds Ratios in Several 2×2 Tables*, unpublished DrPH dissertation, UCLA School Of Public Health, California 1996.

Reis IM, Hirji KF and Afifi AA (1999) Exact and asymptotic tests for homogeneity in several 2×2 tables, *Statistics in Medicine*, **18**: 893–906.

Rice WR (1988) A new probability model for determining exact *p*-values for contingency tables when comparing binomial proportions (with discussion), *Biometrics*, **44**: 1–22.

Robins JM, Breslow NE and Greenland S (1986) Estimators of the Mantel–Haenszel variance consistent in both sparse-data and large strata limiting models, *Biometrics*, **42**: 311–323.

Rosenberger WF and Lachin JM (2002) *Randomization in Clinical Trials: Theory and Practice*, John Wiley & Sons, New York.

Rosenfeld RM (2003) Meta-analysis and systematic literature review (Chapter 4), in RM Rosenfeld and CD Bluestone (editors) (2003) *Evidence-Based Otitis Media*, 2nd edition, BC Decker, Hamilton and London.

Rosner B (1984) Multivariate methods in opthalmology with application to other paired data situations, *Biometrics*, **40**: 1025–1035.

Rosner B (2000) *Fundamentals of Biostatistics*, 5th Edition, Duxbury, Pacific Grove, CA.

Ross SM (2002) *Simulation*, 3rd edition, Academic Press, San Diego, CA.

Rothman KJ (1986a) Significance questing, *Annals of Internal Medicine*, **105**: 445–447.

Rothman KJ (1986b) *Modern Epidemiology*, Little Brown, Boston.

Rothman KJ (2002) *Epidemiology: An Introduction*, Oxford University Press, New York.

Rothman KJ and Greenland S (1998) *Modern Epidemiology*, 2nd edition, Lippincott–Raven, Philadelphia.

Rothman KJ and Poole C (1996) Causation and causal inference, in D Schottenfeld and JF Fraumeni Jr. (editors) (1996) *Cancer Epidemiology and Prevention*, 2nd edition, pages 3–10, Oxford Uinversity Press, New York.

Rothwell PM (2005a) External validity of randomised controlled trials: "To whom do the results of this trial apply?" *The Lancet*, **365**: 82–93.

Rothwell PM (2005b) Subgroup analysis in randomized controlled trials: importance, indications, and interpretation, *The Lancet*, **365**: 176–186.

Roussas G (2003) *Introduction to Probability and Statistical Inference*, Academic Press, Amsterdam.

Routledge RD (1992) Resolving the conflict over Fisher's exact test, *The Canadian Journal of Statistics*, **20**: 201–209.

Routledge RD (1994) Practicing safe statistics with mid-p, *The Canadian Journal of Statistics*, **22**: 103–110.

Routledge RD (1997) P-values from permutation and F-tests, *Computational Statistics and Data Analysis*, **24**: 379–386.

Royall RM (1997) *Statistical Evidence: A Likelihood Paradigm*, Chapman and Hall, London.

Royston P (1993) Exact conditional and unconditional sample size for pair-matched studies with binary outcome: a practical guide, *Statistics in Medicine*, **12**: 699–712.

Rudolph GJ (1967) A quasi-multinomial type of contingency table, *South African Statistical Journal*, **1**: 59–65.

Rustagi JR (1994) *Optimization Techniques in Statistics*, Academic Press, Boston.

Sackett DL (2004) Superiority trials, non-inferiority trials, and prisoners of the 2-sided null hypothesis, *Evidence Based Medicine*, **9**: 38–39.

Sackrowitz H and Samuel–Cahn E (1999) P values as random variables - expected P values, *The American Statistician*, **53**: 326–331.

Sahai H and Khurshid A (1995) On analysis of epidemiological data involving 2×2 contingency tables: an overview of the Fisher's exact test and Yates' correction for continuity, *Journal of Biopharmaceutical Statistics*, **5**: 43–70.

Sahai H and Khurshid A (1996) *Statistics in Epidemiology: Methods, Techniques and Applications*, CRC Press, Boca Raton, FL.

Salvan A and Hirji KF (1991) Asymptotic equivalence of conditional median unbiased and maximum likelihood estimators in exponential families, *Metron*, **49**: 219–232.

Salvia AA (1984) Testing equality of binomial parameters based on inverse sampling, *IEEE Transactions on Reliability*, **R-33**: 377–378.

Samuels SJ, Beaumont JJ and Breslow NE (1991) Power and detectable risk of seven tests for standardized mortality ratios, *American Journal of Epidemiology*, **133**: 1191–1197.

Santner TJ and Duffy DE (1989) *The Statistical Analysis of Discrete Data*, Springer–Verlag, New York.

Santner TJ and Snell MK (1980) Small sample confidence intervals for $p_1 - p_2$ and p_1/p_2 in 2×2 contingency tables, *Journal of the American Statistical Association*, **75**: 386–394.

Satten GA and Kupper LL (1990) Continued fraction representation for exact cell counts of a 2×2 table: a rapid and exact method for conditional maximum likelihood estimation, *Biometrics*, **46**: 217–223.

Schesselman JJ (1982) *Case-Control Studies*, Oxford University Press, New York.

Schinazi RB (2000) The probability of a cancer cluster due to chance alone, *Statistics in Medicine*, **19**: 2195–2198.

Schulz KF and Grimes DA (2005) Sample size calculations in randomized trials: mandatory and mystical, *The Lancet*, **365**: 1348–1353.

Sedgewick R (1988) *Algorithms*, 2nd edition, Addison–Wesley, Reading, Massachusetts.

Selvin S (1995) *Practical Biostatistical Methods*, Duxbury Press, Belmont.

Seneta E, Berry G and Macaskill P (1999) Adjustments to Lancaster's mid-P, *Methodology and Computing in Applied Probability*, **1**: 229–240.

Seneta E and Phipps MC (2001) On the comparison of two observed frequencies, *Biometrical Journal*, **43**: 23–43.

Shaffer JP (1971) An exact multiple comparison test for a multinomial distribution, *British Journal of Mathematical and Statistical Psychology*, **24**: 267–272.

Sheehe PR (1993) A variation on a confidence interval theme (letter), *Epidemiology*, **4**: 185–187.

Shuster JJ (1988) EXACTB and CONF: Exact unconditional procedures for binomial data, *The American Statistician*, **42**: 234.

Shuster JJ (1992) Exact unconditional tables for significance testing in the 2×2 multinomial trial, *Statistics in Medicine*, **11**: 913–922.

Shuster J and van Eys J (1983) Interaction between prognostic factors and treatment, *Controlled Clinical Trials*, **4**: 209–214.

Silvoso GR, Ivey KJ, Butt JH, Lockard OO, Holt SD, Sisk C, Baskin WN, Mackercher PA and Hewett J (1979) Incidence of gastric lesions in patients with rheumatic disease on chronic aspirin therapy, *Annals of Internal Medicine*, **91**: 517–520.

Simpson EH (1951) The interpretation of interaction in contingency tables, *Journal of the Royal Statistical Society, Series B*, **13**: 238–241.

Sinclair JC and Bracken MB (1994) Clinically useful measures of effect in binary analyses of randomized trials, *Journal of Clinical Epidemiology*, **47**: 881–889.

Singh P and Aggarwal AR (1991) Inverse sampling in case control studies, *Environmetrics*, **2**: 293–299.

Skovlund E and Bølviken E (1996) Confidence intervals from Monte Carlo tests, *Journal of The American Statistical Association*, **91**: 1071–1078.

Smith AH and Bates MN (1992) Confidence limits should replace power calculations in the interpretation of epidemiologic studies, *Epidemiology*, **3**: 449–452.

Smith AH and Bates MN (1993) A variation on a confidence interval theme (reply to letter), *Epidemiology*, **4**: 186–187.

Smith PWF, Forster JJ and McDonald JW (1996) Exact tests for square contingency tables, *Journal of the Royal Statistical Society, Series A*, **159**: 309–321.

Smith PWF and McDonald JW (1995) Exact tests for incomplete contingency tables: estimating attained significance levels, *Statistics and Computing*, **5**: 253–256.

Smith PJ, Rae DS, Manderscheid RW and Silbergeld S (1979) Exact and approximate distributions of the chi-square statistic for equiprobability, *Communications in Statistics - Simulations and Computations*, **8**: 131–149.

Smith W and Solow AR (1996) An exact McNemar test for paired binary Markov chains, *Biometrics*, **52**: 1063–1070.

Soares HP, Daniels S, Kumar A, Clarke M, Scott C, Swann S and Djulbegovic B (2004) Bad reporting does not mean bad methods for randomised trials: observational study of randomised controlled trials performed by the Radiation Therapy Oncology Group, *British Medical Journal*, **328**: 22–25.

Soms AP (1989a) Exact confidence intervals, based on the Z statistic, for the difference between two proportions, *Communications in Statistics - Simulation and Computation*, **18**: 1325–1341.

Soms AP (1989b) Some recent results for exact confidence intervals for the difference between two proportions, *Communications in Statistics - Simulation and Computation*, **18**: 1343–1357.

Soper KA and Tonkonoh N (1993) The discrete distribution used for the log-rank test can be inaccurate, *Biometrical Journal*, **35**: 291–298.

Sprott DA and Farewell VT (2000) The analysis of subject specific agreement, in CR Rao and GJ Szekely (editors) (2000) *Statistics for the 21st Century: Methodologies for Applications of the Future*, pages 367–387, Marcel Dekker, New York.

Stern HS (1998) A primer on Bayesian approach to statistical inference, *STATS*, **23**: 3–9.

Sterne TD (1954) Some remarks on confidence or fiducial limits, *Biometrika*, **41**: 275–278.

Sterne JAC and Smith GD (2001) Sifting the evidence - what's wrong with significance tests?, *British Medical Journal*, **322**: 226–231.

Stevens WL (1950) Fiducial limits for the parameters of a discontinuous distribution, *Biometrika*, **37**: 117–129.

Steyn HS (1959) On χ^2 tests for contingency tables of negative multinomial type, *Statistica Neerlandica*, **13**, 433–444.

Stone M (1969) The role of significance testing: some data with a message, *Biometrika*, **56**: 485–493.

Storer BE and Kim C (1990) Exact properties of some exact test statistics for comparing two binomial proportions, *Journal of the American Statistical Association*, **85**: 146–155.

Storer BE, Wacholder S and Breslow NE (1983) Maximum likelihood fitting of general risk models to stratified data, *Applied Statistics*, **32**: 172–181.

Strawderman RL and Wells MT (1998) Approximately exact inference for the common odds ratio in several 2×2 tables (with discussion), *Journal of the American Statistical Association*, **93**: 1294–1320.

Streitberg B and Rohmel J (1984) Exact non-parametrics in APL, *APL Quote Quad*, **14**: 313–325.

Streitberg B and Rohmel J (1986) Exact distributions for permutation and rank tests: an introduction to some recently published algorithms, *Statistical Software Newsletter*, **12**: 10–17.

Suissa S and Salmi R (1989) Unidirectional multiple comparison of Poisson rates, *Statistics in Medicine*, **8**: 757–764.

Suissa S and Shuster JJ (1984) Are uniformly most powerful unbiased tests really best? *The American Statistician*, **38**: 204–206.

Suissa S and Shuster JJ (1985) Exact unconditional sample sizes for the 2×2 binomial trial, *Journal of the Royal Statistical Society, Series A*, **148**: 317–327.

Suissa S and Shuster JJ (1991) The 2×2 matched pairs trial: exact unconditional design and analysis, *Biometrics*, **47**: 361–372.

Sullivan KM and Foster DA (1988) Extension to the use of the *p*-value function (abstract), *American Journal of Epidemiology*, **128**: 907.

Sullivan KM and Foster DA (1990) Use of the confidence interval function, *Epidemiology*, **1**: 39–42.

Sutton AJ, Abrams KR and Jones DR (2001) An illustrated guide to the methods of meta-analysis, *Journal of Evaluation in Clinical Practice*, **7**: 135–148.

Tang ML (1998) Small-sample study of the use of mid-*p* power divergence goodness-of-fit tests, *Journal of Statistical Computation and Simulation*, **62**: 137–180.

Tang ML (1999) Exact power of conditional tests for several binomial populations, *Biometrical Journal*, **41**: 827–839.

Tang ML (2000) On tests of linearity for dose response data: asymptotic, exact conditional and exact unconditional test, *Journal of Applied Statistics*, **27**: 871–880.

Tang ML (2001a) Exact goodness-of-fit test for a binary logistic model, *Statistica Sinica*, **11**: 199–211.

Tang ML (2001b) Exact power computation for stratified dose-response studies, *Statistics and Computing*, **11**: 117–124.

Tang ML and Hirji KF (2002) Simple polynomial multiplication algorithms for exact conditional tests of linearity in a logistic model, *Computer Methods and Programs in Biomedicine*, **69**: 13–23.

Tang ML, Hirji KF and Vollset SE (1995) Exact power computation for dose-response studies, *Statistics in Medicine*, **14**: 2261–2272.

Tang MLN and Tony HK (2004) Comment on: Confidence limits for the ratio of two rates based on likelihood scores: non-iterative method (letter and response), *Statistics in Medicine*, **23**: 685–693.

Tango T (1998) Equivalence test and confidence interval for the difference in proportions for the paired-sample design, *Statistics in Medicine*, **17**: 891–908.

Tarone RE (1985) On heterogeneity tests based on efficient scores, *Biometrika*, **72**: 91–95.

Tarone RE and Gart JJ (1980) On the robustness of combined tests for trend in proportions, *Journal of the American Medical Association*, **75**: 110–116.

The Standards of Reporting Trials Group (1994) A proposal for structured reporting of randomized controlled clinical trials, *Journal of the American Medical Association*, **272**: 1926–1931.

Thomas DG (1971) Algorithm AS 36: Exact confidence limits for the odds ratio in a 2×2 table, *Applied Statistics*, **20**: 105–110.

Thomas DG (1975) Exact and asymptotic methods for the combination of 2×2 tables, *Computers and Biomedical Research*, **8**: 423–446.

Thomas JB (1986) *Introduction to Probability*, Springer–Verlag, New York.

Thomas RG and Conlon M (1992) Sample size determination based on Fisher's exact test for use in 2×2 comparative trials with low event rates, *Controlled Clinical Trials*, **13**: 134–147.

Thomas DG and Gart JJ (1977) A table of exact confidence limits for differences and ratios of two proportions and their odds ratios, *Journal of the American Statistical Association*, **72**: 73–76.

Thomas DG and Gart JJ (1992) Improved and extended exact and asymptotic methods for the combination of 2×2 tables, *Computers and Biomedical Research*, **25**: 75–84.

Thornley B and Adams C (1998) Content and quality of 2000 controlled trials in schizophrenia over 50 years, *British Medical Journal*, **317**: 1181–1184.

Tietjen G (1994) Recursive schemes for calculating cumulative binomial and Poisson probabilities, *The American Statistician*, **48**: 136–137.

Tietjen G (1998) Recursive schemes for calculating common cumulative distributions, *The American Statistician*, **52**: 263–265.

Tillmann HL, van Pelt FNAM, Martz W, Luecke T, Welp H, Dörries F, Veuskens A, Fischer M and Manns MP (1997) Accidental intoxication with methylene dianiline p,p'-diaminodiphenylmethane: Acute liver damage after presumed ecstasy consumption, *Clinical Toxicology*, **35**: 35–40.

Tocher KD (1950) Extension of the Neyman–Pearson theory of tests to discontinuous variates, *Biometrika*, **37**: 130–140.

Tomizawa S (1993) Exact test of uniform association in a 2×3 table, *Biometrika*, **35**: 451–454.

Tsiatis AA, Rosner GL and Mehta CR (1984) Exact confidence intervals following a group sequential test, *Biometrics*, **40**: 797–803.

Upton GJG (1982) A comparison of alternative tests for the 2×2 comparative trial (with discussion), *Journal of the Royal Statistical Society, Series A*, **145**: 86–105.

Upton GJG (1992) Fisher's exact test, *Journal of the Royal Statistical Society, Series A*, **155**: 395–402.

Vail A and Gardner E (2003) Common statistical errors in the design and analysis of subfertility trials, *Human Reproduction*, **18**: 1000–1004.

van de Wiel MA, Di Bucchianico A and van der Laan P (1999) Symbolic computation and exact distributions of nonparametric test statistics, *The Statistician*, **48**: 507–516.

Venables TC and Bhapkar VP (1978) Gart's test for interaction in $2 \times 2 \times 2$ contingency tables for small samples, *Biometrika*, **65**: 669–672.

Vento S, Garofano T, Rezini C, Cainelli F, Casali F, Ghironzi G, Ferraro T and Concia E (1998) Fulminant hepatitis associated with hepatitis A virus superinfection in patients with chronic hepatitis C, *The New England Journal of Medicine*, **338**: 286–290.

Verbeek A and Kroonenberg PM (1985) A survey of algorithms for exact distributions of test statistics in $r \times c$ contingency tables with fixed margins, *Computational Statistics and Data Analysis*, **3**: 159–185.

Vollset SE (1989) *Exact and Asymptotic Inference in a Stratified One Parameter Conditional Logistic Model*, unpublished Dr.PH. Dissertation, Department of Biostatistics, University of California, Los Angeles.

Vollset SE (1993) Confidence intervals for a binomial proportion, *Statistics in Medicine*, **12**: 809–824.

Vollset SE and Hirji KF (1991) A microcomputer program for exact and asymptotic analysis of several 2×2 tables, *Epidemiology*, **2**: 217–220.

Vollset SE and Hirji KF (1992) Confounding of algorithmic efficiency by programming language in exact odds ratio estimation (letter), *Epidemiology*, **3**: 383–384.

Vollset SE and Hirji KF (1994) Paradoxical features of Fisher's exact test as used in medical journals (unpublished manuscript).

Vollset SE, Hirji KF and Afifi AA (1991) Evaluation of exact and asymptotic interval estimators in logistic analysis of matched case-control studies, *Biometrics*, **47**: 1311–1325.

Vollset SE, Hirji KF and Elashoff RM (1991) Fast computation of exact confidence limits for the common odds ratio in a series of 2×2 tables, *Journal of the American Statistical Association*, **86**: 404–409.

Wagner CH (1982) Simpson's paradox in real life, *The American Statistician*, **36**: 46–47.

Wallenstein A (1997) A non-iterative accurate asymptotic confidence interval for the difference between two proportions, *Statistics in Medicine*, **16**: 1329–1336.

Walter SD and Cook RJ (1991) A comparison of several point estimators of the odds ratio in a single 2×2 table, *Biometrics*, **47**: 795–811.

Walter SD and Cook RJ (1992) Response to N Mantel, *Biometrics*, **48**: 1289–1295.

Waxweiler RJ, Stringer W, Wagoner JK, Jones J, Falk H and Coleman C (1976) Neoplastic risk among workers exposed to vinyl chloride, *Annals of the New York Academy of Sciences*, **271**: 41–48.

Weerahandi S (1995) *Exact Statistical Methods for Data Analysis*, Springer–Verlag, New York.

Weinberg CR (1995) How bad is categorization? *Epidemiology*, **6**: 345–347.

Weinberg CR (2001) It's time to rehabilitate the P-value, *Epidemiology*, **12**: 288–290.

Whitehead J (1993) The case for frequentism in clinical trials, *Statistics in Medicine*, **12**: 1405–1415.

Whittle P (1955) Some distribution and moment formulae for the Markov chain, *Journal of the Royal Statistical Society, Series B*, **17**: 235–242.

Wild CJ and Seber GAF (2000) *Chance Encounters: A First Course in Data Analysis and Inference*, John Wiley & Sons, New York.

Williams DA (1988a) Tests for differences between several small proportions, *Applied Statistics*, **37**: 421–434.

Williams DA (1988b) P-values and discontinuity, *Royal Statistical Society News and Notes*, **15**(4): 4–5.

Williams PL (2005) Trend test for counts and proportions, in P Armitage and T Colton (editors-in-chief) *Encyclopedia of Biostatistics*, 2nd edition, volume 7, pages 5516–5527, John Wiley & Sons.

Williamson E and Bretherton MH (1963) *Tables of the Negative Binomial Probability Distribution*, John Wiley & Sons, New York.

Wilson EB (1927) Probable inference, the law of succession, and statistical inference, *Journal of the American Statistical Association*, **22**: 209–212.

Wilson EB (1941) The controlled experiment and the four-fold table, *Science*, **93**: 557–560.

Woodworth GG (2004) *Biostatistics: A Bayesian Introduction*, John Wiley & Sons, New York.

Woolf B (1955) On estimating the relation between blood group and disease, *Annals of Human Genetics*, **19**: 251–253.

Woolf FM (1986) *Meta-Analysis: Quantitative Methods for Research Synthesis*, Sage Publications, Newbury Park, California.

Worcester J (1971) The relative odds in 2^3 contingency tables, *American Journal of Epidemiology*, **93**: 145–149.

Wu CL, Hsu WH, Chiang CD, Kao CH, Hung DZ, King SL and Deng JF (1997) Lung injury related to consuming *Sauropus androgynus* vegetable, *Clinical Toxicology*, **35**: 241–248.

Yao Q and Tritchler D (1993) An exact analysis of conditional independence in several 2×2 tables, *Biometrics*, **49**: 233–236.

Yates F (1934) Contingency tables involving small numbers and the chi-square test, *Supplement to the Journal of the Royal Statistical Society, Series B*, **1**: 217–235.

Yates F (1948) The analysis of contingency tables with grouping based on quantitative characteristics, *Biometrika*, **35**: 176–181.

Yates F (1984) Tests of significance for 2×2 tables (with discussion), *Journal of the Royal Statistical Society, Series A*, **147**: 426–463.

Yusuf S, Collins R and Peto R (1984) Why do we need some large, simple randomized trials, *Statistics in Medicine*, **3**: 409–420.

Yusuf S, Peto R, Lewis J, Collins R and Sleight P (1985) Beta blockade during and after myocardial infarction: an overview of randomized trials, *Progress in Cardiovascular Diseases*, **27**: 335–371.

Zacks S and Solomon H (1976) On testing and estimating the interaction between treatments and environmental conditions in binomial experiments: the case of two stations, *Communications in Statistics - Theory and Methods*, **5**: 197–223.

Zelen M (1971) The analysis of several 2×2 contingency tables, *Biometrika*, **58**: 129–137.

Zelterman D (1999) *Models for Discrete Data*, Clarendon Press, Oxford.

Zelterman D, Chan ISF and Mielke PW (1995) Exact tests of significance in higher dimensional tables, *The American Statistician*, **49**: 357–361.

Zolman JF (1993) *Biostatistics: Experimental Design and Statistical Inference*, Oxford University Press, Oxford.

Index

Acceptability method, *see* Combined tails statistic
Accuracy, *see* Error, systematic
Adjacent categories model, 297, 392, 406, 457
Algorithm
    combinations, 378
    combinatorial, 377
    compositions, 377, 378, 383, 408, 409
    divide and conquer, 345
    exhaustive enumeration, 323–325, 338, 344, 350, 372, 374, 383, 395, 406, 412, 479
    implementation of, 338, 372
    iterative, 88, 107, 113, 301, 335
    Monte–Carlo, xxi, 326–329, 338, 344, 347, 407, 422, 450, 478, 479
    network, xxi, 344, 366–368, 372, 375, 407, 410, 413
    partitions, 376, 377, 380, 408, 409
    properties of, 88
    recursive, xxii, 88, 329–330, 335, 343, 344, 349–351, 363, 367, 372, 374, 376, 383, 384, 395, 399, 401, 407, 408, 410, 411, 413, 421, 422, 442, 444, 480
    shift, xxi, 372
    subsets, 378, 408
Ancillary statistic, 118–121, 140, 152, 166, 251, 458–460, 482, 483
Array storage, 93–95, 339, 351, 372, 373, 430

Backward induction, 246, 355–360, 364, 372, 374, 375, 385, 399, 407, 411
Bayes factor, 452
Bayesian framework, 50, 122, 451–452, 454, 455, 459, 478, 481–483
Bernoulli trial, 2, 42, 432, 453
Bias, *see* Error, systematic
Bimonotonic function, 57, 58, 65, 68, 205, 208, 211, 212, 216, 217, 219
Bisection method, 108, 110, 111, 113, 115
Bonferroni method, 308
Bootstrap method, 450, 478, 480, 482
Boundary point, 80–82, 106, 369
Breslow–Day statistic, *see* Tarone–Breslow–Day statistic

Case-control study, 51, 132, 164, 168, 174, 177–181, 266, 277, 306, 321, 372, 376, 461, 478
Case-control study
    matched, 415, 442–444
Chebychev's inequality, 85
Chisquare statistic, *see* Score statistic
Clinical significance, 41, 76, 469, 480
Clinical trial, 3, 41, 82, 126, 161, 165, 187, 202, 203, 220, 222, 223, 247, 251, 274, 321, 432, 440, 442, 447, 458, 463, 465, 467, 470, 482, 484
Clinical trial
    cross over, 424, 442
    planning of, 466
    quality of, 152, 465, 468, 477, 480, 484
Cochran's rule, 311, 478
Cochran–Armitage test, *see* Cochran–Armitage–Mantel test
Cochran–Armitage–Mantel test, 283, 306, 312
Coefficient
    binomial, 88–92, 99, 101, 102, 112, 113
    factorial, 88–89, 92, 98, 99, 102, 113
    hypergeometric, 88, 91, 92, 113
    multinomial, 88, 91, 92
    negative binomial, 88, 91, 92, 113
    Poisson, 88
Cohort data, *see* Person time data
Cohort design, *see* Person time data
Combined tails statistic, 72, 80, 83, 85, 106, 110, 115, 149, 185, 211, 223, 412
Common odds ratio, 167, 228, 234–237, 248, 254, 256, 268, 313, 334, 338, 343, 344, 346, 347, 369, 375, 376
Complete enumeration, *see* Algorithm, exhaustive enumeration
Computational scale, 96, 111, 113, 373
Conditional analysis
    and case-control study, 168, 174, 177–181, 415
    and discreteness, xxiv, 195
    and exact tests, 148
    and paired data, 442
    and product binomial model, 165, 173–177
    and regression methods, 457, 458, 478
    and relevant subsets, 458
    and statistical models, 461
    and study design, 118, 454

and sufficient statistics, 123
asymptotic, 122, 419, 420, 456
definition of, 118
exact, 122, 139–144, 229–234, 298, 314, 419, 420, 456
for person time data, 174, 176
for stratified data, 227
for trend in odds ratios, 241
in inverse sampling, 153
validity of, 147, 299, 455
Conditional independence test, 229, 237–241, 246–248, 265, 300, 335, 343, 344, 372
Conditional probability, 5
Conditionality principle, 460, 461, 467
Confidence curve, see Evidence function
Confidence interval
and evidence function, xxii, 140
and precision, 76
and random error, 31
and scores, 306
asymptotic, 43, 46, 49, 54, 105, 153, 171, 237, 242, 245, 248, 249, 255, 315, 319, 399, 402, 411, 412, 442–444
combined tails, 73, 80, 85, 105, 115
connected, 42, 58, 195, 219
definition of, 41
exact, 46, 49, 54, 153, 190, 283, 285, 319, 343, 411, 423, 441, 444, 459
for person time data, 132
for random effect model, 270
for relative risk, 220
for risk difference, 121, 156, 220
length of, 205, 224
likelihood ratio, 82, 104, 105, 115, 156, 248
Mantel–Haenszel based, 246, 248
mid-p, 43, 46, 51, 54, 60, 62, 79, 104, 115, 140, 141, 175, 234, 237, 242, 246, 248, 249, 266, 283, 302, 312, 347, 399, 402, 412, 430, 442, 444
Monte–Carlo, 328
nested, 58, 79, 195, 219
nonfinite, 205
one-sided, 43, 44, 46, 49, 53, 54
probability based, 85, 104, 115
randomized, 195, 212
reporting of, 76, 217, 466, 467, 469–471, 477, 481
score, 65, 66, 80, 81, 104, 110, 115, 138, 141, 146, 235, 237, 242, 248, 266, 283, 399, 402, 412, 442
twice the smaller tail, 58, 77, 80, 81, 85, 100, 104, 114, 115, 234, 237, 242, 246, 248, 266, 283, 285
two-sided, 58, 79, 81, 83
unconditional, 220, 224

Wald, 70, 85, 105, 138, 156, 172, 248, 283, 285
weights based, 276
Confidence interval function, see Evidence function
Confounding factor, 161
Continuity correction, 11, 149, 478
Convolution, 8
Coverage function, 206
Coverage level
actual, 79, 204, 206, 211, 212, 217, 224, 249
conditional, 205
guaranteed, 80, 211, 214
nominal, 204, 206, 212
Cox regression, 475, 480
Credible interval, 452, 478
Cressie–Read statistic, 288, 292, 315, 316
Cross product ratio, 263, 265, 275, 457
Cross sectional study, 3, 462
Cumulative distribution function, 5

δ level accuracy, 214
Dependent outcomes, 12, 20, 51, 424, 433, 442, 447, 454, 472
Design and analysis, 47–50, 119, 138, 147–148, 221, 298, 454–462, 477
Distance from center statistic, 71, 86, 235, 283
Distribution
binomial, xxii, 10, 13, 25, 54, 65, 78, 83, 85, 99–100, 112, 114, 222, 326, 428, 433, 454, 470, 482, 484
chisquare, 112, 294, 382, 391, 393, 404
exact, 323, 369, 375, 382, 421, 442, 451, 453
geometric, 14, 25, 27, 48, 53, 84
hypergeometric, xxii, 15, 100–103, 112, 114, 232, 263, 425, 459, 476, 478
multinomial, xxii, 18, 20, 26, 93, 168, 321, 380, 405, 409
negative binomial, xxii, 14, 25, 27, 49, 51, 53, 54, 65, 84, 85, 102, 114, 153, 223, 303, 454, 455
negative trinomial, 21, 26
normal, 11, 14, 24, 25, 43, 451
Poisson, xxii, 12, 24, 25, 27, 45, 52, 65, 83, 85, 86, 97–99, 112, 148, 224, 263, 299, 326, 388, 405, 413, 476, 483
polynomial based, xxii, 6, 17, 23, 24, 26, 33, 35, 42, 52, 79, 84, 96, 108, 115, 121, 123, 125, 129, 131, 135, 139, 140, 143, 148, 157, 167, 174, 195, 198, 208, 224, 230, 231, 234, 266, 314, 328, 440, 442
product binomial, 165, 204, 218, 221, 277, 321, 326, 370, 375, 455, 457, 459, 476, 482–484
product multinomial, xxii, 170, 389

uniform continuous, 194, 217, 224, 326, 339
uniform discrete, 52, 195, 327
Distribution free method, 450
Dose-response analysis, 71, 221, 275, 280, 282, 305, 306, 315, 319, 361, 363, 365, 369, 372, 373, 375, 376, 388, 406, 456, 480
Double blind design, 126
Doubling method, see Twice the smaller tail method
Dynamic programming, see Backward induction

Empirical Bayes method, 452, 484
Empirical logit, 78
Equiprobable multinomial, xxi, 381, 408
Equivalent statistic, 82, 235, 283, 288, 295, 320, 386, 409, 411, 412
Error
   in biomedical studies, 152, 477
   mean squared, 214
   numeric, 87, 89, 90, 94, 96, 102, 107, 113, 201, 338, 343, 345, 372, 386, 468
   random, 31, 159, 160, 204, 480
   systematic, 31, 126, 159, 204, 249, 464, 466, 480
   type I, 39, 80, 196, 211, 219, 277, 451, 459, 471, 474
   type II, 197, 459
Evidence function
   advocacy of, 467
   asymptotic, 37, 44, 52, 65, 66, 138, 184, 190, 247, 283, 316, 319, 320, 412
   combined tails, 72, 84, 154, 156, 184, 211, 248
   definition of, 34
   distance from center method, 71
   exact, 35, 44, 52, 53, 65, 81, 141, 184, 247, 248, 316, 319, 320, 412
   generic, 198, 204
   likelihood ratio, 68, 84, 153, 156, 196, 206
   literature on, 82
   Mantel–Haenszel statistic based, 236
   mid-p, 35, 44, 52, 53, 60, 62, 84, 140, 141, 156, 175, 177, 195, 210, 214, 219, 234, 246–248, 283, 300, 302, 312, 316, 430, 471
   Monte–Carlo, 328
   one-sided, xxii, 195, 212
   other names of, 35
   probability based, 71, 84, 154, 196, 211
   properties of, xxii, 56, 57, 76, 173, 189, 217, 453, 460
   randomized, 195, 196, 210–212
   reporting of, 76, 80, 185, 217, 247, 266, 312, 402, 423, 439, 440, 443, 446, 451, 484
   score, 65, 66, 81, 84, 141, 153, 156, 184, 196, 206, 222, 235, 248, 283, 300, 316
   twice the smaller tail, 58, 59, 80, 84, 154, 156, 210, 211, 234, 283, 300
   two-sided, xxii, 56, 195, 211, 214
   unconditional, 156, 190–192, 194, 217, 224
   variety of, 148
   Wald, 70, 84, 153, 156
Evidentiary consistency, 77, 78, 149, 208, 217, 471
Evidentiary statistic, 64
Exact analysis
   formal, 447, 471, 483
Exact distribution
   and frame of reference, 449
   definition of, 173
   Markov chain, 435, 437
Exact method
   and odds ratio model, 255
   and sparse data, xxi, xxiv
   as a magic solution, 467
   as a recipe, xxi, xxiii
   conventional, 211, 451
   definition of, 450
   integrated framework for, xxii
   misuse of, 462, 476
Exact test
   definition of, 451
   Fisher–Irwin, xxi, 140, 153, 204, 206–210, 221, 223, 451, 452, 462, 464, 468, 472, 476, 477
   Freeman–Halton, 287, 290, 311, 315, 316, 372, 404, 478
   Markov model, 438, 440, 443, 447, 476
   McNemar, 425
   Zelen, 71, 259, 263, 271, 274, 276
Exhaustive enumeration, see Algorithm, exhaustive enumeration
Exponent check, 328, 330–334, 337, 339, 344, 345, 352, 365, 372, 385, 395, 396, 411, 412, 417, 422, 423, 430, 432, 442, 444–446

Factorial design, 183, 187
Factorization theorem, 22, 27, 231, 233, 281
False position method, 109, 115
Fast Fourier transform, xxi, 97, 339–344, 346, 372, 386, 408, 410
Fisher, RA, xxi, 120, 459, 466, 469, 482
Fisherian school, 453
Fixed effect model, 253, 269
Frame of reference, xxiv, 458, 460, 462, 467, 478
Frequentist framework, 453–478, 482, 483

Generalizability, see Validity, external
Generating function, see Generating polynomial
Generating polynomial, 7, 17, 25, 27, 174, 192, 224, 230, 231, 234, 243, 265, 266, 284, 300, 301, 313, 323, 324, 328, 334, 344, 369, 370,

391, 394, 406, 417–419, 421–423, 426, 428, 437, 439, 441, 443
Genetic model, 482
Goodness-of-fit test, 51, 221, 269, 277, 294, 306, 312, 320, 372, 381, 409, 410
Grab sample, 48, 454, 465, 466

Hardy–Weinberg model, 480, 482
Hash function, 94, 115
Heterogeneity, see Interaction
Horner's scheme, 95–98, 100–103, 109, 113, 115, 341

Incidence density design, see Person time data
Independent events, 6, 27, 472
Information matrix, 150, 171, 244, 272, 276, 314
Interaction, 130, 132, 160, 161, 163, 181–183, 228, 229, 242, 247, 255–257, 263, 264, 269, 271, 274, 276, 277, 321, 361, 421, 426, 428, 444, 445, 456, 457
Inverse sampling, 47, 51, 54, 134–135, 138, 152, 153, 157, 188, 224, 266, 274, 277, 303–476

Lexicographic order, 377
Likelihood function, 62, 147, 170, 171, 181, 191, 244, 245, 271, 314, 417, 419, 420, 422, 435, 444, 453, 455, 457, 476
Likelihood ratio statistic, 68, 83–85, 137, 140, 148, 151, 171, 187, 224, 239, 241, 245, 249, 263, 271, 276, 277, 283, 287, 291, 294, 305, 315, 316, 320, 336, 346, 349, 382, 387, 390, 392, 404, 406, 409–411, 417, 435, 443–445, 453, 456
Likelihood ratio statistic
 derivation of, 151
Likelihood school, 50, 122, 453, 455, 478, 482, 483
Linear by linear association model, 405
Log-linear model, 130, 152, 172, 263, 275, 319, 320, 408, 453, 459
Log-rank test, 474, 476
Logistic model, xxi, 126, 152, 160, 183, 205, 221, 228, 254, 255, 280, 296, 297, 300, 305, 306, 312, 316, 319, 345, 376, 393, 408, 415, 424, 425, 440, 450, 459, 474, 478
Logistic regression, 313, 440, 457, 475
Logit model, see Logistic model

Mantel extension test, see
 Cochran–Armitage–Mantel test
Mantel–Haenszel method, 141, 235, 246, 254, 255, 259, 270, 271, 346, 474, 476, 478
Markov chain model, 432–441, 443, 446, 455, 476, 480

Markov inequality, 85
Matched design, 472, 483
Maximum likelihood estimate, 62, 66, 72, 76, 77, 80, 83, 84, 104, 110, 111, 115, 136, 141, 148, 149, 157, 170, 171, 187, 235, 245, 254, 256, 259–261, 263, 269, 273, 276, 283, 285, 298, 301, 315, 345, 399, 402, 419, 420, 435, 444, 456, 457, 459
Mean, definition of, 5
Measures of agreement, 480, 482
Measures of association, 480
Median unbiased estimate, 58, 60, 77, 78, 80, 81, 83, 84, 104, 111, 115, 140, 149, 153, 157, 175, 234, 246, 260, 276, 285, 302, 312, 345, 347, 430, 439
Median, definition of, 5
Memory management, see Array storage
Meta-analysis, 76, 247, 253, 256, 270, 271, 274, 465, 480, 484
Minimum likelihood statistic, see Probability statistic
Missing data, 452, 454
Monte–Carlo method, see Algorithm, Monte–Carlo
Multinomial data, 380, 387

Newton's method, 108, 111, 115
Neyman–Pearson school, 197, 453, 471, 482, 483
Nominal response, 389, 400, 411
Nominal variable, 3
Nonparametric method, xxi, 450, 467, 474, 478
Nuisance parameter, 118, 119, 121, 122, 139, 147, 166, 200, 229, 257, 308, 369, 372, 458

Odds ratio, xxii, 16, 78, 124, 133, 143, 149, 153, 157, 182, 203, 220, 224, 229, 242, 255, 269, 270, 274, 276, 277, 281, 345, 366, 425, 426, 450, 451, 455–459, 472, 476, 479, 480, 482
One hit model, 306
Optimality principle, 355
Ordinal response, 221, 275, 297, 392, 393, 411, 483
Ordinal variable, 3

p-value
 and evidence function, xxii
 and sample size, 41
 as evidence, 31, 35, 38, 40
 asymptotic, 37, 138, 141, 146, 149, 171, 177, 200, 219, 245, 249, 257, 261, 262, 266, 303, 346, 399, 402, 408, 410–412, 442
 combined tails, 73, 74, 80, 222, 223
 critique of, 82, 482, 483
 definition of, 35

distance from center, 72
distribution of, 214, 219, 224
exact, 52, 157, 190, 219, 224, 260, 269, 394, 399, 402, 408, 410–412, 439, 442, 459
interpretation of, 38
likelihood ratio, 68, 75, 144, 222, 241
Mantel–Haenszel based, 237, 248
mid, 36, 50, 60, 61, 73, 77, 79, 140, 149, 175, 177, 219, 223, 225, 234, 237, 242, 244, 246, 248, 266, 283, 284, 302, 312, 346, 347, 394, 399, 402, 411, 412, 442, 470, 471
misuse of, 38, 39, 75, 467, 468, 471, 479, 481
Monte–Carlo, 328
one-sided, 35, 39, 44, 53, 477
probability based, 71, 74, 76, 222, 223, 259, 441
property of, 66
randomized, 194–196, 211–214, 219, 220, 224
score, 65, 66, 74, 77, 80, 141, 144, 149, 151, 177, 200, 222, 235, 237, 240, 242, 245, 248, 266, 284
twice the smaller tail, 57, 59, 61, 73, 74, 76, 77, 80, 83, 140, 144, 177, 200, 222, 234, 237, 242, 249, 266, 283, 284, 346
two-sided, 57, 58, 223, 337
unconditional, 218
varieties of, 148, 221
Wald, 75, 156, 172
p-value function, see Evidence function
Pair-wise comparisons, 308–311, 316, 321, 469
Paired binary outcomes, 424, 442, 445
Pascal triangle, 90, 99
Permutation method, 51, 210, 450, 458, 459, 462, 463, 466
Permutationally equivalent sets, 417
Person time data, 12, 36, 45, 51, 52, 132, 144–147, 152, 169, 176, 224, 263–266, 274, 275, 299–303, 312, 316, 320, 321, 362, 372, 373, 375, 387, 410, 413, 475, 483
Peto statistic, 271
Point estimate, xxii, 31, 57, 190, 196, 214, 247
Poisson model assumptions, 14, 475
Polynomial evaluation, 95, 338
Polynomial multiplication algorithm, see Algorithm, recursive
Polynomial notation, 23
Polytomous regression, 480
Population approach, 210, 462
Posterior probability, 451
Power
  and fixed level, 39
  and study design, 299
  approximate, 198, 203, 204, 307
  conditional, 203, 337, 345, 371
  definition of, 197

exact, 197, 198, 204, 218, 220, 222, 247, 249, 267–269, 271, 277, 306–308, 321, 338, 346, 368–370, 372, 375, 376, 458–460, 480, 481
for other designs, 221
minimal, 197, 203
Monte–Carlo, 369, 370
use of, 79, 466, 470
Power divergence statistic, 287, 387
Power function, 189, 199, 202, 203, 215–217, 219, 223, 460
Power series distribution, see Distribution, polynomial based
Precision, see Error, random
Principles of inference, 460
Prior probability, 451
Probability generating function, 27
Probability statistic, 71, 80, 83, 85, 204, 218, 223, 225, 390, 404, 406, 407, 439, 441
Probit model, 306
Prognostic factor, 126, 161, 466
Proportional odds model, 299, 457

Random allocation, 3, 126, 165, 175, 183, 424, 432, 455, 459, 462–464
Random effect model, 253, 269–271, 274
Random numbers, 326, 339
Random sample, 29, 47, 126, 133, 177, 455, 462, 463, 466
Random variable, 1
Randomization method, 450, 464
Rao, CR, 470
Rare events, see Sparse data
Recall bias, 133
Relative risk, 120, 126, 135, 143, 148, 151, 182, 224, 254, 255, 267, 274, 275, 277, 304, 362, 373, 375, 451, 457, 459
Representative sample, see Validity, external
Resampling method, 450
Risk difference, 120, 126, 147, 151, 156, 182, 224, 253, 255, 274, 275, 451, 457, 459
Risk ratio, see Relative risk
Robins–Breslow–Greenland estimate, 111, 236

Sample size, 50, 203, 218, 220, 224, 454, 463, 464, 466, 467, 469, 480
Sample space, 2, 49, 326, 338, 343, 345, 369, 375, 458
Scale sensing test, 97, 103
Scientific inference, 31, 464, 467
Scientific model, 1, 464
Score statistic, 64, 72, 80, 83–85, 104, 137, 138, 140, 146, 156, 171, 185, 187, 192, 199, 200, 218, 219, 222, 224, 225, 234, 238, 241, 244–246, 259, 263, 271, 273, 276, 277, 283, 288, 294, 300–302, 305, 311, 315, 316, 320,

336, 346, 349, 382, 387, 390, 392, 399, 402, 404, 406, 409–411, 417, 419, 420, 435, 442–445, 454, 456, 457, 474
Score statistic
    derivation of, 151, 301, 314
Scores
    and biologic model, 306
    exponential, 306, 316, 349
    linear, 306, 349
    midpoint, 306
    midrank, 306
    natural, 247, 248, 306, 406, 445
    quadratic, 306, 312, 316, 349
Secant method, 109–111, 115
Significance level
    actual, 39, 197, 203, 206, 210–212, 214, 215, 217, 225, 346
    fixed, 38, 76, 459, 470, 479
    nominal, 38–39, 197, 206, 210, 214, 215, 217, 271
Simpson's paradox, 161, 183
Size, see Error, type I
Sparse data, xxi, xxiv, 5, 12, 25, 44, 79, 190, 219, 246, 274, 306, 337, 343, 382, 475, 481
Standardized mortality ratio, 46, 53
Statistical methods, misuse of, 469
Statistical model, 1, 2, 165, 173, 453
Statistical significance, 38, 40, 41, 76, 255, 469, 481
Statistical software, xxiii, 479
Stern region, 71, 79
Stirling's formula, 89
Stratified analysis, 165, 266, 274–277, 312, 344, 373, 375, 400, 412, 442, 447, 480
Stratified analysis
    literature on, 246
Stratified randomization, 162, 187, 227, 251, 266, 321
Stratum additive statistic, 259
Subgroup analysis, 161
Sufficient statistic, 21–25, 48, 121, 125, 139, 147, 173, 200, 230, 231, 233, 242, 257, 259, 262, 281, 313, 334, 369, 390, 391, 394, 404, 405, 411, 416, 425, 427, 437, 440, 458
Survival data, 472, 476
Synergy, see Interaction
Systematic reviews, see Meta-analysis

Tail area school, 31, 34, 50, 122, 189, 454
Target population, see Frame of reference
Tarone–Breslow–Day statistic, 259, 271
Test
    asymptotic, 271, 274, 376
    biased, 222, 458
    conservative, 196, 218, 219, 271, 459

critical region of, 39, 53, 210, 370
definition of, 39
exact, 274, 372, 376, 440
for equivalence, 126, 157, 459
for noninferiority, 40, 459
inversion of, 43, 469
liberal, 196
most powerful, 216, 220
one-sided, 30–40, 51, 207, 468–470, 477
randomization, 328
randomized, 212
two-sided, 30–40, 57, 58, 140, 207, 469, 479
unbiased, 216, 220
Test for linearity, see Goodness-of-fit test
Test for trend, see Dose-response analysis
Test statistic
    definition of, 64
    derivation of, 136, 149–151, 171, 313
Transition probabilities, 433, 438, 439, 455
Trend in odds ratios, 228, 241–248, 250, 261–263, 273, 282, 375
Trend in rate ratios, 300
Trend in relative risk, 277
Twice the smaller tail method, 59, 80, 103, 185

Unconditional analysis
    and additive risk model, 121, 123
    and case-control study, 181
    and conditional analysis, 148, 184, 218, 221, 277, 442, 456, 458, 460, 482, 483
    and study design, 118
    approximate, 220
    as the only valid method, 457–462
    asymptotic, 122, 135–138, 170–173, 245, 314
    computational issues, 194, 220, 459
    definition of, 118
    exact, 79, 122, 189–194, 274, 276, 458, 478
    features of, 217–218
    for a $2 \times 2 \times 2$ table, 170
    for a 2×2 table, 149, 220
    for a $2 \times K$ table, 283, 298
    for interaction, 260
    for Markov chains, 434
    for odds ratio, 144
    for person time data, 145
    for several 2×2 tables, 239
    in inverse sampling, 144, 277
    literature on, 219
Unrelated statistic, 118

Validity
    and exclusive dichotomies, 467
    external, 30, 126, 129, 455, 458, 462, 464–467, 483
    internal, 465, 466, 484

Variance, definition of, 5

Wald statistic, 69–70, 76, 82–85, 105, 106, 137, 140, 156, 171, 187, 249, 263, 276, 277, 283, 320
Weights based analysis, 255

Zelterman statistic, 288, 289, 293, 315, 316
Zero event data, 44, 53, 275